Colloids and Interface Science Series
Volume 2

Colloid Stability

Edited by
Tharwat F. Tadros

1807–2007 Knowledge for Generations

Each generation has its unique needs and aspirations. When Charles Wiley first opened his small printing shop in lower Manhattan in 1807, it was a generation of boundless potential searching for an identity. And we were there, helping to define a new American literary tradition. Over half a century later, in the midst of the Second Industrial Revolution, it was a generation focused on building the future. Once again, we were there, supplying the critical scientific, technical, and engineering knowledge that helped frame the world. Throughout the 20th Century, and into the new millennium, nations began to reach out beyond their own borders and a new international community was born. Wiley was there, expanding its operations around the world to enable a global exchange of ideas, opinions, and know-how.

For 200 years, Wiley has been an integral part of each generation's journey, enabling the flow of information and understanding necessary to meet their needs and fulfill their aspirations. Today, bold new technologies are changing the way we live and learn. Wiley will be there, providing you the must-have knowledge you need to imagine new worlds, new possibilities, and new opportunities.

Generations come and go, but you can always count on Wiley to provide you the knowledge you need, when and where you need it!

William J. Pesce
President and Chief Executive Officer

Peter Booth Wiley
Chairman of the Board

Colloids and Interface Science Series
Volume 2

Colloid Stability

The Role of Surface Forces – Part II

Edited by
Tharwat F. Tadros

BICENTENNIAL 1807 WILEY 2007 BICENTENNIAL

WILEY-VCH Verlag GmbH & Co. KGaA

The Editor

Prof. Dr. Tharwat F. Tadros
89 Nash Grove Lane
Wokingham, Berkshire RG40 4HE
Great Britain

Library of Congress Card No.: applied for

British Library Cataloguing-in-Publication Data
A catalogue record for this book is available from the British Library

Bibliographic information published by the Deutsche Nationalbibliothek
The Deutsche Nationalbibliothek lists this publication in the Deutsche Nationalbibliografie; detailed bibliographic data are available in the Internet at http://dnb.d-nb.de

Printed in the Federal Republic of Germany
Printed on acid-free paper

Composition K+V Fotosatz GmbH, Beerfelden
Printing betz-druck GmbH, Darmstadt
Bookbinding Litges & Dopf Buchbinderei GmbH, Heppenheim
Wiley Bicentennial Logo Richard J. Pacifico

ISBN 978-3-527-31503-1

Contents

Colloids and Interface Science Series, Vol. 2
Colloid Stability: The Role of Surface Forces, Part II. Edited by Tharwat F. Tadros

ISBN: 978-3-527-31503-1

Preface

This is a new series of reviews that are aimed at identifying the role of colloid and interface science in various fields. The first two volumes describe some aspects of colloid stability with special reference to the role of surface forces. Several reviews with different scopes are written by leading scientists from all over the world. They cover topics such as the thermodynamic criteria of colloid stability, the role of surface forces, hydrophobic interaction, long-range forces, nanoparticles, colloid stability using polymeric surfactants, etc. A great deal of emphasis is given to foam and emulsion films, which are used fundamentally to investigate the role of surface forces in the stabilization of such films. Some other aspects covered are wetting films, both static and dynamic, and emulsion stability. The reviews are not given in any specific order and they are published on the basis of the order of receiving the manuscripts. These reviews are comprehensive, with many references, and they should be extremely useful for those engaged in fundamental studies of colloid stability and the role of surface forces both in academia and in industry. The first two volumes are dedicated to Professors Dotchi Exerowa and Dimo Platikanov (on the occasion of their 70th birthdays), both from the famous school of colloid science that was led by the late Professor A. Scheludko in Bulgaria. Photographs, biographies and lists of selected papers published by both scientists are given directly after this preface.

I would like to thank all the authors for their dedication in producing these excellent reviews, which made my editing task fairly easy. I would like also to thank the staff of Wiley-VCH for producing these two volumes quickly.

Wokingham, October 2006

Tharwat Tadros
Editor of the Series

Colloids and Interface Science Series, Vol. 2
Colloid Stability: The Role of Surface Forces, Part II. Edited by Tharwat F. Tadros

ISBN: 978-3-527-31503-1

Professor Dotchi Exerowa, DrSc, Academician

Dotchi Exerowa was born on 20 May 1935 in Varna, Bulgaria. After finishing secondary school in Sofia (1953), she studied chemistry at the University of Sofia. In 1958 she obtained the degree of Diploma-Chemist (equal to an MSc). Her thesis was in the area of colloid science, carried out in the Department of Physical Chemistry. Her scientific career began at the Institute of Physical Chemistry of the Bulgarian Academy of Sciences, which has continued to be her workplace up to the present. As a junior research associate she completed her PhD thesis in 1969, supervised by the great Bulgarian colloid scientist Professor A. Scheludko. She advanced her research work and in 1972 was awarded Habilitation, which entitled her to a senior position at the Institute. In 1983 she succeeded Professor Scheludko as Head of the Department of Colloid and Interface Science. Her DrSc thesis was successfully defended in 1987 and a year later she became Professor in Physical Chemistry. Parallel to her main duties at the Institute of Physical Chemistry she has lectured on Physicochemical Methods in Biology, being a Visiting Professor at the University of Sofia. She has also supervised many PhD and MSc students and postdoctoral fellows, thus bringing many young people into scientific research.

Dotchi Exerowa has published about 200 papers in the scientific literature. In addition, she has written a monograph, *Foam and Foam Films*, co-authored with Peter Kruglyakov (published by Elsevier, Amsterdam, 1998). Owing to her extensive publications, Dotchi Exerowa has become internationally well known in the field of colloid and interface science. She has been invited many times to lecture at international conferences and seminars in leading scientific institutions. She has been a member of many scientific committees of conferences and an editorial board member of four international scientific journals. In 1997 she was co-chairman of the 9th International Conference on Surface and Colloid Science. She was twice elected a member of the Council of the International Association of Colloid and Interface Scientists. In 2004, Professor Exerowa was elected to the Bulgarian Academy of Sciences, receiving the title Academician.

The contributions of Dotchi Exerowa are mainly in the field of thin liquid films, surfactants, foams, liquid interfaces, lung surfactant systems, etc. Many of the results obtained were on aspects that have stimulated new directions in the development of knowledge in the field of thin liquid films and also the physics and chemistry of interfacial phenomena.

In the 1960s, Dotchi Exerowa, together with her teacher Professor Scheludko, developed a unique experimental method for the study of thin liquid films based on the very useful model of a microscopic (radius ca. 100 μm) horizontal thin liquid film. This allowed the measurement of important parameters characterizing their properties: equilibrium thickness, critical thickness of rupture, disjoining pressure, contact angle film/bulk liquid, etc. The method and equipment for microscopic thin liquid film investigations are known as the Scheludko–Exerowa micro-interferometric technique. Special attention has been given to the direct measurement of the interaction forces in microscopic liquid films, the disjoining pressure/thickness isotherm and the transition from long- to short-range molecular interactions. This is now referred to as the thin liquid film pressure balance technique, and is widely used in many laboratories all over the world.

Extensive studies on surface forces in thin liquid films have been performed. A quantitative assessment of the main theory of colloid stability, the DLVO theory, was made. A new vision of the electrostatic interactions in liquid films has been developed. For the first time, the values of the diffuse electric layer potential at the liquid/air interface and the isoelectric points at this interface have been determined. These parameters are very informative for gaining an understanding of the charge nature and the electrostatic interaction, respectively. Barrier and non-barrier transitions in the disjoining pressure isotherm of foam films from liposome suspensions of phospholipids have been obtained experimentally.

DLVO and non-DLVO surface forces in liquid films from amphiphilic block copolymers (PEO–PPO–PEO type) have been determined. The transition from electrostatic to steric stabilization has been elucidated by determination of the critical electrolyte concentration, which divides the two types interactions. It was found that the electrostatic repulsion arises from the charge at the water/air interface due to preferential adsorption of OH^- ions. For the non-DLVO surface forces, brush-to-brush contact was established and the disjoining pressure isotherm followed the de Gennes scaling theory.

A new approach to amphiphile bilayers, the thinnest Newton black films, has been developed. A microscopic theory of the formation and stability of amphiphile bilayers was created. The rupture of bilayers was considered on the basis of a fluctuation mechanism of formation of nanoscopic holes in the bilayers. The hole formation was treated as a nucleation process of a new phase in a two-dimensional system with short-range intermolecular forces. Free rupture and deliberate rupture (by α-particles) of bilayers have been described. The role of bulk surfactant concentration for the formation and stability of amphiphile bilayers was demonstrated. A number of important parameters, the binding energy of an amphiphile molecule in the bilayer and the specific hole linear energy, which are important characteristics of the short-range surface forces in bilayers, were determined. Also for the first time the equilibrium surfactant concentration has been found, at which the bilayer (in contact with the bulk phase) is thermodynamically stable. In that way the ruptured, the metastable and the stable amphiphile bilayers can be clearly differentiated.

A two-dimensional chain-melting phase transition in foam bilayers was established for the first time. The binding energy of two neighboring phospholipid molecules was determined for the gel and liquid crystalline state of the bilayers from several phospholipids. It is to be expected that foam bilayers from phospholipids could be used as a model for the investigation of short-range forces in biological structures, of interactions between membranes, etc.

A new theoretical vision of polyhedric foams has been developed. It was verified experimentally for solutions of different surfactants, amphiphilic polymers and natural and technological mixtures. Methods for the differentiation of the processes connected with the syneresis (drainage) and stability of foams by creating a pressure gradient in the liquid phase have been developed. In that way the processes and factors acting in liquid channels and foam films are distinguished. On this basis, new methods and equipment have been developed, for foam stability determination at constant capillary pressure (foam pressure drop technique), rapid foam rupture, effective foam concentration and separation, water purification from surfactants, foam elimination in waste materials of nuclear fuel processing, effective foam formation at surfactants with high self-stabilizing ability during oil recovery, etc.

The lung surfactant system has been studied on the basis of theoretical and experimental investigations of amphiphile bilayers from amniotic fluid and alveolar surfactant. A new *in vitro* model for studying alveolar surface and stability, namely the microscopic foam film, has been introduced under the conditions in the lung alveolus: capillary pressure, radius, electrolyte concentration and temperature. It was shown that under these conditions, a foam bilayer stabilized by short-range interaction forces was formed and new parameters characterizing its formation and stability were introduced. On this basis, new clinical methods for the diagnosis of lung maturity and assessment of surfactant lung maturity of newborns have been created. The very good fit of the clinical results and the parameters of the *in vitro* model have allowed a new hypothesis to be created for the structure of the alveoli, namely an ordered structure in contrast to the widely accepted "monolayer" model. The most significant feature of the offered new model for the alveolar structure is that its stability is determined by the lateral short-range interactions in addition to the normal interactions between the ordered molecules. This gives the possibility of the quantitative study of the lung surfactant system and the processes related to the main physiological process – breathing.

The newly created method for fetal lung maturity assessment has a number of advantages: high precision (90%), a small quantity of liquid used (1 cm^3) and speed – the result is ready in about 20 minutes. The method for lung maturity diagnosis has also been very successfully developed for therapy control, i.e. defining the action of therapeutic surfactants, which cure the respiratory distress syndrome. This creates the possibility of looking for the most effective medicines to influence the lung surfactant system in respiratory distress.

From the above description, it is clear that Professor Exerowa has made significant original contributions in the field of colloid and interface science, for which she has been awarded the highest possible scientific position in Bulgaria, namely an Academician. A list of her most important publications is provided.

Selected Publications

1 Foam and Foam Films. Monograph in the Series *Studies in Interface Science*. D. Exerowa, P. M. Kruglyakov. Elsevier Science, Amsterdam, 1998, pp. 796.

2 Über den elektostatischen und van der Waalsschen zusätzlichen Druck in wässeringen Schaumfilmen. A. Scheludko, D. Exerowa, *Kolloid-Z., 168* (**1960**) 24.

3 Effect of Adsorption, Ionic Strength and pH on the Potential of the Diffuse Electric Layer. D. Exerowa, *Kolloid-Z., 232* (**1969**) 703.

4 Some Techniques for the Investigation of Foam Stability. D. Exerowa, Kh. Khristov, I. Penev, in: *Foams* (R. Akers, Ed.), Academic Press, London, 1976, p. 109.

5 Influence of the Pressure in the Plateau–Gibbs Borders on the Drainage and the Foam Stability. Khr. Khristov, P. M. Kruglyakov, D. Exerowa, *Colloid Polym. Sci., 257* (**1979**) 506.

6 Nucleation Mechanism of Rupture of Newtonian Black Films. I. Theory. D. Kashchiev, D. Exerowa, *J. Colloid Interface Sci., 77* (**1980**) 501.

7 Common Black and Newton Film Formation. D. Exerowa, A. Nikolov, M. Zacharieva, *J. Colloid Interface Sci., 81* (**1981**) 419.

8 Newtonian Black Films Rupture by Irradiation with α-Particles. I. Stochastic Model of the Phenomenon. I. Penev, D. Exerowa, *J. Colloid Interface Sci., 87* (**1982**) 5.

9 Influence of the Type of Foam Films and the Type of Surfactants on Foam Stability. Khr. Khristov, D. Exerowa, P. M. Kruglyakov, *Colloid Polym. Sci., 261* (**1983**) 265.

10 Nucleation Mechanism of Rupture of Newtonian Black Films. II. Experiment. D. Exerowa, B. Balinov, D. Kashchiev, *J. Colloid Interface Sci., 94* (**1983**) 45.

11 Bilayer Lipid Membrane Permeation and Rupture Due to Hole Formation. D. Kashchiev, D. Exerowa, *Biochim. Biophys. Acta, 732* (**1983**) 133.

12 Hole-mediated Stability and Permeability of Bilayers. D. Exerowa, D. Kashchiev, *Contemp. Phys., 27* (**1986**) 429.

13 Method for Assessment of Fetal Lung Maturity. D. Exerowa, Zdr. Lalchev, B. Marinov, K. Ognianov, *Langmuir, 2* (**1986**) 664.

14 Bilayer and Multilayer Foam Films – Model for Study of the Alveolar Surface and Stability. D. Exerowa, Zdr. Lalchev, *Langmuir, 2* (**1986**) 668.

15 Direct Measurement of Disjoining Pressure in Black Foam Films. I. Films from an Ionic Surfactant. D. Exerowa, T. Kolarov, Khr. Khristov, *Colloids Surf., 22* (**1987**) 171.

16 Newtonian Black Films Stabilized with Insoluble Monolayers Obtained by Adsorption from the Gas Phase. D. Exerowa, R. Cohen, A. Nikolova, *Colloids Surf., 24* (**1987**) 43.

17 Stability and Permeability of Bilayers. D. Exerowa, D. Kashchiev, D. Platikanov, *Adv. Colloid Interface Sci., 40* (**1992**) 201.

18 Phase Transitions in Phospholipid Foam Bilayers. D. Exerowa, A. Nikolova, *Langmuir, 8* (**1992**) 3102.

19 Foam Bilayer from Amniotic Fluid: Formation and Phase State. A. Nikolova, D. Exerowa, *Langmuir, 12* (**1996**) 1846.

20 DLVO and Non-DLVO Surface Forces in Foam Films from Amphiphilic Block Copolymers.

R. Sedev, D. Exerowa, *Adv. Colloid Interface Sci., 83* (**1999**) 111.

21 Structure and Surface Energy of the Surfactant Layer on the Alveolar Surface.
D. Kashchiev, D. Exerowa, *Eur. Biophys. J., 30* (**2001**) 34.

22 Chain-melting Phase Transition and Short-range Molecular Interactions in Phospholipid Foam Bilayers.
D. Exerowa, *Adv. Colloid Interface Sci., 96* (**2002**) 75.

23 Foam Films as Instrumentation in the Study of Amphiphile Self-assembly.
E. Mileva, D. Exerowa, *Adv. Colloid Interface Sci., 100–102* (**2003**) 547.

24 Foam and Wetting Films: Electrostatic and Steric Stabilization.
D. Exerowa, N. Churaev, T. Kolarov, N. E. Esipova, N. Panchev, Z. M. Zorin, *Adv. Colloid Interface Sci., 104* (**2003**) 1.

25 Thin Liquid Films from Phospholipids: Formation, Stability and Phase Transitions.
D. Exerowa, *Prog. Colloid Polym. Sci., 128* (**2004**) 135.

26 Amphiphile Bilayers from DPPC: Bilayer Lipid Membranes (BLM) and Newton Black Films (NBF).
D. Exerowa, R. Todorov, L. Nikolov, *Colloids Surf. A, 250* (**2004**) 195.

Professor Dimo Platikanov, PhD, DrSc

Dimo Platikanov was born on 2 February 1936 in Sofia, Bulgaria. After finishing secondary school in Sofia (1953), he studied chemistry at the University of Sofia. In 1958 he obtained the degree of Diploma-Chemist (equal to an MSc). His thesis was in the area of colloid science, performed in the Department of Physical Chemistry. His scientific career began at the Department of Physical Chemistry of the University of Sofia, which has continued to be his workplace up to the present. As a Junior Assistant Professor he completed his PhD thesis in 1968, supervised by the great Bulgarian colloid scientist Professor A. Scheludko. He became Associate Professor in 1970. He spent the academic year 1973–74 at the University of Munich, Germany, as Alexander von Humboldt Foundation Research Fellow. In 1989 he succeeded Professor Scheludko as Head of the Department of Physical Chemistry at the University of Sofia. His DrSc thesis was successfully defended also in 1989 and a year later he became Professor in Physical Chemistry. During the last 30 years he had lectured in the main course of Physical Chemistry to students in the Faculty of Chemistry at the University of Sofia. He has also been supervisor of many PhD and MSc students and postdoctoral fellows, thus bringing many young people into scientific research.

The scientific results of Dimo Platikanov have been published in about 120 papers in the scientific literature. He also published two extensive chapters together with Dotchi Exerowa: “Thin Liquid Films”, in Fundamentals of Interface and Colloid Science, edited by J. Lyklema (Elsevier, 2005), and “Symmetric Thin Liquid Films with Fluid Interfaces”, in *Emulsions and Emulsion Stability*, edited by J. Sjoblom (Taylor and Francis, 2005). Owing to his original scientific work and publications, Dimo Platikanov has been invited many times to lecture at international conferences and seminars in leading scientific institutions. He has been a member of many scientific committees of conferences and an editorial board member of four international scientific journals. In 1997 he was co-chairman of the 9th International Conference on Surface and Colloid Science. He was elected a member of the Council and later President (2000–2003) of the International Association of Colloid and Interface Scientists (IACIS), and is currently a member of the Standing Committee and the Council of IACIS. In the past 15 years he had been member of the Standing Committee of the European Chemistry at Interfaces Conferences and since 2004 he has been member of the Physical and Biophysical Chemistry Division Committee of IUPAC. He has also been President of the Humboldt Union in Bulgaria since 2002.

The contributions of Dimo Platikanov are mainly in the field of thin liquid films, liquid interfaces, three-phase contact, wetting, etc. The many original results obtained have stimulated new directions in the development of knowledge in the field of thin liquid films and also the physical chemistry of the interfacial phenomena.

Most of the scientific results obtained by Dimo Platikanov are experimental. In most cases a unique experimental method has been developed, especially for corresponding studies. Such methods allowed the measurement of important parameters characterizing the system studied (wetting film, foam film, black film, etc.) and its properties. About 20 unique experimental techniques have been developed in his experimental work.

An original experimental cell developed for the investigation of microscopic, circular, wetting liquid films on solid surfaces proved to be very effective and it has been used by many researchers in different countries. The Scheludko–Exerowa micro-interferometric technique has been extended in conjunction with this cell. Using this method, the shape of the three-phase contact gas/liquid film + meniscus/solid surface has been studied in detail under dynamic and equilibrium conditions. The experimental data for the development of the 'dimpling' in the initial stages of the formation of a wetting film were subsequently used for elaborating the hydrodynamic theory of this phenomenon. The equilibrium profile of the transition zone between a wetting film and the bulk liquid has been experimentally obtained and data for the 'contact thickness' have been calculated using a disjoining pressure theory.

Studies on surface forces in thin wetting films have also been performed. The disjoining pressure/thickness isotherm measured for a wetting benzene film on mercury showed a complicated character of the molecular interactions in this system. Aqueous wetting films (without or with amphiphilic PEO–PPO–PEO block copolymers added) on surfaces of fused quartz and of silicon carbide have been studied in detail. The electrolyte concentration and the solid surface pretreatment strongly influenced their stability. At electrolyte concentrations where the electrostatic disjoining pressure was fully suppressed, the disjoining pressure/thickness isotherm was measured using the dynamic method. It has been interpreted by the superposition of a negative van der Waals component and a positive steric component (due to brush-to-brush contact) of the disjoining pressure, hence the electrostatic and steric stabilization of wetting films have been distinguished. Surface forces in thin, non-aqueous foam films have also been studied. The disjoining pressure/thickness isotherms for films from benzene and chlorobenzene were measured using the dynamic method. The effect of electromagnetic retardation of the dispersion molecular interactions has been experimentally established and the Hamaker constant and London wavelength calculated.

Extensive investigations on common and Newton black foam films (CBFs and NBFs) have been performed using a number of unique experimental methods. Through deformation of a black film in electric field, the reversibility of the

CBF/NBF transition and vice versa and also the electrical neutrality of the thinnest NBFs have been proved. The measured longitudinal electrical conductivity and the transport numbers of ions in black films gave information about their structure: a three-layer structure for the CBF whereas the NBF is a bilayer of amphiphilic molecules. Other new methods allowed the measurement of the film tension and the line tension of NBFs. The film tension of NBFs from sodium dodecyl sulfate was found to be constant over wide range of static and dynamic conditions; this was not the case with NBFs from phospholipids. The values of the line tension of NBFs from sodium dodecyl sulfate have been determined – positive at low and negative at high electrolyte concentrations.

The gas permeability through foam films has been determined for several cases using two newly developed methods. The gas permeability coefficient of NBFs depends strongly on the surfactant concentration. This dependence was in good agreement with the nucleation theory of fluctuation formation of nanoscopic holes responsible for the bilayer stability and permeability. A very interesting result obtained is that the gas permeability coefficient of thicker CBFs is 2–3 times larger than that for the thinnest NBFs. Another important result is that the coefficient of the CBFs increases with decreasing electrolyte concentration (increasing film thickness), passing through a maximum.

Dynamic contact angles and gas permeability coefficients of NBFs from aqueous dispersions of phospholipids have been measured by an original method. The results for two types of solutions, (1) liposome suspension and (2) ethanol + water solution of phospholipids, were found to be very different. The contact angles in case (1) vary strongly under dynamic conditions whereas in case (2) they remain almost constant. The gas permeability is larger in case (2) than in case (1). The results were discussed in connection with the thickness and structure of the NBFs from the two types of solutions, taking into account the solubility (or insolubility) and the hydration of the adsorption layers of phospholipid molecules.

Extensive investigations of black films from aqueous protein solutions showed more complicated behavior. A dynamic hysteresis of the contact angles has been established and studied. The results have been interpreted in connection with the rheological properties of the protein adsorption layers.

A combination of the Langmuir–Blodgett technique and neutron activation analysis has been used to determine the stoichiometry of the interaction between arachidic acid monolayers and cadmium or barium ions dissolved in the subsolution. The stability constants of the corresponding arachidic soaps formed in the monolayer have been calculated from the experimental data. Equations for equilibrium constants of arachidic acid monolayer–subsolution counterion ion exchange were also derived. The interaction of octadecylamine monolayers with the subsolution phosphate counterions at different pH and ionic strength have been studied by the same combination of techniques and the stability constant of octadecylammonium hydrogenphosphate has been estimated. A series of experiments on the elasticity of soluble and non-soluble monolayers on a liquid substrate have also been performed.

From the above description, it is clear that Professor Platikanov has made significant original scientific contributions in the field of colloid and interface science. His publications allowed him to become internationally known and for this reason he has been elected President of the International Association of Colloid and Interface Scientists. A list of his most important papers is provided.

Selected Publications

1 Untersuchung dünner flüssiger Schichten auf Quecksilber.
A. Scheludko, D. Platikanov, *Kolloid-Z.*, *175* (**1961**) 150.

2 Experimental Investigation on the "Dimpling" of Thin Liquid Films.
D. Platikanov, *J. Phys. Chem.*, *68* (**1964**) 3619.

3 Disjoining Pressure in Thin Liquid Films and the Electro-Magnetic Retardation Effect of the Molecular Dispersion Interactions.
A. Scheludko, D. Platikanov, E. Manev, *Discuss. Faraday Soc.*, *40* (**1965**) 253.

4 Electrical Conductivity of Black Foam Films.
D. Platikanov, N. Rangelova, in *Research in Surface Forces* (B.V. Derjaguin, Ed.), Vol. 4, Consultants Bureau, New York, 1972, p. 246.

5 Orientation of Nonionic Surfactants on Solid Surfaces: n-Alkyl Polyglycol Ethers on Montmorillonite.
D. Platikanov, A. Weiss, G. Lagaly, *Colloid Polym. Sci.*, *255* (**1977**) 907.

6 Free Black Films of Proteins.
D. Platikanov, G.P. Yampolskaya, N. Rangelova, Zh. Angarska, L.E. Bobrova, V. N. Izmailova, *Colloid J. USSR*, *42* (**1980**) 753; *43* (**1981**) 149; *52* (**1990**) 442.

7 Line Tension of Newton Black Films.
D. Platikanov, M. Nedyalkov, A. Scheludko, V. Nasteva, *J. Colloid Interface Sci.*, *75* (**1980**) 612, 620.

8 Interaction of Octadecylamine Monolayers with Phosphate Counterions.
J.G. Petrov, I. Kuleff, D. Platikanov, *J.Colloid Interface Sci.*, *109* (**1986**) 455.

9 Equilibrium Constants of Ion Exchange Reactions between Fatty Acid Monolayers and Dissolved Counterions.
J.G. Petrov, D. Platikanov, *Colloid Polym. Sci.*, *265* (**1987**) 65.

10 The Transition Region between Aqueous Wetting Films on Quartz and the Adjacent Meniscus.
Z. Zorin, D. Platikanov, T. Kolarov, *Colloids Surf.*, *22* (**1987**) 147; *51* (**1990**) 37.

11 Method for Direct Measurement of the Film Tension of Black Foam Films.
D. Platikanov, M. Nedyalkov, N. Rangelova, *Colloid Polym. Sci*, *265* (**1987**) 72; *269* (**1991**) 272.

12 On the Curvature Dependence of the Film Tension of Newton Black Films.
D. Platikanov, M. Nedyalkov, A. Scheludko, B.V. Toshev, *J.Colloid Interface Sci*, *121* (**1988**) 100.

13 Line Tension in Three-phase Equilibrium Systems.
B.V. Toshev, D. Platikanov, A. Scheludko, *Langmuir*, *4* (**1988**) 489.

14 Method for Direct Measurement of Film Tension of Newton Black Films on a Diminishing Bubble.
M. Nedyalkov, G. Schoepe, D. Platikanov, *Colloids Surf.*, *47* (**1990**) 95.

15 Disjoining Pressure, Contact Angles and Line Tension in Free Thin Liquid Films.
B.V. Toshev, D. Platikanov, *Adv. Colloid Interface Sci.*, *40* (**1992**) 157.

16 On the Mechanism of Permeation of Gas through Newton Black Films at Different Temperatures.
M. Nedyalkov, R. Krustev, A. Stankova, D. Platikanov, *Langmuir*, *8* (**1992**) 3142; *12* (**1996**) 1688.

17 Permeability of Common Black Films to Gas.
R. Krustev, D. Platikanov, M. Nedyalkov, *Colloids Surf.*, *79* (**1993**) 129; *123/124* (**1997**) 383.

18 Linear Energy with Positive and Negative Sign.
D. Exerowa, D. Kashchiev, D. Platikanov, B.V. Toshev, *Adv. Colloid Interface Sci.*, *49* (**1994**) 303.

19 Thin Liquid Films from Polyoxyethylene–Polyoxypropylene Block Copolymer on the Surface of Fused Quartz.
B. Diakova, M. Kaisheva, D. Platikanov, *Colloids Surf. A, 190* (**2001**) 61.

20 Thin Wetting Films from Aqueous Electrolyte Solutions on SiC/Si Wafer.
B. Diakova, C. Filiatre, D. Platikanov, A. Foissy, M. Kaisheva, *Adv. Colloid Interf. Sci., 96* (**2002**) 193.

21 Phospholipid Black Foam Films: Dynamic Contact Angles and Gas Permeability.
D. Platikanov, M. Nedyalkov, V. Petkova, *Adv. Colloid Interf. Sci., 101* (**2003**) 185; *104* (**2003**) 37.

22 Physico-chemical Background of the Foaming Protein Separation for Waste Minimization.
D. Platikanov, V. N. Izmailova, G P. Yampolskaya, in *Role of Interfaces in Environmental Protection* (S. Barany, Ed.), *NATO Science Series, IV/24* (**2003**) 507.

23 Thin Wetting Films from Aqueous Solutions of a Polyoxyethilene–Polyoxypropylene Block Copolymer on Silicon Carbide Surface.
B. Diakova, D. Platikanov, R. Atanassov, M. Kaisheva, *Adv. Colloid Interface Sci., 104* (**2003**) 25.

24 Thin Liquid Films.
D. Platikanov, D. Exerowa, in *Fundamentals of Interface and Colloid Science,* Vol. 5 (J. Lyklema, Ed.), Elsevier, Amsterdam, 2005, Chap. 6, p. 6.1.

25 Symmetric Thin Liquid Films with Fluid Interfaces.
D. Platikanov, D. Exerowa, in *Emulsions and Emulsion Stability,* 2nd edn. (J. Sjöblom, Ed.), CRC Press, Taylor and Francis, New York, 2006, Chap. 3, p. 127.

List of Contributors

Zbigniew Adamczyk
Institute of Catalysis and Surface Chemistry
Polish Academy of Sciences
ul. Niezapominajek 8
30-239 Kraków
Poland

Idit Amar-Yuli
Casali Institute of Applied Chemistry
The Hebrew University of Jerusalem
Jerusalem 91904
Israel

Makoto Aratono
Department of Chemistry
Faculty of Sciences
Kyushu University
Hakozaki 6-10-1, Higashiku
Fukuoka 812-8581
Japan

Jakub Barbasz
Institute of Catalysis and Surface Chemistry
Polish Academy of Sciences
ul. Niezapominajek 8
30-239 Kraków
Poland

James K. Beattie
School of Chemistry
University of Sydney
Sydney, NSW 2006
Australia

Nicolay V. Churaev
A.N. Frumkin Institute of Physical Chemistry and Electrochemistry
Russian Academy of Science (IPCE RAS)
Leninsky Prospect 31
119991 Moscow
Russia

Andrei S. Dukhin
Electrokinetic Technology
Goldens Bridge
New York, NY 10526
USA

Stanislav S. Dukhin
Electrokinetic Technology
Goldens Bridge
New York, NY 10526
USA

Michele Ferrari
Consiglio Nazionale delle Ricerche
Istituto per l'Energetica e le Interfaci
via De Marini, 6
16149 Genova
Italy

Colloids and Interface Science Series, Vol. 2
Colloid Stability: The Role of Surface Forces, Part II. Edited by Tharwat F. Tadros

ISBN: 978-3-527-31503-1

Nissim Garti
Casali Institute of Applied Chemistry
The Hebrew University of Jerusalem
Jerusalem 91904
Israel

Katarzyna Jaszczółt
Institute of Catalysis and Surface Chemistry
Polish Academy of Sciences
ul. Niezapominajek 8
30-239 Kraków
Poland

Pyotr M. Kruglyakov
Penza State University of Architecture and Building
44028 Penza
Russia

Libero Liggieri
Consiglio Nazionale delle Ricerche
Istituto per l'Energetica e le Interfaci
via De Marini, 6
16149 Genova
Italy

Hiroki Matsubara
Department of Chemistry
Faculty of Sciences
Kyushu University
Hakozaki 6-10-1, Higashiku
Fukuoka 812-8581
Japan

Aneta Michna
Institute of Catalysis and Surface Chemistry
Polish Academy of Sciences
ul. Niezapominajek 8
30-239 Kraków
Poland

Francesca Ravera
Consiglio Nazionale delle Ricerche
Istituto per l'Energetica e le Interfaci
via De Marini, 6
16149 Genova
Italy

Jarl B. Rosenholm
Department of Physical Chemistry
Åbo Akademi University
Porthansgatan 3
20500 Åbo (Turku)
Finland

Brian Saunders
School of Materials
University of Manchester
Grosvenor Street
Manchester, M1 7HS
United Kingdom

Vladimir D. Sobolev
A.N. Frumkin Institute of Physical Chemistry and Electrochemistry
Russian Academy of Science (IPCE RAS)
Leninsky Prospect 31
119991 Moscow
Russia

Victor M. Starov
Department of Chemical Engineering
Loughborough University
Loughborough LE11 3TU
United Kingdom

Natalia G. Vilkova
Penza State University of Architecture and Building
44028 Penza
Russia

Brian Vincent
School of Chemistry
University of Bristol
Cantock's Close
Bristol, BS8 1TS
United Kingdom

Maria Zembala
Institute of Catalysis and Surface Chemistry
Polish Academy of Sciences
ul. Niezapominajek 8
30-239 Kraków
Poland

1
Wetting of Surfaces and Interfaces: a Conceptual Equilibrium Thermodynamic Approach

Jarl B. Rosenholm

Abstract

Owing to the focus on molecular engineering of intelligent materials, growing interest has been focused on the specific interactions occurring at molecular distances from a surface. Advanced experimental techniques have been developed including instruments able to measure directly interactions at nanometer distances and to identify special structural features with comparable perpendicular and lateral resolution. With this new information at hand, the theories of molecular interactions have been re-evaluated and developed further to encompass specific interactions such as Lewis and Brønsted acidity and basicity. However, the new theories are based on critical approximations, making the upscaling to macroscopic condensed systems uncertain. Therefore, the aim of this study was to evaluate some recent models by introducing macroscopic work functions (cohesion, adhesion, spreading and immersion) of wetting of solid surfaces within the proper conceptual thermodynamic (macroscopic) framework. The properties of binary and ternary systems are discussed with the focus on four non-ideal inorganic (SiO_2 and TiO_2) model substrates. The results obtained after applying the recent and more traditional models for dispersive and specific (polar) interactions are compared with those utilizing simplifying assumptions. The sources of uncertainties are sought, e.g. from the contribution of surface pressure determined from contact angle and adsorption isotherms. Finally, the influence of chemical and structural heterogeneities and also external stimuli on wetting is briefly discussed.

1.1
Introduction

Owing to the focus on molecular engineering of intelligent materials, growing interest has been focused on the specific interactions occurring at molecular distances from a surface. Advanced experimental techniques have been developed including instruments able to measure directly interactions at nanometer dis-

Colloids and Interface Science Series, Vol. 2
Colloid Stability: The Role of Surface Forces, Part II. Edited by Tharwat F. Tadros

ISBN: 978-3-527-31503-1

tances and identify special structural features with the same perpendicular and lateral resolution. Consequently, the theories developed for the dispersive interaction of hydrocarbons has been re-evaluated and developed further to encompass specific interactions such as both Lewis and Brønsted acidity and basicity. However, the new theories are based on molecular properties for which the up-scaling to macroscopic condensed systems includes a number of critical approximations.

This chapter takes the opposite, conceptual approach. The basic thermodynamic functions are chosen as the basic framework by introducing macroscopic work functions (cohesion, adhesion, spreading and immersion) of wetting of solid surfaces. The properties and wetting processes characterizing binary and ternary systems are discussed with reference to recent molecular models and more traditional models for dispersive and specific (polar) interactions including surface pressure and adsorption approaches. The Lewis and Brønsted acid–base interaction is carefully separated, since they diverge both in strength and distance. This fact is frequently disregarded. Care is taken to use systematic and clear indexing.

A number of new simplifying experimental procedures to utilize the models developed are suggested. Again, the different frameworks that the models represent are kept apart and later compared mutually with key properties. In order to facilitate the comparison of the data presented, the analysis is focused on four non-ideal solid samples of silica and titania at equilibrium. Finally, the influence of chemical and structural heterogeneities and also external stimuli on wetting is placed within the same thermodynamic conceptual framework.

1.2
Thermodynamic Reference Parameters

The state of a system is defined by its internal energy (U or E), which equals the sum of heat (Q) and work (W). For a spontaneous reaction to occur U is expected to be negative. Using the conventions for heat and work we may write the differential equation

$$dU = dQ + dW \tag{1}$$

It should be noted that U is a state function and is thus dependent solely on the initial and final state ($dU \approx \Delta U$), whereas the work (dW) is dependent on the path between these two states. IUPAC recommends that both are positive when there is an increase in energy of the system. For reversible processes the heat exchange is customarily exchanged for the entropic work ($dQ = TdS$). The change of work is defined as the product of two conjugative properties, the change being given by the extensive property multiplied by its intensive conjugated pair (Table 1.1) [1]. All the work functions are exchangeable and must be considered when defining the internal energy.

Table 1.1 Typical internally exchangeable conjugated extensive and intensive work variables [1].

Process	Extensive variable	Intensive variable
Thermal	Entropy	Temperature
P–V	Volume	Pressure
Chemical	Amount	Chemical potential
Surface	Surface area	Surface tension (energy)
Electric	Amount of charge	Electric potential
Magnetic	Magnetization	Magnetic field strength

Selecting the most typical set of parameters for a system containing two homogeneous phases (α and β) separated by one flat interface, we obtain for the differentials of the internal energy (U) of each phase

$$\mathrm{d}U = T\mathrm{d}S - P\mathrm{d}V + \gamma\mathrm{d}A + \mu\mathrm{d}n \tag{2}$$

As shown, the change is directed only to the extensive properties, while the intensive variable are kept constant ($\mathrm{d}W = -P\mathrm{d}V + \gamma\mathrm{d}A + \mu\mathrm{d}n$). The differentials of the intensive state variables have been omitted as stated specifically by the Gibbs-Duhem relationship:

$$S\mathrm{d}T - V\mathrm{d}P + A\mathrm{d}\gamma + n\mathrm{d}\mu \equiv 0 \tag{3 a}$$

The Helmholtz energy (F or A), the enthalpy (H) and the Gibbs free energy (G) are all related to the internal energy:

$$\mathrm{d}F = \mathrm{d}(U - TS) = -S\mathrm{d}T - P\mathrm{d}V + \gamma\mathrm{d}A + \mu\mathrm{d}n \tag{3 b}$$

$$\mathrm{d}H = \mathrm{d}(U + PV) = T\mathrm{d}S + V\mathrm{d}P + \gamma\mathrm{d}A + \mu\mathrm{d}n \tag{3 c}$$

$$\mathrm{d}G = \mathrm{d}(H - TS) = -S\mathrm{d}T + V\mathrm{d}P + \gamma\mathrm{d}A + \mu\mathrm{d}n \tag{3 d}$$

Each of them considers different dependences on the working state variables, T, P, V and S. An extended evaluation of these relationships is presented elsewhere [1].

If more than one component (liquid mixture) or if more than one surface (crystal facets) is accounted for, a summation over these variables must be considered separately. Since non-relaxed solid surfaces are included in the considerations, it is advisable to distinguish specifically the relaxed surface tension of the liquid components denoted $\gamma/\mathrm{mN\ m^{-1}}$ (force) from the (strained) surface energies of the solids denoted $\sigma/\mathrm{mJ\ m^{-2}}$ (energy). Assuming that a pure liquid (L) is placed in contact with a smooth and homogeneous solid surface (S), maintaining the temperature and pressure and composition constant, we may, in the absence of other work functions, derive the Young equation [2] in the following way (Fig. 1.1):

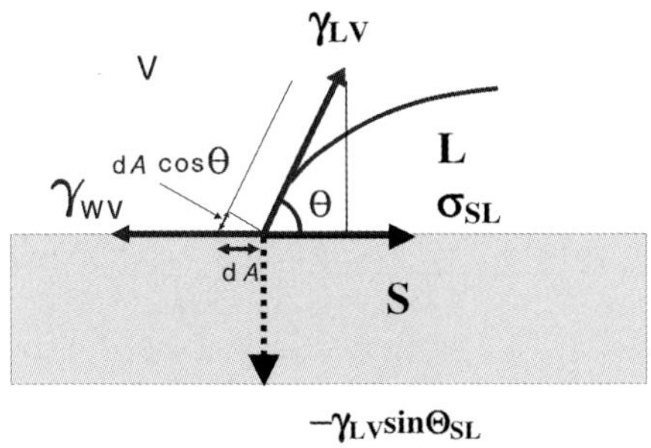

Fig. 1.1 The contact angle of a sessile drop on an ideally smooth and homogeneous surface is defined by the vectorial stress laid upon the three-phase (solid–liquid–vapor) contact line (tpcl, Young equation). The transverse component ($-\gamma_{LV}\sin\Theta_{SL}$) may be considered to represent the frictional pinning of the tpcl.

$$dG_{SL} = \Sigma_i \gamma_i dA_i = \sigma_{SV} dA - \gamma_{LV}(dA\cos\Theta_{SL}) - \sigma_{SL} dA \tag{4 a}$$

where the subscript i represents V=vapor, L=liquid and S=solid. At equilibrium:

$$dG_{SL}/dA \approx \Delta G_{SL}/A = 0 = \sigma_{SV} - \gamma_{LV}\cos\Theta_{SL} - \sigma_{SL} \tag{4 b}$$

Hence under these circumstances (but only then) the work function equals the Gibbs free energy per unit area with an opposite sign.

The change in free surface energy may also be expressed by the Dupré equation [3] for the work of adhesion:

$$dG_{SL}/dA \approx \Delta G_{SL}/A = \Delta G_A^s = -W_A = \sigma_{SL} - \gamma_{LV} - \sigma_{SV} \tag{5}$$

The contact angle can be determined graphically, geometrically by assuming that the drop is represented by a hemisphere and by deriving it from the expression for the Laplace pressure [4]. Note that the pinning of the three-phase contact line (tpcl) may be represented by the frictional surface tension vector directed perpendicular to the surface.

Combining the Young and Dupré equations, the work done at the interface may be defined as four key wetting (work) functions (omitting the differential sign):

$$\text{Cohesion:} \quad W_C = C_{LL} = 2\gamma_{LV} \quad \text{or} \quad C_{SS} = 2\sigma_{SV} \tag{6 a}$$

$$\text{Adhesion:} \quad W_A = W_{SL} = \sigma_{SV} + \gamma_{LV} - \sigma_{SL} = \gamma_{LV}(\cos\Theta_{SL} + 1) \tag{6 b}$$

$$\text{Spreading:} \quad W_S = S_{SL} = \sigma_{SV} - \gamma_{LV} - \sigma_{SL} = W_{SL} - C_{LL} = \gamma_{LV}(\cos\Theta_{SL} - 1) \tag{6 c}$$

$$\text{Immersion:} \quad W_I = I_{SL} = \sigma_{SV} - \sigma_{SL} = \gamma_{LV}\cos\Theta_{SL} \tag{6 d}$$

However, as shown, the work functions are defined for the separation of the interconnected phases (work done by the system), while the Gibbs free energy for adhesion is usually defined as uniting the surfaces. Hence for a spontaneous process they have opposite signs!

If the processes occurring at the sharp solid–liquid interface (S) alone are considered, then the Gibbs dividing plane may be applied, being characterized by a zero volume and a zero surface excess of the liquid. For a single surface, when the two homogeneous phases have been subtracted ($V^s=0$) (α and β), we find [5]

$$dU^s = dU - dU^\alpha - dU^\beta = TdS^s + \gamma dA + \mu dn^s \tag{7 a}$$

$$dF^s = d(U^s - TS^s) = -S^s dT + \gamma dA + \mu dn^s \tag{7 b}$$

$$dH^s = d(U^s - \gamma A) = TdS^s - Ad\gamma + \mu dn^s \tag{7 c}$$

$$dG^s = d(H^s - TS^s) = -S^s dT - Ad\gamma + \mu dn^s \tag{7 d}$$

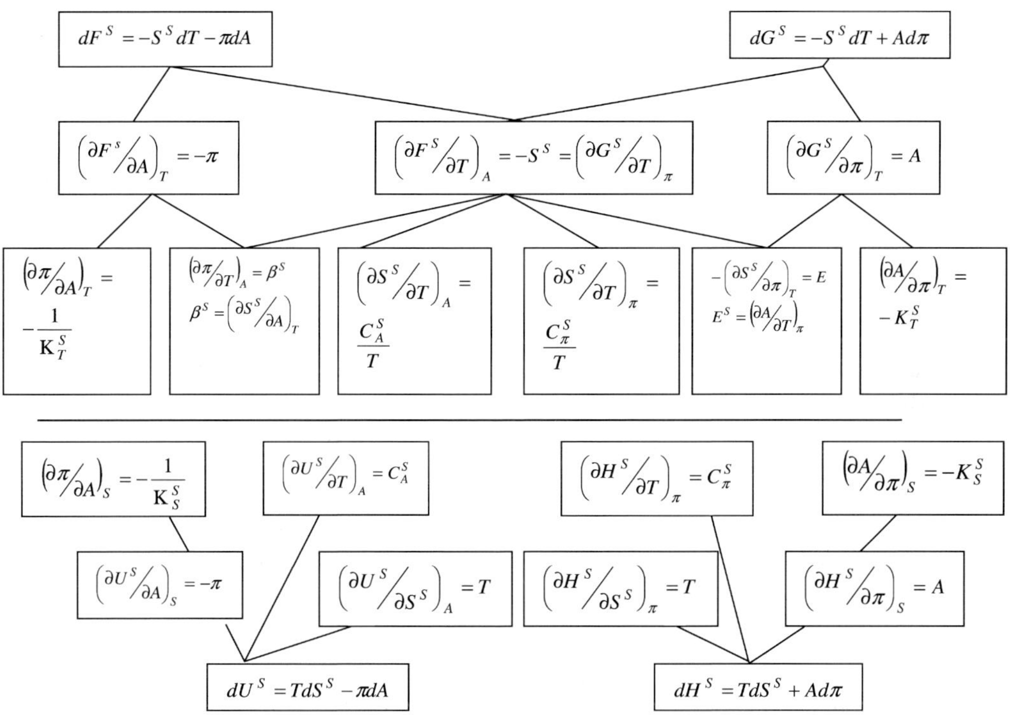

Fig. 1.2 The interrelationships between the first- and second-order partial derivatives according to the working state variables *V*, *T*, *P* and *S* and the free energies *F* and *G* on the one hand and U and *H* on the other for processes at the Gibbs surface.

Note that if the IUPAC recommendation for the surface enthalpy is followed, then neither the Gibbs free energy nor the enthalpy can be used to derive the Young equation. If the PV work is excluded ($V^s=0$), then it would appear to be more appropriate to equate the internal energy and the enthalpy in order to maintain the symmetry of the equations. However, if both the absolute value and the difference in the surface pressure is taken to be opposite to that of the surface tension (energy), then PdV terms and the VdP of the bulk systems correspond to the πdA and the $Ad\pi$ terms of the surface system, respectively [6].

It should be particularly pointed out that a conceptual analysis of the hierarchy of thermodynamic parameters (the Thermodynamic Family Tree [1]) both for bulk systems and for interfaces [6] reveals that the traditional "work" state parameters (T–S and P–V) are only sensitive to dramatic changes, providing information on the order of the phase transitions (Fig. 1.2). On the other hand, the next class of partial derivative parameters (heat capacities C_P, C_V and compressibilities K_T, K_S) and their cross-derivatives (their ratio γ, cubic expansion coefficient α and pressure coefficient β) are sensitive to higher order interactions, such as hydrogen bonding or Lewis acidity and basicity. In particular, they provide information on the extension of the interaction such as the cooperativity of molecular association (cf. lambda transitions). This has not been fully understood by those relating particular molecular properties to the macroscopic thermodynamic network (Fig. 1.2).

1.3
Wetting in Idealized Binary Systems

Viewed from the point of view of thermodynamics, the models for dispersive (hydrocarbon) interactions are usually based on the van der Waals gas law. The van Laar model for hydrocarbon liquids considers the components to be mixed in the ideal gaseous state and the non-ideality is averaged geometrically. These considerations form a base for the modeling of the dispersive interaction parameters.

1.3.1
Models for Dispersive Solid–Liquid Interactions

Traditionally, medium to long chain length (C_6–C_{16}) saturated hydrocarbon liquids have been utilized as standards for the fully dispersive (London) interactions with solids. Thus, the standard method of Zisman [7] relies on a range of contact angles measured for such hydrocarbons on a solid. When plotted as $\cos\Theta_L$ against the surface tension of the liquids, the extrapolation to $\cos\Theta_{SL}=1$ for experimental points falling on a straight line gives the critical surface tension (not surface energy), which is considered to correspond to the surface energy of the solid ($\sigma_{SV} \approx \gamma_{crit} \approx \lim\gamma_{LV}$). In Fig. 1.3, $\cos\Theta_{SL}$ is plotted as a function of the surface tension of seven probe liquids on silica [8].

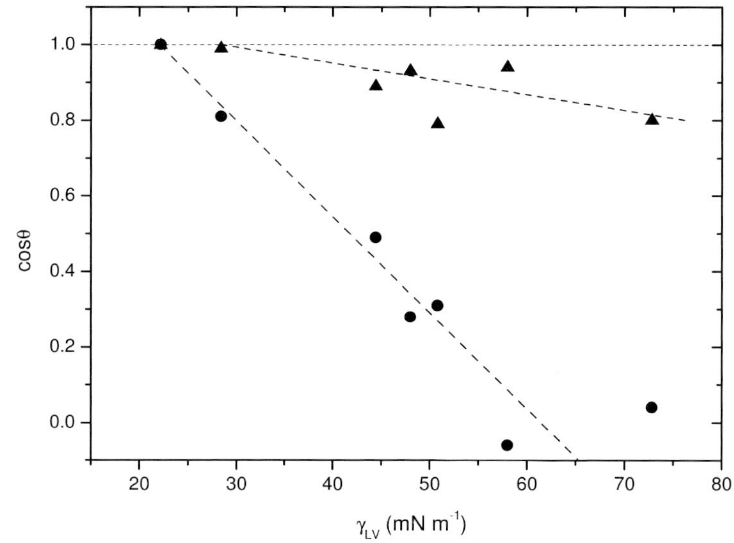

Fig. 1.3 Zisman plot for silanized hydrophobic silica (spheres) and neat hydrophilic silica (triangles). The probe liquids are in increasing order of the surface tension; octane, hexa-decane, α-bromonaphthalene, ethylene glycol, diiodomethane, formamide and water. The contact angles have been determined in air using the Laplace approach (data partially from [8, 42]).

It is seen that octane represents the critical liquid (γ_{crit}) for the hydrophobic silica, but its surface tension is too small for the determination of the surface tension of hydrophilic silica. Hexadecane may be considered to represent the critical liquid for the hydrophilic silica. On the other hand, water seems to have a too high surface tension in order to comply with the trend of the other probe liquids on hydrophobic silica. It thus appears that there exist a frame within the surface tensions and the surface energies are sufficiently close in order to provide physically relevant information. Outside this range the results extracted may be seriously distorted.

Instead of using the total surface tension, Fowkes subtracted a dispersive part and evaluated the excess as the surface pressure of a vapor film [9]. Later, Zettlemoyer, for example, identified this fraction as a polar ($\gamma^{p}_{LV}=\gamma_{LV}-\gamma^{d}_{LV}$) component [10]. The polar interaction should be understood as a specific (molecularly arresting) interaction without any specific nature (e.g. dipolar) in mind. This procedure enabled experimentalists also to use a broader range of probe liquids. Two straightforward alternatives for averaging the work of adhesion have been proposed based on the dispersive component of the surface tension/energy [9, 10]:

$$W_{SL} = 2(\sigma^{d}_{SV}\gamma^{d}_{LV})^{1/2} \quad \text{geometric average} \tag{8a}$$

$$W_{SL} = \sigma^{d}_{SV} + \gamma^{d}_{LV} \quad \text{arithmetic average} \tag{8b}$$

Combining these models with the previous definition of the work of adhesion gives the following expressions for the interfacial tension assuming that $\gamma_{LV} \approx \gamma_{LV}^{d}$ and $\sigma_{SV} \approx \sigma_{SV}^{d}$:

$$\sigma_{SL} = \sigma_{SV} + \gamma_{LV} - 2(\sigma_{SV}^{d}\gamma_{LV}^{d})^{1/2} \approx (\sqrt{\sigma_{SV}^{d}} - \sqrt{\gamma_{LV}^{d}})^2 \tag{9 a}$$

$$\sigma_{SL} = \sigma_{SV} + \gamma_{LV} - (\sigma_{SV}^{d} + \gamma_{LV}^{d}) \approx 0 \tag{9 b}$$

Owing to the square of the term in parentheses, the geometrically averaged interaction is always attractive, but for the critical surface tension ($\sigma_{S}^{d} \approx \gamma_{L}^{d} \approx \gamma_{crit}$) the interfacial surface tension vanishes. For the arithmetic mean the interfacial tension is irrespectively zero. This is inherent to the definition.

In the Scatchard-Hildebrand and the Regular Solution models the solution interaction is documented as the solubility parameter (δ). For the relationship with the surface tension ($\gamma \approx k_{\delta} V^{1/3} \delta^{y}$) Hildebrand [11] suggested k_{δ}=0.0376 and y=2.326, and Beerbower [12] arrived at k_{δ}= 0.0714 and y=2.000. The interfacial energy in the last case thus be related to the solubility parameters as

$$\sigma_{SL} \approx (k_S\delta_S - k_L\delta_L)^2 \tag{10}$$

where $k_i = \sqrt{k_{\delta}}$=0,2763.

Girifalco and Good [13] offered a geometric mean procedure involving the total surface tensions/energies corrected by a ratio (Φ). For London-van der Waals interactions it is represented by the ratio of arithmetic averaging to geometric averaging of the work of adhesion. For non-specific interactions this ratio is close to unity even when the substances are appreciably different, thus claiming equal validity for the geometric and arithmetic averaging procedures. Wu presented a harmonic mean model for the dispersive and polar interactions equaling the ratio when the ionization potentials are replaced by the surface tension/energy components [14]. Neumann and Sell suggested an equation of state model with an exponential dependence on the surface tension/energy components [15]. These methods have found only limited use and are not dealt with here.

Fowkes related the Hamaker constant to the GG ratio. For a solid–liquid contact he arrived at the following relationship [9]:

$$A_{SL} = \sqrt{A_{SV}^2} - 2\Phi\sqrt{A_{SV}}\sqrt{A_{LV}} + \sqrt{A_{LV}^2} \tag{11 a}$$

Only when Φ is not too far from unity (or when the ionization potentials are not too different), the equation reduces to

$$A_{SL} = (\sqrt{A_{SV}} - \sqrt{A_{LV}})^2 \tag{11 b}$$

in symmetry with geometric averaging of the dispersive interfacial energy (Eq. 9 a). The van der Waals gas law constants have also been related to the molecular Hamaker constant and the surface tension for liquids and surface energies for solids [16–18]:

$$A_{LL} = k_A \gamma_{LV} d_0^2 \tag{12 a}$$

$$A_{SS} = k_A \sigma_{SV} d_0^2 \tag{12 b}$$

where $k_A \approx 75.40$ for van der Waals liquids, 100.5 for pure hydrocarbons and 10.47 for semi-polar liquids. The equilibrium distance between the molecules in the condensed medium (d_0) has been reported to be 0.22 ± 0.05 nm [16] and $0.13 < d_0 < 1.7 \pm 0.01$ nm [17, 18]. It is therefore obvious that the specific (polar) interactions should be considered as an excess from the ideal gaseous state.

Owing to the direct relationship between the surface tension/energy and the work of cohesion (C) on the one hand and the Hamaker constant (A) on the other for fully or nearly dispersive substances, we can relate the work of cohesion and the Hamaker constant to the contact angle:

$$C_{SS} = 0.25 C_{LL} (\cos \Theta_{SL} + 1)^2 \tag{13 a}$$

$$A_{SS} = 0.25 A_{LL} (\cos \Theta_{SL} + 1)^2 \tag{13 b}$$

Conversely, the latter equation offers the possibility of estimating $\cos \Theta_L$ from tabulated data.

Fowkes [9] developed a method for evaluating the work of adhesion from geometric averaging (G) of the dispersive components of the surface tension or surface energy. Combining the Young-Dupré and Fowkes geometric average models for the work of adhesion gives the Young-Dupré-Fowkes (YDF) equation ($\gamma_{LV} = \gamma_{OV} \approx \gamma_{OV}^d$):

$$W_{SO}(G) = \gamma_{OV} (\cos \Theta_{SO} + 1) = 2(\sigma_{SV}^d \gamma_{OV}^d)^{1/2} \tag{14 a}$$

$$\sigma_{SV}^d (G) \approx 0.25 (\gamma_{OV})(\cos \Theta_{SO} + 1)^2 \tag{14 b}$$

Rewritten in terms of $\cos \Theta_{SO}$, we obtain

$$\cos \Theta_{SL} = -1 + 2(\sigma_{SV}^d)^{1/2} [(\gamma_{LV}^d)^{1/2} / \gamma_{LV}] \approx -1 + 2[(\sigma_{SV}^d)/(\gamma_{LV}^d)]^{1/2} \tag{14 c}$$

If $\cos \Theta_{SL}$ is plotted against $[1/(\gamma_{OV}^d)^{1/2}]$ for fully dispersive liquids or $(\gamma_{LV}^d)^{1/2}/\gamma_{LV}$ for polar liquids, a straight line should be obtained with slope $2(\sigma_{SV}^d)^{1/2}$. For geometric averaging the line should moreover cross the ordinate at –1. The latter requirement is a crucial intrinsic standard to ensure that polar interactions do not seriously distort the slope providing the dispersive component of the solid.

In Fig. 1.4 it is shown that the expectation is fulfilled only for the silanized hydrophobic silica sample [8]. The more polar the surfaces are the smaller is the slope. For hydrophilic titania both octane and hexadecane have too small

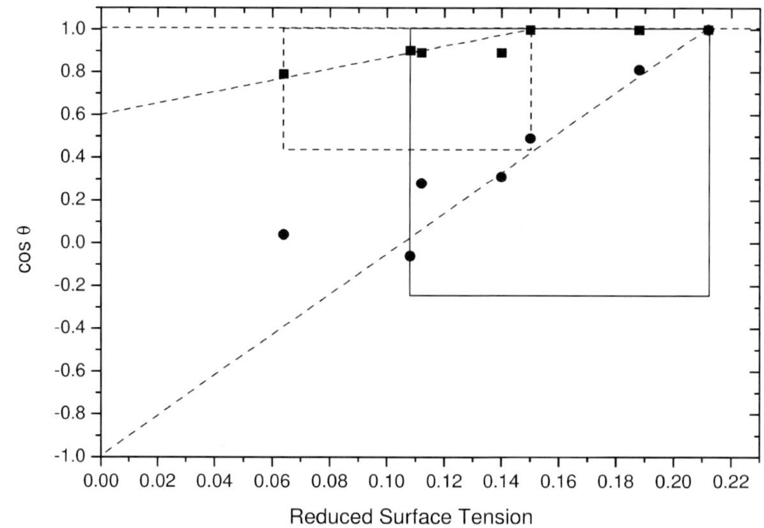

Fig. 1.4 YDF-plot of cos Θ_{SL} of the probe liquids (see Fig. 1.3) plotted against the reduced surface tension of hydrophobic silica (spheres) and hydrophilic titania (squares). The frames encompass the probe liquids having matching surface tensions with the surface energies of the solids [8, 42].

surface tensions for the determination of but α-bromonaphthalene has a matching surface tension. As compared with Fig. 1.3 the requirement of matching surface tensions with surface energies become again evident. For some time the deviation from –1 was used as a measure for the polarity of the solid surface or as an indication of the presence of a vapour film on the surface [9]. Thus water (and EG) seems to be unsuitable for the characterisation of hydrophobic silica.

Combining the Young-Dupré and the arithmetic average (A) equations for a fully dispersive system, we obtain the Young-Dupré-Zettlemoyer (YDZ) equation ($\gamma_{LV} = \gamma_{OV} \approx \gamma^{d}_{OV}$) [10]:

$$W_{SO}(A) = \gamma_{OV}(\cos\Theta_{SO} + 1) = (\sigma^{d}_{SV} + \gamma^{d}_{OV}) \tag{15 a}$$

$$\sigma^{d}_{SV}(A) \approx \gamma_{OV}\cos\Theta_{SO} \tag{15 b}$$

Obviously, the dispersive surface energy [$\sigma^{d}_{SV}(A)$] equals the work of immersion when $\Theta_{SO} \to 0$, since $\sigma^{d}_{SL} \approx 0$. The arithmetic and geometric surface energies and the GG ratio for our model substrates (cf. Fig. 1.4) determined with the probe liquids are collected in Table 1.2 [8].

The influence of non-matching liquids on the determination of surface energies is obvious. With octane and in most cases hexadecane the surface energy is independent of the properties of the solids, equalling the surface tension of the

Table 1.2 Arithmetic (A) and geometric (G) averaging of the dispersive surface energy σ_{SV}^{d} and the ratio Φ for four solid model surfaces with octane, hexadecane, α-bromonaphthalene and diiodomethane [8].

	σ_{SV}^{d} (A) (mJ m^{-2})	σ_{SV}^{d} (G) (mJ m^{-2})	Φ (°)
Octane			
SiO_2–hydrophobic	22.2	22.2	1.00
SiO_2–hydrophilic	22.2	22.2	1.00
TiO_2–hydrophobic	22.2	22.2	1.00
TiO_2–hydrophilic	22.2	22.2	1.00
Hexadecane			
SiO_2–hydrophobic	23.1	23.3	0.99
SiO_2–hydrophilic	28.1	28.1	1.00
TiO_2–hydrophobic	28.4	28.4	1.00
TiO_2–hydrophilic	28.4	28.4	1.00
α-Bromonaphthalene			
SiO_2–hydrophobic	21.6	24.6	0.88
SiO_2–hydrophilic	39.4	39.6	0.99
TiO_2–hydrophobic	37.8	38.0	0.99
TiO_2–hydrophilic	44.3	44.3	1.00
Diiodomethane			
SiO_2–hydrophobic	15.6	21.7	0.72
SiO_2–hydrophilic	40.2	40.7	0.99
TiO_2–hydrophobic	39.6	40.2	0.99
TiO_2–hydrophilic	45.2	45.3	1.00

(completely wetting) liquid. Only ABN and DIM provide a surface energy which reflects the expectations. On the other hand all probe liquids give a comparable (low) surface energy of the hydrophobic silica. Zettlemoyer [10] compared the arithmetic mean with the geometric mean for a more polar metallic mercury surface and found that the former method was much more sensitive to show salient polarities of unsaturated and halogenated hydrocarbons unavailable with the geometric averaging (Table 1.3).

Obviously, if one component is (nearly) zero, the geometric average vanishes whereas the arithmetic remains significant. This insensitivity of the geometric mean to respond to specific interactions has also been widely noted using the Regular Solution model [19].

Table 1.3 Interfacial mercury–organic interfacial tension ($mN\ m^{-1}$) calculated using arithmetic and geometric averaging [10].

Liquid	Geometric average	Arithmetic average
Hexane	0	0
Benzene	0	13
Toluene	0	17
Bromobenzene	0	25
1,2-Dibromomethane	0	29
Aniline	0	34

The work of immersion offers a sensitive mean to determine the dispersive component of the surface tension/energy. If we relate the Gibbs free surface energy from the Gibbs-Helmholtz equation with the surface tension/energy, we may write

$$\Delta G^{s}_{Si} = \sigma_{Si} = \Delta H^{s}_{Si} + T[\partial(\Delta G^{s}_{Si})/\partial T]_{P} = \Delta H^{s}_{Si} + T[\partial(\sigma_{Si})/\partial T]_{P,n} \tag{16}$$

where i = V, L. Since $\Delta G_{I} = \Delta G_{SL}/A = -I_{SL} = \sigma_{SL} - \sigma_{SV}$, the equation takes the form

$$\Delta H^{s}_{I} = \Delta G^{s}_{I} - T(\mathrm{d}\Delta G^{s}_{I}/\mathrm{d}T) \tag{17 a}$$

$$\Delta H^{s}_{I} = \sigma_{SL} - T[\partial(\sigma_{SL})/\partial T]_{P,n} - \sigma_{SV} + T[\partial(\sigma_{SV})/\partial T]_{P,n} \tag{17 b}$$

$$\Delta H^{s}_{I} = \sigma_{SL} - \sigma_{SV} - T[\partial(\sigma_{SL} - \sigma_{SV})/\partial T]_{P,n} \tag{17 c}$$

where $\Delta H^{s}_{I} = \Delta H^{s}_{SL}/A = (H^{s}_{SL} - H^{s}_{SV})/A$. Substituting Young's equation, we obtain

$$\Delta H^{s}_{I} = -\gamma_{LV}\cos\Theta_{SL} + T[\partial(\gamma_{LV}\cos\Theta_{SL})/\partial T]_{P,n} \tag{18 a}$$

$$\Delta H^{s}_{I} = -\gamma_{LV}\cos\Theta_{SL} + T\cos\Theta_{SL}[\partial(\gamma_{LV})/]_{P,n} + T\gamma_{LV}[\partial(\cos\Theta_{SL})/\partial \mathrm{T}]_{P,n} \tag{18 b}$$

$$\Delta H^{s}_{I} + I_{SL} = T\cos\Theta_{SL}[\partial(\gamma_{LV})/\partial T]_{P,n} + T\gamma_{LV}[\partial(\cos\Theta_{SL})/\partial \mathrm{T}]_{P,n} \tag{18 c}$$

It is therefore possible to relate the enthalpy of immersion to the temperature dependence of the surface tension of the test liquid and the change of its contact angle with the solid. In Fig. 1.5, the immersion enthalpy is calculated using Eq. (18 b) and compared with those determined using calorimetry [20].

However, as discussed in the Introduction, the relationships introduced correspond to the two phases in equilibrium with the interface, which is different from the process occurring solely at the interface. For hydrocarbons on hydrophobic surfaces (implied for all equations), the temperature dependence of the surface tension is usually small [21] $\{(-\Delta S^{s}_{LV})^{d} = [\partial(\gamma^{d}_{LV})/\partial T]_{P,n} \approx -0.1\ \mathrm{mJ\ m^{-2}\ K^{-1}}\}$, which simplifies the calculation.

We may use the geometric averaging of the interfacial energy to express the Gibbs free energy of immersion:

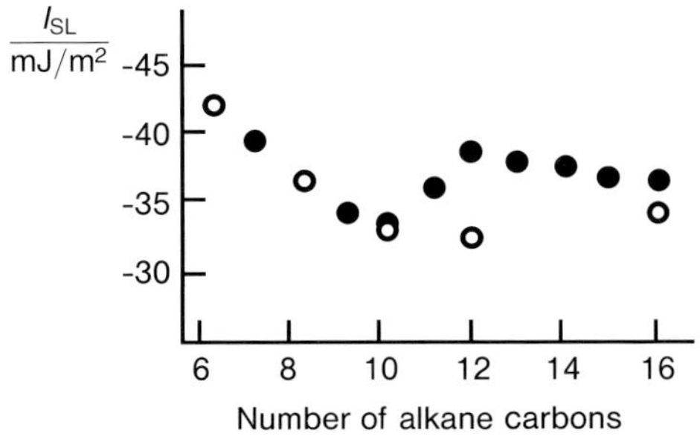

Fig. 1.5 Immersion enthalpy for Teflon in various alkanes is determined calorimetrically and calculated using Eq. (18 b) (from [20], with permission).

$$\Delta G_I^s = \Delta G_{SL}/A = -I_{SL} = \sigma_{SL} - \sigma_{SV} = \gamma_{LV} - 2(\sigma_{SV}^d \gamma_{LV}^d)^{1/2} \tag{19 a}$$

$$\Delta H_I^s = \gamma_{LV} - 2(\sigma_{SV}^d \gamma_{LV}^d)^{1/2} - T\{[\partial(\gamma_{LV})/\partial T]_{P,n} - 2\sqrt{\sigma_{SV}^d}[(\partial\sqrt{\gamma_{LV}^d})/\partial T]_{P,n} - 2\sqrt{\gamma_{LV}^d}[(\partial\sqrt{\sigma_{SV}^d})/\partial T]_{P,n}\} \tag{19 b}$$

The expression is again dramatically simplified if the arithmetic mean is applied for $\gamma_{LV} \approx \gamma_{LV}^d$:

$$\Delta G_I^s = \Delta G_{SL}/A = -I_{SL} = \sigma_{SL} - \sigma_{SV} = \gamma_{LV} - (\sigma_{SV}^d + \gamma_{LV}^d) \tag{20 a}$$

$$(\Delta G_I^s)^d \approx -\sigma_{SV}^d \tag{20 b}$$

$$\Delta H_I^s = -\sigma_{SV}^d + T[(\partial(\sigma_{SV}^d)/T)_{P,n}] \tag{20 c}$$

For a dispersive probe liquid (O) on low-energy surfaces, we find at room temperature that [10]

$$(\Delta H_I^S)^d \approx -\sigma_{SV}^d + 298(-0.07) \approx -\sigma_{SV}^d + 21 \tag{21}$$

Note that Eq. (21) states that the temperature dependence (entropy contribution) is constant when considering the dispersive surface energy of the solid.

Van Oss et al. [17, 18] introduced an extended scale for the non-specific interaction by choosing halogenated hydrocarbons as probe (oil, O) liquids. The most popular of them, diiodomethane (DIM; $\gamma_{OV}^{LW} \approx \gamma_{OV} = 51$ mN m^{-1}) and α-bromonaphthalene (ABN; $\gamma_{OV}^{LW} \approx \gamma_{OV} = 44$ mN m^{-1}) have a cohesive energy ($C_{OO} = 2\gamma_{OV}^{LW}/$ mN m^{-1}) which is about double that of dispersive probe hydrocarbon liquids ($18 < \gamma_{OV}^d < 26$ mN m^{-1}) [22]. This reduces the contribution of the specific interaction to only a fraction of its "polar" value and distorts the scaling published previously in the literature. The designation Lifshitz-van der Waals (LW) component emphasizes the fact that the contribution is considered to include the semi-polar (Debye and Keesom) interactions of the slightly acidic halocarbons. As discussed, these are largely erased in geometric averaging, but remain observable using arithmetic averaging [10].

However, as discussed in relation to Figs. 1.3 and 1.4 the high surface tension of ABN and DIM are superior for high energy surfaces, while O and HD are applicable to low energy surfaces. In the past most model studies were made on polymers with a rather low surface energy. In conclusion, we may define two different alternative scales for the surface tension/energy:

$$\gamma_{LV} = \gamma_{LV}^{d} + \gamma_{LV}^{p} \quad \text{and} \quad \sigma_{SV} = \sigma_{SV}^{d} + \sigma_{SV}^{p} \tag{22 a}$$

$$\gamma_{LV} = \gamma_{LV}^{LW} + \gamma_{LV}^{AB} \quad \text{and} \quad \sigma_{SV} = \sigma_{SV}^{LW} + \sigma_{SV}^{AB} \tag{22 b}$$

It should be pointed out that maintaining the traditional designation "polar" for non-dispersive interactions does not mean that they are solely dipolar in origin, but non-specific in a more general chemical sense.

1.3.2
Contribution from the Surface Pressure of (Gaseous) Molecules and Spreading of Liquid Films

The ideal solid surface model assumes a smooth homogeneous surface structure, resembling that of a liquid. The model is suitable for the definition of the wetting of (mainly) hydrophobic surfaces. In reality, the surface of semipolar or polar surfaces is always covered by an adsorbed, condensed layer of vapor which reduces the surface energy considerably [9, 10, 23–25]. As shown in Fig. 1.6 the contribution of the surface pressure is quite dramatic on polar surfaces, but small on hydrophobic surfaces. We therefore consider the influence of an adsorbed vapor layer in equilibrium with its own liquid (drop). The surface pressures are defined as

$$\pi_{L(L)} = \gamma_{LV} - \gamma_{L(L)} \approx \gamma_{L}^{o} - \gamma_{L(L)} \approx 0 \tag{23 a}$$

$$\pi_{S(L)} = \sigma_{SV} - \sigma_{S(L)} \neq \sigma_{S}^{o} - \sigma_{S(L)} \tag{23 b}$$

The contribution of an adsorbed vapor layer to the surface tension of the liquids may be considerable, but has been found to be negligible for our model systems [8]. In the previous equations the vapor (V) denotes the adventitious adsorption of ever-existing vapor from the environment. Hence in practice the work of adhesion represents the displacement of V by L. This vapor dramatically lowers the surface energies. The Dupré equation takes the form {note the designation $W_{S(L)L} = W_{[S(L)]L}$, i.e. preadsorbed (L)}:

$$W_{S(L)L} = \sigma_{S(L)} + \gamma_{L(L)} - \sigma_{SL} \approx \sigma_{SV} - \pi_{S(L)} + \gamma_{LV} - \sigma_{SL} \tag{24 a}$$

$$W_{S(L)L} = W_{SL} - \pi_{S(L)} \Longrightarrow \pi_{S(L)} = W_{SL} - W_{S(L)L} \tag{24 b}$$

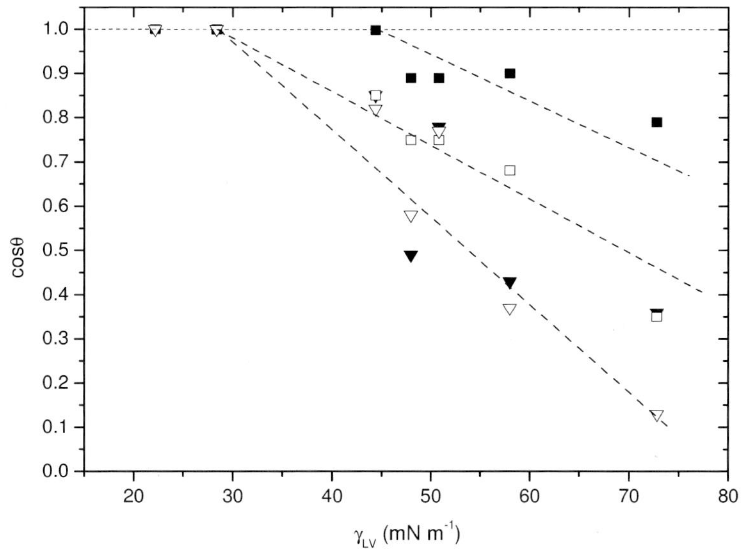

Fig. 1.6 The cos Θ_{SL} measured in air (filled symbols) and in saturated probe liquid vapour (open symbols) on hydrophobic (inversted triangles) and hydrophilic (squares) titania, plotted as a function of the surface tension of the probe liquids [8, 42].

We now consider the Young equation for these two limiting states:

$$\sigma_{S(L)} - \sigma_{SL} = \gamma_{LV} \cos \Theta_{S(L)L} = \sigma_{SV} - \pi_{S(L)} - \sigma_{SL} \tag{25 a}$$

$$\sigma_{SV} - \sigma_{SL} = \gamma_{LV} \cos \Theta_{SL} \tag{25 b}$$

$$\pi_{S(L)} = \gamma_{LV} \left[\cos \Theta_{SL} - \cos \Theta_{S(L)L}\right] \tag{25 c}$$

since $\gamma_{LV} \approx \gamma_{L(L)}$. Obviously, the surface pressure of polar liquid vapors on the solid surface can be determined from the change in contact angle on adsorption of the liquid vapor replacing the air. As shown in Fig. 1.6, this contribution can be substantial for polar surfaces [8].

A more consistent value for the surface pressure may be determined from the adsorption isotherm of liquid vapors on evacuated (powder) samples. For the surface pressure of the monomolecular film we may write [10]

$$\pi^0_{S(L)} = \sigma^0_S - \sigma_{S(L)} = RT \int_{\Gamma_L=0}^{\Gamma_L(\text{mon})} \Gamma_L \mathrm{d}\ln(P_L/P^\bullet_L) \tag{26 a}$$

$$\pi_{S(L)} = (RT/M_L A_w) = \int_{\Gamma_L=0}^{\Gamma_L(\text{mon})} w_L \mathrm{d}\ln(P_L/P^\bullet_L) \tag{26 b}$$

where $\pi^0_{S(L)}$ and σ^0_S (0 = reference state vapor free surface in vacuum), Γ_L = surface excess ($\approx w_L/M_L A_w$), $P^\bullet_L$ = (partial) pressure (• = at standard pressure = 1 bar), A_w = specific surface area of the sample, w_L = weight ratio of adsorbate to substrate and M_L = molar mass of adsorbent. Note that for monolayer coverage P_{mon} corresponds to $\Gamma_{L,mon}$. It should be noted that this equation is restricted to the monolayer adsorption and it should be revised for multilayer and infinitely thick films equaling the work of spreading. For an infinitely thick (duplex) film, we find [10]

$$S^0_{SL} = \sigma^0_S - \gamma_{L(L)} - \sigma_{SL} = RT \int_{\Gamma_L=0}^{\infty} \Gamma_L \mathrm{d}\ln(P_L/P^\bullet_L) \tag{27}$$

These two states are illustrated in Fig. 1.7.

The liquid close to the tpcl can therefore microscopically be divided into three distinct regions: the bulk liquid, the unstable transition region and the vapor film. The thickness of the liquid collar has been related to the work of spreading [26]:

$$h = (3\gamma_{LV} d_0^2/2S_{SL})^{1/2} = (3A_{LL}/2k_A S_{SL})^{1/2} \tag{28}$$

where d_0 again denotes the mean distance between the molecules, $0.18 < d_0 < 0.26$ nm, in the substance. The development of a contact line tension τ ($\approx 10^{-11}$ -10^{-10} J m^{-1}) contributes to the macroscopic ($\sigma_{SL} - \sigma_{SV}$) surface energy balance only for very small droplets, i.e. the line tension is of the order τ/σ_{SL} ($\approx 10^{-11}$ J m^{-1}/10^{-2} J m^{-2} ≈ 1 nm). However, line tensions up to 10^{-5} J m^{-1} have been reported [27], corresponding to experimentally verified liquid collars in the millimeter range [29]. The contribution of the line tension can be introduced as a corrected contact angle over the Young contact angle:

$$\cos\Theta^*_{SL} = \cos\Theta_{SL} - (\tau\kappa/\gamma_{LV}) = \cos\Theta_{SL} - (\tau/r_c\gamma_{LV}) \tag{29}$$

where γ_{LV} is the surface tension of the liquid and κ is the curvature of the contact line [27] or the local radius of curvature r_c of the contact line [28]. A linear dependence is thus expected of $\cos\Theta^*_{SL}$ upon the local curvature of the contact line. In kinetic considerations, a hydrodynamic model is usually applied for the unstable phase condensed liquid halo whereas a molecular kinetic model is applied on the gaseous vapor film [30].

The difference between these two integrals is represented by the shaded area (van der Waals loops) [10]. Subtraction gives the work of spreading on a vapor-covered surface:

$$S_{S(L)L} = S^0_{S(L)} - \pi^0_{S(L)} = \sigma_{S(L)} - \gamma_{L(L)} - \sigma_{SL} = RT \int_{\Gamma_{L(mon)}}^{\infty} \Gamma_L \mathrm{d}\ln(P_L/P^\bullet_L) \tag{30}$$

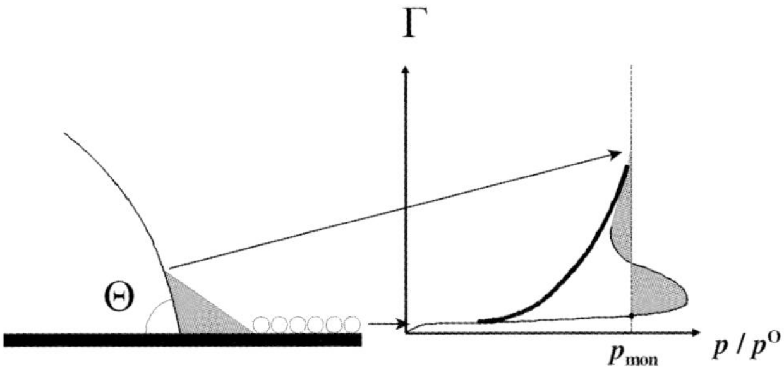

Fig. 1.7 The three-phase-contact line can in reality be subdivided into three regions, bulk liquid, condensation range and gaseous molecular film. Plotted as the surface excess against the relative pressure, the phase condensation is represented by the van der Waals loops. In most cases only the adsorption (bold line) is observed.

The amount adsorbed can also be evaluated from a simple experiment. The sample plate is hooked on to a balance and placed in an almost closed sample cell above, but not in contact with the liquid. Then the change in weight is monitored as a function of time. The surface of the sample plate is known ($A = A_{w*}w_S$, specific surface area times the weight), as is the molecular surface area of water ($a_L = 3V_L/r_L N_A = A/N_m^S$). The number of molecules at the saturated surface is $N_m^S = n_m^S A = A/aN_A$ and the surface excess for monomolecular coverage is $\Gamma_{L(mon)} = N_m^S/AN_A = n_m^S/A$. Multiplying by RT, we may therefore calculate the surface pressure of the liquid vapor, $\pi_{S(L)} = \sigma_{SV} - \sigma_{S(L)}$, from gravimetric data. Alternatively, the amount adsorbed on a flat surface can be determined utilizing, e.g., ellipsometry [29].

Introducing the work of cohesion, $C_{L(L)L} = 2\gamma_{L(L)} \approx C_{LL}$, we find for the work of spreading of the bulk liquid:

$$S_{S(L)L} = W_{S(L)L} - C_{L(L)L} = \sigma_{S(L)} - \gamma_{L(L)} - \sigma_{SL} \tag{31 a}$$

$$S_{S(L)L} = \sigma_{SV} - \pi_{S(L)} - \gamma_{LV} - \sigma_{SL} \tag{31 b}$$

$$S_{S(L)L} = S_{SL} - \pi_{S(L)} \tag{31 c}$$

These equations apply since the original reference state (σ_S^0) is subtracted from the equation. Following the formalism introduced, the work of adhesion of the wetting liquid in equilibrium with its own vapor $W_{S(L)L}$ may thus be subdivided into the work of adhesion (W_{SL}) in the absence of an adsorbed film and the surface pressure of the vapor film on the solid [$\pi_{S(L)}$]:

$$W_{S(L)L} = \gamma_{L(L)}[\cos\Theta_{S(L)L} + 1] = \gamma_{LV}(\cos\Theta_{SL} + 1) - \pi_{S(L)} \tag{32 a}$$

$$S_{S(L)L} = \gamma_{L(L)}[\cos\Theta_{S(L)L} - 1] = \gamma_{AV}(\cos\Theta_{SL} - 1) - \pi_{S(L)} \tag{32 b}$$

At the critical point of complete wetting ($\cos \Theta_{SL}=1$), the work of adhesion equals the difference between the work of cohesion of the liquid and the surface pressure of the film [$W_{S(L)L}=C_{LL}-\pi_{S(L)}$]. Then, the negative surface pressure equals the work of spreading: $S_{S(L)L}=-\pi_{S(L)}$.

Table 1.4 collects the work of immersion, adhesion and spreading when the reference state was a vapor-free surface in vacuum. The contribution from the surface pressure of adsorbed vapor films is, depending on the system, considerable (unequal probe liquids) or predominant (most polar surfaces). It is therefore not surprising that wetting vapors and liquids are used to lower the surface energy when machining, drilling or grinding polar solids. Note that $W^0_{SL} \approx I^0_{SL}+\gamma_{LV}$ and $S^0_{SL} \approx I^0_{SL}-\gamma_{LV}$ (cf. Eq. 6 b–d).

Since the work of spreading from the monomolecular film to a condensed surface is zero [$S_{S(L)L}=0$, Eq. (30); $\Theta_{SL}=0$], obviously the surface pressure equals $S^0_{SL} = \pi^0_{S(L)}$.

The surface pressure of the probe liquids used as examples in Fig. 1.6 were determined from the contact angle difference in air and under saturated vapor. The results of these calculations are collected in Table 1.5.

For TiO_2 (anatase) the following saturation surface pressures/work of adhesion have been found [25]: 46/86 (*n*-heptane), 108/154 (*n*-propyl alcohol) and 196/340 (water). Although the surface pressure of the monomolecular film (Eq. 26) is only a fraction of this saturation value (Eq. 27), the values reported in Table 1.4 are almost negligible. ABN and DIM produce, however, a significant surface pressure on hydrophilic TiO_2. As shown, the vapor surface pressure of the polar liquids is considerably greater on all sample surfaces. The surface pressure on hydrophilic TiO_2 is greatest, whereas the other varies in an irrational way from positive to negative values. The latter are again due to $\Theta_{S(L)L}<\Theta_{SL}$. The influence of adventitious vapors competing for the adsorption sites is obvious, i.e. $\pi_{S(L)} \ll \pi^0_{S(L)}$.

Combining the Young and Dupré equations with Fowkes' geometric averaging gives the Young-Dupré-Fowkes (YDF) equation including the contribution from the surface pressure:

$$W_{S(L)L} = \gamma_{L(L)}[\cos \Theta_{S(L)L} + 1] = 2(\sigma^d_{SV}\gamma^d_{LV})^{1/2} - \pi_{S(L)} \tag{33 a}$$

$$\cos \Theta_{S(L)L} = -1 + [2(\sigma^d_S\gamma^L_L)^{1/2} - (\pi_{S(L)})]/\gamma_{L(L)} \tag{33 b}$$

Hence the deviation of the slope of the line through the experimental points plotted as a function of $1/\gamma_{L(L)}$ (Fig. 1.8) may be interpreted in part as a contribution from the surface film pressure. As an intrinsic consistency test the line should pass through –1 at $1/\gamma_{L(L)}=0$.

In the absence of a surface pressure the equation equals the YDF equation (14 c).

Note that the contact angle measured in air and in saturated probe liquid vapor differ considerably in particular for hydrophilic surfaces. This observation

Table 1.4 Work of immersion, adhesion and spreading (mJ/m^2) for silica and titania-liquid vapor pairs determined from vacuum [24].

Solid	Vapor	I^0_{SL}	W^0_{SL}	S^0_{SL}
SiO_2	Water	316	388	244
SiO_2	n-Propanol	134	158	110
SiO_2	Benzene	81	110	52
SiO_2	n-Heptane	59	79	38
TiO_2	Water	300	370	228
TiO_2	n-Propanol	114	138	90
TiO_2	Benzene	85	114	56
TiO_2	n-Heptane	58	78	38

Table 1.5 Surface pressure of octane, hexadecane, α-bromonaphthalene, diiodomethane, ethylene glycol and formamide and water on different solid substrates calculated from the measured contact angle in air and in saturated vapor (Eq. 25 c) [8, 42].

	$\pi_{S(L)}$		$\pi_{S(L)}$
Octane		*Ethylene glycol*	
SiO_2–hydrophobic	0.0	SiO_2–hydrophobic	3.3
SiO_2–hydrophilic	0.0	SiO_2–hydrophilic	5.7
TiO_2–hydrophobic	0.0	TiO_2–hydrophobic	–4.6
TiO_2–hydrophilic	0.0	TiO_2–hydrophilic	7.1
Hexadecane		*Formamide*	
SiO_2–hydrophobic	0.2	SiO_2–hydrophobic	–6.8
SiO_2–hydrophilic	1.7	SiO_2–hydrophilic	7.0
TiO_2–hydrophobic	0.0	TiO_2–hydrophobic	3.1
TiO_2–hydrophilic	0.0	TiO_2–hydrophilic	12.7
α-Bromonaphthalene		*Water*	
SiO_2–hydrophobic	–0.7	SiO_2–hydrophobic	8.9
SiO_2–hydrophilic	–1.1	SiO_2–hydrophilic	6.3
TiO_2–hydrophobic	1.5	TiO_2–hydrophobic	19.1
TiO_2–hydrophilic	6.4	TiO_2–hydrophilic	31.7
Diiodomethane			
SiO_2–hydrophobic	–3.2		
SiO_2–hydrophilic	0.0		
TiO_2–hydrophobic	0.3		
TiO_2–hydrophilic	6.3		

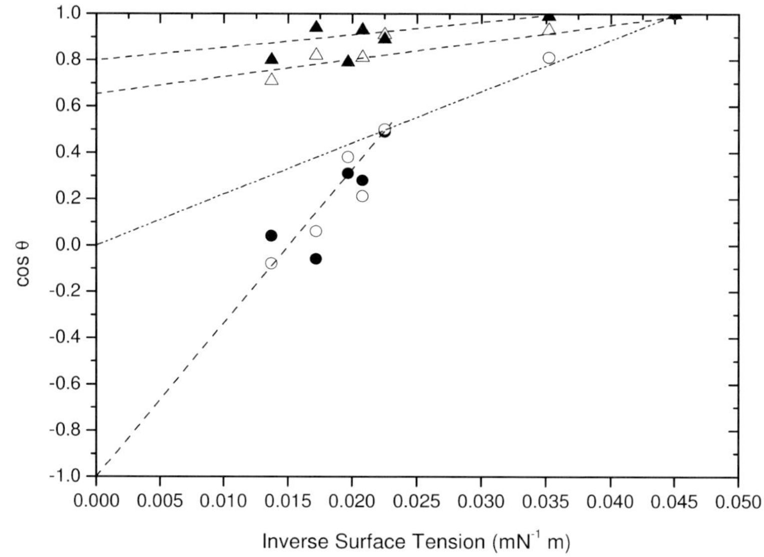

Fig. 1.8 A comparison of the dependency cos Θ_{SL} (in air, filled symbols) and cos $\Theta_{S(L)L}$ (in saturated vapor, open symbols) for hydrophobic (spheres) and hydrophilic (triangles) silica on the inverse surface tension of the solid-probe liquid systems [8, 42].

was discussed previously (Fig. 1.6). However, the plot for hydrophobic silica breaks into two parts at ABN, the low energy liquids extrapolating to octane at $\cos\Theta_{SL}=1$ and the high-energy liquids, including water to the expected intercept -1 at $1/\gamma_{LV}=0$. The surface pressure obviously does not explain the deviation of hydrophilic silica from this point, but should rather be devoted to specific interactions.

Recalling the Young-Dupré-Zettlemoyer (YDZ) equation, we may express the work of adhesion in terms of the arithmetic averaging:

$$W_{S(L)L} = \gamma_{L(L)}[\cos\Theta_{S(L)L} + 1] = (\sigma_{SV}^{d} + \gamma_{LV}^{d}) - \pi_{S(L)} \tag{34 a}$$

$$\cos\Theta_{S(L)L} = -1 + [\sigma_{SV}^{d} + \gamma_{LV}^{d} - \pi_{S(L)}]/\gamma_{LV} \tag{34 b}$$

As an intrinsic consistency test the line should pass through -1 at $1/\gamma_{L(L)}=0$, which is obeyed by ABN, DIM, EG, FA and W. For hydrocarbons $[\gamma_{L(L)} \approx \gamma_{LV} \approx \gamma_{LV}^{d}]$ we have

$$\cos\Theta_{S(L)L} \approx (\sigma_{SV}^{d} - \pi_{S(L)})/\gamma_{LV} \tag{34 c}$$

for which the line should pass through zero at $1/\gamma_{L(L)}=0$. This expectation is fulfilled for the O-HD-ABN branch of hydrophobic silica in Fig. 1.8. ABN is thus represented in both liquid series. Obviously Eq. (34 c) may be used to evaluate

Table 1.6 Values of σ_S^d and $\pi_{S(H)}$ at 25 °C for a number of solid surfaces using heptane as probe liquid [10].

Solid	$\pi^0_{S(H)}$	σ_S^d (arithmetic mean)	σ_S^d (geometric mean)	Φ
Copper	29	49	60	0.82
Silver	37	57	74	0.77
Lead	49	69	99	0.70
Tin	50	70	100	0.70
Iron	53	73	108	0.68
SiO_2	39	59	78	0.76
TiO_2 (anatase)	46	66	92	0.72
SnO_2	54	74	111	0.67
Fe_2O_3	54	74	107	0.69
Graphite	56	76	115	0.66

the applicability of the arithmetic averaging for a range of probe liquids. When $\gamma_{LV} \neq \gamma_{LV}^d$ the YDZ Eq. (15 b) does not apply. We cannot deduce from Fig. 1.8 the preference between Eqs. (33 b) and (34 c). Table 1.6 reports on the difference between arithmetic and geometric averaging of σ_S^d using heptane ($\gamma_H \approx \gamma_H^d$) as probe liquid.

For polar solids, the arithmetic and geometric averaging provide divergent values for the dispersive component of the solids. One reason is obviously the fairly large surface pressure. However, the ratio between the geometric and arithmetic components, equaling the Girifalco–Good ratio (Φ), remains fairly constant in the range 0.6–0.8, being substantially above unity assumed for dispersive interaction.

1.3.3
Models for Specific Polar (Lewis) Interactions

Starting from the truly dispersive (London) interactions of hydrocarbons, there is a broad range of molecular interactions of Lewis nature. However, as discussed, the Debye and Keesom interactions diverge from the traditional van der Waals range to the Coulomb range when molecularly arrested or "frozen" upon adsorption on the surface sites [31].

In the classical treatment of surface interactions, the total contribution is subdivided into a dispersive part and a polar part. The latter should be understood as specific (molecularly arresting) interactions without any particular (e.g. dipolar) interaction in mind [10]:

$$\gamma_{LV} = \gamma_{LV}^d + \gamma_{LV}^p \tag{35 a}$$

$$\sigma_{SV} = \sigma_{SV}^d + \sigma_{SV}^p \tag{35 b}$$

Introducing arithmetic and geometric averaging of the dispersive component for the solid, we may now calculate the specific (polar) interaction as $\gamma_{LV} \neq \gamma_{OV} \approx \gamma^{d}_{OV}$ [10]:

$$\sigma^{p}_{SV}(A,O) = \sigma_{SV} - \gamma_{OV} \cos \Theta_{SO} \tag{36 a}$$

$$\sigma^{p}_{SV}(G,O) = \sigma_{SV} - 0.25(\gamma_{OV})(\cos \Theta_{SO} + 1)^2 \tag{36 b}$$

However, since the surface energy of the solid (σ_{SV}) is unknown, we cannot utilize these simple equations. Instead, we may calculate the specific (polar) work of adhesion by using the dispersive surface energy of the solid determined with the reference oils to calculate the polar work of adhesion (W^{p}_{SL}) in the following way:

$$W_{SL}(I) = \gamma_{LV}(\cos \Theta_{SL} + 1) = W^{d}_{SL} + W^{p}_{SL} \tag{37 a}$$

$$W^{p}_{SL}(A) = W_{SL} - [\sigma^{d}_{SV}(A, O) + \gamma^{d}_{LV}] \tag{37 b}$$

$$W^{p}_{SL}(G) = W_{SL} - 2[\sigma^{d}_{SV}(G, O)\gamma^{d}_{LV}]^{1/2} \tag{37 c}$$

Note that the separate determination of $\sigma^{d}_{SV}(O)$ using a fully dispersive liquid (oil, O) is specifically indicated.

Assuming that also the polar contributions are additive, the arithmetic averaging available to resolve the polar component of the surface energy of the solid may be represented as

$$\sigma_{SV}(A)^{p} = W^{p}_{SL}(A) - \gamma^{p}_{LV} \tag{38 a}$$

$$\sigma_{SV}(A)^{p} = \gamma_{LV} \cos \Theta_{SL} - \sigma^{d}_{SV}(A, O) = \gamma_{LV} \cos \Theta_{SL} - \gamma_{OV} \cos \Theta_{SO} \tag{38 b}$$

Although it has been considered unacceptable to apply geometric averaging to any interaction of a specific nature, we also consider this option (Eq. 37 c):

$$\sigma_{SV}(G, I)^{p} = [W^{p}_{SL}(G)]^2 / 4\gamma^{p}_{LV} \tag{39}$$

If the dispersive surface tension of the probe liquid (L) and oil are almost equal ($\gamma^{d}_{LV} \approx \gamma^{d}_{OV} \approx \gamma_{OV}$), we may also assume that the reference oil fully represents the dispersive interaction and write considering Eq. (38 b):

$$W^{p}_{SL}(II) = W_{SL} - W^{d}_{SL} = \gamma_{LV}(\cos \Theta_{SL} + 1) - \gamma_{OV}(\cos \Theta_{SO} + 1) \tag{40 a}$$

$$\sigma^{p}_{SV}(G, II) = [\sigma^{p}_{SV}(A) + \gamma^{p}_{LV}]^2 / 4\gamma^{p}_{LV} \tag{40 b}$$

The specific (polar & AB) surface energy components (Eq. 38 b) of hydrophobic silica is plotted as a function of the surface tension of the probe oils (γ_{OV}) in Fig. 1.9 (for details, see Ref. [42]).

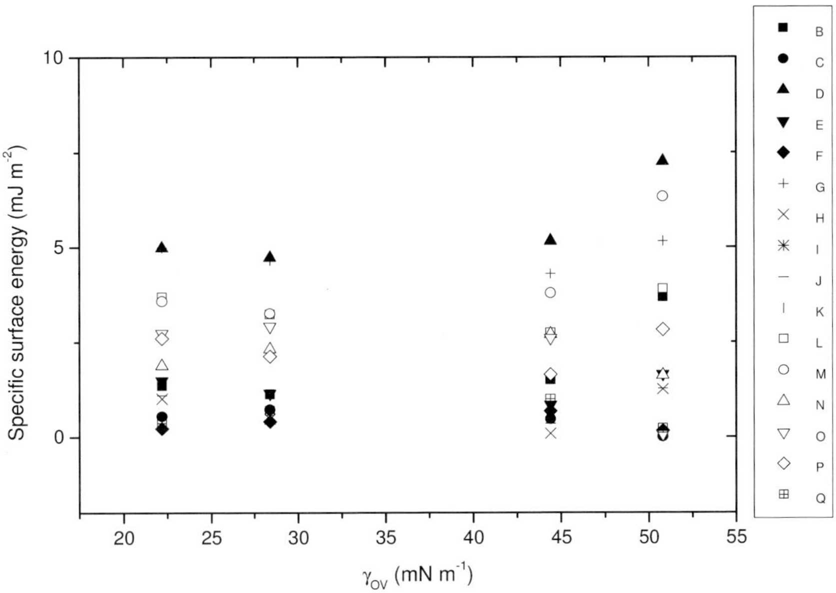

Fig. 1.9 A comparison of the specific (polar and AB) component of the surface energy of hydrophobic silica calculated with: Eq. (38 b) (B–D), Eq. (39) (E–G), Eqs. (46) and (47) (H–K), Eqs. (49) and (50) (L–O) and Eqs. (49) and (52) (P, Q), respectively plotted as a function of the surface tension of the probe liquids (for details see text and [42]).

As shown the specific component is quite consistent for all probe oils ranging from 0 to 5 mJ/m^2. A slightly larger scatter is found for DIM.

Most new methods rely on a geometric (product) averaging of the directional specific interactions. When applying arithmetic averaging, maintaining the previous symmetry, all terms cancel out, making it unavailable for this purpose. One may collectively express the Lewis interaction models representing monodentate acid–base bimolecular pairing with each (monodentate) site possessing both an electrostatic and a covalent binding character:

$$Y - D_1 D_2 = X_1^A X_2^B + Z_1^A Z_2^B \tag{41}$$

where Y is a generalized property or a state variable (typically G, F or H), D is the dispersive interaction, the A and B interaction between the molecular pairs 1 and 2. Such models are, for example, those of Drago and Gutmann (see Table 1.7).

As shown, none of these models considers in particular the non-specific interactions which in later developments of the models are subtracted from the generalized property Y. Consequently, most of the scales refer to poorly solvating solutions (or gas mixtures) of acidic and basic molecules. The latter are based on enthalpy alone, which was shown to be a tentative property and does not

Table 1.7 The *D*, *X* and *Z* terms of different models[a)] for specific interactions.

Property (*Y*)	*D* term	*X* term	*Z* term	Additional term
$\text{Log}K^{AB}$	N/A	S(trength)	σ (soft)	N/A
$\text{Log}K^{AB}$	N/A	S(trength)	σ (soft)	s(teric)
$\text{Log}K^{AB}$	N/A	E(lectrostatic)	C(ovalent)	d(esolvation)
$-\Delta H^{AB}$	N/A	E(lectrostatic)	C(ovalent)	D(ispersive)
$-\Delta H^{AB}$	N/A	e(lectrostatic)	c(ovalent)	t(ransfer)
$-\Delta G^{AB}$	N/A	A(cidic)	B(asic)	N/A

a) The models refer to the models of Gutmann et al. [33, 34], Maria and Gal [35], Handcock and Marciano [36], Drago and Wayland [32], Kroeger and Drago [37] and Della Volpe and Siboni [38].

include the entropic contribution to the Gibbs free energy. As discussed in a previous section, none of the properties are related to the second derivatives sensitive to salient interactions such as the Lewis interactions.

In a recent paper, Peterson made a conceptual analysis of these types of models and made a matrix transformation producing the following constants [39]:

$$X^{+} = 1/\sqrt{2}(X_1^A + X_2^B) \quad \text{and} \quad X^{-} = 1/\sqrt{2}(X_1^A - X_2^B) \tag{42 a, b}$$

$$Z^{+} = 1/\sqrt{2}(Z_1^A + Z_2^B) \quad \text{and} \quad Z^{-} = 1/\sqrt{2}(Z_1^A - Z_2^B) \tag{42 c, d}$$

He concluded that the positive diagonal X^+ and Z^+ matrix terms represent the concept of "like strengths attract their like". The negative X^- and Z^- terms represent the situation when "opposite sites attract each other"; the larger the difference, the greater is the attraction. He suggested that all types of interactions could contribute to the wetting phenomena. The analysis is interesting since it makes it possible to evaluate the terms separately. However, depending on the instrumental method utilized, the empirical constants obtained from fitting to experimental data refer to enthalpic or free energy surface components.

1.3.4
Partial Acid and Base Components

The method of van Oss et al. (vOCG) [17, 18] is related to the geometric derivation procedure of the Lifshitz-van der Waals (LW) contribution [$\sigma_{SV}^{d}(G) = \sigma_{SV}^{LW}(G)$]. As for the division into dispersive and polar components, the LW forces are particularly considered and subtracted from the total surface tension/energy to give the specific acid–base (AB) component [$\sigma_{SV}^{p}(G) = \sigma_{SV}^{AB}(G)$]. However, in the vOCG model, each probe molecule and surface are assigned both acidic and basic sites (bidentacy), which interact with their counterparts independently. Also, the intrinsic interaction between these sites is allowed for. Hence for bidentate (one acidic and one basic site) probe liquids we find

Table 1.8 Surface tension components of the specific probe liquids (~ 20 °C) [17, 18].

Probe liquid	γ_{LV}	γ_{LV}^{LW}	γ_{LV}^{AB}	$\gamma_{LV}^{\oplus}$	$\gamma_{LV}^{\ominus}$
Water	72.8	21.8	51.0	25.5	25.5
Ethylene glycol	48.0	29.0	19.0	1.92	47.0
Formamide	58.0	39.0	19.0	2.28	39.6

$$\gamma_{LV}^{AB} = \gamma_{LV} - \gamma_{LV}^{LW} = 2(\gamma_{LV}^{\oplus}\gamma_{LV}^{\ominus})^{1/2} \tag{43 a}$$

$$\sigma_{SV}^{AB} = \sigma_{SV} - \sigma_{SV}^{LW} = 2(\sigma_{SV}^{\oplus}\sigma_{SV}^{\ominus})^{1/2} \tag{43 b}$$

where vOCG denote $\gamma_{LV}^{\oplus}/\sigma_{SV}^{\oplus}$ the acid (electron acceptor) component and $\gamma_{LV}^{\ominus}/\sigma_{SV}^{\ominus}$ the base (electron donor) component of the surface tension/energy. For comparability with the literature values, the values in Table 1.8 were assumed to be representative. The AB work of cohesion is as usually twice this value, $C_{LL}^{AB} = 2\gamma_{LV}^{AB}$ and $C_{SS}^{AB} = 2\sigma_{SV}^{AB}$. For monodentate (one acid or one base site), the AB term vanishes for the respective component.

Similarly, the work of adhesion is taken as the sum of the LW and AB contributions, the latter being defined as [17, 18]

$$W_{SL}^{AB} = W_{SL} - W_{SL}^{LW} = 2[(\sigma_{SV}^{\oplus}\gamma_{LV}^{\ominus})^{1/2} + (\sigma_{SV}^{\ominus}\gamma_{LV}^{\oplus})^{1/2}] \tag{44 a}$$

$$W_{SL}^{AB} = \sigma_{SV}^{AB} + \gamma_{LV}^{AB} - \sigma_{SL}^{AB} \tag{44 b}$$

Again, if one of the AB pairs is monodentate, the term involving this surface tension/energy vanishes. Combining the AB work of adhesion we obtain the AB component of the interfacial energy:

$$\sigma_{SL}^{AB} = \sigma_{SV}^{AB} + \gamma_{LV}^{AB} - 2[(\sigma_{SV}^{\oplus}\gamma_{LV}^{\ominus})^{1/2} + (\sigma_{SV}^{\ominus}\gamma_{LV}^{\oplus})^{1/2}] \tag{45 a}$$

$$\sigma_{SL}^{AB} = 2[(\sigma_{SV}^{\oplus}\sigma_{SV}^{\ominus})^{1/2} + (\gamma_{LV}^{\oplus}\gamma_{LV}^{\ominus})^{1/2}] - 2[(\sigma_{SV}^{\oplus}\gamma_{LV}^{\ominus})^{1/2} + (\sigma_{SV}^{\ominus}\gamma_{LV}^{\oplus})^{1/2}] \tag{45 b}$$

$$\sigma_{SL}^{AB} = 2[(\sqrt{\sigma_{SV}^{\oplus}} - \sqrt{\gamma_{SV}^{\oplus}})(\sqrt{\sigma_{SV}^{\ominus}} - \sqrt{\gamma_{LV}^{\ominus}})] \tag{45 c}$$

The last equation indicates that the AB interaction is repulsive if $\gamma_{SV}^{\oplus} > \sigma_{SV}^{\oplus}$ and $\gamma_{LV}^{\ominus} < \sigma_{SV}^{\ominus}$ or if the reverse is true. In practice, the bimolecular bidentate interaction has the symmetry of a three-phase (non-specific) liquid contact (see below).

If the LW component of the solid has been determined with the LW probe oils (O) according to one of the methods indicated above, we may write according to the vOCG model for the work of adhesion for two bidentate AB probe liquids K and L [17, 18]

$$W_{SK}^{AB} = W_{SK} - 2[\sigma_{SV}^{LW}(G, O)\gamma_{KV}^{LW}]^{1/2} = C\sqrt{\sigma_{SV}^{\oplus}}(K) + D\sqrt{\sigma_{SV}^{\ominus}}(K) \tag{46 a}$$

$$W_{SL}^{AB} = W_{SL} - 2[\sigma_{SV}^{LW}(G,O)\gamma_{LV}^{LW}]^{1/2} = E\sqrt{\sigma_{SV}^{\oplus}}(L) + F\sqrt{\sigma_{SV}^{\ominus}}(L) \quad (46\,b)$$

where $W_{SK} = \gamma_{KV}(\cos\Theta_{SK}+1)$, $W_{SL} = \gamma_{LV}(\cos\Theta_{SL}+1)$, $C = 2\sqrt{\gamma_{KV}^{\ominus}}$, $D = 2\sqrt{\gamma_{KV}^{\oplus}}$, $E = 2\sqrt{\gamma_{LV}^{\ominus}}$ and $F = 2\sqrt{\gamma_{LV}^{\oplus}}$. Since $\sigma_{SV}^{\oplus}(K) = \sigma_{SV}^{\oplus}(L)$ and $\sigma_{SV}^{\ominus}(K) = \sigma_{SV}^{\ominus}(L)$, the acid and base components of the solid may then be obtained according to [17, 18]

$$\sqrt{\sigma_{SV}^{\oplus}} = (W_{SK}^{AB}F - W_{SL}^{AB}D)/(CF - DE) \quad (47\,a)$$

$$\sqrt{\sigma_{SV}^{\ominus}} = (W_{SL}^{AB}C - W_{SK}^{AB}E)/(CF - DE) \quad (47\,b)$$

The σ_{SV}^{AB} values calculated from Eqs. (45) and (47) are included in Fig. 1.9. The $\sigma_{SV}^{\oplus}$ and $\sigma_{SV}^{\ominus}$ components are presented in Fig. 1.10.

The problem with this method is that it may produce negative values for $\sqrt{\sigma_S^{\oplus}}$ and $\sqrt{\sigma_S^{\ominus}}$, which are squared artificially to positive numbers. In order to control this problem, the intrinsic self-consistency check of recalculating W_{SK}^{AB} and W_{SL}^{AB} from the evaluated $\sigma_{SV}^{\oplus}$ and $\sigma_{SV}^{\ominus}$ using Eqs. (12 a) and (12 b), respectively, must be applied! If either of the acid–base adhesions does not agree with those calculated from the equations

$$W_{SK}^{AB} = (\cos\Theta_{SK} + 1) - 2[\sigma_{SV}(G,O)^{LW}\gamma_{KV}^{LW}]^{1/2} \quad (48\,a)$$

$$W_{SL}^{AB} = (\cos\Theta_{SL} + 1) - 2[\sigma_{SV}(G,O)^{LW}\gamma_{LV}^{LW}]^{1/2} \quad (48\,b)$$

they must be disregarded as intrinsically inconsistent.

Apart from the problems related to the use of halogenated hydrocarbons as LW references, the polar liquids utilized are of a predominantly basic character. It therefore seems rational to consider one or all of the liquid(s) as almost pure monodentate base(s). This is in accord with vOCG [17, 18], who state: "If either the acidic or basic property is negligible and the other property is appreciable, the substance is termed monopolar". The "degree of monodentacy" of the probe liquids is given in Table 1.9.

As shown, water has the largest AB component, but a zero base dominance. Glycerol has the second largest AB contribution, the largest absolute base com-

Table 1.9 Degree of basicity of the non-aqueous probe liquids suggested by vOCG [42].

Probe liquid	γ_{LV}	γ_{LV}^{LW}	γ_{LV}^{AB}	$\gamma_{LV}^{\oplus}$	$\gamma_{LV}^{\ominus}$	γ_{LV}^{AB} (%) [a]	$\Delta\gamma_{LV}^{\ominus}/\gamma_{LV}^{\ominus}$ (%) [a]
Water	72.8	21.8	51	25.5	25.5	70.1	0.0
Glycerol	64	34	30	3.92	57.4	46.9	93.2
Ethylene glycol	48	29	19	1.92	47.0	39.6	95.9
Formamide	58	39	19	2.28	39.6	32.8	94.2
Dimethyl sulfoxide	44	36	8	0.50	32	18.2	98.4

a) γ_{LV}^{AB} (%) = $(\gamma_{LV}^{AB}/\gamma_{LV}) \times 100$ and $\Delta\gamma_{LV}^{\ominus}/\gamma_{LV}^{\ominus}$ (%) = $[(\gamma_{LV}^{\ominus} - \gamma_{LV}^{\oplus})/\gamma_{LV}^{\ominus}] \times 100$.

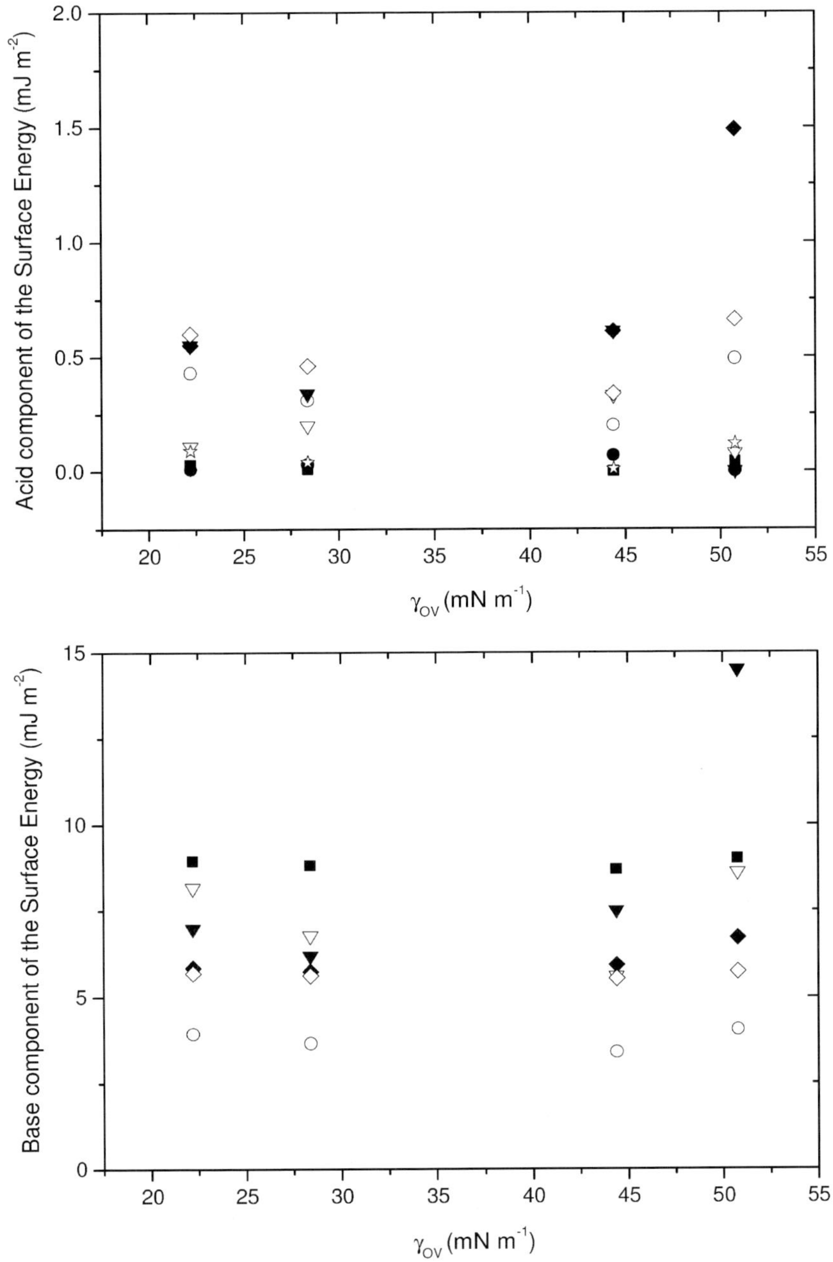

Fig. 1.10 The acid (left) and base (right) surface energy components of hydrophobic silica calculated with the models defined in Fig. 1.9 plotted as a function of the surface tension of the probe liquids (for details see text and [42]).

ponent, but the second smallest base dominance. Ethylene glycol seems to have the overall most favorable properties. With reference to the Girifalco–Good quotient it is expected that the arithmetic/total work of adhesion should not be too different from the geometric work of adhesion ($\Phi \approx 1$). The partial contributions should then add up to the total values. However, the base components exceed the AB-component and even equal the total surface tension (EG). Although mathematically consistent, the geometric averaging seems to produce rather unphysical values. Instead of becoming involved in the complicated vOCG averaging discussed above, a monodentacy is considered instead. Thus, since $\gamma_{KV}^{\oplus} \approx 0$ we may write the AB work of adhesion as [42]

$$W_{SK}^{AB} = W_{SK} - W_{SK}^{LW} \approx W_{SK}^{A(B)} = 2[\sigma_{SV}^{A}(K)\gamma_{KV}^{\ominus}]^{1/2} \tag{49 a}$$

$$\sigma_{SV}^{A}(K) = [W_{SK}^{A(B)}]^{2}/4\gamma_{KV}^{\ominus} \tag{49 b}$$

The superscript A is used in order to distinguish the σ_{SV}^{A} component from the $\sigma_{SV}^{\oplus}$ component. The problem is, however, that no probe liquid has been reported with as pure acidity as the nearly pure basic solutions suggested for the vOCG model. Therefore, we choose another way of calculating the $\sigma_{SV}^{B}(L)$ component. The acid–base contribution to the adhesion can be calculated by accepting that the LW probe (O) oil can fully represent the non-specific interaction (cf. Eq. 40 a). A matching of the LW component of the AB liquid (L) is then advisable ($\gamma_{LV}^{LW} \approx \gamma_{OV}^{LW} \approx \gamma_{OV}$). In Eq. (38 b) it was assumed that $\gamma_{LV}^{AB} \approx \gamma_{LV} - \gamma_{OV}^{LW}$. In this case we may write [42]

$$W_{SL}^{AB}(I) = W_{SL} - W_{SO}^{LW} = \gamma_{LV}(\cos\Theta_{SL} + 1) - \gamma_{OV}(\cos\Theta_{SO} + 1) \tag{50 a}$$

$$W_{SL}^{AB}(II) = \gamma_{LV}\cos\Theta_{SL} - \gamma_{OV}^{LW}\cos\Theta_{SO}^{LW} + \gamma_{LV}^{AB} \tag{50 b}$$

$$W_{SL}^{AB} = 2[(\sigma_{SV}^{A}(K)\gamma_{LV}^{\ominus})^{1/2} + (\sigma_{SV}^{B}(L)\gamma_{LV}^{\oplus})^{1/2}] \tag{50 c}$$

Rewritten in terms of the basic solid component, we find [42]

$$\sigma_{SV}^{B}(L) = \{W_{SL}^{AB} - 2[\sigma_{SV}^{A}(K)\gamma_{LV}^{\ominus})^{1/2}]^{2}/4\gamma_{LV}^{\oplus} \tag{50 d}$$

Depending on the way in which the work of adhesion is calculated (Eq. 50 a and b) for introduction into Eq. (50 d) the indexing is $\sigma_{SV}^{B}(I)$ or $\sigma_{SV}^{B}(II)$. These A and B components are compared with the corresponding parameters in Fig. 1.10. As shown the values produced are rather consistent for each probe liquid, but with an enhanced spread for DIM, being dependent on the model used. Yet another way is to derive a computational $\cos\Theta_{SL}^{AB}$:

$$W_{SL}^{AB} = \gamma_{LV}^{AB}(\cos\Theta_{SL}^{AB} + 1) \tag{51 a}$$

$$W_{SL}^{AB} = \gamma_{LV}(\cos\Theta_{SL} + 1) - \gamma_{OV}^{LW}(\cos\Theta_{SO}^{LW} + 1) \tag{51 b}$$

Table 1.10 Weighted surface tension components of the polar probe liquids suggested by Della Volpe and Siboni [38].

Probe liquid	γ_{LV}	γ_{LV}^{LW}	γ_{LV}^{AB}	γ_{LV}^{a}	γ_{LV}^{b}
Water	72.8	21.8	51.0	65.0	10.0
Ethylene glycol	48.0	31.4	16.4	1.58	42.5
Formamide	58.0	35.6	22.6	1.95	65.7

Testing the values for intrinsic consistency the acid base components agree, but the LW component for EG should read (31.6 mN/m) and for FA (35.4 mN/m), respectively.

Table 1.11 Weighted surface tension components and the degree of acidity or basicity of the polar probe liquids suggested by Della Volpe and Siboni [38].

Probe liquid	γ_{LV}	γ_{LV}^{LW}	γ_{LV}^{AB}	γ_{LV}^{a}	γ_{LV}^{b}	γ_{LV}^{AB} (%) a)	$\Delta\gamma_{LV}^{a,b}/\gamma_{LV}^{a,b}$ (%) a)
Water	72.8	21.8	51.0	65.0	10.0	70.1	84.6
Ethylene glycol	48.0	31.4	16.4	1.58	42.5	34.2	96.3
Formamide	58.0	35.6	22.6	1.95	65.7	39.0	97.0

a) See Table 1.10.

$$\cos\Theta_{SL}^{AB} = (\gamma_{LV}\cos\Theta_{SL} - \gamma_{OV}^{LW}\cos\Theta_{SO})/\gamma_{LV}^{AB} \tag{51c}$$

$$\sigma_{SV}^{B}(III) = \{\gamma_{LV}^{AB}(\cos\Theta_{SL}^{AB} + 1) - 2[\sigma_{SV}^{A}(K)\gamma_{LV}^{\ominus}]^{1/2})\}^2/4\gamma_{LV}^{\oplus} \tag{51d}$$

However, $\sigma_{SV}^{B}(III)$ (Eq. 51d) equals $\sigma_{SV}^{B}(II)$ (Eq. 50d). It should be noted that if a faction of the surface were assigned both to the dispersive and the polar (AB) surface, Eq. (51c) would equal the well known Cassie equation for chemical heterogeneous surfaces discussed below!

In the calculations the AB contribution of water has artificially been divided into two equal contributions (25.5 mN m^{-1}, Table 1.4). The reference to water is rational since all acid–base scales have been related to some particular property of water. However, in order to better reflect the true balance between the acid–base character for water, a strongly weighted acid contribution has been suggested on experimental and theoretical grounds by a number of authors [40, 41]. The most extreme balance suggested by Della Volpe and Siboni [38] is given in Table 1.10.

As the most extreme case, we may again use the strongly weighted acid contribution suggested for water by Della Volpe and Siboni [38] and compare its predominant acidity with the predominant basicity of the other vOCG probe liquids (Table 1.11).

Although the A/B balance is slightly changed also for ethylene glycol and formamide, their nearly pure basicity remains. The strongly dominant acidity of water suggests that water may be taken as an acidic probe. The base component is then obtained simply as (cf. Eq. 49 b) [42]

$$W_{SL}^{AB} \approx W_{SL}^{B(A)} = 2[\sigma_{SV}^{B}(W)\gamma_{WV}^{a}]^{1/2} \tag{52 a}$$

$$\sigma_{SV}^{B}(W) = [W_{SW}^{B(A)}]^{2}/4\gamma_{WV}^{a} \tag{52 b}$$

The use of water is specifically denoted. We may then proceed to calculate the total and AB component of the surface energy [42]:

$$\sigma_{SV} = \sigma_{SV}^{LW} + \sigma_{SV}^{AB} = \sigma_{SV(O)}^{LW} + 2[\sigma_{SV}^{A}(K)\sigma_{SV}^{B}(W)]^{1/2} \tag{53}$$

The σ_{SV}^{A} and σ_{SV}^{B} components are compared with the corresponding parameters in Fig. 1.10, and the σ_{SV}^{AB} component is reproduced in Fig. 1.9 (for details, see [42]). The surface energy components calculated with each model introduced for the model surfaces introduced in Fig. 1.3 are plotted as a function of the surface tension of the probe liquids in Fig. 1.11. It is particularly rewarding to note that the mono-bi and mono-mono dentate models closely agree with the bi-bi dentate vOCG model, when corrected for the DVS acid-base balance.

The geometric averaging principle can be maintained for total surface energies and surface tensions by applying the GG ratio Φ. The work of adhesion is then

$$W_{SL} = 2[(\Phi_{SV}\sigma_{SV})(\Phi_{LV}\gamma_{LV})]^{1/2} = 2\Phi_{SL}(\sigma_{SV}\gamma_{LV})^{1/2} \tag{54}$$

i.e. $\Phi_{SL} = (\Phi_{SV}\Phi_{LV})^{1/2}$. However, since the numerical value of the GG ratio is mostly unknown, it does not provide any advantages over the vOCG model. We can instead use the arithmetic equivalent for the separation of the acid (a) and base (b) components from the total specific (polar) part of the surface tension and surface energy (cf. Eq. 4) [42]:

$$\gamma_{LV}^{p} = \gamma_{LV} - \gamma_{LV}^{d} = \gamma_{LV}^{a} + \gamma_{LV}^{b} \tag{54 a}$$

$$\sigma_{SV}^{p} = \sigma_{SV} - \sigma_{SV}^{d} = \sigma_{SV}^{a} + \sigma_{SV}^{b} \tag{54 b}$$

Considering Eqs. (6 a) and (6 b), we may now define the work of acid–base interaction:

$$W_{SL}^{p} = W_{SL} - W_{SL}^{LW} = \gamma_{LV}(\cos\Theta_{SL} + 1) - [\sigma_{SV}^{d}(A, O) + \gamma_{LV}^{d}] \tag{55 a}$$

$$W_{SL}^{p} = (\sigma_{SV}^{a} + \gamma_{LV}^{b}) + (\sigma_{SV}^{b} + \gamma_{LV}^{a}) \tag{55 b}$$

$$W_{SL}^{p} = \sigma_{SV}^{p} + \gamma_{LV}^{p} - \sigma_{SL}^{p} = (\sigma_{SV}^{a} + \sigma_{SV}^{b}) + (\gamma_{LV}^{a} + \gamma_{LV}^{b}) - \sigma_{SL}^{p} \tag{55 c}$$

Combining the last two equations, it is obvious that the arithmetic averaging being symmetric with the vOCG model predicts that the interfacial tension

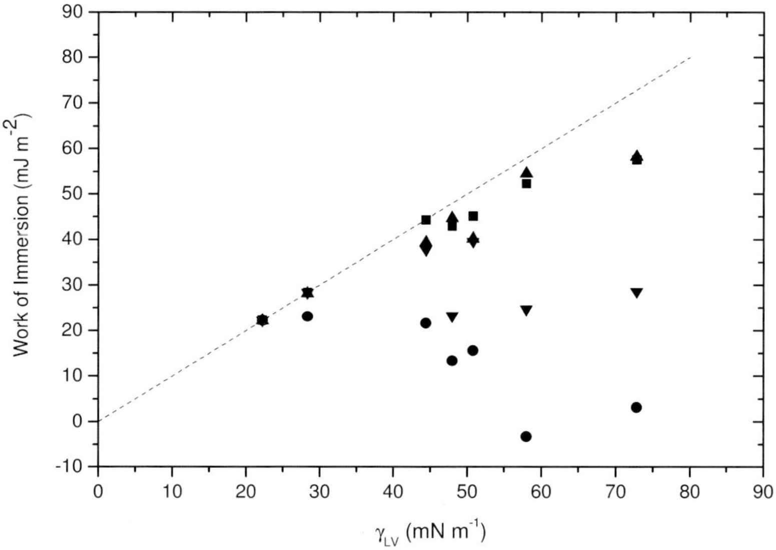

Fig. 1.11 The work of immersion for silica and titania (symbols as in previous figures, air, filled symbols and saturated vapour, open symbols) plotted as a function of the surface tension of the probe liquid-solids pairs (from [42]).

(σ_{SL}^{p}) is always zero at equilibrium. Thus, the arithmetic averaging does not lend itself for calculations of interfacial energies. With reference to the Hard–Soft Acid–Base (HSAB) principle, we define that if one of the sites of the acid–base pair is absent the other cannot interact and the entire parentheses vanish. This principle for monodentacy is also applied in the vOCG model. Combining Eqs. (21 a) and (21 b) and rearranging we find an exceedingly simple expression for the surface energy:

$$\sigma_{SV} = (\sigma_{SV}^{a} + \sigma_{SV}^{b}) + \sigma_{SV}^{d}(A, O) = \gamma_{LV} \cos \Theta_{SL} = I_{SL} \tag{56}$$

This expression equals the work of immersion which agrees with Eq. (15 b) for fully dispersive liquids. Note that $\sigma_{SV}^{b} \approx \sigma_{SV}$. For dispersive systems, the equation may be tested for consistency by extrapolating $\gamma_{LV} \cos \Theta_{SL}$ to $\Theta_{SL} = 0$:

$$\lim(\gamma_{LV}^{d} \cos \Theta_{SL}) = \gamma_{LV}^{d} \approx \gamma_{crit}^{d} \approx \sigma_{SV}^{d} \tag{57}$$

Obviously a Zisman like plot is recovered. Replotted as the work of immersion against the surface tension of the liquids Neumann et al. have identified the surface energy of the solid as the maximum crossing point of the lines defined by $\Theta_{SL} = 0(\gamma_{LV} < \sigma_{SV})$ and $\sigma_{SV} \approx$ constant $(\gamma_{LV} > \sigma_{SV})$ [13].

As shown the break point may, indeed, be identified for the hydrophobic surfaces. For the hydrophilic surfaces the break point is less clear.

When considering enthalpic relationships, the state of the surface film must also be kept in mind. For polar liquids adsorbing on polar surface sites on an otherwise hydrophobic surface, it has been found that the enthalpy of immersion may rise with the degree of vapor coverage (cf. Fig. 1.7) [10]. This means that the film has a greater affinity for the vapor molecules than the bare surface. The enthalpy of (mono-molecular) adsorption ($\Delta H^{\mathrm{s}}_{\mathrm{ads}}$) may be related to the enthalpy of adhesion ($\Delta H^{\mathrm{s}}_{\mathrm{A}}$) of multilayer adhesion (or spreading):

$$\Delta H^{\mathrm{s}}_{\mathrm{ads}} = \Delta H_{\mathrm{ads}}/A = (\Delta H^{\mathrm{s}}_{\mathrm{I(L)}} - \Delta H^{\mathrm{s}}_{\mathrm{I(V)}}) \tag{58 a}$$

$$\Delta H^{\mathrm{s}}_{\mathrm{A}} = \Delta H^{\mathrm{s}}_{\mathrm{ads}} + \Delta H^{\mathrm{s}}_{\mathrm{liq}} \tag{58 b}$$

where the molar enthalpy of liquefaction is $\Delta H_{\mathrm{m,liq}}$ ($\Delta H^{\mathrm{s}}_{\mathrm{liq}} = \Gamma_{\mathrm{L}} \Delta H_{\mathrm{m,liq}}$). The first part ($\Delta H^{\mathrm{s}}_{\mathrm{ads}}$) relates to the adsorbed (gaseous) film while ($\Delta H^{\mathrm{s}}_{\mathrm{A}}$) accounts also for the condensation of the multiplex film. All the terms are negative. The enthalpy of immersion for the vapor covered and the precoated sample is:

$$\Delta H^{\mathrm{s}}_{\mathrm{I(V)}} = \Delta H^{\mathrm{s}}_{\mathrm{SL}}/A = (H_{\mathrm{SL}} - H_{\mathrm{SV}})/A \tag{59 a}$$

$$\Delta H^{\mathrm{s}}_{\mathrm{I(L)}} = \Delta H^{\mathrm{s}}_{\mathrm{S(L)L}}/A = (H_{\mathrm{SL}} - H_{\mathrm{S(L)}})/A \tag{59 b}$$

The difference between immersion enthalpy of vapor and liquid covered sample is:

$$\Delta H^{\mathrm{s}}_{\mathrm{ads}} = (\Delta H^{\mathrm{s}}_{\mathrm{I(V)}} - \Delta H^{\mathrm{s}}_{\mathrm{I(L)}}) = (H^{\mathrm{s}}_{\mathrm{S(L)}} - H^{\mathrm{s}}_{\mathrm{SV}})/A \tag{60 a}$$

The enthalpy of adsorption is typically determined utilizing the Clausius–Clapeyron relation for the isosteric heat of adsorption (Q_{st}):

$$\Delta H^{0}_{\mathrm{ads}} = \int_{\Gamma_{\mathrm{L}=0}}^{\Gamma_{\mathrm{L}}} Q_{\mathrm{st}} \,\mathrm{d} \ln \Gamma_{\mathrm{L}} \tag{61}$$

Hence the change in the immersion enthalpy may be expressed as

$$\Delta H^{\mathrm{s}}_{\mathrm{I(V)}} - \Delta H^{\mathrm{s}}_{\mathrm{I(L)}} \approx \Gamma_{\mathrm{L}} (\Delta H_{\mathrm{m,ads}} - \Delta H_{\mathrm{m,liq}}) \tag{62}$$

In this way it expresses the energy change of the adsorbate in moving from the bulk liquid to the solid surface if the solid is negligibly perturbed and lateral interactions are similar in the adsorbed film to those in the bulk liquid.

Fowkes and Mostafa [43] suggested relating the work of adhesion to the (exothermic) enthalpy per mole of acid–base adduct formation at the interface ($-\Delta H^{\mathrm{ab}}$) with a function f supposed to convert the enthalpic quantity $[\Gamma^{\mathrm{ab}}(-\Delta H^{\mathrm{ab}})]$ to the Gibbs free energy for AB interaction:

$$W^{\mathrm{AB}}_{\mathrm{SL}} = W_{\mathrm{SL}} - W^{\mathrm{LW}}_{\mathrm{SL}} = -\Delta G^{\mathrm{AB}}_{\mathrm{I}} = f[\Gamma^{\mathrm{ab}}(-\Delta H^{\mathrm{ab}})] \tag{63}$$

where $\Gamma^{ab}(= N^s/N_A A)$ is the number of moles of accessible acid or base functional groups per unit area of the solid surface determined by colorimetric titration or adsorption isotherms of the Lewis sites. Assuming that the temperature dependence does not influence the division into LW and AB contributions, we recall Eqs. (19) and (20):

$$\Delta G_I^S = \Delta G_{SL}^S/A = \Delta H_I^S + T[(d\Delta G_I^S)/dT]_{P,n} = \Delta G_I^{LW} + \Delta G_I^{AB} \quad (64)$$

Then, introducing the Fowkes model:

$$\Delta H_I^{AB} = \Delta H_{SL}^{AB}/A = \Gamma^{ab}\Delta H^{ab} = (1/f)\Delta G_I^{AB} \quad (65\,a)$$

$$\Delta H_I^{AB} = \Delta G_I^{AB} - T[(d\Delta G_I^{AB})/dT]_{P,n} \quad (65\,b)$$

we may derive the condition for the equality of ΔG_I^{AB} as [44]

$$1/f = 1 - [d(\Delta \ln G_I^{AB})/d\ln T]_{P,n} \quad (66)$$

Alternatively, we assume that the arithmetic averaging of the surface energy components applies ($\sigma_{SL}^{AB}=0$). Then, as shown in Eq. (65 a) we may write:

$$\Delta G_I^{AB} = -W_{SL}^{AB} = -\sigma_{SV}^{AB} = \Delta H_I^{AB} - T(d(\sigma_{SV}^{AB})/dT)_{P,n} \quad (67)$$

We may thus rewrite the f-factor in the form:

$$f = -\sigma_{SV}^{AB}/\Delta H_I^{AB} = 1 - \{T[d(\sigma_{SV}^{AB})/dT]_{P,n}/\Delta H_I^{AB}\} \quad (68\,a)$$

$$f = 1 + T(\Delta S_I^{AB})/(\Delta H_I^{AB}) \quad (68\,b)$$

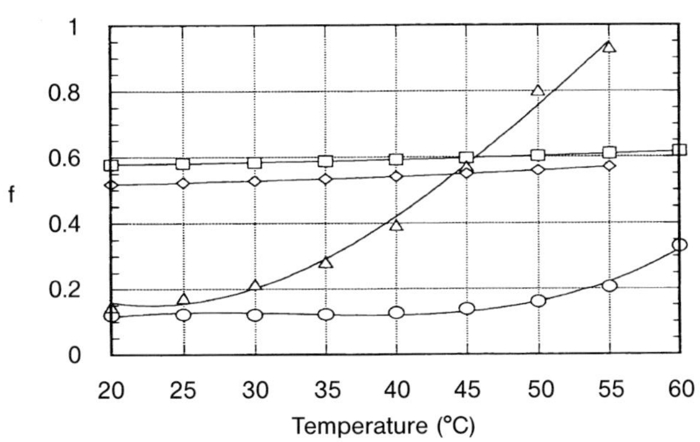

Fig. 1.12 The Fowkes f factor as a function of temperature for bromoform–poly(methyl methacrylate) (circles), dimethyl sufloxide (DMSO)–poly(vinyl chloride) (squares), DMSO–[polyethylene/poly(acrylic acid), 5%] (triangles) and DMSO–[polyethylene/poly(acrylic acid), 20%] (diamonds) (from [44], with permission)

Fig. 1.12 shows that the f factor is substantially less than unity in most cases and increases with temperature. Obviously, for these systems the prediction of a straightforward relationship does not exist between the enthalpy and free energy for the acid–base interaction.

It should be noted that the Drago model [32, 33] refers to the enthalpy of formation of one-to-one molecular adducts in the gas phase and in poorly coordinating solvents. In the latter case, the enthalpy of the non-specific interactions must be subtracted to obtain ΔH_{SL}^{AB}, which, according to Eq. (21), can be approximated as

$$\Delta H_I^{AB} = \Delta H_I - \Delta H_I^{LW} \approx \Delta H_I + \sigma_{SV}^d - 21 \tag{69}$$

Alternatively, we may recall the enthalpy of immersion (Eq. 17):

$$\Delta H_I^{AB} = \sigma_{SL}^{AB} - T[\partial(\sigma_{SL}^{AB})/\partial T]_{P,n} - \sigma_{SV}^{AB} + T[\partial(\sigma_{SV}^{AB})/\partial T]_{P,n} \tag{70 a}$$

$$\Delta H_I^{AB} = \sigma_{SL}^{AB} - \sigma_{SV}^{AB} - T[\partial(\sigma_{SL}^{AB} - \sigma_{SV}^{AB})/\partial T]_{P,n} \tag{70 b}$$

$$\Delta H_I^{AB} = -\gamma_{LV}^{AB} \cos \Theta_L^{AB} + T[\partial(\gamma_{LV}^{AB} \cos \Theta_L^{AB})/\partial T]_{P,n} \tag{70 c}$$

$$\Delta H_I^{AB} = -\gamma_{LV}^{AB} \cos \Theta_L^{AB} + T\gamma_{LV}^{AB}[\partial(\cos \Theta_L^{AB})/\partial T]_{P,n} + T \cos \Theta_L^{AB}[\partial(\gamma_{LV}^{AB})/\partial T]_{P,n} \tag{70 d}$$

where $\cos \Theta_L^{AB}$ may be derived in the way described above. This approach seems, however, to be an unacceptably tedious approach. Douillard and Médout-Marère extended the vOCG division of the components to the enthalpic contribution [45]:

$$H_{SL}^s = H_{SV}^s + H_{LV}^s - 2(H_{SV}^{LW} H_{LV}^{LW})^{1/2} - 2[(H_{SV}^{\oplus} H_{LV}^{\ominus})^{1/2} + (H_{SV}^{\ominus} H_{LV}^{\oplus})^{1/2}] \tag{71}$$

where the components of the heat of immersion are

$$\Delta H_I^s = (H_{SL}^s - H_{SV}^s)/A = \Delta H_I^{LW} + \Delta H_I^{AB} \tag{72 a}$$

$$\Delta H_I^{LW} = H_{LV}^{LW} - 2(H_{SV}^{LW} H_{LV}^{LW})^{1/2} \tag{72 b}$$

$$\Delta H_I^{AB} = H_{LV}^{AB} - 2[(H_{SV}^{\oplus} H_{LV}^{\ominus})^{1/2} + (H_{SV}^{\ominus} H_{LV}^{\oplus})^{1/2}] \tag{72 c}$$

If the immersion is first done in non-specific liquids, then ΔH_I^{LW} can be subtracted from the total heat of immersion for acidic and basic probes to give the acid–base components. Alternatively, the acidic or basic probes are titrated to the solid dispersed in the non-specific liquid displacing the LW molecules from the AB sites. Douillard and Médout-Marère suggested using the Fowkes f function to convert the enthalpies further to surface energy components of the vOCG model discussed previously. As shown above, the latter suggestion is bound to fail.

The conversion into arithmetic averaging maintaining the symmetry of the vOCG model is not possible, since the subtraction of the terms involving the acid–base components cancels the interfacial terms altogether. However, since the extensive experimental material analyzed by Drago is enthalpic in origin, we may rewrite his E and C constants as partial enthalpies:

$$\Delta H_{\mathrm{I}}^{\mathrm{AB}} = \Delta H_{\mathrm{I}} - \Delta H_{\mathrm{I}}^{\mathrm{LW}} = (H_{\mathrm{SV}}^{\mathrm{EA}} H_{\mathrm{LV}}^{\mathrm{EB}}) + (H_{\mathrm{SV}}^{\mathrm{CA}} H_{\mathrm{LV}}^{\mathrm{CB}}) \tag{73}$$

In this way, all the accumulated data are made available for monodentate acid–base reactions immediately. Thus, the immersion is first done in pure LW liquids to give the $\Delta H_{\mathrm{I}}^{\mathrm{AB}}$ contribution.

Alternatively, we may, in line with the Douillard and Médout-Marère (DMM) geometric model, apply the much simpler arithmetic averaging model:

$$H_{\mathrm{SL}}^{\mathrm{s}} = H_{\mathrm{SV}}^{\mathrm{s}} + H_{\mathrm{LV}}^{\mathrm{s}} - (H_{\mathrm{SV}}^{\mathrm{LW}} + H_{\mathrm{LV}}^{\mathrm{LW}}) - [(H_{\mathrm{SV}}^{\mathrm{A}} + H_{\mathrm{LV}}^{\mathrm{B}}) + (H_{\mathrm{SV}}^{\mathrm{B}} + H_{\mathrm{LV}}^{\mathrm{A}})] \tag{74}$$

The components of the heat of immersion are by symmetry

$$\Delta H_{\mathrm{I}}^{\mathrm{s}} = (H_{\mathrm{SL}}^{\mathrm{s}} - H_{\mathrm{SV}}^{\mathrm{s}})/A = \Delta H_{\mathrm{I}}^{\mathrm{LW}} + \Delta H_{\mathrm{I}}^{\mathrm{AB}} \tag{75 a}$$

$$\Delta H_{\mathrm{I}}^{\mathrm{LW}} = H_{\mathrm{LV}}^{\mathrm{LW}} - (H_{\mathrm{SV}}^{\mathrm{LW}} + H_{\mathrm{LV}}^{\mathrm{LW}}) \tag{75 b}$$

$$\Delta H_{\mathrm{I}}^{\mathrm{AB}} = H_{\mathrm{LV}}^{\mathrm{AB}} - [(H_{\mathrm{SV}}^{\mathrm{A}} + H_{\mathrm{LV}}^{\mathrm{B}}) + (H_{\mathrm{SV}}^{\mathrm{B}} + H_{\mathrm{LV}}^{\mathrm{A}})] \tag{75 c}$$

In all these cases, the state molecular gaseous film should be kept apart from the condensed liquid film including the enthalpy of condensation and the behavior of the bulk liquid (see Fig. 1.7).

As discussed in the Introduction, rather than aiming for the free energies (F and G) as done in the discussion above, one should relate the enthalpy to heat capacity instead. For the free energies, all interactions are balanced against each other and thence only a break point is recorded for free energies at first-order phase transitions. For enthalpy this produces a sudden jump to a new level, which is sharper the more extensive the phase transition is. However, the salient interactions are sensitively reflected only for the second order derivative properties, such as heat capacities, expansivities and compressibilities [1, 6]. Consider the distribution of a probe between acidic (or neutral) solution (state A) and basic surface sites (state B):

$$\mathrm{A} \leftrightarrow \mathrm{B} \Rightarrow K = x_{\mathrm{B}}/x_{\mathrm{A}} = x_{\mathrm{B}}/(1 - x_{\mathrm{B}}) \tag{76}$$

where the system is considered ideal, i.e. the activity coefficients have been set equal to unity. The heat capacity of such a system will first contain contributions from the probe in each state and may be written as [46]

$$C_{\mathrm{P}}^{\mathrm{intra}} = x_{\mathrm{A}} C_{\mathrm{P}}^{\mathrm{A}} + x_{\mathrm{B}} C_{\mathrm{P}}^{\mathrm{B}} \tag{77}$$

The latter is termed “intra” in order to distinguish it from another possible contribution, which can arise from shifts in the equilibrium populations of each site with temperature. If there is an enthalpy difference between states A and B (ΔH^{AB}), then the equilibrium shift is obtained through

$$d(\ln K)/dT = [d(\ln K)/dx_B]dx_B/dT = \Delta H^{AB}/RT^2 \quad (78)$$

After derivation and rearranging, we may write

$$dx_B/dT = x_A x_B \Delta H^{AB}/RT^2 \quad (79)$$

Now, the heat adsorbed for this equilibrium shift will contribute and “inter-state” heat capacity, defined as [46]

$$C_P^{inter} = dH/dT = (dH/dx_B)(dx_B/dT) = \Delta H^{AB}(dx_B/dT) \quad (80)$$

Insertion of (dx_B/dT) gives

$$C_P^{inter} = x_A x_B (\Delta H^{AB})^2/RT^2 \quad (81)$$

Hence the total heat capacity has been related to both Gibbs free energy ($\ln K$) and the enthalpy change of the acid–base site binding (ΔH^{AB}) [46]:

$$C_P = C_P^{intra} + C_P^{inter} = x_A C_P^A + x_B C_P^B + x_A x_B (\Delta H^{AB})^2/RT^2 \quad (82)$$

Corresponding relationships can also be written for the other second derivatives from the Gibbs free energy, i.e. the expansivity and the isothermal compressibility [46]:

$$E = E^{intra} + E^{inter} = x_A E^A + x_B E^B + x_A x_B (\Delta H_V^{AB} \Delta V^{AB})/RT^2 \quad (83)$$

$$K_T = K_T^{intra} + K_T^{inter} = x_A K_T^A + x_B K_T^B + x_A x_B (\Delta V^{AB})^2/RT^2 \quad (84)$$

where the cubic expansion coefficient $\alpha = E/V$ and the isothermal compressibility coefficient $\kappa_T = K_T/V$ [1].

The interstate contribution will be maximum at $x_A = x_B = 1/2$ (i.e. $K = 1, \Delta G^{AB} = 0$) and its maximum will depend on $(\Delta H^{AB})^2$, $(\Delta H^{AB} \Delta V^{AB})$ or $(\Delta V^{AB})^2$. When squared the inter-state contribution is always positive, regardless of the sign of the enthalpy change. It must be acknowledged that in this system, x_A and x_B cannot be varied at will, except by changing the temperature or pressure. The limiting cases where the inter-state contribution is small, i.e. $x_A \to 0$ or $x_B \to 0$, originate from either very large or very small equilibrium constants. It is now understood that very weak (van der Waals) interactions will yield rather small contributions to C_P^{inter}. On the other hand, such interactions are sensitively reflected in K_T^{inter}. Intermediate Lewis acid–base interactions such as hydrogen bonding with energies several times RT are expected to produce greater effects on C_P^{inter}. The expansivity (E^{inter}) is expected to be sensitive to both types of interaction. The real potential of (C_P^{inter}, E^{inter}, K_T^{inter}) measurements may thus be fully appreciated as a selective emphasis on contributions which correspond to a particular range of interaction energies.

1.4
Wetting in Idealized Ternary Systems

In order to rationalize the concepts, we describe the processes of wetting using half-spheres having unit target area directed towards the dividing plane. The work of cohesion and adhesion are illustrated in Fig. 1.13.

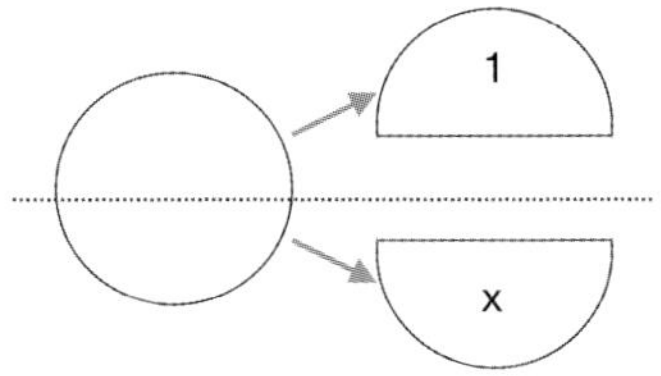

Fig. 1.13 The work of cohesion ($X = 1$) represents the separation of the same phase and the work adhesion ($X = 2$) two phases (half droplets) in contact, thus bringing them in contact with (their) vapor.

We consider specifically the three phases discussed previously, i.e. the liquids (K and L) and the solid (S) (Fig. 1.14). We may now easily derive the work of spreading for each pair of phases, disregarding the third phase (vapor, V) at the three-phase contact-line (tpcl). For non-condensed (vapor) phases the surface tension is negligible. In the indexing the most condensed phase is written first:

$$\left.\begin{aligned} W_{KL} &= \gamma_{KV} + \gamma_{LV} - \gamma_{KL} \\ S_{KL} &= \gamma_{LV} - \gamma_{KV} - \gamma_{KL} \end{aligned}\right\} S_{KL} = W_{KL} - C_{KK} \tag{85 a}$$

$$\left.\begin{aligned} W_{SK} &= \sigma_{SV} + \gamma_{KV} - \sigma_{SK} \\ S_{SK} &= \sigma_{SV} - \gamma_{KV} - \sigma_{SK} \end{aligned}\right\} S_{KL} = W_{SK} - C_{KK} \tag{85 b}$$

$$\left.\begin{aligned} W_{SL} &= \sigma_{SV} + \gamma_{LV} - \sigma_{SL} \\ S_{SL} &= \sigma_{SV} - \gamma_{LV} - \sigma_{SL} \end{aligned}\right\} S_{KL} = W_{SL} - C_{LL} \tag{85 c}$$

In all cases the upper phase is like a reversed process considered to spread on the lower one, until the work of adhesion and work of cohesion are equal. We shall make use of these binary systems when considering the phase equilibrium in three-component systems.

We may expand the optional work of adhesion in terms of the surface tensions of two liquids (K = 1 and L = 2) previously discussed in contact with a solid (S = 3), assuming that they are fully immiscible with each other (more condensed phase first):

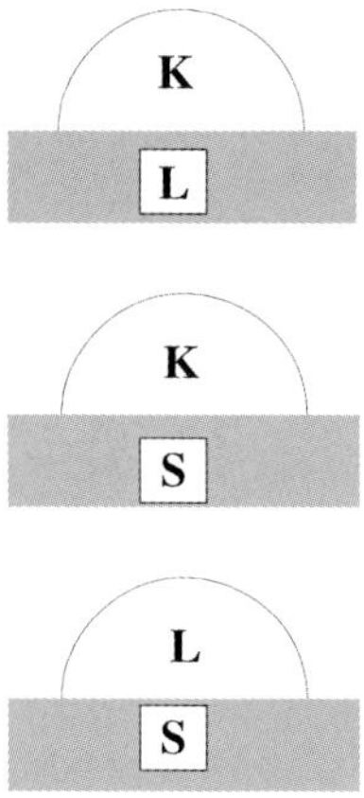

$$\left.\begin{array}{l} W_{KL} = \gamma_{KV} + \gamma_{LV} - \gamma_{KL} \\ S_{KL} = \gamma_{LV} - \gamma_{KV} - \gamma_{KL} \end{array}\right\} S_{KL} = W_{KL} - C_{KK}$$

$$\left.\begin{array}{l} W_{SK} = \sigma_{SV} + \gamma_{KV} - \sigma_{SK} \\ S_{SK} = \sigma_{SV} - \gamma_{KV} - \sigma_{SK} \end{array}\right\} S_{SK} = W_{SK} - C_{KK}$$

$$\left.\begin{array}{l} W_{SL} = \sigma_{SV} + \gamma_{LV} - \sigma_{SL} \\ S_{SL} = \sigma_{SV} - \gamma_{LV} - \sigma_{SL} \end{array}\right\} S_{KL} = W_{SL} - C_{LL}$$

Fig. 1.14 The work of spreading may be expressed as the difference between the work of adhesion and the work of cohesion (Eq. 89). The liquid phase is described as a half droplet in contact with the solid being in contact with the vapor.

$$W_{KL} = \gamma_{KV} + \gamma_{LV} - \gamma_{KL} \Rightarrow W_{12} = \gamma_1 + \gamma_2 - \gamma_{12} \tag{86 a}$$

$$W_{SK} = \sigma_{SV} + \gamma_{KV} - \sigma_{SK} \Rightarrow W_{13} = \sigma_3 + \gamma_1 - \sigma_{13} \tag{86 b}$$

$$W_{SL} = \sigma_{SV} + \gamma_{LV} - \sigma_{SL} \Rightarrow W_{23} = \sigma_3 + \gamma_2 - \sigma_{23} \tag{86 c}$$

When three phases are brought into contact, the situation is rendered much more complex. In addition to the binary contact area we have to consider a three-phase contact point (tpcp) (Fig. 1.15). Assuming that the outer curved lines remain excluded from the considerations, the following options for the processes appear reasonable (no particular indexing order):

If only one third (phase 1) is separated, we find for the work of adhesion (process I)

$$W(\mathrm{I})_{123} = 2\gamma_1 + \gamma_2 + \sigma_3 - \gamma_{12} - \sigma_{13} - \sigma_{123} \tag{87 a}$$

$$W(\mathrm{I})_{123} = W_{12} + W_{13} - \sigma_{123} \tag{87 b}$$

When phase 1 is immersed in phases 2 and 3 the interfacial contacts 1–2 and 1–3 remain and the work of adhesion is dramatically simplified (should be 3 in 1 and 2, process II):

$$W(\mathrm{II})_{123} = \gamma_2 + \sigma_3 \tag{88 a}$$

$$W(\mathrm{II})_{123} = W_{23} + \sigma_{23} \tag{88 b}$$

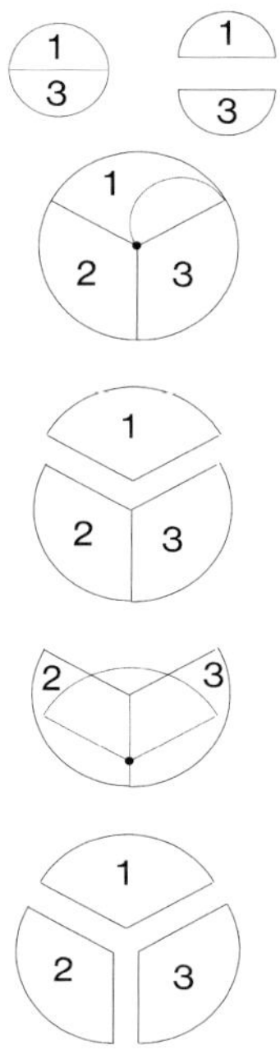

Fig. 1.15 The work of adhesion represented by the separation of one to three phases bringing them in contact with (their) vapor (Eqs. 91–93).

If all three phases are separated simultaneously, we find (process III)

$$W(\mathrm{III})_{123} = 2\gamma_1 + 2\gamma_2 + 2_3 - \gamma_{12} - \sigma_{13} - \sigma_{23} - \sigma_{123} \tag{89 a}$$

$$W(\mathrm{III})_{123} = W_{12} + W_{13} + W_{23} - \sigma_{123} \tag{89 b}$$

The process considered is obviously of prime importance for the surface tension–surface energy balance found. The energy balance at the tpcp should equal zero at equilibrium.

Two processes are offered as a standard for the work of adhesion in textbooks on surface and colloid chemistry [16, 31]. The reason for the particular averaging scheme is probably to maintain the symmetry of the geometric averaging rule.

First we consider that a liquid (L) and a solid (S) initially in contact are separated from each other and brought into cohesive contact (Fig. 1.16):

$$W(\mathrm{IV})_{\mathrm{SLS}} = C_{\mathrm{LL}} + C_{\mathrm{SS}} - 2W_{\mathrm{SL}} \tag{90}$$

Written in terms of interfacial tensions, this equation reduces to

$$W(\mathrm{IV})_{\mathrm{SL}} = 2\sigma_{\mathrm{SL}} \tag{91}$$

The second is the separation between two phases (S and K) initially in contact with the medium (L) to form a contact with the two phases and the third phase internally.

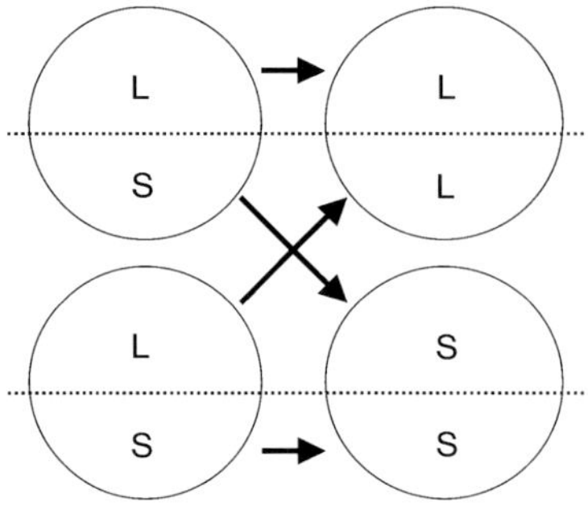

Fig. 1.16 The work of adhesion between a solid and a liquid represented by the separation of the initially interconnected phases and joining each phase to a unified phase, represented by the work of cohesion.

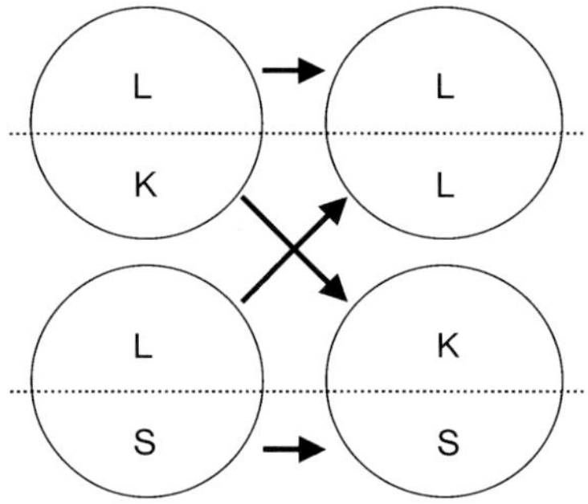

Fig. 1.17 The work of adhesion between liquid–liquid and solid–liquid phases represented by the separation of the initially interconnected phases and joining dispersion liquid to a unified phase, represented by the work of cohesion and the two dispersed phases represented by the work of adhesion.

Converted into binary work of cohesion and adhesion (Fig. 1.17) [16, 31]:

$$W(\mathrm{V})_{\mathrm{SLK}} = C_{\mathrm{LL}} + W_{\mathrm{SK}} - W_{\mathrm{KL}} - W_{\mathrm{SL}} \tag{92}$$

Written in terms of interfacial tensions:

$$W(\mathrm{V})_{\mathrm{SLK}} = \gamma_{\mathrm{KL}} + \sigma_{\mathrm{SL}} - \sigma_{\mathrm{SK}} \tag{93}$$

The latter process is illustrated in terms of the triangular sphere in Fig. 1.18.

As shown, the latter two-phase $[W(\mathrm{IV})_{\mathrm{SL}}]$ and three-phase $[W(\mathrm{V})_{\mathrm{SKL}}]$ work of adhesion correspond to only a fraction of the total work of adhesion $[W(\mathrm{III})_{\mathrm{PLK}}]$. They therefore all represent different thermodynamic realities.

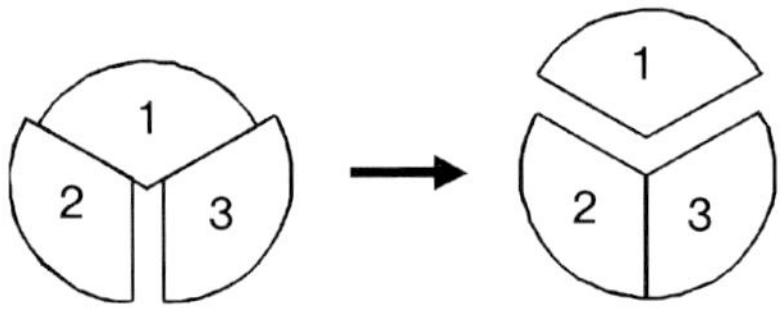

Fig. 1.18 The work of adhesion between liquid–liquid and solid–liquid phases represented by the separation of the initially interconnected phase (1 = L) and joining of the separated phases (2 = K and 3 = S).

1.4.1
Preferential Spreading at Three-component Interfaces

Dispersing a solid (S) and a liquid (K) in small amounts in a immiscible liquid (L) may lead to a full dispersion (*rejection*) of all the phases or an *engulfment* (preferential wetting) of the solid into the K liquid. The intervening situation when all phases partially wet each other is denoted a *funicular state*. In order to determine these limiting states, we derive the ternary work of adhesion and spreading denoting the dispersion medium between the dispersed phases in the lower index. We select the traditional process (Eq. 86) and permute the liquid (K) and the dispersion medium (L) maintaining the solid (S).

With reference to the binary processes (Fig. 1.14), we may write the ternary work of spreading and the work of adhesion in terms of interfacial tensions. For the first case, we find

$$\left.\begin{aligned} S_{SLK} &= \sigma_{SL} - \gamma_{LK} - \sigma_{SK} \\ W_{SLK} &= \sigma_{SL} + \gamma_{LK} - \sigma_{SK} \end{aligned}\right\} S_{SLK} = W_{SLK} - 2\gamma_{LK} \tag{94}$$

Considering L as the dispersed liquid and K as the dispersion medium, we find

$$\left.\begin{aligned} S_{SKL} &= \sigma_{SK} - \gamma_{KL} - \sigma_{SL} \\ W_{SKL} &= \sigma_{SK} + \gamma_{KL} - \sigma_{SL} \end{aligned}\right\} S_{SKL} = W_{SKL} - 2\gamma_{KL} \tag{95}$$

The ternary work of spreading may thus be expressed as the difference between the ternary work of adhesion and the two times the interfacial tension between the liquids (Fig. 1.19).

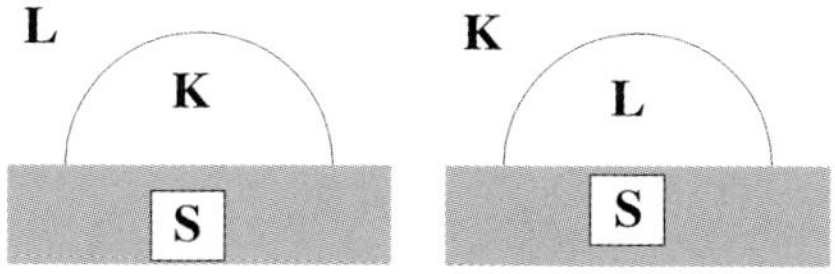

Fig. 1.19 The preferential wetting of two non-miscible liquids on a solid may be expressed by the ternary work of spreading expressed in terms of interfacial tensions. The work of spreading represents the difference between the ternary work of adhesion and two times the interfacial tension between the liquids.

Likewise as for the binary case, the spreading coefficient is expected to be positive (negative Gibbs free energy) for spontaneous preferential spreading to occur. Three limiting cases can be distinguished:

1. The dispersed liquid (K) cannot spread on the solid (S) since the dispersion liquid (L) preferentially wets the particles, i.e. $S_{SLK} \geq 0$, but $S_{SKL} < 0$.
2. The dispersed liquid (K) partially forms (liquid bridges between) the solids (S) if both $S_{SLK} < 0$ and $S_{SKL} < 0$.

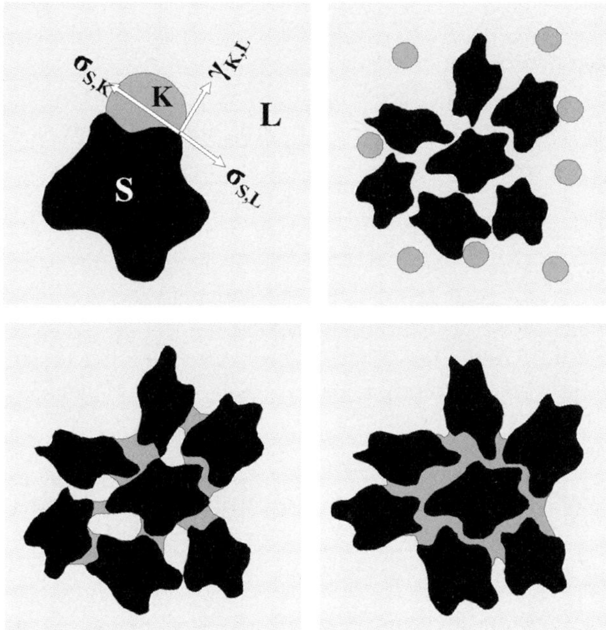

Fig. 1.20 Preferential wetting of oil on hydrophobic particles dispersed in an aqueous dispersion. The particles are the more efficiently removed the smaller the interfacial tension between the water and the oil is and the larger the difference $\sigma_{SK} > \sigma_{SL}$ grows.

3. The dispersed liquid (K) preferentially wets the solid particles (S), displacing the dispersion liquid (L), if $S_{SKL} \geq 0$, but $S_{SLK} < 0$.

Note that the interfacial tension between the liquids determine whether the ternary work of spreading has a positive or negative sign (Fig. 1.20).

We may measure directly the work of adhesion by the introduction of a ternary Young-Dupré equation:

$$W_{SLK} = \gamma_{KL}(\cos \Theta_{SLK} + 1) \tag{96 a}$$

$$W_{SKL} = \gamma_{LK}(\cos \Theta_{SKL} + 1) \tag{96 b}$$

In the former case the contact angle between the solid (S) and the droplet (K) is measured immersed in liquid (L) and in the latter case the liquids are reversed. Owing to density differences, one measurement is usually made from a sessile drop and the other from a pendant drop.

The ternary work of adhesion can be related to the binary work of adhesion as discussed previously [8]:

$$W_{SLK} = C_{LL} + W_{SK} - W_{SL} - W_{LK} \tag{97 a}$$

$$W_{SKL} = C_{KK} + W_{SL} - W_{SK} - W_{KL} \tag{97 b}$$

The liquid (interfacial) tensions are measured as usual and the binary work of adhesion for the solid as

$$W_{SL} = \gamma_{LV}(\cos\Theta_{SL} + 1) \tag{98 a}$$

$$W_{SK} = \gamma_{KV}(\cos\Theta_{SK} + 1) \tag{98 b}$$

Since $C_{LL} - W_{LK} = -S_{LK}$ and $C_{KK} - W_{KL} = -S_{KL}$, we may rearrange the equations:

$$W_{SLK} = \gamma_{KL}(\cos\Theta_{SLK} + 1) = \gamma_{KV}(\cos\Theta_{SK} + 1) - \gamma_{LV}(\cos\Theta_{SL} + 1) - S_{LK} \tag{99 a}$$

$$\gamma_{KL}\cos\Theta_{SLK} = \gamma_{KV}\cos\Theta_{SK} - \gamma_{LV}\cos\Theta_{SL} \tag{99 b}$$

$$W_{SKL} = \gamma_{LK}(\cos\Theta_{SKL} + 1) = \gamma_{LV}(\cos\Theta_{SL} + 1) - \gamma_{KV}(\cos\Theta_{SK} + 1) - S_{KL} \tag{99 c}$$

$$\gamma_{LK}\cos\Theta_{SKL} = \gamma_{LV}\cos\Theta_{SL} - \gamma_{KV}\cos\Theta_{SK} \tag{99 d}$$

This so-called Bartell-Osterhof equation [47] shows that the ternary contact angle (solid–liquid–liquid) may be related to the binary one (solid–liquid–vapor) in a straightforward way. It may be considered as a Cassie equation for a multicomponent system.

1.4.2
Models for Dispersive Solid–Liquid–Liquid Interaction

When considering the two standard processes for the work of cohesion and adhesion, we introduce the geometric average of the dispersive component:

$$C^{d}_{SS} = 2(\sqrt{\sigma^{d}_{SV}}\sqrt{\sigma^{d}_{SV}}) \text{ and } C^{d}_{LL} = 2(\sqrt{\gamma^{d}_{LV}}\sqrt{\gamma^{d}_{LV}}) \tag{100}$$

$$W^{d}_{SL} = 2(\sigma^{d}_{SV}\gamma^{d}_{LV})^{1/2} \tag{101}$$

Then we may write the work of adhesion for the extended (SLS) binary system defined by the equation [8, 17, 18]

$$W^{d}_{SLS} = 2[(\gamma^{d}_{LV}\gamma^{d}_{LV})^{1/2} + (\sigma^{d}_{SV}\sigma^{d}_{SV})^{1/2} - 2(\gamma^{d}_{LV}\sigma^{d}_{SV})^{1/2}] \tag{102 a}$$

$$W^{d}_{SLS} = 2(\sqrt{\sigma^{d}_{SV}} - \sqrt{\gamma^{d}_{LV}})^{2} = 2\sigma^{d}_{SL} \tag{102 b}$$

As shown, the extended binary system produces, as expected from the process considerations, a double dispersive interfacial tension. For both versions the equation shows that only when $\sqrt{\sigma^{d}_{SV}} = \sqrt{\gamma^{d}_{LV}}$ does $W^{d}_{SLS} = 0$. Otherwise W^{d}_{SLS} is always positive. For the three-component system we may write for the work of adhesion

$$W^{d}_{SLK} = 2[(\gamma^{d}_{LV}\gamma^{d}_{LV})^{1/2} + (\sigma^{d}_{SV}\gamma^{d}_{KV})^{1/2} - (\sigma^{d}_{SV}\gamma^{d}_{LV})^{1/2} - (\gamma^{d}_{KV}\gamma^{d}_{LV})^{1/2}] \tag{103 a}$$

$$W^{d}_{SLK} = 2[(\sqrt{\gamma^{d}_{KV}} - \sqrt{\gamma^{d}_{LV}})(\sqrt{\sigma^{d}_{SV}} - \sqrt{\gamma^{d}_{LV}})] \tag{103 b}$$

The equation indicates that the dispersive interaction is repulsive if $\sigma^{d}_{SV} > \gamma^{d}_{LV} > \gamma^{d}_{KV}$ or $\sigma^{d}_{SV} < \gamma^{d}_{LV} < \gamma^{d}_{KV}$.

Since $\gamma \approx A_{H}/(k_{A}d_{0}^{2})$, where $k_{A} \approx 75.40$ for van der Waals liquids, 100.5 for pure hydrocarbons and 10.47 for semi-polar liquids, we may also write the equations in terms of Hamaker constants (A_{H}) [16, 31]:

$$A_{SLK} = A_{LL} + A_{SK} - A_{SL} - A_{LK} \tag{104 a}$$

$$A_{SKL} = A_{KK} + A_{SL} - A_{SK} - A_{KL} \tag{104 b}$$

The same geometric averaging rules have been applied to these interfacial Hamaker constants. Owing to the definition of the work of adhesion for three-component systems without the tpcl(p), the interfacial tension cannot be derived in a straightforward way.

The three-phase systems offer an interesting alternative to measure contact angles of, e.g., a solid (S) immersed in a hydrocarbon (oil, O). If a drop of water (W) is placed as a sessile drop on the solid immersed in the oil, we may write for the ternary Young equation [8]

$$\sigma_{SO} = \sigma_{SW} + \gamma_{OW} \cos \Theta_{SOW} \tag{105}$$

We assume that the hydrocarbon interacts with the solid solely through London-van der Waals forces and write the interfacial energy in terms of the Dupré-Fowkes equation:

$$\sigma_{SO} = \sigma_{SV} + \gamma_{OV} - 2(\sigma^{d}_{SV}\gamma^{d}_{OV})^{1/2} \tag{106}$$

On the other hand, the σ_{SW} component is assumed to be polar, hence the interaction is both dispersive and specific (polar) in origin ($W_{SW} = W^{d}_{SW} - W^{p}_{SW}$) [48]:

$$\sigma_{SW} = \sigma_{SV} + \gamma_{WV} - 2(\sigma^{d}_{SV}\gamma^{d}_{WV})^{1/2} - W^{p}_{SW} \tag{107}$$

Inserted into Eq. (105), we obtain the Schultz equation [49]:

$$\gamma_{WV} - \gamma_{OV} + \gamma_{OW} \cos \Theta_{SOW} = 2\sqrt{\sigma^{d}_{SV}}(\sqrt{\gamma^{d}_{WV}} - \sqrt{\gamma^{d}_{OV}}) + W^{p}_{SW} \tag{108}$$

A Schultz plot of $\gamma_{WV} - \gamma_{OV} + \gamma_{OW} \cos \Theta_{SOW}$ against $\sqrt{\gamma_{WV}^{d}} - \sqrt{\gamma_{OV}^{d}}$ is expected to give a straight line with slope $2\sqrt{\sigma_{SV}^{d}}$ and intercept W_{SW}^{p}. The extraction of W_{SW}^{p} can be improved by choosing octane as the immersion liquid, since its surface tension equals the dispersion component of water. The accuracy of the measurement is frequently fairly low, but it can be confirmed by measuring the contact angle from the pendant hydrocarbon drop against the solid immersed in water [8, 48].

1.4.3
Contribution from the Surface Pressure of a Monomolecular (Gaseous) Film

The preferential spreading of a liquid (K) dispersed in small amounts in an immiscible liquid (L) on an equally dispersed solid (S) may lead to a phenomenon similar to preferential adsorption of one component from a mixed solvent. However, as discussed before, the state of the film considered should be specified, i.e. whether it is a molecular vapor (given by the surface pressure), an unstable intermediate or a bulk (immiscible) liquid (given by the work of spreading). Since $\pi_{S(K)} \approx \sigma_{SV} - \sigma_{S(K)}$, it is realized that the anticipated process is correct. It is obvious from Eqs. (31) and (32) that the spreading coefficient equals the negative surface pressure of a duplex film, i.e. when the liquid fully wets the solid surface as a duplex film. Exchanging the vapor (gas) for the liquid (K), we obtain the following permutative ternary spreading coefficients between K, component L (liquid) and S (substrate), i.e. S_{SLK} and S_{SKL}, respectively.

The preferential spreading of the liquid probe (K) on the solid (S) displacing the dispersion liquid (L) may be assumed to occur via an intermediate state where both liquids are preadsorbed on the solid represented by the surface pressures before immersion in the other liquid. Neglecting the surface pressures on the liquids, we obtain

$$\pi_{S(K)L} = \sigma_{SL} - \sigma_{S(K)L} \tag{109 a}$$

$$\pi_{S(L)K} = \sigma_{SK} - \sigma_{S(L)K} \tag{109 b}$$

We obtain for the work of adhesion for competing surface pressures (Eq. 12 a)

$$W_{S(K,L)LK} = C_{LL} + W_{S(L)K} - W_{S(K)L} - W_{LK} \tag{110 a}$$

$$= 2\gamma_{LV} + [\sigma_{SV} + \gamma_{KV} - \sigma_{S(L)K}] - [\sigma_{SV} + \gamma_{LV} - \sigma_{S(K)L}] - (\gamma_{LV} + \gamma_{KV} - \gamma_{LK}) \tag{110 b}$$

$$W_{S(K,L)LK} = \gamma_{LK} + \sigma_{SL} - \sigma_{SK} - \pi_{S(K)L} + \pi_{S(L)K} \tag{111}$$

where $C_{LL} = C_{L(L)} = 2\gamma_{L(L)} = \gamma_{LV}$. We find the following work of spreading:

$$S_{S(K,L)LK} = W_{S(K,L)LK} - 2\gamma_{LK} = \sigma_{SL} - \sigma_{SK} - \gamma_{LK} - \pi_{S(K)L} - \pi_{S(L)K} \tag{112}$$

The preferential spreading of the liquid (K) occurs in parallel with the retreat of liquid L (negative spreading, opposite signs). Clearly, the preferential wetting may be treated as preferential adsorption from a mixed solvent system to produce a film pressure! For non-miscible liquids the basic phenomena is, however,' more favorably described by an adsorption isotherm.

1.4.4
Models for Lewis (Polar) Solid–Liquid–Liquid Interaction

When considering the two standard processes for the work of cohesion and adhesion, we introduce the geometric average of the acid–base component:

$$W_{ij}^{AB} = W_{ij} - W_{ij}^{LW} = 2[(\gamma_i^{\oplus}\gamma_j^{\ominus})^{1/2} + (\gamma_i^{\ominus}\gamma_j^{\oplus})^{1/2}] \\ = 2[(\sqrt{\gamma_i^{\oplus}}\sqrt{\gamma_j^{\ominus}}) + (\sqrt{\gamma_i^{\ominus}}\sqrt{\gamma_j^{\oplus}})] \quad (113)$$

Additionally, for bidentate (one acidic and one basic site) probe liquids, we find

$$\sigma_{Si}^{AB} = \sigma_{Si} - \sigma_{Si}^{LW} = 2(\sigma_{Si}^{\oplus}\gamma_{Si}^{\ominus})^{1/2} = 2(\sqrt{\sigma_{Si}^{\oplus}}\sqrt{\sigma_{Si}^{\ominus}}) \quad (114\,a)$$

$$\gamma_{Li}^{AB} = \gamma_{Li} - \gamma_{Li}^{LW} = 2(\gamma_{Li}^{\oplus}\gamma_{Li}^{\ominus})^{1/2} = 2(\sqrt{\gamma_{Li}^{\oplus}\gamma_{Li}^{\ominus}}) \quad (114\,b)$$

The symmetry rule also applies for the work of cohesion, being $W_{ii}^{AB} = C_{ii}^{AB} = 2\gamma_i^{AB}$. The work of adhesion and cohesion indicated above thus takes the form [8, 17, 18]

$$W_{SLK}^{AB} = W(4)_{SLK}^{AB} = C_{SS}^{AB} + W_{LK}^{AB} - W_{SL}^{AB} - W_{SK}^{AB} \quad (115\,a)$$

$$W_{SLK}^{AB} = 2\{2(\sqrt{\sigma_{SV}^{\oplus}}\sqrt{\sigma_{SV}^{\ominus}}) + [(\sqrt{\gamma_{LV}^{\oplus}}\sqrt{\gamma_{KV}^{\ominus}}) + (\sqrt{\gamma_{LV}^{\ominus}}\sqrt{\gamma_{KV}^{\oplus}})] - [(\sqrt{\sigma_{SV}^{\oplus}}\sqrt{\gamma_{LV}^{\ominus}}) \\ + (\sqrt{\sigma_{SV}^{\ominus}}\sqrt{\gamma_{LV}^{\oplus}})] - [(\sqrt{\sigma_{SV}^{\oplus}}\sqrt{\gamma_{KV}^{\ominus}}) + (\sqrt{\sigma_{SV}^{\ominus}}\sqrt{\gamma_{KV}^{\oplus}})]\} \quad (115\,b)$$

$$W_{SLK}^{AB} = 2\{[(\sqrt{\sigma_{SV}^{\oplus}} - \sqrt{\gamma_{LV}^{\oplus}})(\sqrt{\sigma_{SV}^{\ominus}} - \sqrt{\gamma_{LV}^{\ominus}})] + [(\sqrt{\sigma_{SV}^{\oplus}} - \sqrt{\gamma_{KV}^{\oplus}})(\sqrt{\sigma_{SV}^{\ominus}} - \sqrt{\gamma_{KV}^{\ominus}})] \\ - [(\sqrt{\gamma_{LV}^{\oplus}} - \sqrt{\gamma_{KV}^{\oplus}})(\sqrt{\gamma_{LV}^{\ominus}} - \sqrt{\gamma_{KV}^{\ominus}})]\} \quad (115\,c)$$

$$W_{SLK}^{AB} = 2\{[(\sqrt{\gamma_{KV}^{\oplus}} - \sqrt{\gamma_{LV}^{\oplus}})(\sqrt{\sigma_{SV}^{\ominus}} - \sqrt{\gamma_{LV}^{\ominus}})] + [(\sqrt{\gamma_{KV}^{\ominus}} - \sqrt{\gamma_{LV}^{\ominus}})(\sqrt{\sigma_{SV}^{\oplus}} - \sqrt{\gamma_{LV}^{\oplus}})]\} \quad (115\,d)$$

The LW and AB interactions in the three-component system may be written in a more illustrative way [8]:

$$W_{SLK}^{LW} = W_{SK}^{LW} - 2[\sqrt{\gamma_{LV}^{LW}}(\sqrt{\sigma_{SV}^{LW}} + \sqrt{\gamma_{KV}^{LW}} - \sqrt{\gamma_{LV}^{LW}})] \quad (116\,a)$$

$$W_{SLK}^{AB} = W_{SK}^{AB} - 2[\sqrt{\gamma_{LV}^{\oplus}}(\sqrt{\sigma_{SV}^{\ominus}} + \sqrt{\gamma_{KV}^{\ominus}} - \sqrt{\gamma_{LV}^{\ominus}})] - 2[\sqrt{\gamma_{LV}^{\ominus}}(\sqrt{\sigma_{SV}^{\oplus}} + \sqrt{\gamma_{KV}^{\oplus}} - \sqrt{\gamma_{LV}^{\oplus}})] \quad (116\,b)$$

In both cases the binary work of adhesion between the dispersed components may be separated from the interaction between the medium liquid and the dispersed components. Thus the LW component of the medium liquid interacts with (is multiplied with) the dispersed S and K components, while the interaction with itself is subtracted from the balance. In a similar way, the acidic site interacts with the basic sites of the dispersed S and K components, while the interaction between the acidic and basic sites of the liquid is subtracted from the balance. Conversely, the basic sites of the liquid interact with the acidic sites of the S and K components, while the interactions with its own acidic sites are subtracted from the balance.

An interesting opportunity to evaluate the work of ternary interaction is provided by the atomic force microscope (AFM), utilized as such or as a colloidal probe [50]. According to the Derjaguin-Muller-Toporov (DMT) theory [51] for a small-radius solid (tip, T) interacting with a flat solid (S) in a liquid (L), the force of adhesion is given by

$$F_A = 2\pi R W_{TLS} \tag{117}$$

where R is the radius of curvature of the tip (or colloid). Since

$$W_{TLS} = C_{LL} + W_{TS} - W_{TL} - W_{SL} \tag{118}$$

on combining the equations we obtain

$$W_{TS} = (F_A/2\pi R) + W_{TL} + W_{SL} - C_{LL} \tag{119}$$

Now W_{TL} and W_{SL} may be determined from contact angle measurements and $C_{LL} = 2\gamma_{LV}$. Using standard vOCG liquids, the surface energy components were determined for a number of solid substrates using an Si_3N_4 AFM tip [50]. However, in the colloidal probe procedure a roughly spherical particle (ca. 1 μm) is glued on the cantilever and then just about any combination of T–L–S and T–S interactions can be measured.

1.5 Adsorption from Solution

As indicated earlier, the preferential adsorption of a liquid component from a mixture cannot be determined from wetting experiments, e.g. from the work of spreading. However, the determination of the adsorption of probe molecules on the sites offers a straightforward way to determine the molecular surface pressure. In this section we discuss the Lewis type and the Brønsted type of acid–base interaction separately, since the mechanism and energy involved differ. We shall also distinguish between adsorption from an undefined medium resembling the gas adsorption. The difference is, however, that in a hydrocarbon solution the dispersive interaction with the surface is neutralized whereas it remains in gas adsorption. Alternatively, a competing adsorption is considered separately when probe molecules displace, e.g., solvent molecules from the surface sites.

1.5.1
Determination of Lewis (Polar) Interactions with Surface Sites

Adsorption provides the proper mean to evaluate the surface states of the solid. In the first model the process involves two steps. First, the adsorption of probe molecules to the surface which is determined separately providing the number of sites. Second, the ability to transfer electrons from the adsorbed basic absorbate to the acidic surface sites provides the strength of the surface sites. Assuming that the adsorbed probe molecule is an almost pure Lewis base (B^s) reacting on the surface with acidic surface sites (A^s) to form an adduct (AB^s) we may write the equilibrium and the equilibrium constant as [52]:

$$A^s + B^s \leftrightarrow (AB)^s \Rightarrow 1/K_B^s = K_{ads}^s = (a_{AB}^s)/(a_A^s a_B^s) = (x_{AB}^s/x_A^s)(f_{AB}^s/f_A^s a_B^s) \quad (120)$$

where a = activity, x = mole fraction, f = activity coefficient, s = surface and b = (equilibrium) bulk solution. We may now introduce the Hammett function (H_0):

$$H_0 = -\log K_B^s + \log(x_B^s/x_{AB}^s) = -\log(a_A^s) - \log(f_B^s/f_{AB}^s) \quad (121\,a)$$

$$H_0 = pK_B^s + \log(x_B^s/x_{AB}^s) = p(a_A^s) - \log(f_{AB}^s/f_A^s) \quad (121\,b)$$

When $x_{AB}^s = x_B^s$ then $H_0 = pK^s$. Since a mole fraction ratio is considered, it may be exchanged for any other concentration scale. A slight excess of acidic and basic indicator probes has been adsorbed on solids of opposite nature dispersed in a saturated hydrocarbon solvent. After equilibration, the indicators are desorbed using even stronger acids and bases. The amount acid and base needed for changing the color of the adsorbed indicator ($x_{AB}^s = x_A^s$) gives the number of sites and then $H_0 = pK^s = pK_a$ (indicator).

The fraction of acidic surface sites (A) occupied by the basic probe molecules (B) dispersed in indifferent oil (O) for low surface site occupancy (surface coverage) may be related to the surface film pressure [53]:

$$\pi_{S(B)O} = \pi_{S(K)L} = \sigma_{SO} - \sigma_{S(B)O} = (RT/A_m)\ln(a_{AB}^s/a_A^s) \approx (RT/A_m)(x_{AB}^s/x_A^s) \quad (122)$$

where the molar surface area $A_m = N_A a$ and a is the surface area occupied by each B or rather each site area. The number of surface sites $N_m^s = N_A(n_{AB}^s)_m$ and the area occupied by one site $a = A/N^s = w_S A_w/N_m^s$. The monolayer surface excess is $\Gamma_m = (n_{AB}^s)_m/A = N_m^s/AN_A$. In this calculation, it is assumed that the solvent is a fully inert oil (O) and that there is no (surface or concentration) potential against which the adsorption occurs. It may not be possible to

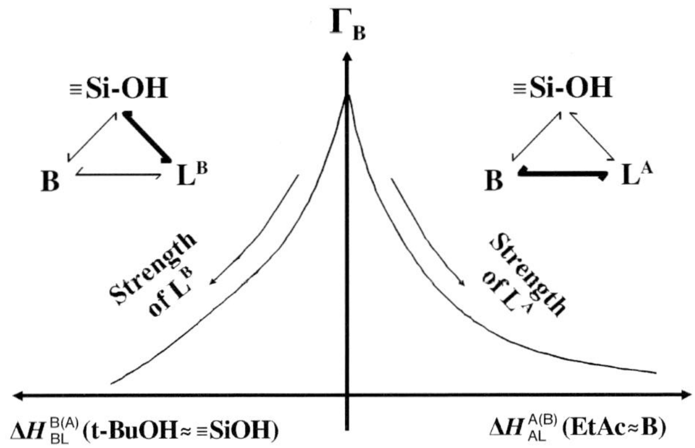

Fig. 1.21 Schematic illustration how basic probe molecules adsorb on the acidic surface silanol (Si–OH) groups of silica. The surface excess is greatest from neutral solvents (middle) but is reduced when the basicity (L^B, left branch) or the acidity (L^A, right branch) of the solvent molecules increases due to SiOH–solvent complexation (left) or probe–solvent complexation (right).

identify $(n^s_A)_m$, the end-point, but rather the equivalence point where $x^s_{AB}/x^s_A = 1$. This can conveniently be identified, e.g. from spectroscopic measurements (spectral or color changes).

For ideal surfaces, the term in Eq. (121) involving activity coefficients can be omitted. The relationship with the energy exchange upon adsorption can be confirmed with the Boltzmann equation:

$$n^s_{AB} = n^b_B \exp(\Delta\mu_{S(B)O}/RT) \tag{123}$$

Introduced in Eq. (117) we obtain:

$$\pi_{S(B)O} = \sigma_{SO} - \sigma_{S(B)O} = (RT/A_m)(\Delta\mu_{S(B)O}/RT) = (\Delta\mu_{S(B)O}/A_m) \tag{124}$$

since $n^s_{AB}/n^b_B = x^s_{AB}/x^b_B$.

Thus, for dilute solutions depletion measurements may be used.

Calorimetry can also be used to determine the degree of adsorption. Figure 1.21 illustrates the amount of an adsorbed basic (probe) molecule on acidic silica silanol (Si–OH) groups plotted against increasing basicity (left) and increasing acidity (right) of a number of solvents. The adsorption is considered to be dependent on the relative degree of (specific) interaction, being greatest from neutral solvents. The strength of the basic solvent is plotted as $\Delta H^{B(A)}_{BL}$, the heat of interaction of the basic solvents with *t*-BuOH. The choice of this alcohol is due to the assumption that it has acidic properties similar to those of the surface Si–OH sites of silica [43]. The strength of the acidic solvents is plotted as

$\Delta H_{AL}^{A(B)}$, the heat of interaction of the solvents with ethyl acetate. It is considered as an oxygen base which models the oxygen basicity of the basic probe.

The left side of Fig. 1.21 shows that as the solvent becomes more basic (e.g. aromatic or oxygenated substances) it forms acid–base complexes with the Si–OH sites. A smaller amount of basic probes is then adsorbed because of reduced exothermicity of adsorption to the SiOH–solvent complex sites [43]. Similarly, the right side of Fig. 1.21 shows that as the solvent becomes more acidic (e.g. halogenated hydrocarbons) and forms acid–base complexes with the basic probe molecule, less probe–solvent complex is adsorbed on the acidic Si–OH surface sites of silica owing to decreased exothermicity of adsorption.

1.5.2
Determination of Brønsted (Charge) Interactions with Surface Sites

A Brønsted acid–base interaction is activated if the Lewis interaction is strong enough, e.g. for hydrogen bonds a protolysis occurs. Then in water both the adsorbate (probe) and the adsorbent (solid substrate) become charged. The Brønsted acidity and basicity thus interlink the Lewis electron acceptor and donor activity into true Coulomb charge interactions. Since the distance over which this interaction is active supersedes the extension of the van der Waals interactions by orders of magnitude, they should be kept apart. However, the considerations of proton and electrolyte distributions as a function of the distance from the surface (given, e.g., by the DLVO theory) is not considered here since the discussion is focused on the surface properties alone.

The clear difference between Lewis and Brønsted acid–base interactions has, however, not always been recognized when selecting molecules for surface probing. In order to avoid complications, nearly ideal polymers are then used as model surfaces. However, in particular when using water as a vOCG probe liquid on inorganic polar surfaces, the Brønsted activity must be considered.

It may be difficult to detect proton transfer at surface sites if the surface area is not sufficient for detectable adsorption to occur. In the simplest form, the adsorption of a proton (acid) on a basic surface site may be described by the Hammett parameter (H_0) [52]:

$$AH^s + B^s \leftrightarrow A^s + (BH^+)^s \tag{125 a}$$

$$1/K_a^s = K_{ads}^s = (a_A^s)(a_{BH}^s)/(a_{AH}^s a_B^s) = (a_A^s x_{BH}^s / a_{AH}^s x_B^s)(f_{BH}^s / f_B^s) \tag{125 b}$$

where a = activity, x = mole fraction, f = activity coefficient, s = surface. Assuming the surface to be ideal (a_A^s=1) we obtain:

$$H_0 = -\log K_a^s + \log(x_B^s / x_{BH}^s) = -\log(a_{(A)H}^s) - \log(f_B^s / f_{BH}^s) \tag{126}$$

$$H_0 = pK_a^s + \log(x_B^s / x_{BH}^s) = p(A)H^s - \log(f_B^s / f_{BH}^s)$$

where $p(A)H^s$ represents the proton activity at the surface sites. At the equivalence point when $x^s_{BH} = x^s_B$ then $H_0 = pK^s_a$, the acidity constant.

The reaction does not necessarily have to be in water. It is sufficient that a proton exchange between the surface and the basic probe (indicator) molecules occur. Figure 1.22 illustrates the titration of titania (anatase and rutile) powders dispersed in cyclohexane using n-butylamine as the base titrant for acidic surface sites and trichloroacetic acid as the acidic titrant for basic surface sites [54].

The indicator probe molecules chosen for the acidic surface sites have increasing, but low, pK^s_a. They are all weaker bases than n-butylamine. The strength of the surface sites is determined by $H_0 < pK^s_a$ and the number of sites is determined by the amount of n-butylamine consumed in order to reach the equilibrium point (color change of indicator). For the basic surface sites, indicators with a rather high pK^s_a are used and trichloroacetic acid is used to desorb these indicators from the surface until the equivalence point. As shown in Fig. 1.22, the titania samples have both acidic and basic sites which can be identified both in number and in (H_0) strength.

In water, neglecting the activity coefficients (ideal surface conditions), the equation may be rewritten in the form

$$H_0 - p(A)H^s = pK^s_a + \log(x^s_{BH}/x^s_B) \tag{127}$$

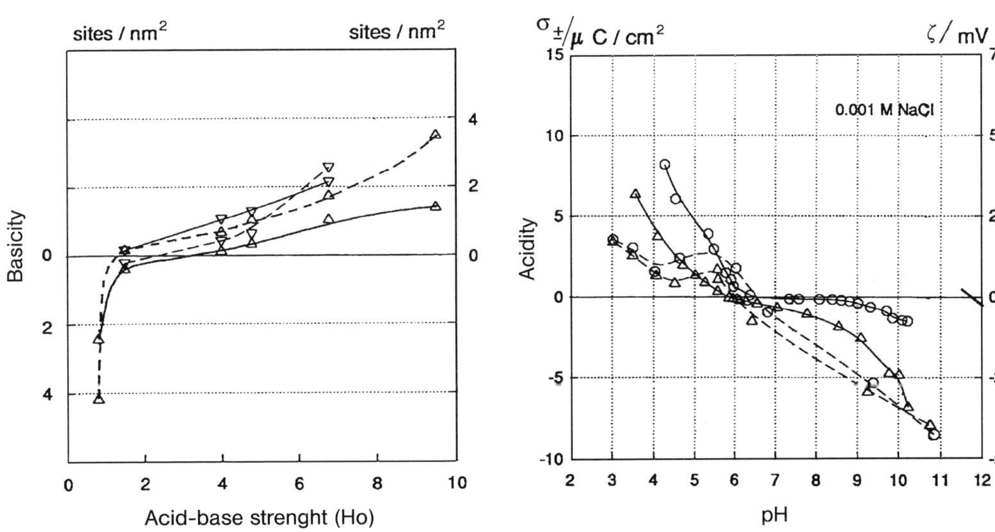

Fig. 1.22 Number (per nm^2) and strength of acidic and basic surface sites of titania powders expressed as the Hammett function (H_0); anatase (down triangles) and rutile (up triangles) both as delivered (broken line) and washed (full drawn line). For comparison, the surface charge density (left axis, full line) and zeta potential (right axis, broken line) are given for anatase (triangles) and rutile (circles) determined in 0.001 mol dm^{-3} NaCl at 25 °C (from [54], with permission).

It is obvious that the Hammett function may be expressed as a corrected pH scale. However, in non-aqueous solvents the pH concept is not clear. All critical values $H_{0,max}$, pH_{PZC} and pH_{IEP} usually match (~6.2 for anatase and ~5.3 for rutile) fairly well, but as shown they diverge in the present case [54]. This is indicative of surface impurities. The corresponding equation may be written for the adsorption of the hydroxyl ions on acidic surface sites. Since the surface is charged in aqueous dispersions, it is customary to relate the pH of the solution with the surface concentration with the Boltzmann relation:

$$n^{s}_{H} = n^{b}_{H} \exp(-F\psi_0/RT) \qquad (128)$$

where $x^{s}_{H}/x^{b}_{H} = n^{s}_{H}/n^{b}_{H}$.

The previous equations do not make any explicit consideration of the proton exchange equilibrium at the surface. According to the Partial Charge Model, the degree of hydrolysis (h) of a cation can be estimated from [55]

$$h = [1/(1+0.41\,\mathrm{pH})]\{[(1.36z - N_C)(0.236 - 0.08\,\mathrm{pH})] - [(2.621 - 0.02\,\mathrm{pH} - \chi^{*}_{M})/\sqrt{\chi^{*}_{M}}]\} \qquad (129\,a)$$

where h = the number of protons spontaneously released by the coordinated complex $[M(OH_2)_N]^{z+}$ in solution, z = valency (charge number), N_C = coordination number and $\overset{*}{\chi}_M$ = Mulliken type electronegativity of the metal. At pH = 0, the equation reduces to [55]

$$h = (1.36z - 0.24N_C) - [(2.621 - \chi_M*)/\sqrt{\chi^{*}_{M}}] \qquad (129\,b)$$

and at pH = 14 to [55]

$$h = (1.14z - 0.25N_C) - \{0.836[(2.341 - \chi^{*}_{M})]/\sqrt{\chi^{*}_{M}}\} \qquad (129\,c)$$

The most important parameter is the formal valency (z, charge number) of the metal cation; N_C and (χ^{*}_{M}) are of lesser importance. The type of coordination can be approximated as the z–pH dependence (Fig. 1.23).

Assuming initially that the metal maintains its coordination complex at the surface, the ligands may reside in the oxo (M–O–), the hydroxo (M–OH) and the aquo ($M\text{–}OH_2^+$) form. The charging of the surface is then due to a single type of (average metal) surface sites. The hydrolysis may then be expressed in terms of the surface charge density:

$$\sigma_{\pm} = F(\Gamma_H - \Gamma_{OH}) = (F/w_S A_w)[(n_H - n_{OH}) - (n^{b}_{H} - n^{b}_{OH})] \qquad (130)$$

where F = Faraday constant, w_S = mass of the solid powder sample, A_w = specific surface area, n_i = acid or base added to the suspension and n^{b}_{i} = acid or base added to the separated supernatant. Of course, other acids and bases may com-

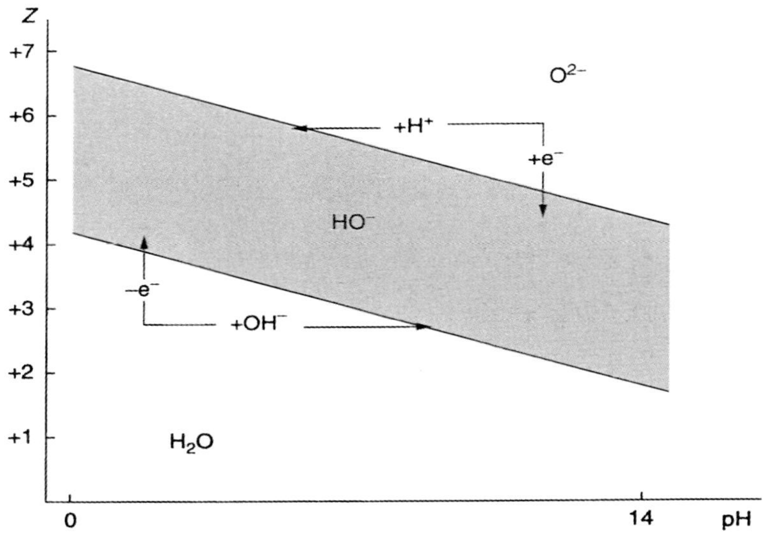

Fig. 1.23 Predominance of aquo, hydroxo and oxo ligands coordinated to metal cations of formal charge *z* as a function of solution pH (from [55], with permission).

pete for the adsorption sites. The point where the charges are neutralized ($\sigma_0 = 0$) is denoted the point of zero charge, pH_{PZC}.

The pH_{PZC} values can be used to determine the ratio of OH groups attached to hydrolyzable surface species (metals) such as Al or Ti. The surface site dissociation can be written as [56]

$$M{-}OH_2^+ \leftrightarrow M{-}OH + H^+ \quad K^{int}_{M,2} = [M{-}OH][H^+]/[M{-}OH_2^+] \tag{131 a}$$

$$M{-}OH \leftrightarrow M{-}O^- + H^+ \quad K^{int}_{M,1} = [M{-}O^-][H^+]/[M{-}OH] \tag{131 b}$$

where $K_{X,m}$ is the equilibrium constant for the metal oxide (M) and m = number of protons. $[H^{+,s}]$ denotes the activity of the protons at the Brønsted surface sites, which is related to the bulk proton activity ($[H^{+,b}]$) through the Boltzmann relation (Eq. 128):

$$[H^{+,s}] = [H^{+,b}]\exp(-F\psi_0/RT) \tag{132}$$

where ψ_0 = surface potential, R = gas constant and T = absolute temperature.

At pH_{PZC}, when $\psi_0 = 0$, Eq. (132) states that $[H^{+,s}] = [H^{+,b}]$, i.e. the intrinsic constant, $K^{int}_{n.m}$ represents the proton equilibrium constant (acidity constant of the surface sites) in a chargeless environment. They are assumed to be independent of the concentrations of the species and the surface potential.

The site number ($N^{s}_{\pm}$) and the total number of OH groups at the surface sites (N^{s}_{m}) is given by

$$N^{s}_{m} = [\text{M–OH}] + [\text{M–OH}_2^+] + [\text{M–O}^-] \tag{133}$$

Using these site number definitions, the surface charge density can be defined as

$$\sigma_{\pm} = (F/A)N^{s}_{m}\{([\text{M–OH}_2^+]-[\text{M–O}^-])/([\text{M–OH}] + [\text{M–OH}_2^+] + [\text{M–O}^-])\} \tag{134}$$

Introducing the equilibrium constants [56]:

$$\begin{aligned}\sigma_{\pm} = F(N^{s}_{m}/A)\{([\text{H}^{+,b}]\exp(-F\psi_0/RT)/K^{\text{int}}_{\text{M},2}) - (K^{\text{int}}_{\text{M},1}\exp(-F\psi_0/RT)/[\text{H}^{+,b}])\}/\\ \{1 + ([\text{H}^{+,b}]\exp(-F\psi_0/RT)/K^{\text{int}}_{\text{M}}) + (K^{\text{int}}_{\text{M},1}\exp(-F\psi_0/RT)/[\text{H}^{+,b}])\}\end{aligned} \tag{135}$$

The total number of OH per nm^2 can be determined by, e.g., titration or adsorption experiments. The site density is very dependent on the experimental method and model of analyzing the data [56].

Formally, the charged surfaces are subdivided into non-polarizable and polarizable surfaces. The polarizable surfaces do not share potential-determining ions (PDIs) with the liquid. Non-polarizable surfaces are characterized by one common species for the surface matrix and the intervening solution. This is typical for most solids where potential-determining cations dissolve partially from the surface, thus determining the surface charge (ψ_0). Assuming the metal oxide surface to be fully polarizable (insoluble) at constant ionic strength (i.e. neglecting the ion contribution), we may relate the electrochemical potential to the interfacial energy by adding the electrical work to the Gibbs-Duhem equation (Eq. 3a). At constant T and P we obtain

$$(n^{s}_{\text{H}}\text{d}\mu_{\text{H}} + n^{s}_{\text{OH}}\text{d}\mu_{\text{OH}}) + A\text{d}\sigma_{\text{SL}} + N^{s}_{\pm}\text{d}\psi_0 = 0 \tag{136 a}$$

$$(\Gamma^{s}_{\text{H}} - \Gamma^{s}_{\text{OH}})\text{d}\mu_{\text{H}} + \text{d}\sigma_{\text{SL}} + \sigma_{\pm}\text{d}\psi_0 = (\sigma_{\pm}/F)\text{d}\mu_{\text{H}} + \text{d}\sigma_{\text{SL}} + \sigma_{\pm}\text{d}\psi_0 = 0 \tag{136 b}$$

where $\sigma_{\pm}$ is the surface charge density. This is the Lippmann equation:

$$-\text{d}\sigma_{\text{SL}} = \sigma_{\pm}\text{d}\psi_0 + (\sigma_{\pm}/F)\text{d}\mu_{\text{H}} = \sigma_{\pm}\text{d}\psi_0 + (\Gamma^{s}_{\text{H}} - \Gamma^{s}_{\text{OH}})\text{d}\mu_{\text{H}} \tag{136 c}$$

This quasi-thermodynamic relation defines an electrical and a chemical contribution to the interfacial energy. The difference between the solid and liquid phases may be varied by an externally applied electrical potential V. The electrical potential difference $U = \Delta V$ can be used to replace $\Delta\psi_0$. Deriving the surface charge density with respect of the surface potential at constant chemical potential for the protons, μ_{H}:

$$-(\text{d}\sigma_{\text{SL}}/\text{d}V)_{\text{P,T},\mu} = (\sigma_{\pm})_{\text{L}} = -(\sigma_{\pm})_{\text{S}} \Rightarrow (\sigma_{\pm})_{\text{L}} = F(\Gamma^{s}_{\text{H}} - \Gamma^{s}_{\text{OH}}) \tag{137}$$

This is the electrocapillarity equation (Fig. 1.23), which shows that when $d\sigma_{SL}/dV > 0$, then $\Gamma^s_{OH} > \Gamma^s_H$ and when $d\sigma_{SL}/dV < 0$, then $\Gamma^s_H > \Gamma^s_{OH}$. A maximum surface energy occurs at pH_{PZC}, when $d\sigma_{SL}/dV = 0$:

$$(d\sigma_{SL}/d\psi_0)_{P,T,\mu} = 0 \Rightarrow \Gamma^s_H = \Gamma^s_{OH} \tag{138}$$

Thus, at the pH_{PZC}, $\sigma_\pm = 0$ and thence $\Gamma^s_H = \Gamma^s_{OH}$ as expected. In the absence of specifically adsorbing ions, σ_{SL} is at a maximum at pH_{PZC}. A shift from this position due to the presence of specifically adsorbing ions is denoted the Esin-Markov effect [57]. The chemical contribution may be evaluated assuming Langmuir isotherm conditions for the chemical potential (e.g. for protons alone) [58]:

$$d\mu_H = RT d\ln[\theta/(1-\theta)] \tag{139 a}$$

Again, assuming that only M–O$^-$ and M–OH_2^+ sites exist, we may write

$$\theta = (\sigma_{max} + \sigma_\pm)/2\sigma_{max} \tag{139 b}$$

where $\sigma_{max} = 2F\Gamma_{PZC}$ and Γ_{PZC} is the surface excess at pH_{PZC} ($\sigma\pm = 0$ and $\psi_0 = 0$) and $\Gamma_H = \Gamma_{OH} = 1/2\Gamma_{max}$. The equation shows that the interfacial tension is maximum at pH_{PZC} and that both chemical and charge factors contribute to the decrease in surface energy from PZC. Since the adsorption causes a decrease in interfacial energy and since spontaneous dispersion of the system occurs for $\sigma_{SL} < 0$, a point of zero interfacial tension pH_{PZIT} may be identified. In the presence of PDIs, two pH_{PZIT} may be identified on both sides of the maximum. In the presence of electrolytes, a range ($pH > pH_{PZIT}$) has been identified where the surface charge becomes saturated [58].

In addition to a complete account of electrolytes in the double layer provided by the Gouy-Chapman approach [16, 31], Stol and de Bruyn offer the following simplified solution to the integrated interfacial energy [58]:

$$\Delta\sigma_{SL} \approx 2(RT/F)(\sigma_- - \sigma_+) - (RT/F)(\sigma^2_\pm/\sigma_{max}) \tag{140}$$

where σ_- and σ_+ are the charge densities of the anions and cations, respectively, in the diffuse part of the double layer.

Recalling equation (136 c) ($d\mu_H = RT d\ln a_H \approx 2.3RT dpH$), the Lippmann equation may be rewritten in the form

$$d\sigma_{SL} = -\sigma_\pm d\psi_0 + 2.3\sigma_\pm(RT/F)dpH \tag{141}$$

Barthes-Labrousse and Joud derived two limiting conditions from this equation [48].

First, when the pH of the aqueous solution is close to pH_{PZC} of the metal oxide surface, a parabolic dependence of the integrated σ_{SL} on pH is observed (Fig. 1.24):

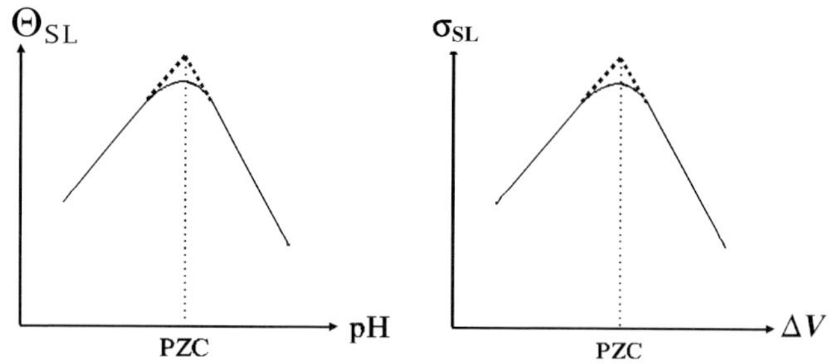

Fig. 1.24 Electrocapillary curve (Eq. 137) and the corresponding variation of the contact angle (Θ_{SL}) as a function of pH (Eq. 144 c) in contact with an amphoteric metal oxide surface. The dependence around the maximum is parabolic. The maximum at pH_{PZC} can be deduced from the positive and negative linear slopes.

$$\Delta\sigma_{SL} = -8RT(c_{1:1}/\kappa_d)(pH_{PZC} - pH)^2 \quad (142)$$

where $c_{1:1}$ is the concentration of possible supporting aqueous 1:1 electrolyte solution and $1/\kappa_d$ is the Debye screening length of the Gouy-Chapman equation. The ratio expresses the amount of ions per unit surface area.

Second, if it is assumed that the mineral acid or base used to adjust the pH does not influence the components (e.g. σ_{SO} and γ_{OW}) other than the solid in contact with water, they may derive the Young equation (cf. Eq. 106):

$$d\sigma_{SW}/dpH = -\gamma_{OW}(d\cos\Theta_{SOW})/dpH = \gamma_{OW}\sin\Theta_{SOW}(d\Theta_{SOW}/dpH) \quad (143)$$

In the presence of electrolytes, γ_{OW} is, however, expected to change but is easily measurable. When the surface charge density is close to the maximum value (σ_{max}), the interfacial surface energy is linearly dependent on pH (cf. Eq. 141):

$$d\sigma_{SW}/dpH = 4.6(RT/F)(\sigma_{\pm})_{max} \quad (144\,a)$$

$$(\sigma_{\pm})_{max} = -(F/4.6RT)\gamma_{WO}(d\cos\Theta_{SOW}/dpH) \quad (144\,b)$$

$$(\sigma_{\pm})_{max} = (F/4.6RT)\gamma_{WO}\sin\Theta_{SOW}(d\Theta_{SOW}/dpH) \quad (144\,c)$$

Note that, in order to avoid the development (adsorption and spreading) of an aqueous surface film [$\pi_{S(W)}$], the measurements were made in a hydrocarbon (O) liquid. Hence γ_{OW} represents the interfacial tension between an aqueous solution (W) and a hydrocarbon (oil, O) and Θ_{SOW} the contact angle of the sessile aqueous drop on the solid (S) immersed in the hydrocarbon.

The influence of the electrolyte concentration and of pH on the surface tension of water and on its contact angle with the hydrophilic silica in air (Table 1.12) may be related to the previous equations by replacing the oil medium by air:

$$\sigma_{\pm}(c_{1:1}, \mathrm{pH}) = -(F/4.6RT)\gamma_{\mathrm{WV}}(\mathrm{d}\cos\Theta_{\mathrm{SW}}/\mathrm{dpH}) \tag{145}$$

Assuming that $\mathrm{pH_{PZC}} \approx 2$ and that the number of surface sites were $N^{\mathrm{s}}/A \approx 1.5\ \mathrm{nm}^2$ [56], the surface charge density was calculated using Eq. (145) and the change in surface energy using Eq. (142), where $RT(c_{1:1}/\kappa_{\mathrm{d}}) \approx kT(N^{\mathrm{s}}/A)$. The results were not realistic and are not listed in Table 1.12. The change in contact angle and the change in pH (Eq. 146 b) also did not balance. Obviously, there must be some other effects that are not included in the models offered. One reason may be the enhanced surface pressure of electrolyte solution or that the $\mathrm{pH_{PZC}}$, assumed to be 2, is shifted upon electrolyte addition. Therefore, the calculations were repeated assuming the $\mathrm{pH_{PZC}}$ to be 6, but this did not improve the results. The underlying assumption that the concepts of a fully polarizable surface can be applied is not supported by the experiments.

In the absence of a water film on the surface, the equations should apply also in the absence of the hydrocarbon liquid. The slopes are proportional to σ_{max}. Bain and Whitesides related the contact angle to the $\mathrm{p}K_{\mathrm{a}}$ values of the carboxyl groups in a film [60]. The model can be modified to apply to the surface M–OH groups with a single acid constant:

$$\cos\Theta_{\mathrm{SL}} = \cos\Theta_{\mathrm{SL}}(\mathrm{PZC}) - (N^{\mathrm{s}}kT/A\gamma_{\mathrm{LV}})\ln[c_{\mathrm{H}}/(c_{\mathrm{H}} - K_{\mathrm{a}})] \tag{146 a}$$

Since $\mathrm{pH_{PZC}} = \mathrm{p}K_{\mathrm{a}}$, the equation takes the form ($c_{\mathrm{H}} << K_{\mathrm{a}}$)

$$\gamma_{\mathrm{LV}}[\cos\Theta_{\mathrm{SL}} - \cos\Theta_{\mathrm{SL}}(\mathrm{PZC})] = 2.3kT(N^{\mathrm{s}}/A)(\mathrm{pH} - \mathrm{pH_{PZC}}) \tag{146 b}$$

Obviously, when $\mathrm{pH} = \mathrm{pH_{PZC}} = K^{\mathrm{int}}_{\mathrm{M,1}} = \mathrm{p}K_{\mathrm{a}}$ then $\cos\Theta_{\mathrm{SL}} = \cos\Theta_{\mathrm{SL}}(\mathrm{PZC})$ and the contact angles of water showed the typical maximum. The derivative $[(\gamma_{\mathrm{LV}}\cos\Theta_{\mathrm{SL}}) - (\gamma_{\mathrm{LV}}(\mathrm{PZC})\ \cos\Theta_{\mathrm{SL}}(\mathrm{PZC}))]/2.3kT(\mathrm{pH} - \mathrm{pH_{PZC}})$ should thus give the change in the number of surface sites (N^{s}/A) produced by the change in pH

Table 1.12 Change in surface tension of the electrolyte solution and the (cosine) contact angle for the electrolyte solution on the hydrophilic silica at pH 2, 6, 10 and 22 ± 1 °C (from [59], with permission).

[NaCl] (mol dm^{-3})	Parameter	Value		
0	pH	2.05	6.33	10.03
	γ_{WV} (mN m^{-1})	72.43	72.31	72.57
	Θ_{SW} (°)	46.47	36.88	38.02
0.1	pH	2.04	6.20	9.92
	γ_{WV} (mN m^{-1})	72.79	72.75	73.03
	Θ_{SW} (°)	41.71	42.15	35.51
0.5	pH	2.04	6.38	10.04
	γ_{WV} (mN m^{-1})	73.48	73.55	73.68
	Θ_{SW} (°)	45.31	47.99	40.30

(Table 1.12). Assuming that $pH_{PZC} \approx 2$ and that the number of surface sites were originally $(N^s/A) \approx 4/\text{nm}^2$ the change in surface charge density calculated using Eq. (146 b) is of the order of only 1–5%.

1.5.3
Adsorption Isotherms for Competitive Interaction at Surface Sites

The adsorption isotherms discussed so far can be extended to include the displacement of e.g. previously adsorbed liquid (solvent, L) molecules by base molecules (B) from solution:

$$(AL)^s + B^b \leftrightarrow (AB)^s + L^b \tag{147 a}$$

$$K_L = (a^s_{AB} a^b_L)/(a^s_{AL} a^b_B) = [(x^s_{AB}/x^s_{AL})(x^b_L/x^b_B)(f^s_{AB} f^b_L/f^s_{AL} f^b_B)] \tag{147 b}$$

where $K_L = K^s_{ads}$ in previous equations.

Assuming ideal conditions both at the surface and in the bulk (dilute) solution, we may omit the term including the activity coefficients. If we additionally introduce $x^s_{AL} = 1 - x^s_{AB}$ and $x^b_L/x^b_B = c^b_L/c^b_B$ (molar concentrations) into the equilibrium constant, then

$$K_L = [x^s_{AB}/(1 - x^s_{AB})](c^b_L/c^b_B) \tag{147 c}$$

Rearrangement gives

$$x^s_{AB} = [c^b_B K_L/(c^b_L + c^b_B K_L)] \tag{148 a}$$

This is one form of the Langmuir adsorption isotherm where x^s_{AB} represents the surface site occupancy (θ) of equal non-communicating ideal surface sites. Note that c^b_L has been maintained in the equation in order to keep K_L dimensionless.

Two limiting cases can be anticipated:

1. If $a^b_B << 1$, i.e. when $c^b_L >> c^b_B$, we find $\theta = x^s_{AB} = (c^b_B/c^b_L)K_L$, representing dilute solutions.
2. If $a^b_B >> 1$, i.e. when $c^b_L << c^b_B$, we find $\theta = x^s_{AB} = 1$, representing a monolayer of adsorbate molecules.

The experimental data are linearized by inverting the equation:

$$1/x^s_{AB} = [c^b_L/(c^b_B K_L)] + 1 \tag{148 b}$$

A plot of $1/x^s_{AB}$ against $1/c^b_B$ should result in a straight line with intercept equal to unity and a slope of c^b_L/K_L.

Recognizing that $x^s_{AB} = n^s_{AB}/(n^s_{AB})_m$ (m = maximum number of surface sites) and multiplying both sides by c^b_B, we obtain

$$c^b_B/n^s_{AB} = c^b_L/(n^s_{AB})_m K_L L + c^b_B/(n^s_{AB})_m \tag{148 c}$$

which is the typical form used (Fig. 1.25). A plot of c^b_B/n^s_{AB} against c^b_B is expected to give a straight line with an intercept $c^b_L/(n^s_{AB})_m K_L$ and a slope of $1/(n^s_{AB})_m$. Since the total volume (V) is known for dilute solutions, $n^s_{AB} = n^i_B - c^b_B V$ and $n^b_L = c^b_L V$. The number of surface sites is $N^s_m = N_A(n^s_{AB})_m$ and the area occupied by one site is $a = A/N^s_m = w_S A_w/N^s_m$. In this calculation, it is assumed that there is no (surface or concentration) potential against which the adsorption occurs.

The molar Gibbs free energy of adsorption is given by

$$\Delta G^s_{m,ads} = -RT \ln K_L = \Delta H^s_{m,ads} - T\Delta S^s_{m,ads} \tag{149}$$

For spontaneous adsorption, $\Delta G^s_{m,ads}$ must be negative. $\Delta S^s_{m,ads}$ is also negative since the motion in three dimensions is restricted to bound molecules in two dimensions. Self-evidently, $\Delta H^s_{m,ads}$ must be negative, corresponding to an exothermic process.

Suppose that the solid is homogeneous and that the dispersion solvent (L) and the base (B) molecules adsorb on acidic surface sites (A) of equal average molar surface areas $A_m (= N_A a$, where a is the average area occupied by each site). Suppose, further, that both the bulk and the surface phases are ideal. Under these circumstances, at adsorption equilibrium, we find for the fractional adsorption from dilute solutions of B in indifferent (non-adsorbing) liquid L

$$\pi_{S(B)L} = \sigma_{SL} - \sigma_{S(B)L} = (RT/A_m)\ln(x^s_{AB}/x^b_B) \tag{150 a}$$

Correspondingly, we find for the fractional adsorption of L in non-adsorbing B

$$\pi_{S(L)L} = \sigma_{SL} - \sigma_{S(L)L} = (RT/A_m)\ln(x^s_{AL}/x^b_A) \tag{150 b}$$

The latter equation serves as an illustration of the partial coadsorption of the indifferent dispersion medium. Obviously $(x^s_{AB} + x^s_A) - (x^s_{AB})_{max} = (x^s_A)_{max}$. We may therefore write formally for the replacement of adsorbed L by B in an indifferent media:

$$\pi_{S(B,L)L} = \pi_{S(B)L} - \pi_{S(L)L} = \sigma_{S(L)L} - \sigma_{S(B)L} = RT/A_m \ln(x^s_{AB}x^b_L/x^s_{AL}x^b_B) \tag{150 c}$$

We define the molar Gibbs free energy of adsorption of monodentate bases on monodentate acid surface sites thus replacing preadsorbed solvent molecules in terms of surface pressure for the adsorbed molecules:

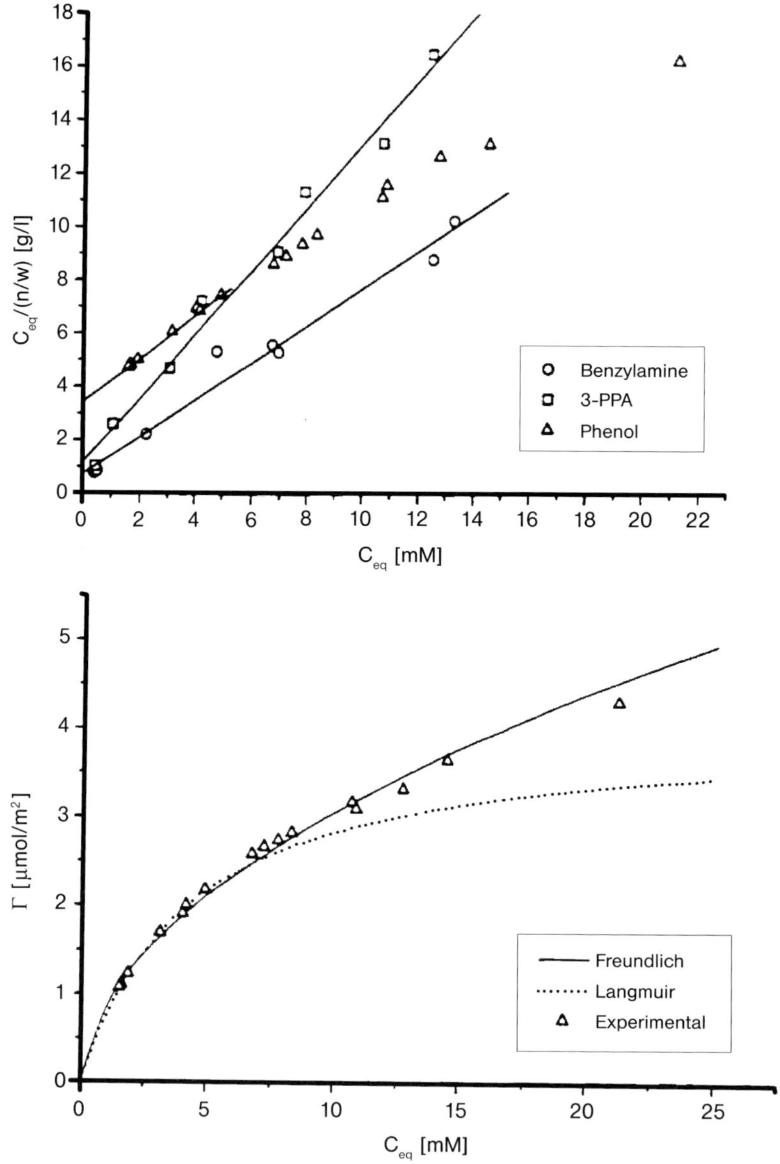

Fig. 1.25 Linearized Langmuir adsorption isotherms for phenol, benzylamine and 3-phenylpropionic acid (3-PPA) on porous silica powder from cyclohexane. Fits using both the Freundlich and the Langmuir isotherms to draw the line through the experimental points for phenol are also shown (from [61], with permission).

$$\Delta G^{s}_{m,ads} = -RT \ln K^{s}_{ads} = -RT \ln[(x^{s}_{AB}/x^{s}_{AL})(x^{b}_{L}/x^{b}_{B})] \tag{151 a}$$

$$\Delta G^{s}_{m,ads} = -(RT/A_{m})[\ln(x^{s}_{AB}/x^{b}_{B}) + \ln(x^{b}_{L}/x^{s}_{AL})] \tag{151 b}$$

$$\Delta G^{s}_{m,ads} = -\pi_{S(B)L} + \pi_{S(L)L} = \Delta\mu_{S(B)L} - \Delta\mu_{S(L)L} \tag{151 c}$$

Utilizing Boltzmann equation the overall Gibbs free energy can thus be subdivided into its components.

The surface pressure can be determined $[x^{s}_{AB} = n^{s}_{AB}/(n^{s}_{AB})_{max} = x^{s}_{AB}/(x^{s}_{AB})_{max}]$ as a depletion from the solution or an adsorption on the surface with different (e.g. spectroscopic) techniques. We define a bidentate surface (S^{AB}) with predominantly acidic ($S^{A(B)} = A^{s}$) and basic ($S^{(A)B} = B^{s}$) interactions, respectively. Likewise, for a bidentate probe (P^{AB}) we denote ($P^{A(B)} = P^{A}$) and ($P^{(A)B} = P^{B}$). For a bidentate probe adsorbing from an indifferent solution onto a bidentate surface we may then write:

$$(P^{B})^{b} + (P^{A})^{b} + A^{s} + B^{s} \leftrightarrow (P^{B}A)^{s} + (P^{A}B)^{s} \tag{152 a}$$

$$K_{ads} \approx [(x^{s}_{PBA}x^{s}_{PAB})/(x^{b}_{PB}x^{b}_{PA})(a^{s}_{A}a^{s}_{B})] \tag{152 b}$$

Assuming that the surface may be assumed ideal ($a^{s}_{A} = a^{s}_{B} = 1$) we find for the fractional adsorption of P^{A} and P^{B} in the indifferent liquid (L):

$$\pi_{S(PB)L} = \sigma_{SL} - \sigma_{S(PB)L} = (RT/A_{m}) \ln(x^{s}_{PBA}/x^{b}_{PB}) \tag{153 a}$$

$$\pi_{S(PA)L} = \sigma_{SL} - \sigma_{S(PA)L} = (RT/A_{m}) \ln(x^{s}_{PAB}/x^{b}_{PA}) \tag{153 b}$$

The sum of these reactions in indifferent media gives:

$$\pi_{S(PA,PB)L} = \pi_{S(PA)L} + \pi_{S(PB)L} = \sigma_{S(L)L} + \sigma_{S(B)L} + 2\sigma_{SL} \tag{154 a}$$

$$\pi_{S(PA,PB)L} = RT/A_{m}[\ln(x^{s}_{PAB}/x^{b}_{PB}) + \ln(x^{s}_{PBA}/x^{b}_{PA})] \tag{154 b}$$

The molar Gibbs free energy of adsorption then takes the form:

$$\Delta G^{s}_{m,ads} = -RT \ln K^{s}_{ads} = -RT \ln[(x^{s}_{PBA}x^{s}_{PAB})/(x^{b}_{PB}x^{b}_{PA})] \tag{155 a}$$

$$\Delta G^{s}_{ads} = -\pi_{S(PA)L} - \pi_{S(PB)L} = \Delta\mu_{S(B)L} + \Delta\mu_{S(L)L} \tag{155 b}$$

where all terms may be determined experimentally as discussed previously. The mass balance is recovered when summing the contributions:

$$n_{P} = n_{PA} + n_{PB} = (n^{s}_{PBA} + n^{s}_{PAB}) + (n^{b}_{PA} + n^{b}_{PB}) \tag{156 a}$$

$$n_{P} = n^{b}_{PA}(k^{s}_{PAB} + 1) + n^{b}_{PB}(k^{s}_{PBA} + 1) \tag{156 b}$$

since $k^{s}_{PBA} = (n^{s}_{PBA}/n^{b}_{PB})$ and $k^{s}_{PAB} = (n^{s}_{PAB}/n^{b}_{PA})$.

There is another way of evaluating the competitive adsorption of basic molecules with the solvent molecules or with a competing base for the same surface sites [62–64]. The preferential adsorption m_B^s (= surface excess, n_B^s, per unit mass of solid substrate powder, w_S) from a mixed solvent on a solid surface (specific surface area A_w) may also be quantified by considering the following limiting cases [65]:

1. The total amount of competing components (B and L = solvent) before adsorption: $n_O = n_{BO} + n_{LO}$.
2. The total amount of competing components in bulk solution at adsorption equilibrium: $n_T^b = n_B^b + n_L^b$.
3. The total number of competing components adsorbed per unit mass of solid at adsorption equilibrium: $m_{AT}^s = m_{AB}^s + m_{AL}^s$, where $m_B^s = n_B^s / w_S$ and $n_B = w_B / M_B$ and $c_B = n_B / V$.

Note that $n_{AB}^s = n_{BO} - n_B^s$. Consequently, $n_O = (n_B^b + w_S m_B^s) + (n_L^b + w_S m_L^s)$. Since $n_B^b / n_L^b = x_B^b / x_L^b$, the mass balance may be described by

$$n_{BO} = n_L^b x_B^b / x_L^b + w_S m_B^s \tag{157 a}$$

$$n_{LO} = n_A^b x_L^b / x_A^b + w_S n_L^s \tag{157 b}$$

Multiplication with x_L^b and x_B^b, respectively, gives

$$n_{BO} x_L^b = n_L^b x_B^b + w_S m_B^s x_L^b \tag{158 a}$$

$$n_{LO} x_B^b = n_B^b x_L^b + w_S m_L^s x_B^b \tag{158 b}$$

Subtraction of Eq. (158 b) from Eq. (158 a) and considering that $n_L^b x_B^b = n_B^b x_L^b$ yields

$$n_{BO} x_L^b - n_{LO} x_B^b = w_S (m_B^s x_L^b - m_L^s x_B^b) \tag{159}$$

Substitution of $x_L^b = 1 - x_B^b$ and $n_{LO} = n_O - n_{BO}$ and recalling that $n_{BO} = n_O x_{BO}$ gives the equation for the surface excess isotherm [62–64]:

$$n_O \Delta x_B^b / w_S = m_B^s x_L^b - m_L^s x_B^b \tag{160}$$

where $\Delta x_B^b = x_{BO} - x_B^b$, which is frequently determined as depletion, e.g. through the change in refractive index or by separate sampling using gas chromatography. The plot of $n_O \Delta x_B^b / w_S$ against x_B indicates, for positive values (i.e. when $x_{BO} > x_B$), a preferential adsorption of component B over L on the solid S. It then follows that a negative plot of $n_O \Delta x_L^b / w_S$ indicates that the liquid component L is negatively adsorbed and the surface phase is less rich than the bulk in L. Figure 1.26 shows the subdivision of the composite isotherm into two general classes.

Provided that a linear section can be identified in the composite isotherm, the thickness of the adsorbed layer can be estimated in the following way. Exchanging $x_L = 1 - x_B$ in Eq. (160) gives

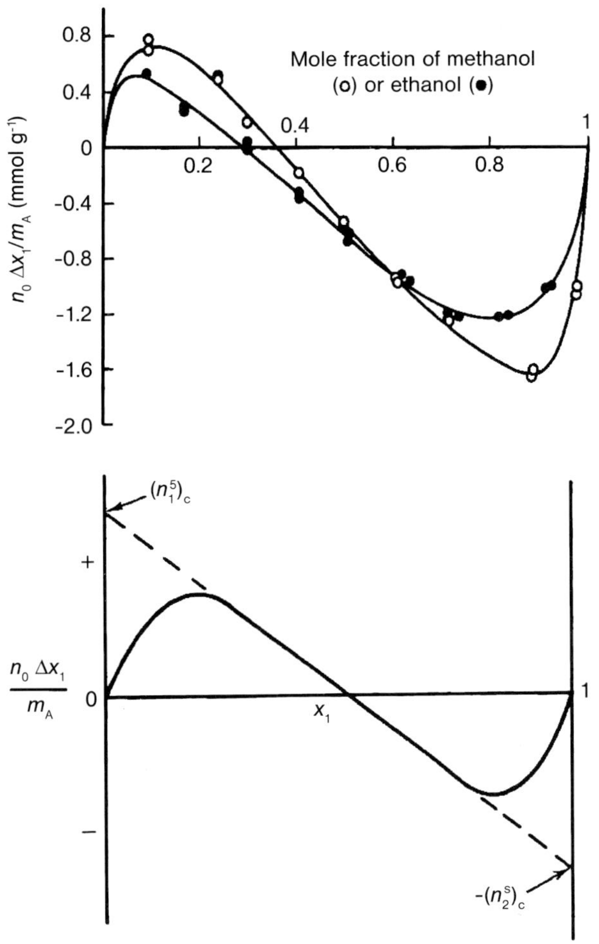

Fig. 1.26 Schematic linearization of composite adsorption isotherms showing concentration regions of excess adsorption of one component over the other (1 = B, 2 = L) with benzene–methanol (open circles) and benzene–ethanol (filled circles) as examples (from [64, 65], with permission).

$$n_O \Delta x_B^b / w_S = m_B^s - (m_B^s + m_L^s) x_B^b \tag{161}$$

Evidently the extrapolated value at $x_B^b = 0$ gives $(m_B^s)_{xL=1}$ and the extrapolated value at $x_B^b = 1$ gives $-(m_L^s)_{xB=1}$. Monolayer adsorption is indicated if [63]

$$(m_B^s)_{crit}/(m_B^s)_{max} + (m_L^s)_{crit}/(m_L^s)_{max} = 1 \tag{162}$$

Assuming that the probe molecules change surface tension and that all depleted molecules adsorb on the solid surface, the Gibbs adsorption equation can be used to determine the surface excess:

$$\Gamma_B = (n_B^s/A) = (m_B^s/A_w) = (d\gamma_{LV}/d\mu_B) \approx -1/RT(\partial\gamma_{LV}/d\partial \ln c_B^b) \quad (163)$$

The influence of the composition on the surface energy (σ_{SV}) tension of the mixed liquid may also be derived in the following way. Since $n_B^s = n_{BO} - n_B$, the composite isotherm Eq. (160) may be written in the form

$$n_O \Delta x_B^b / A = \Gamma_B x_L^b - \Gamma_L x_B^b \quad (164\,a)$$

Introducing $x_L^b = 1 - x_B^b$, we obtain

$$n_O \Delta x_B^b / A = \Gamma_B - (\Gamma_B + \Gamma_L) x_B^b \approx \Gamma_B^{(L)} x_L^b \quad (164\,b)$$

$$\Gamma_B^{(L)} = (n_O \Delta x_B^b / A x_L^b) \quad (164\,c)$$

The latter equality assumes that the phase boundary is normalized to erase the surface excess of the dispersion liquid and to account only for the surface excess of the solute probe (B). Replacing $\Gamma_B^{(L)}$ with the Gibbs adsorption equation [64]

$$-x_L^b(1/RT)[d\sigma_{S(B)L}/d\ln a_B] = \Gamma_B - (\Gamma_B + \Gamma_L)x_B^b = \Gamma_B - \Gamma_{max} x_B^b \quad (165)$$

It is thus possible to quantify the preferential adsorption on the L/V surface by simply measuring the surface tension of the liquid as a function of the composition. Since $n_O \Delta x_B / A$ represents the change in surface tension with the base probe concentration (B), a positive deviation should indicate preferential adsorption and a negative deviation competitive desorption of component (B). For U-shaped isotherms, the component with the lower surface tension (B) adsorbs preferentially on the liquid surface (positive surface excess). In dilute solutions, $x_B = 0$ and hence $\Gamma_B = \Gamma_B^{(L)}$.

The model introduced may be combined with the adsorption equilibrium, i.e. when a liquid (L) in contact with the solid (S) is replaced by the basic component B [64]:

$$\pi_{S(B)L} = \sigma_{SL} - \sigma_{S(B)L} = (RT/A) \int_{a_B=1}^{a_B} (n_O \Delta x_B^b / x_L) d\ln a_B^b \quad (166)$$

Δ plot of $(n_O \Delta x_B^b / x_L)$ against $\ln a_B^b$ may thus be graphically integrated to give $\pi_{S(B)L}$.

We now have all the elements required to relate the key parameters of adsorption to the molecular models described in Table 1.7: $\Delta G_{ads}^s = -RT \ln K_L$ should be proportional to $\ln K_{AB}$ of the Edwards, Maria and Gal, Handcock and Marciano models, and ΔH_{ads} should equal ΔH_{AB} of the Drago and Wayland and Kroeger and Drago models. The implicit condition is that the equilibrium constant refers to adsorption from a fully dispersive solvent and that the enthalpy of wetting the solid with the dispersive solvent has been subtracted from the total enthalpy of adsorption to give ΔH_{AB}. No consideration has been given, however, to whether the process is considered to occur solely at the interface or at an interface in equilibrium with the bulk solution. Tentatively, the molecular models relate to interactions at surfaces, whereas the adsorption concerns an equilibrium of the probe molecules between the bulk solution and the surface. The thermodynamic function relating to the surface alone is the surface pressure. Moreover, since no *PV* work has been considered, the proper state function would, as shown, be the internal energy. If salient specific (AB) interactions are aimed for, the third=order derivatives from the free energies (*F* or *G*) should probably be superior state functions for the correlation.

1.6
Contributions from Surface Heterogeneities

The surface of a solid substrate differs considerably from that of a liquid in that the heterogeneities are not equilibrated by the rapid molecular motions. In reality the solid surface is not molecularly smooth, but consists of surface heterogeneities, such as asperities, dislocations (steps, kinks adatoms and vacancies) and different crystal habits (crystal planes) and other physico-chemical surface heterogeneities (Fig. 1.27). At each heterogeneous site an energy is stored (e.g. as broken bonds) providing the surface with specific binding sites, which influence the wetting phenomena.

Two cases are considered which influence the properties of the system. First, the surface may be of a chemically heterogeneous character, for which the Brønsted interaction and adsorption isotherm are discussed. Self-evidently, there is an even more extensive influence on the Lewis interaction sites, exemplified, e.g., by the dislocations found at the molecular level on the surface. Second, the surface may at the macroscopic level be structurally rough, which influences the wetting for extremely hydrophobic and hydrophilic surfaces. In the latter case, the effect of line tension must be considered.

Fig. 1.27 Solid surfaces may be discontinuous on both the molecular and the macroscopic scale. At the molecular level the dislocations (steps, kinks, adatoms and vacancies) liberate energetic bonds, the number of which depends on the direction of cleavage.

1.6.1
Non-ideal Solid–Liquid Brønsted (Charge) Interactions

Jolivet strongly criticizes the naivistic treatment of solid surfaces described previously and claims that no group may in reality exhibit an amphoteric character, in spite of the fact that they may be positively or negatively charged as a function of pH [55]. Instead, a single equilibrium occurs for each site with $pK_{X,m}^{int} = pH_{PZC}$. This means that the successive involvement of two protons on the same surface site appears completely unrealistic. Instead, for the equilibria in Eq. (131) to apply, they must all be assigned to different surface groups. The latter view is fully possible, since the surfaces contain in reality imperfections representing different crystal planes with the surface elements being involved in different degrees of binding [66]. Following the multisite complexation (MUSIC) model and the formal bond valence concept ($\nu = z/N_C$, N_C = coordination number), we extend our focus on silica and titania to include alumina in order to illustrate the capabilities of this model. For alumina, having the a valency $z = +3$ and coordination numbers 4 and 6, we find the following sequence for ν and the formal charge $\delta = n\nu + z(O^{2-}) + mz(H^+)$ where for singly coordinated OH ligands $n = 1$ and $m = 1$ [55, 56]:

$$\mathrm{Al{-}OH} \quad \nu = 3/4 = 3/4 \quad \delta = 3/4 - 2 + 1 = -1/4 \quad \mathrm{Al(4){-}OH^{1/4-}}$$

$$\mathrm{Al{-}OH} \quad \nu = 3/6 = 1/2 \quad \delta = 1/2 - 2 + 1 = -1/2 \quad \mathrm{Al(6){-}OH^{1/2-}}$$

However, there are also doubly and triply coordinated OH ($n = 2, 3$; $m = 1$) [55]. The dissociation equilibria should therefore be written in the generalized form (cf. Eq. 132):

$$\mathrm{Al_n{-}OH^{(n\nu-1)}} \leftrightarrow \mathrm{Al_n{-}O^{(n\nu-2)}} + \mathrm{H^+} \quad K_{n,1}^{int} = [\mathrm{Al_n{-}O^{(n\nu-2)}}][\mathrm{H^+}]/[\mathrm{Al_n{-}OH^{(n\nu-1)}}] \tag{167 a}$$

$$\mathrm{Al_n{-}OH_2^{(n\nu)}} \leftrightarrow \mathrm{Al_n{-}OH^{n\nu-1)}} + \mathrm{H^+} \quad K_{n,2}^{int} = [\mathrm{Al_n{-}OH^{(n\nu-1)}}][\mathrm{H^+}]/[\mathrm{Al_n{-}OH_2^{(n\nu)}}] \tag{167 b}$$

For gibbsite, $Al(OH)_3$, the large 001 faces (13.8 OH per nm^2) are characterized by doubly coordinated OH groups ($n = 2$) while the sides of the platelets (*hk*0 faces) contain singly (9.6 OH per nm^2) and doubly (4.8 OH per nm^2) coordinated OH groups (Fig. 1.28).

For these groups, Jolivet [55] estimates ($N_C = 6$) that for the dissociations

$$\mathrm{Al{-}OH_2^{1/2+}} \leftrightarrow \mathrm{Al{-}OH^{1/2-}} + \mathrm{H^+} \quad pK_{1,2}^{int} = 10 \tag{168 a}$$

$$\mathrm{Al_2{-}OH^0} \leftrightarrow \mathrm{Al_2{-}O^-} + \mathrm{H^+} \quad pK_{2,1}^{int} = 12.3 \tag{168 b}$$

$$\mathrm{Al_2{-}OH_2^+} \leftrightarrow \mathrm{Al_2{-}OH} + \mathrm{H^+} \quad pK_{2,2}^{int} = -1.5 \tag{168 c}$$

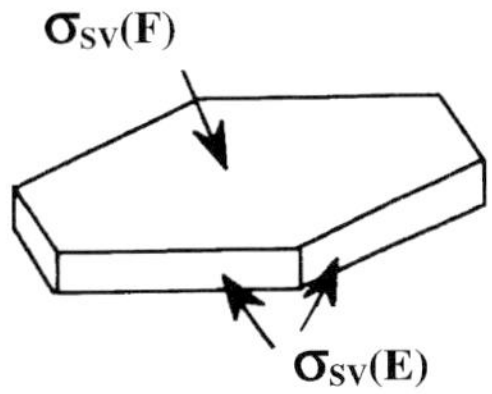

Fig. 1.28 The hexagonal structure of gibbsite particles and the dimensions given as assumed maximal cross-section (ϕ = 0.90 nm), length (l = 0.78 nm), width (w = 0.45 nm) and variable thickness (t). The flat surface is indexed 001 (n = 2) and the sides 010 and 001, the last two being characterized by n = 1 and 2.

The $pK_{1,m}^{int}$ of the $[Al–OH]^{1/2-}/[Al–O^{-}]$ equilibrium ($pK_{1,1}^{int}$ = 23.88) is very high and the singly coordinated groups are therefore present only as $Al–OH^{1/2-}$ or $Al–OH_2^{1/2+}$, depending on the pH. James and Parks report (C. P. Huang, PhD Thesis, 1971) [67] for γ-Al_2O_3 $pK_{1,2}^{int}$ = 7.89 (0.42 OH per nm^2) and $pK_{1,1}^{int}$ = 9.05 (0.39 OH per nm^2) and for α-Al_2O_3 pK_{a1}^{int} = 8.50 and pK_{a2}^{int} = 9.70 (2.7 OH per nm^2) in the presence of 0.1 mol dm^{-3} NaCl. Doubly coordinated groups exist only as Al_2–OH within the normal pH range. For Al_2O_3, the following dissociation constants $pK_{n,1}^{int}$ were calculated: $pK_{1,1}^{int}$ ($-Al–OH^{1/2-}$) = 24, $pK_{2,1}^{int}$ ($-Al_2–OH^0$) = 12.3 and $pK_{3,1}^{int}$ ($-Al_3–OH^{1/2+}$) = 1.6. The increased acidity or weakened strength of the O–H bond with increased degree of coordination of the hydroxyl ligand is clearly reflected.

If one or more of the $pK_{M,m}^{int}$ is outside the available pH range, only a single equilibrium occurs where $pH_{PZC} = pK_{M,m}^{int}$. Jolivet [55] also compared the experimental pH_{PZC} with those calculated with the MUSIC model: Al_2O_3 (z=3, CN=6, ν=1/2), pH_{PZC}(calc.) = 9.1. This compares well with the experimentally found value (9.1) and with those reported by James and Parks for γ-Al_2O_3 (pH_{PZC}=8.47) and for α-Al_2O_3 (pK_{PZC}=9.10). According to Jolivet, we may write [55, 56]

$$\Sigma pK_X^{int} - 2p[H^+] = pK_{1,1}^{int} + pK_{1,2}^{int} - 2p[H^+] = \log([XOH_2^+]/[XO^-]) \qquad (169)$$

At the pH_{PZC} where $\psi_0 = 0$, $[H^+] = [H^{+,b}]$ and $[XOH_2^+] \approx [XO^-]$, we find

$$pH_{PZC} = 1/2(pK_{1,1}^{int} + pK_{1,2}^{int}) = 1/2\Sigma pKx^{int} \qquad (170\,a)$$

$$pH_{PZC}(Al_2O_3) = 1/2\Sigma pK_{Al}^{int} = 1/2(pK_{Al,1}^{int} + pK_{Al,2}^{int}) = 9.1 \qquad (170\,b)$$

Obviously, the pH_{PZC} values compare favorably with the values found experimentally and calculated with the MUSIC model. The equilibria can also be written in the following way:

$$\Delta pK_{Al}^{int} = pK_{1,1}^{int} - pK_{1,2}^{int} = \log([AlOH]^2/[AlOH_2^+]/[AlO^-]) = 1.2 \qquad (171)$$

We may draw the following general conclusions regarding any single metal (M) surface sites [55]:

- If $\Delta pK_X^{int} > 4$ (high), then $[XOH] >> [XOH_2^+] \approx [XO^-]$ and the acid $[XOH_2^+]$ is much stronger than the acid [XOH] and the base $[XO^-]$ is much stronger than the base [XOH]. The predominant species is XOH and the number of ionized species is very small.

- If $\Delta pK_X^{int} < 4$ (small), the XOH_2^+ and XOH acids and XO^- and XOH bases, respectively, have similar strengths. Then the number of charged groups XOH_2^+ and XOH^- is large.

These predictions seem, however, not to be met by a number of acidic and basic groups listed by James and Parks [67], indicating a variable hydroxyl group density for oxides. A more refined analysis of the MUSIC model shows that the dissociation equilibrium must be considered separately for each surface group. The pH at which the net charge is zero depends on the relative fractions of each type of group, and also on their respective pK_1^{int}. For many oxides, cancellation of the global charge may take place through compensation. Moreover, the influence of neighboring hydroxyl groups must be taken into account. The hydroxyl groups decrease linearly with increase in temperature [68]. However, when the communication between the –OH groups ceases, the dependence on temperature is strongly reduced. In this case the rehydroxylation becomes much slower. In porous matrices doubly (geminal) and triply coordinated hydroxyl groups exist [66] which are not described by the MUSIC model and they are only fractionally available for chemical reactions.

1.6.2
Surface Energy of Coexisting Crystal Planes

Since the crystal habit of gibbsite is fairly symmetrical (Fig. 1.28), we may derive a simple model to resolve the surface energy for the two dominant crystal planes, i.e. the surface (face = F) and the sides (edge = E) in equilibrium with the saturated solution. The surface area, the volume and the thickness of the particle may be expressed in terms of the width (w) and a constant k as follows:

$$A = 2kw^2 + 6tw \tag{172 a}$$

$$V = kw^2t \tag{172 b}$$

$$t = V/kw^2 \tag{172 c}$$

where $k = (3/2)\sqrt{3} = 2.598$. The density ($\rho$) and molar mass ($M$) of gibbsite are known and we define the molar surface energy G_m^s of the particle as

$$G_m^s = 2/3\{[2kw^2\sigma_{SV}(F) + 6wt\sigma_{SV}(E)]M\}/V\rho \tag{173 a}$$

Introducing the thickness, the equation takes the form

$$G_m^s = 2/3\{2[kw^2\sigma_{SV}(F)] + 6[V\sigma_{SV}(E)/kw)]M\}/V\rho \tag{173 b}$$

Keeping V and ρ constant, we derive the molar surface energy with respect to w:

$$dG^s_m/dw = 2/3\{4[kw\sigma_{SV}(F)] - 6[V\sigma_{SV}(E)/kw^2)]M\}/V\rho \tag{173 c}$$

Since also the second derivative is positive, the extreme point is a minimum. For this equilibrium, we set the free energy derivative equal to zero and reintroduce the thickness:

$$4[kw\sigma_{SV}(F)] = 6[V\sigma_{SV}(E)/kw^2)] = 6t\sigma_{SV}(E) \tag{174 a}$$

$$\sigma_{SV}(E) = 2kw\sigma_{SV}(F)/3t \tag{174 b}$$

We reintroduce this relationship into the original equation for the molar free energy and eliminate V:

$$\left.\begin{array}{l}\sigma_{SV}(F) = G^s_m t\rho/4M \\ \sigma_{SV}(E) = kwG^s_m\rho/6M\end{array}\right\} \Rightarrow \sigma_{SV}(E)/\sigma_{SV}(F) = 2kw/3t = 3.46 \tag{175}$$

The ratio has been found to agree with electron microscope-measured ratios. For intermediate aging times, $\sigma_{SV}(F) = 140 \pm 24$ mJ m^{-2} and $\sigma_{SV}(E) = 483 \pm 84$ mJ m^{-2}, which agree with experimentally determined values. It has been shown that the ratio of a variable mixture of hydrophilic and hydrophobic particles was linearly related to the contact angle determined with the Cassie model [69, 70]. Consequently, the measured average surface energy determined for gibbsite particles can be subdivided into the surface energy contributions of each crystal plane knowing the fraction of each partial surface. The opposite is also true: from the total surface energy the contributions of each crystal plane can be calculated.

1.6.3
Competing Multi-site Adsorption

The nature of these surface sites has been characterized using attributes such as polar, acid, basic, etc. In medium and high dielectric media, the surface sites may develop charges which are enhanced by surface reactions and isomorphic substitutions of the constituent atoms (ions). The driving force for the adsorption is to neutralize the excess energy of these surface sites. For adsorption on these surface sites a "generalized Langmuir equation" (GL) has been developed from the localized Langmuir (L) isotherm [71]:

$$n^s_A/(n^s_A)_m = [(K_{GL}n_A)^q/(1 + K_{GL}n_A)^q]^{r/q} \tag{176}$$

The constants q and r are assumed to characterize the width of the distribution function and lie within the range 0–1. The constant q characterizes the distribution widening in the direction of the lower adsorption energies and r represents this widening towards higher adsorption energies. For $q = r = 1$ the equation is transformed back into the local Langmuir equation for adsorption on homogeneous solids. However, when $q = 1$ and r $(= 1/k)$ is between 0 and 1, a generalized Freundlich isotherm is obtained. This corresponds to an asymmetric, quasi-Gaussian energy distribution, with widening occurring in the direction of lower adsorption energies. When both q and r lie between 0 and 1, the equation describes the Langmuir-Freundlich isotherm, having a symmetrical quasi-Gaussian energy distribution of the sites. Alternatively, the adsorption energy determined with the Freundlich isotherm can be considered to be dependent on the degree of surface coverage (the lateral interaction between the adsorbate molecules). Empirically, the Freundlich adsorption equation is expressed as [16, 31]

$$m^{s}_{AB} = k_{F}(n^{b}_{B})^{1/k} \tag{177 a}$$

$$\ln m^{s}_{AB} = \ln k_{F} + 1/k \ln n^{b}_{B} \tag{177 b}$$

where k_F and k are experimental constants. The logarithm of the amount adsorbed per unit mass of solid $(m^{s}_{AB} = n^{s}_{AB}/w_{S})$ is plotted against the logarithm of equilibrium concentration $(\ln n^{b}_{B})$ to give the constant $1/k$ as the slope and the constant $\ln k_F$ as the intercept. The enhanced fit over the Langmuir adsorption isotherm is illustrated for phenol on silica in Fig. 1.24.

For uniform sites ($k = 1$), the Freundlich isotherm is comparable to the Langmuir isotherm for dilute solutions:

Freundlich isotherm: $n^{s}_{AB} = k_{F}(n^{b}_{B})^{1/k} \Rightarrow k_{F} = n^{s}_{AB}/n^{b}_{B}$

Langmuir isotherm: $n^{s}_{AB} = (n^{s}_{AB})_{m}(n^{b}_{B}/n^{b}_{L})K_{L} \Rightarrow K_{L} = n^{s}_{AB}/[(n^{s}_{AB})_{m}(n^{b}_{B}/n^{b}_{L})]$

Consequently, the constants used are related to each others as
$k_{F} = K_{L}[(n^{s}_{AB})_{m}/n^{b}_{L})]$.

According to Perkel and Ullman, the adsorption saturation $[(m^{s}_{AB})_{max}]$ for polymers fits the equation [72]

$$(w^{s}_{AB}/w_{S})_{max} = k_{F}M^{k} \tag{177 c}$$

The exponent can be related to the conformation that the polymer takes at the surface: $k = 0$ (in plane), $k = 1$ (upright), $k = 0.5$ (tangled and intertwined), $0 < k < 0.1$ (spherical threads).

The indifferent adsorption in excess of the Langmuir adsorption can be accounted for by setting the exponential constant of the Freundlich isotherm equal to one ($k = 1$). The resulting isotherm is then called a Henry isotherm. The Langmuir-Henry isotherm then takes the form [61]

$$(m^{s}_{A})_{LH} = (m^{s}_{A})_{exp} - k_{H}n_{A} = (m^{s}_{A})_{LH}(K_{LH}n^{b}_{A})/(1 + K_{LH}n^{b}_{A}) \tag{178}$$

The Freundlich equation is particularly usable for the characterization of the concurrent adsorption on the different crystal planes. It may also be used for characterizing the adsorption in multilayers. Since the degree of adsorption may continue infinitely with increasing n_B^b, the equation is not suitable for high coverages. Assume that the probe molecules first adsorbed form a monolayer (k=0) on the surface. In such cases, the multilayer adsorption may be described by the following equation:

$$m_{AB}^s = (k_F)_m + k_F n_B^{1/k} \tag{179}$$

The constant $(k_F)_m$ represents the monolayer adsorption (k = 0) and the last term $k_F n_B^{1/k}$ the multilayer adsorption.

1.6.4
Structural Heterogeneities of the Surface

The dependence of the contact angle on chemical heterogeneities at the surface was studied by Cassie and Baxter [70]. As discussed previously, the contribution of known crystal planes to the average total surface energy is linearly dependent on their fractional surface area. This observation can be rationalized with the well-known Cassie equation being related to the work of adhesion as

$$W_{SLV} = -(\Delta G/\Delta A) = f_1(\sigma_{SL1} - \sigma_{SV}) + f_2(\sigma_{SL2} - \sigma_{SV}) \tag{180}$$

where the function f_i denotes the probability of finding (fractional) surface area with the property i characterized by the contact angle Θ_{SLi}. Applying the Young equation, we may write for the contact angle

$$\cos\langle\Theta_{SL}\rangle = f_1 \cos\Theta_{SL1} + f_2 \cos\Theta_{SL2} \tag{181 a}$$

where $\langle\Theta_{SL}\rangle$ is the average contact angle. Multiplying each term by the surface tension, one obtain the arithmetic surface energy dependence on the fractional surface energy of each diverging property. For porous surfaces, the fractional porc area covered by the liquid (e.g. f_2) may be accounted for by setting $\Theta_{SL2} = 0$ to give

$$\cos\langle\Theta_{SL}\rangle = f_1 \cos\Theta_{SL1} + f_2 \tag{181 b}$$

Wenzel reported on the interdependence of wettability and surface roughness for polar surfaces as early as 1936 [73]. The so-called Wenzel equation can be derived from the Young equation [74]:

$$dG = \Sigma_i \gamma_i dA_i = \sigma_{SV} dA - \gamma_{LV} dA \cos\Theta_{SL} - \sigma_{SL} dA \tag{182}$$

where i represents V = vapor, L = liquid, S = solid. Realizing that the real surface area is much greater than the projected (Young) surface area $q = A_{real}/A_{projected}$, we obtain the work of adhesion:

$$W_{\mathrm{SLV}} = -(\Delta G/\Delta A) = q(\sigma_{\mathrm{SL}} - \sigma_{\mathrm{SV}}) + \gamma_{\mathrm{LV}} \cos\Theta^*_{\mathrm{SL}} \qquad (183)$$

Introducing the Young equation $(\sigma_{\mathrm{SL}} - \sigma_{\mathrm{SV}}) = -\gamma_{\mathrm{LV}} \cos\Theta_{\mathrm{SL}}$ at equilibrium $(\Delta G/\Delta A = 0)$ we obtain the well-known Wenzel equation:

$$\cos\Theta^*_{\mathrm{SL}} = q\cos\Theta_{\mathrm{SL}} \qquad (184)$$

where $q \geq 1$ relates the Young contact angle (Θ_{SL}) to the effective contact angle (Θ^*_{SL}) of the real surface.

Four basic wetting behaviors of a corrugated surface can be identified:

1. the Imbition range
3. the Wenzel range
4. the Cassie range
6. the Lotus range

and two transition zones (2) and (5) (Fig. 1.29).

Above the limit of imbition, the liquid is modeled to be sucked into the porous surface structure leaving the top of the asperities in contact with air [75]. The surface area fraction in contact with air is denoted $f_1 = \phi^{\mathrm{s}}_{\mathrm{SV}}$. However, we also have to consider the extended real surface by multiplying the solid surface by q, but subtracting the fraction of surface not in contact with the liquid $(\phi^{\mathrm{s}}_{\mathrm{SV}})$. Finally, as discussed above, we find a fraction of the surface area covered with liquid for which $f_2 = 1 - \phi^{\mathrm{s}}_{\mathrm{SV}}$. The work of adhesion thus takes the form $(\cos\Theta_{\mathrm{LV}} = 1)$

$$W_{\mathrm{SLV}} = -(\Delta G/\Delta A) = q(\sigma_{\mathrm{SL}} - \sigma_{\mathrm{SV}}) - \phi^{\mathrm{s}}_{\mathrm{SV}}(\sigma_{\mathrm{SL}} - \sigma_{\mathrm{SV}}) + \gamma_{\mathrm{LV}}(1 - \phi^{\mathrm{s}}_{\mathrm{SV}}) \qquad (185)$$

Introducing Young equation at equilibrium, we find

$$\cos\Theta^*_{\mathrm{SL}} = q\cos\Theta_{\mathrm{SL}} = \phi^{\mathrm{s}}_{\mathrm{SV}}\cos\Theta_{\mathrm{SL}} + (1 - \phi^{\mathrm{s}}_{\mathrm{SV}}) \qquad (186)$$

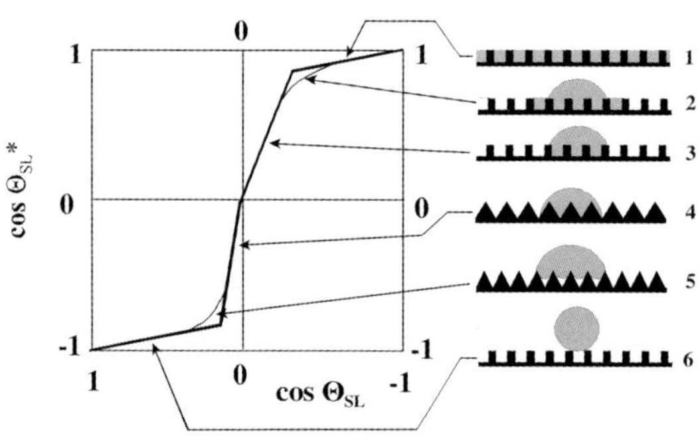

Fig. 1.29 Relationship between the effective contact angle $(\cos\Theta^*_{\mathrm{SL}})$ and the ideal (Young) contact angle $(\cos\Theta_{\mathrm{SL}})$ can be divided schematically into four basic ranges (1, 3, 4 and 6) and two transition zones (2 and 5). The influence of the surface structure on wetting is illustrated for each range.

Therefore, if $\cos\Theta^*_{SL}$ is plotted against $\cos\Theta_{SL}$ for the imbition range, the line extrapolates to $1-\phi^s_{SV}$ with a slope ϕ^s_{SV} when $\cos\Theta_{SL}=0$. The critical contact angle for imbition is [75]

$$\cos\Theta^I_c = (1-\phi^s_{SV})/(q-\phi^s_{SV}) \tag{187}$$

which may be found as a break point above the $\cos\Theta$ axis. Imbition occurs when $\Theta_{SL} < \Theta_c$, i.e. the surface roughness is flooded by the liquid. For a flat surface $q = 1$, i.e. $\cos\Theta_c = 1$. Common for both the Wenzel and Imbition ranges is that $\Theta_{SL} < 90°$. When $\Theta_{SL} > \Theta_c$ the rough surface remains dry ahead of the drop.

Considering the surface asperities as a powder applied as a layer on a plate for the wicking experiments, but assuming full contact of all surfaces with the wetting liquid, White derived an expression for the suction (Laplace like) pressure due to wetting [76]:

$$\Delta P = [2(\sigma_{SV}-\sigma_{SL})/R_{eff}] + (\rho_V-\rho_L)gh = [2\gamma_{LV}\cos\Theta_{SL}/R_{eff}] + (\rho_V-\rho_L)gh \tag{188}$$

where $R_{eff} = 2(1-\phi_S)/(\rho_S A_w \phi_S)$, ρ_S the density of the particles, A_w the specific surface area of the particles and ϕ_S the volume fraction of the particles in the wetting space considered.

The transition from complete to partial wetting of the surface as a function of changing shape of the solid surface (slope of the surface asperities) was demonstrated by Wapner and Hoffman [77]. Their paper actually demonstrates how certain topographical features may give rise to the birth of air pockets and thereby, for example, explains the formation of nanobubbles when such a surface is covered by a liquid. This transition is opposite to the Wenzel range, but now the air pockets form below the drop on top of the asperities. Hence maintaining the surface fractions, the same the Cassie range is defined by [78]

$$W_{SLV} = -(\Delta G/\Delta A) = q(\sigma_{SL}-\sigma_{SV}) - \phi_{SL}(\sigma_{SL}-\sigma_{SV}) - \gamma_{LV}(1-\phi_{SL}) \tag{189}$$

since $\cos\Theta_{LV} = -1$ on the liquid-air pockets. Hence, at equilibrium, the Young equation gives

$$\cos\Theta^*_{SL} = q\cos\Theta_{SL} = -1 + \phi_{SL}(\cos\Theta_{SL}+1) = (\phi_{SL}-1) + \phi_{SL}\cos\Theta_{SL} \tag{190}$$

Therefore, if $\cos\Theta^*_{SL}$ is plotted against $\cos\Theta_{SL}$ for the imbition range, the line extrapolates to $\phi_{SL}-1$ with slope ϕ_{SL} when $\cos\Theta_{SL}=0$. The critical contact angle for the Lotus surface is [79]

$$\cos\Theta^L_c = (\phi_{SL}-1)/(q-\phi_{SL}) \tag{191}$$

Hence it is through the interdependence of topography and enhanced hydrophobicity that a surface turns from "normal" hydrophobic to superhydrophobic.

Nature utilizes this phenomenon, for example, in self-cleaning plant leaves (the so-called Lotus effect) [78, 79]. Bico et al. assumed a discontinuity between the Wenzel and Lotus (i.e. the Cassie) ranges; however, this was not supported by the experimental results presented [79], but is more an artifact due to the assumption of only purely vertical surface asperities. For hemispherical asperities, Bico et al. derived the equation [78]

$$\cos \Theta_{SL}^{*} = -1 + \phi_{B}^{s} (\cos \Theta_{SL} + 1)^{2} \tag{192}$$

where ϕ_{B}^{s} is the ratio of the base over the total surface area of the asperities. This type of representation is in agreement with experimental findings and may probably be used to represent the transition ranges (2) and (5). For example, in the transition range between the Imbition (1) and Wenzel (3) ranges a liquid collar may be found around the drop in the surface heterogeneities as described by Apel-Paz and Marmur [80].

Tsujii's group has demonstrated that even molecular-scale topography contributes to contact angle hysteresis [81]. Topographical characterization has to be carried out with high resolution and at scales of different lengths, which is characteristic for the surface structure [81]:

$$\cos \Theta_{SL}^{*} = (L/l)^{D-2} \cos \Theta_{SL} \tag{193}$$

where L and l are the upper and lower limit step lengths, respectively, over which the fractal dimension was analyzed and D ($2 < D < 3$) is the fractal dimension of the surface (Fig. 1.30). However, in their presentation they made a fit only to the data within the Wenzel range, leaving the analysis of the Cassie range uncompleted [82].

Compared with artificial textured surfaces, the description of the form and shape of real surfaces with complex topography sets high requirements on sur-

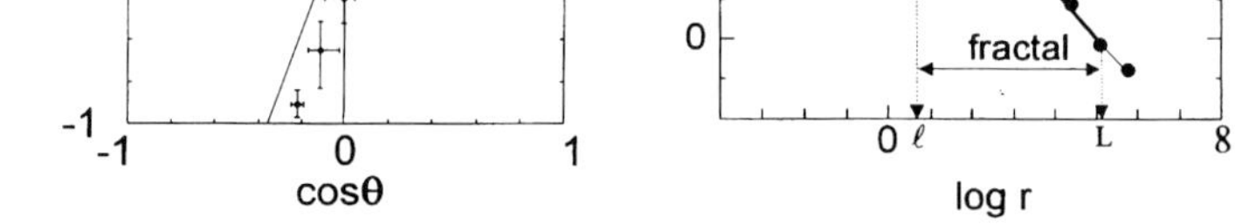

Fig. 1.30 Plot of $\cos \Theta_{SL}^{*}$ against $\cos \Theta_{SL}$ for an alkylketene dimer (AKD)/dialkyl ketone (DAK) surface. The line drawn according to Eq. (193) is characteristic only for part of the surface structure (bold line) relating to the Wenzel region (from [81], with permission).

face microscopy and especially image analysis. Obviously, root mean square (RMS) roughness or peak-to-valley height parameters may be regarded as only indicative when considering, for example, surface porosity or topography-corrected wetting behavior. Peltonen et al. applied a set of topographical parameters for the description of amplitude, spatial and hybrid properties of surfaces for a versatile 3D surface characterization of sol–gel samples with different topography [83]. It was demonstrated that different sets of parameters describe and identify surfaces of different character. They also demonstrated the topography-dependent functionality of the studied surfaces. The 3D image data were captured by atomic force microscopy (AFM). The challenge to be met is to quantify a real surface not only by RMS roughness, but also, for example, by the effective surface area, height asymmetry, surface porosity and the number, size and form of local maxima. In this way, the understanding of the role of topography in phenomena such as wetting, adsorption/precipitation and liquid penetration can be considerably enhanced.

1.7 Contributions from External Stimuli

Many wetting experiments are performed under uncontrolled conditions, e.g. under the influence of external electrostatic potentials and intensive light. It may therefore be of value to consider some not immediately obvious dependences of wetting on external stimuli. The discussion on the relationship between the surface energy and the surface potential showed that electromagnetism may have a profound influence on the surface states of a solid. This is particularly true for semiconductors, in which the electron distribution within the space charge region may be considered to be a (negative) mirror image of the ions in the aqueous double layer outside the surface.

1.7.1 External Electrostatic Potential

For conductive polarizable or non-polarizable surfaces, the distribution of excess cations or anions near the surface is represented by the surface charge density. The Lippmann or electrocapillarity equation discussed previously shows that the surface energy may be adjusted by applying an external electrostatic potential (E) over the surface (Eq. 137):

$$\Delta\sigma_{\mathrm{SL}} = \int_0 \mathrm{d}\sigma_{\mathrm{SL}} = -\int_0 \sigma_{\pm}\mathrm{d}V \tag{194}$$

From the Young equation, we obtain

$$\cos\Theta_{\mathrm{SL}} = (1/\gamma_{\mathrm{LV}})[(\int_0 \sigma_{\pm}\mathrm{d}V) + \sigma_{\mathrm{SV}}] \tag{195}$$

Alternatively, the electrowetting can be derived by way of the minimum free energy requirement for thermodynamic equilibrium conditions for a droplet of conducting liquid (e.g. water, W) on a solid (S) immersed in a insulating oil (O) [84]:

$$\cos\Theta_{\mathrm{SOW}} = \cos\Theta^0_{\mathrm{SOW}} + (\varepsilon_r\varepsilon_0\Delta V^2/2\gamma_{\mathrm{OW}}\delta) \tag{196 a}$$

$$\gamma_{\mathrm{OW}}\cos\Theta_{\mathrm{SOW}} = \sigma_{\mathrm{SO}} - \sigma_{\mathrm{SW}} + (\varepsilon_r\varepsilon_0\Delta V^2/2\delta) \tag{196 b}$$

where Θ^0_{SOW} is the contact angle in the absence and Θ_{SOW} that in the presence of the applied external electrical potential (ΔV), ε_r $(=\varepsilon_{\mathrm{SW}})$ the relative dielectric constant of the conductive liquid close to the solid and δ is the thickness of the dielectric layer. The latter is typically of the order of micrometers whereas size of the droplet is of the order of millimeters.

Kang identified the Maxwell stress (F_e) and the perpendicular force components acting at the three-phase contact line (tpcl) as (Fig. 1.31) [84]

$$F_e = (\varepsilon_r\varepsilon_0\Delta V^2/2\delta)\,\mathrm{cosec}\Theta_{\mathrm{SOW}} \tag{197 a}$$

$$F_{ex} = (\varepsilon_r\varepsilon_0\Delta V^2/2\delta) \tag{197 b}$$

$$F_{ey} = (\varepsilon_r\varepsilon_0\Delta V^2/2\delta)\cot\Theta_{\mathrm{SOW}} \tag{197 c}$$

It is interesting that the horizontal force component is independent of the contact angle. This very localized point force would pull the tpcl until it balances with the dragging force of surface tension. The macroscopic balance of the horizontal force components is given by the electrocapillary equation (Eq. 137).

In simple terms, the surface potential gradient is related to the surface charge density $\sigma_\pm(= N^s_\pm/A)$

$$(\mathrm{d}\psi/\mathrm{d}x) \approx (\Delta\psi/\delta) = \sigma_\pm/\varepsilon_r\varepsilon_0 \tag{198}$$

$\Delta V = \Delta\psi = \psi_0 - \psi_\delta$ is the potential drop between two plates at a distance δ with a medium characterized by the relative dielectric constant ε_r and ΔV is the

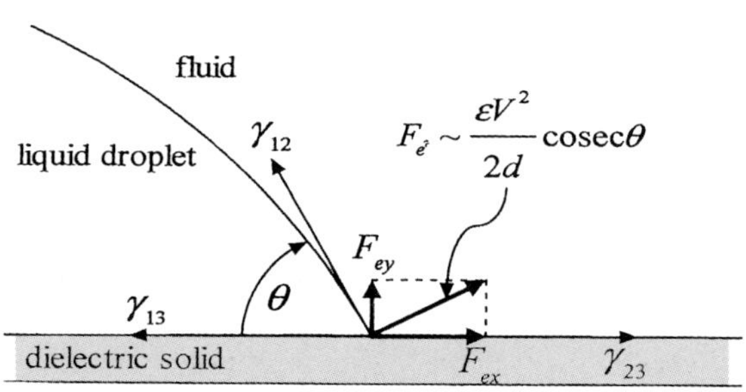

Fig. 1.31 Electrostatic force and its influence of horizontal and vertical balance of forces acting on the three-phase contact line (tpcl) (from [84], with permission).

externally applied potential. We may now proceed to relate the electrostatic potential across the interface, being represented by the differential capacity of the double layer in the liquid phase using the electrocapillary equation:

$$C_{SL} = (d\sigma_{\pm}/dV)_{P,T,\mu} = -(d^2\sigma_{SL}/dV^2)_{P,T,\mu} \tag{199}$$

where $\varepsilon_r\varepsilon_0 = \delta C_{SL}$. The major difference between differential capacity and the capacity of an electric condenser is that it depends on the potential across the double layer, whereas that of the electric condenser does not. For the purpose of integration the potential difference is referred to the electrocapillary maximum (U_{ECM}), which for pH scales occurs at pH_{PZC} (see Fig. 1.24).

1.7.2 External Illumination

Let $V_{h\nu}$ be the electrical potential generated by photonic energy. Then the potential difference in relation to an arbitrary reference potential ($\Delta V_{h\nu} = V_{h\nu} - V_0$) is [85]

$$d\sigma_{SL}/d\psi \approx d\sigma_{SL}/dV_{h\nu} \tag{200}$$

We have defined the work of immersion at constant P, T and μ (Eq. 6 d), which is also termed the adhesion tension:

$$I_{SL} = \sigma_{SV} - \sigma_{SL} = \gamma_{LV}\cos\Theta_{SL} \tag{201}$$

The Young equation is utilized to define the surface charge density difference between the SV and the SL interfaces [85]:

$$\Delta\sigma_{\pm} = d(I_{SL})/dV_{h\nu} = d(\sigma_{SV} - \sigma_{SL})/dV_{h\nu} \tag{202 a}$$

$$\Delta\sigma_{\pm} = \gamma_{LV}[d(\cos\Theta_{SL})/dV_{h\nu}] = -\gamma_{LV}\sin\Theta_{SL}(d\Theta_{SL}/dV_{h\nu}) \tag{202 b}$$

The latter expression is symmetrical with Eq. (144 c) and shows that the change in the contact angle on illumination is proportional to the charge density difference.

Table 1.13 illustrates the dependence of the contact angle on the illumination. As shown, the contact angle decreases slowly with time, indicating that the solid may store the optical energy.

As shown in Fig. 1.32, the hydrophilization (reduction in contact angle) is related to the degradation time of stearic acid reflected as a reduced absorbance which is characteristic for hydrocarbon groups [87]. It is assumed that the hydrophilization is due to the (radical?) formation of hydroxyl groups at the surface [86]. This activation is obviously also related to the degradation mechanism. When energy-rich UV light was used, the time needed to reduce the contact

Table 1.13 Dependence of the contact angle of water on mesoporous titania (anatase) surfaces on the UV and fluorescent visible illumination as a function of time (from [86, 87], with permission).

	Time (min)						
	0	**15**	**30**	**60**	**180**	**360**	**540**
Fluorescent visible light							
Contact angle, Θ_{SW} (°)	33			30	24	18	17
Ultraviolet light							
Contact angle, Θ_S (°)	49	32	26	9			
$E^{s}_{h\nu}$ (kJ m^{-2})	0	10	20	40			

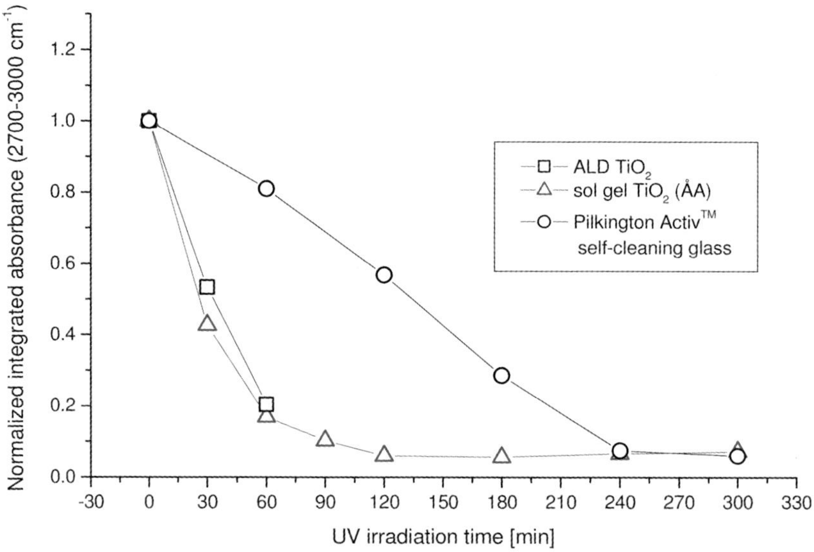

Fig. 1.32 Reduction in the characteristic infrared peaks for stearic acid as a function of time due to catalytic degradation by titania. Pilkington glass, atomic layer deposition and (sol–gel) surfaces under illumination (from [87], with permission).

angle was shortened. The UV irradiance was measured [87] as 1.1 mW cm^{-2} = 11 J m^{-2} s^{-1}, which when multiplied by the irradiation time gives the illumination energy at the surface ($E^{s}_{h\nu}$) reported in Table 1.13.

By differentiation, we obtain the differential capacity difference between the SV and SL interfaces [85]:

$$\Delta C_{SL} = d^2(I_{SL})/d(V_{h\nu})^2 = d(\Delta\sigma_{\pm})/dV_{h\nu} \tag{203}$$

where I_{SL} is denoted adhesion tension. Recalling that the process is carried out at constant P, T and μ, we replace $\sigma_{\pm}$ and C_{SL} by $\Delta\sigma_{\pm}$ and ΔC_{SL} (Eqs. 199 and 203), respectively, to give [85]

$$\Delta\psi_0 = \Delta V_{h\nu} = (\Delta\sigma_{\pm}/\Delta C_{SL})_{T,P,\mu} \tag{204}$$

A plot equivalent to the electrocapillary curve should be obtained on plotting I_{SL} against $\Delta\psi_0$ or $\Delta V_{h\nu}$ (Fig. 1.33).

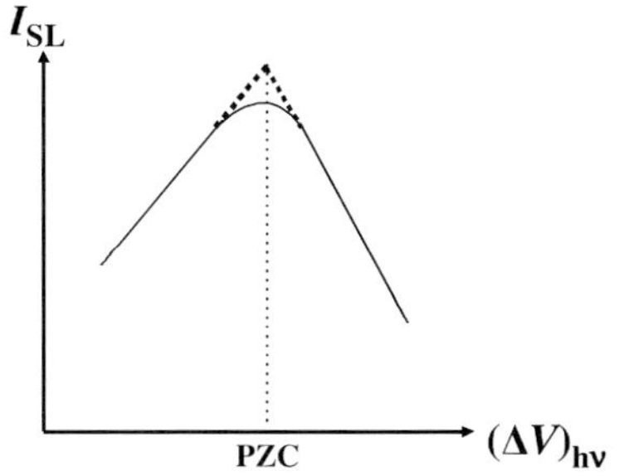

Fig. 1.33 Schematic illustration of the dependence of the work of immersion (I_{SL}) on the photonic potential ($V_{h\nu}$), resembling an electrocapillary curve.

The left-hand side from the maximum refers to n-type semiconductor solids and the right-hand side to p-type. The maximum corresponds to $V_{h\nu} = V_0 \propto U_{ECM}$, i.e. the point of zero charge (cf. pH_{PZC}) or the flat band potential. It is assumed that at PZC both the space charge layer and the electrical double layer vanish.

The energy gap of intrinsic semiconductors, $\langle E_g \rangle$, approximately equals twice the equivalent heat of formation ($-\Delta H^f_{eq} \propto -\Delta G^f_{eq}$) for several binary inorganic substances. For defect-free (intrinsic) semiconductors, we find that [88]

$$1/2\langle E_g \rangle \approx -\Delta H^f_{eq} = E_F = \langle E_b \rangle - k \tag{205}$$

where $\langle E_b \rangle$ is the average bond energy and k = 2.7 eV for a large number of solids. Using this equation, Vijh [88] found an energy gap of 4.7 eV for titania, which is not far from the values reported below.

The band gap can also be determined spectroscopically through the transmittance (Fig. 1.34). First the refractive index (n) and extinction coefficient (k) are determined. The following relationship has been defined for the photonic energy [89]:

$$E_{h\nu} = h\nu = \kappa_e/kn \tag{206}$$

where κ_e is the conductivity of the solid. The absorption coefficient (a) is then determined as [90]

$$ah\nu = l(h\nu - \langle E_g \rangle)^m \tag{207}$$

where l is a model-dependent constant relating to the transition considered and is determined by $\langle E_g \rangle$ and m. The exponent m equals 1/2 for allowed direct, 2 for allowed indirect, 3/2 for forbidden direct and 3 for forbidden indirect transitions. Extrapolating the linear part of $(ahv)^m$ against hv (both in eV) to the abscissa gives m as the slope and $\langle E_g \rangle$ at $(ahv)^m = 0$ [87].

The bulk value for rutile is 3.0 eV and for anatase 3.2 eV, which correspond favorably with the values found (3.3–3.4 eV) in Fig. 1.34.

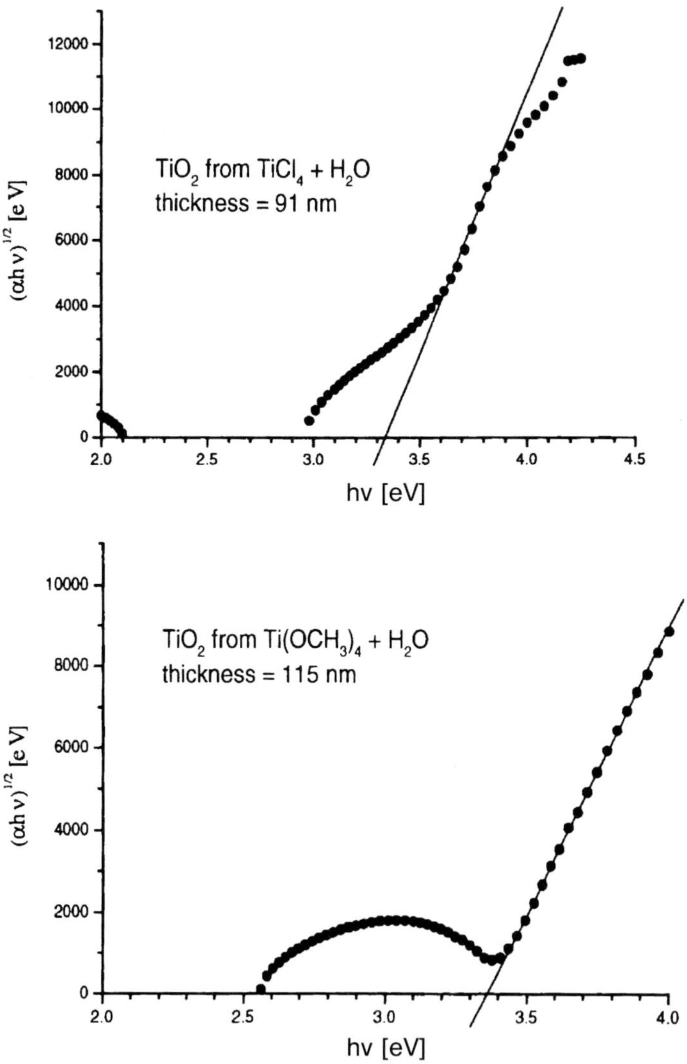

Fig. 1.34 Band gap $\langle Eg \rangle$ determinations of TiO_2 films for an allowed direct transition ($m = 1/2$) (from [87], with permission).

Since the Hamaker constant is also described in terms of refractive indices and dielectric constants, including a characteristic frequency in the ultraviolet region ($\nu \approx 8 \times 10^{15}\ s^{-1}$), the relationship between the photonic energy parameters is obvious [16, 31]. Converting the band gap energies to thermodynamic energy units, we obtain for the 3.0–3.4 eV range $(48.1–54.5) \times 10^{-20}$ J, which is not far from published Hamaker constant values for titania in air (vacuum) [90]: $(15.5–16.8) \times 10^{-20}$ J, representing the cohesive energy of the solid (Eq. 12b).

1.8 Conclusions

It has been demonstrated that nearly all phenomena occurring at solid surfaces can be related to and investigated by the use of different work of wetting. This holds true for the adsorption isotherms, which can be considered as representing the surface pressure in liquids competing for surface sites.

Despite their sensitivity to salient molecular interactions, higher order thermodynamic parameters, such as heat capacity, expansivity and compressibility, are not determined. In contrast, first-order parameters such as enthalpy are (unsuccessfully) related to the Gibbs free energy.

The relationships between wetting models currently considered have been presented and evaluated. Although the absolute values vary between the models, the overall trends remain the same, frequently being dependent on the probe liquids used. Since the choice is to a large extent a matter of convenience, simplification of most current models is suggested.

The equations presented are restricted to equilibrium conditions excluding considerations on dynamic wetting, hysteresis effects and differences in advancing and retreating (receding) contact angles. Moreover, only the interfaces between pure solids, liquids and probe molecules are evaluated. The properties of aqueous electrolyte solutions at a distance from the solid surfaces is discussed only to a limited extent.

The influence of chemical and structural heterogeneity of the surface has been considered with a few examples. Moreover, it has been shown that external stimuli influence the surface energy and contact angle. All models can, however, be rationalized in a straightforward way.

Acknowledgments

This chapter is written in support for the PINTA-ShinePro and NETCOAT-MOL-PRINT projects financed by the National Technology Agency of Finland. The author is grateful to Dr. Folke Eriksson for figure drawings and to Dr. Markku Leskelä, Dr. Mikko Ritala, Dr. Sami Areva, Björn Granqvist and Mikael Järn for permission to use unpublished data.

References

1 Rosenholm, J. B., *Fresenius' Z. Anal. Chem.*, *321* (**1985**) 731.
2 Young, T., *Philos. Trans. R. Soc. London*, (**1805**) 84.
3 Dupré, A., *Théorie Méchnique de la Chaleur.* Gauthier-Villars, Paris, p. 369 (**1869**).
4 Makkonen, L., *Langmuir, 18* (**2000**) 7669.
5 Lyklema, J., *Fundamentals of Interface and Colloid Science*, Vol. 1. Academic Press, London, Appendix 5 (**1991**).
6 Rosenholm, J. B., Ihalainen, P., Peltonen, J., *Colloids Surf. A, 228* (**2003**) 119.
7 Zisman, W. A., *Ind. Eng. Chem., 55* (**1963**) 19.
8 Järn, M., Granqvist, B., Lindfors, J., Kallio, T., Rosenholm, J. B., *Adv. Colloid Interface Sci.*, in press.
9 Fowkes, F. M., *Ind. Eng. Chem., 56* (**1964**) 40.
10 Zettlemoyer, A. C., in *Hydrophobic Surfaces*, Fowkes, F. M. (ed.). Academic Press, New York, pp. 1–27 (**1969**).
11 Hildebrand, J. H., *J. Am. Chem. Soc., 51* (**1929**) 66.
12 Beerbower, A., *J. Colloid Interface Sci., 35* (**1971**) 46.
13 Girifalco, I. A., Good, R. J., *J. Phys. Chem., 61* (**1957**) 904.
14 Wu, S., *J. Polym. Sci. C, 34* (**1971**) 19.
15 Neumann, A. W., Sell, P. J., *Z. Phys. Chem. (Leipzig), 227* (**1964**) 187.
16 Hiemenz, P. C., *Principles of Colloid and Surface Chemistry*, 2nd edn. Marcel Dekker, New York, Ch. 11.8, p. 649 (**1986**).
17 van Oss, C. J., Chadhury, M. J., Good, R. J., *Adv. Colloid Interface Sci., 28* (**1987**) 35.
18 van Oss, C. J., Chadhury, M. J., Good, R. J., *Chem. Rev., 88* (**1988**) 927.
19 Shinoda, K., Becher, P., *Principles of Solution and Solubility.* Marcel Dekker, New York (**1978**).
20 Neumann, A. W., *Adv. Colloid Interface Sci., 4* (**1974**) 105.
21 Schonhorn, H., *J. Phys. Chem., 70* (**1966**) 4086.
22 Harkins, W. D., Feldman, A., *J. Am. Chem. Soc., 44* (**1922**) 2665.
23 Boyd, G. E., Livingston, H. K., *J. Am. Chem. Soc., 64* (**1942**) 2383.
24 Harkins, W. D., Livingston, H. K., *J. Chem. Phys., 10* (**1942**) 342.
25 Loesner, E. H., Harkins, W. D., Twiss, S. B., *J. Phys. Chem., 57* (**1953**) 251.
26 Garnier, G., Bertin, M., Smrckova, M., *Langmuir, 15* (**1999**) 7863.
27 Marmur, A., *Adv. Colloid Interface Sci., 19* (**1983**) 75.
28 Pompe, T., Fery, A., Herminghaus, S., in *Contact Angle, Wettability and Adhesion*, Mittal, K.L. (ed.), Vol. 2. VSP, Utrecht, p. 377 (**2002**).
29 Vilette, S., Valignat, M. P., Cazabat, A. M., Jullien, L., Tiberg, F., *Langmuir, 12* (**1996**) 825.
30 Hayes, R. A., Ralston, J., *J. Colloid Interface Sci., 159* (**1993**) 429.
31 Hunter, R. J., *Foundations of Colloid Science*, Vol. 1. Clarendon Press, Oxford, (**1991**).
32 Drago, R. S., Wayland, B., *J. Am. Chem. Soc., 87* (**1965**) 3571.
33 Drago, R. S., Vogel, G. C., Needham, T. E., *J. Am. Chem. Soc., 93* (**1971**) 6014.
34 Gutmann, V., Steininger, A., Wychera, E., *Monatsh. Chem., 106* (**1966**) 460.
35 Maria, P. C., Gal, J. F., *J. Phys. Chem., 89* (**1985**) 1296.
36 Handcock, R. D., Marciano, F., *S. Afr. J. Chem., 33* (**1980**) 77.
37 Kroeger, M. K., Drago, R.S., *J. Am. Chem. Soc., 103* (**1981**) 3250.
38 Della Volpe, C., Siboni, S., *J. Colloid Interface Sci., 195* (**1997**) 121.
39 Peterson, I. R., *Surf. Coat. Int. B, 88* (**2005**), 1.
40 Lee, L.-H., *Langmuir, 12* (**1996**) 1681.
41 Shen, Q., *Langmuir, 16* (**2000**) 4394.
42 Granqvist, B., Järn, M., Rosenholm, J. B., *Colloids Surfaces* (submitted).
43 Fowkes, F. M., Mostafa, M. A., *Ind. Eng. Chem. Prod. Res. Dev., 17* (**1978**) 3.
44 Vrbanac, M. D., Berg, J., in *Acid–Base Interactions*, Mittal, K. L., Anderson, H. R., Jr. (eds.). VSP, Utrecht, p. 67 (**1991**).
45 Douillard, J.-M., Médout-Marère, V., in *Acid–Base Interactions*, Vol. 2, Mittal, K. L. (ed.). VSP, Utrecht, p. 317 (**2000**).
46 Jolicoeur, C., *Methods Biochem. Anal., 27* (**1981**) 171.
47 Bartell, F. E., Osterhof, H. J., *Ind. Eng. Chem., 19* (**1927**) 1277.

48 Barthes-Labrousse, M.-G., Joud, J.-C., in *Acid–Base Interactions*, Vol. 2, Mittal, K.L. (ed.). VSP, Utrecht, p. 453 (**2000**).

49 Schultz, J., Tsutsumi, K., Donnet, J. B., *J. Colloid Interface Sci., 59* (**1977**) 272.

50 Jacquot, C., Takadoum, J., *J. Adhesion Sci. Technol., 15* (**2001**) 681.

51 Derjaguin, B. V., Muller, V. M., Toporov, Y. P., *J. Colloid Interface Sci., 53* (**1975**) 314.

52 Tanabe, K., *Solid Acids and Bases*. Academic Press, New York (**1970**).

53 Good, R. J., *J. Adhesion Sci. Technol., 6* (**1992**) 1269.

54 Pettersson, A. B. A., Byman-Fagerholm, H., Rosenholm, J. B., in *Ceramic Materials and Components for Engines*, Carlsson, R., Johansson, T., Kahlman, L. (eds.). Elsevier Applied Science, Barking, p. 260 (**1992**).

55 Jolivet, J.-P., *Metal Oxide Chemistry and Synthesis*. Wiley, Chichester, Appendix 3, p. 195 (**2000**).

56 Rosenholm, J. B., Rahiala, H., Puputti, J., Stathopoulos, V., Pomonis, P., Beurroies, I., Backfolk, K., *Colloids Surf. A, 250* (**2004**) 289.

57 Lyklema, J., *J. Colloid Interface Sci., 99* (**1984**) 109.

58 Stol, R. J., de Bruyn, P. L., *J. Colloid Interface Sci., 75* (**1980**) 185.

59 Granqvist, B., Järn, M., Rosenholm, J. B., unpublished results.

60 Bain, C. D., Whitesides, G. M., *Langmuir, 5* (**1989**) 1370.

61 Bäckman, J., Eklund, T., Rosenholm, J. B., in *Acid–Base Interactions*, Vol. 2, Mittal, K. L. (ed.). VSP, Utrecht, p. 465 (**2000**).

62 Brunauer, S., Deming, L. S., Deming, W. E., Teller, E., *J. Am. Chem. Soc., 62* (**1940**) 1723.

63 Jaycock, M. J., in *Dispersion of Powders in Liquids*, 2nd edn, Parfitt, G. D. (ed.). Applied Science, London, Ch. 2, p. 44 (**1973**).

64 Aveyard, R., Haydon, D. A., *An Introduction to the Principles of Surface Chemistry*. Cambridge University Press, Cambridge, Ch. 6, p. 195 (**1973**).

65 Gasser, C. G., Kipling, J. J., *J. Phys. Chem., 64* (**1960**) 710.

66 Czuryskiewicz, T., Rosenholm, J., Kleitz, F., Rosenholm, J. B., Lindén, M., *Colloids Surfaces* (submitted).

67 James, R. O., Parks, G. A., *Surf. Colloid Sci.*, 12 (**1982**) 119.

68 Kytökivi, A., Growth of ZrO_2 and CrO_x on high surface area oxide supports by atomic layer epitaxy. Thesis, Helsinki University of Technology (**1997**).

69 Diggins, D., Fokkink, L. G. J., Ralston, J., *Colloids Surf. A, 44* (**1990**) 299.

70 Cassie, A. B. D., Baxter, S., *Trans. Faraday Soc., 40* (**1944**) 546.

71 Derylo-Marczewska, A. W., Jaroniec, M., *Surf. Colloid Sci., 14* (**1987**) 301.

72 Perkel, R., Ullman, R., *J. Polym. Sci., 54* (**1961**) 127.

73 Wenzel, R. N., *Ind. Eng. Chem., 28* (**1936**) 988.

74 Wenzel, R. N., *J. Phys. Colloid Chem., 53* (**1949**) 1466.

75 Bico, J., Tordeux, C., Quéré, D., *Europhys. Lett., 55* (**2001**) 214.

76 White, L. R., *J. Colloid Interface Sci., 90* (**1982**) 536.

77 Wapner, P. G., Hoffman, W. P., *Langmuir, 18* (**2002**) 1225.

78 Bico, J., Marzolin, C., Quéré, D., *Europhys. Lett., 47* (**1999**) 220.

79 Bico, J., Thiele, U., Quéré, D., *Colloids Surf. A, 206* (**2002**) 41.

80 Apel-Paz, M., Marmur, A., *Colloids Surf. A, 146* (**1999**) 273.

81 Onda, T., Shibuichi, S., Satoh, N., Tsujii, K., *J. Phys. Chem., 100* (**1996**) 19512.

82 Onda, T., Shibuichi, S., Satoh, N., Tsujii, K., *Langmuir, 12* (**1996**) 2125.

83 Peltonen, J., Järn, M., Areva, S., Lindén, M., Rosenholm, J. B., *Langmuir, 20* (**2004**) 9428.

84 Kang, K. H., *Langmuir, 18* (**2002**) 10318.

85 Vera-Graziano, R., Torres, F. R., Ordónez-Medrano, A., in *Contact Angle, Wettability and Adhesion*, Vol. 2, Mittal, K. L. (ed.). VSP, Utrecht, p. 239 (**2002**).

86 Areva, S., unpublished results.

87 Pore, V., Heikkilä, M., Ritala, M., Leskelä, M., Areva, S., *J. Photochem. Photobiol.*, A 177 (**2006**) 68.

88 Vijh, A. K., in *Oxides and Oxide Films*, Vol. 2, Diggle, J. W. (ed.). Marcel Dekker, New York, p. 1 (**1973**).

89 Bhattacharrya, D., Chaudhuri, S., Pal, A. K., *Vacuum, 43* (**1992**) 313.

90 Larson, I., Drummond, C. J., Chan, D. Y. C., Grieser, F., *J. Am. Chem. Soc., 115* (**1993**) 11885.

2
Surface Forces and Wetting Phenomena

Victor M. Starov

2.1
Wetting and Neumann-Young's Equation

Why do droplets of different liquids deposited on identical solid substrates behave so differently? Why do identical droplets, e.g. aqueous droplets, deposited on different substrates behave so differently?

Note that the contact angle is always measured inside the liquid phase (Fig. 2.1). A mercury droplet does not spread out on a glass substrate at all but forms a spherical cap with a contact angle larger than $\pi/2$ (Fig. 2.1 a). An aqueous droplet deposited on the same glass substrate spreads out only partially down to some contact angle, θ, which is between 0 and $\pi/2$ (Fig. 2.1 b). However, an oil droplet (hexane or decane) deposited on the same glass substrate spreads out completely (Fig. 2.1 c), and the contact angle decreases with time to zero. These three cases are referred to as non-wetting, partial wetting and complete wetting, respectively.

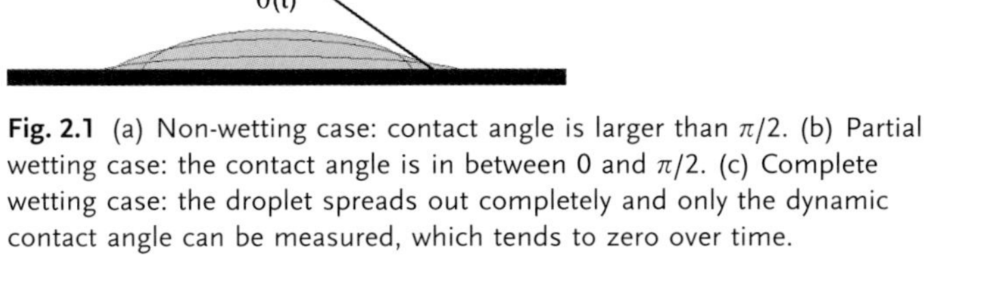

Fig. 2.1 (a) Non-wetting case: contact angle is larger than $\pi/2$. (b) Partial wetting case: the contact angle is in between 0 and $\pi/2$. (c) Complete wetting case: the droplet spreads out completely and only the dynamic contact angle can be measured, which tends to zero over time.

Colloids and Interface Science Series, Vol. 2
Colloid Stability: The Role of Surface Forces, Part II. Edited by Tharwat F. Tadros

ISBN: 978-3-527-31503-1

The same liquid can spread out completely or does not spread at all depending on the nature of the solid substrate. For example, water partially wets a glass surface and does not wet the Teflon substrate.

It is obvious that complete wetting, partial wetting and non-wetting behaviors are determined by the nature of both the liquid and the solid substrate.

The well-known Neumann-Young rule allows the connection of three interfacial tensions, γ_{sl}, γ_{sv} and γ, with the value of the equilibrium contact angle, θ (Fig. 2.2), where γ_{sl}, γ_{sv} and γ are solid–liquid, solid–vapor and liquid–vapor interfacial tensions, respectively:

$$\cos\theta_{NY} = (\gamma_{sv} - \gamma_{sl})/\gamma \tag{1}$$

Note that we denoted the equilibrium contact angle in Eq. (1) θ_{NY}, and we shall see below that there is a good reason for that.

Let us deduce Eq. (1) based on the consideration of the excess free energy of the system presented in Fig. 2.2. Let us assume that the excess free energy of a small droplet (gravity is neglected) is

$$F = \gamma S + PV + \pi\Re^2(\gamma_{sl} - \gamma_{sv}) \tag{2}$$

where S is the area of the liquid/air interface, $P = P_a - P_l$ is the excess pressure inside the liquid, P_a is the pressure in the ambient air and P_l is the pressure inside the liquid; the last term on the right-hand side gives the difference between the energy of the part of the bare surface covered by the liquid drop and the energy of the same solid surface without the droplet. Note that according to the definition of the excess pressure, P, it is negative in the case of liquid droplets (concave liquid/air interface) and positive in the case of a meniscus in partially or completely wetted capillaries (convex liquid/air interface).

Let $h(r)$ be the unknown profile of the liquid droplet, then Eq. (2) can be rewritten as

$$F = 2\pi \int_0^{\Re} r\left[\gamma\sqrt{1 + (h')^2} + Ph + \gamma_{sl} - \gamma_{sv}\right] dr \tag{3}$$

Under equilibrium conditions the excess free energy, F, should reach its minimum value. The mathematical expressions for this requirement are the following four conditions: (1) the first variation of the free energy, ∂F, should become zero; (2) the second variation, $\partial^2 F$, should be positive; and (3) the transversality condition (see below) at the droplet three-phase contact line, that is, at $r = \Re$, should be satisfied. At the moment we ignore the fourth condition because it is possible to check that this condition is always satisfied in the case of the excess free energy given by Eq. (3). However, below we consider this fourth condition, which is usually ignored.

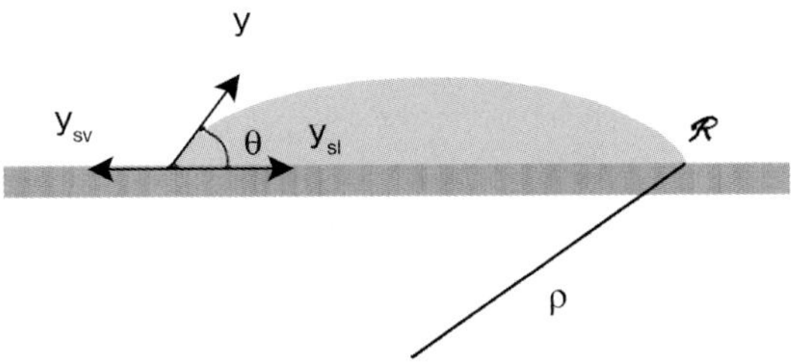

Fig. 2.2 Interfacial tensions at the three-phase contact line. $\Re$ is the radius of the droplet base and ρ is the radius of the droplet. The droplet is small enough and gravity is neglected.

Conditions (1) and (2) are identical with those for a minimum of a regular function of one variable. Condition (3) is usually deduced using a different consideration.

Condition (1) results in the well-known Euler equation, which gives an equation for the drop profile:

$$\frac{\partial \Phi}{\partial h} - \frac{\mathrm{d}}{\mathrm{d}r}\frac{\partial \Phi}{\partial h'} = 0$$

where $\Phi = r(\gamma\sqrt{1+h'^2} + Ph + \gamma_{\mathrm{sl}} - \gamma_{\mathrm{sv}})$, or

$$\frac{\gamma}{r}\frac{\mathrm{d}}{\mathrm{d}r}\left[r\frac{h'}{(1+h'^2)^{\frac{1}{2}}}\right] = P \tag{4}$$

Solution of the latter equation is part of the sphere of radius $\rho = 2\gamma/P$ (Fig. 2.2).

The second condition, (2), gives

$$\frac{\partial^2 \Phi}{\partial h'^2} > 0 \quad \text{or} \quad \frac{\gamma}{(1+h'^2)^{\frac{3}{2}}} > 0$$

which is always satisfied. The latter means that Eq. (4) really gives a minimum value to the excess free energy, Eq. (3).

Now the third, transversality condition (3) is as follows:

$$\left[\Phi - h'\frac{\partial \Phi}{\partial h'}\right]_{r=\Re} = 0$$

or

$$\left[r(\gamma\sqrt{1+h'} + Ph + \gamma_{\mathrm{sl}} - \gamma_{\mathrm{sv}}) - h'\frac{r\gamma h'}{\sqrt{1+h'^2}}\right]_{r=\Re} = 0$$

Taking into account that $h|_{r=\Re} = 0$ we conclude from the latter equation that

$$\left(\frac{\gamma}{\sqrt{1+h'^2}} + \gamma_{sl} - \gamma_{sv}\right)_{r=\Re} = 0 \tag{5}$$

Figure 2.2 shows that $h'|_{r=\Re} = -\tan\theta_{NY}$. Substitution of the latter expression into Eq. (5) results in Eq. (1).

It is easy to check that Eq. (1) also expresses the absence of any force in the tangential direction on the solid substrate at the three-phase contact line, which is the line where liquid, air and solid meet.

According to Fig. 2.2, the complete wetting case corresponds to the case when all forces cannot be compensated in the tangential direction at any contact angle, that is, if $\gamma_{sv} > \gamma_{sl} + \gamma$. The partial wetting case according to Eq. (1) corresponds to $0 < \cos\theta_{NY} < 1$, and, finally, the non-wetting case corresponds to $-1 < \cos\theta_{NY} < 0$. That is, Eq. (1) reduces complete wettability, partial wettability and non-wettability cases to the determination of three interfacial tensions, γ_{sl}, γ_{sv} and γ.

2.2
When is the Neumann-Young Equation Valid?

It is necessary to keep in mind that the Neumann-Young equation was deduced under certain conditions, the main ones being (a) the requirement for thermodynamic equilibrium in the system under consideration and (b) that the free energy of the system is given by Eq. (2). Does it include all required energy excesses?

The requirement for thermodynamic equilibrium means (1) an equilibrium of the droplet with the vapor of the same liquid in the surrounding air, (2) the absence of any flow in the droplet itself and in the vicinity of the droplet three-phase contact line and (3) the equilibrium of the vapor with the solid substrate.

Note that the radius of the droplet, ρ, and the radius of the droplet base, $\Re$, are left arbitrary according to the previous derivation and also according to the Neumann-Young equation, Eq. (1). However, according to the requirement for thermodynamic equilibrium it should be a unique equilibrium value of the radius of the droplet, ρ, or the radius of the droplet base, $\Re$. The latter does not agree with Eq. (1).

It is well known that liquid droplets on a solid substrate can be at equilibrium only with the oversaturated vapor of the same liquid. Equilibration process between the droplet and the vapor includes the following steps:

- The solid substrate under investigation should be kept in the atmosphere of the oversaturated vapor until equilibrium adsorption of vapor on the solid substrate is reached.

- After that, a droplet of the size which should be at the equilibrium with the oversaturated vapor should be deposited and kept until the equilibrium is reached.

The characteristic time-scale of the latter processes depends on the liquid volatility and viscosity and is, in general, hours in magnitude. We are not aware of such experiments in the atmosphere of an oversaturated vapor. Can we infer that the latter means that equilibrium liquid droplets of volatile liquids have never been observed experimentally? If so, it means that the applicability of the Neumann-Young equation in this case is ill-supported.

The latter means that Neumann-Young's equation can supposedly be used only in the case of non-volatile liquids. Do truly non-volatile liquids exist? Usually low volatility means liquids with large molecules, and the latter means high viscosity and correspondingly a longer characteristic time-scale of the equilibration process with the oversaturated vapor. In spite of that, let us assume that the liquid is non-volatile. In the case of partial wetting at equilibrium, liquid droplets cannot be in equilibrium with a bare solid surface. There is always an adsorption layer of liquid molecules on the solid substrate in front of the droplet. If the liquid is volatile, then this layer is created by means of evaporation–adsorption. However, if the liquid is non-volatile, the same layer should be created by means of flow from the droplet edge on to the solid substrate. As a result, the solid substrate is covered at equilibrium by an equilibrium liquid layer of thickness h. The thickness of the equilibrium liquid film, h, is determined by the potential of action of surface forces. The characteristic time-scale of this process is hours, because it is determined by the flow in the narrowest part in the vicinity of the three-phase contact line, where the viscous resistance is very high. During these hours, evaporation of the liquid from the droplet cannot be ignored and we go back to the problem of volatility.

Let us assume, however, that the equilibrium film after all forms in front of the liquid droplet and we have waited long enough for equilibrium. However, now we have three interfacial tensions, γ, γ_{sl} and γ_{vh}, namely liquid–vapor, solid–liquid and solid substrate covered with the liquid film of thickness h–vapor interfacial tensions, respectively. The interfacial tension γ_{vh} can in principle be expressed via surface forces [1]. However, no direct attempts have been made to measure γ_{vh}, which is obviously very different from the interfacial tension of the bare surface, γ_{sv}.

The latter means that the Neumann-Young equation actually should be rewritten as follows:

$$\cos\theta = \frac{\gamma_{hv} - \gamma_{sl}}{\gamma} \tag{6}$$

This equation shows that even in the case of non-volatile liquids the applicability of the Neumann-Young equation, Eq. (1), still remains questionable.

2.3
Hysteresis of Contact Angle

The previous consideration is important but there is a phenomenon that is even more important, namely the contact angle hysteresis.

The above derivation of Eq. (1) shows that this equation determines only one unique equilibrium contact angle. Contact angle hysteresis results in an infinite number of "equilibrium positions" and "equilibrium contact angles" of the drop on the solid surface; not a unique contact angle, θ, but a whole range of contact angles, $\theta_r<\theta<\theta_a$, where θ_r and θ_a are static receding and advancing contact angles, respectively, which are "equilibrium contact angles". Are all those angles equilibrium angles?

Let us explain the meaning of static advancing and receding contact angles. For this purpose, let us consider a liquid droplet on a horizontal substrate. The liquid is slowly pumped in or out through an orifice in the substrate (Fig. 2.3). Let us assume that in some way an initial contact angle is in the range $\theta_r<\theta<\theta_a$. Let us start carefully and slowly pump the liquid through the orifice in the center. The contact angle will grow, but the radius of the drop base will not change until a critical value of the contact angle, θ_a, is reached. Further pumping will result in drop spreading.

If we start from the same initial contact angle as above and begin to pump out the liquid through the same orifice, then again the contact angle will decrease but the droplet will not shrink until the critical contact angle, θ_r, is reached.

For example, in the case of water droplets on a smooth homogeneous glass surface, $\theta_r \approx 0$–5°, whereas θ_a is in the range 40–60°.

It is usually believed that the hysteresis of the contact angle is determined by the surface roughness and heterogeneity. However, the hysteresis is present even on smooth and homogeneous surfaces. Below we consider the contact angle hysteresis only on smooth and homogeneous solid surfaces.

The previous considerations show that the following dependence of the contact angle on the velocity of the three-phase contact line can be adopted (see Fig. 2.4). In Fig. 2.4, θ_r and θ_a are static receding and advancing contact angles, respectively.

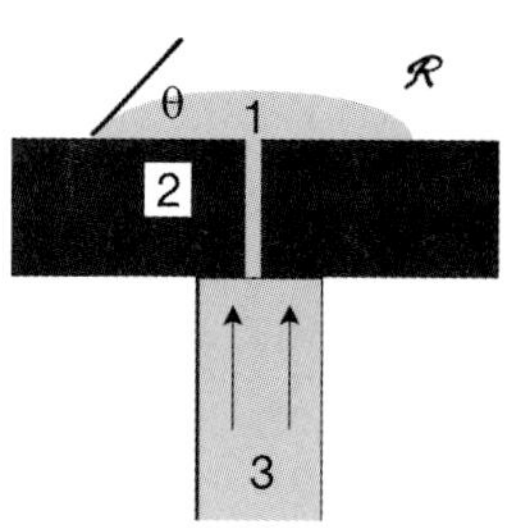

Fig. 2.3 Schematic presentation of a liquid droplet on a horizontal solid substrate, which is slowly pumped through the liquid source in the drop center. $\Re$, radius of the drop base; θ, contact angle. (1) Liquid drop; (2) solid substrate with a small orifice in the center; (3) liquid source (syringe).

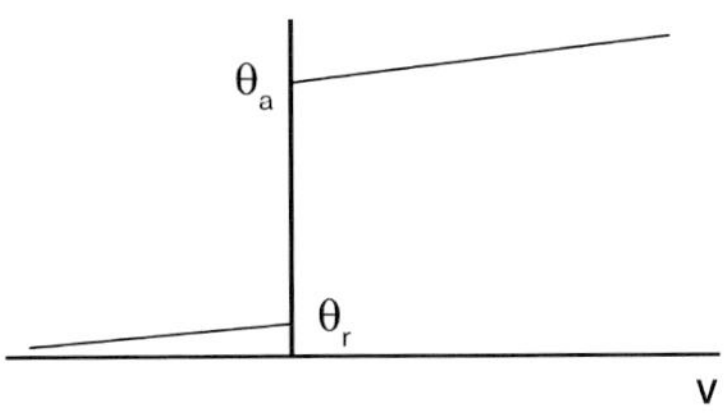

Fig. 2.4 Dependence of the contact angle on the velocity of advancing ($\nu > 0$) or receding ($\nu < 0$).

The latter picture, however, is in severe contradiction with thermodynamics, which requires a unique equilibrium contact angle, θ_e. The latter means that at any other contact angle, θ, in the range $\theta_r < \theta < \theta_a$ the liquid droplet cannot be at equilibrium but in a state of very slow motion. More detailed observations and theoretical considerations show [2, 3] that at any contact angle different from the equilibrium value, θ_e, the liquid droplet is in a state of slow "microscopic motion", which is located in a tiny vicinity of the apparent three-phase contact line. The latter motion abruptly becomes "macroscopic motion" after a critical contact angle θ_a or θ_r is reached. The latter means that the dependence presented in Fig. 2.4 should be replaced by a more complicated but realistic dependence as shown in Fig. 2.5.

The presence of contact angle hysteresis shows that the actual equilibrium contact angle is very difficult to obtain experimentally even if we neglect the equilibrium with vapor and solid substrate.

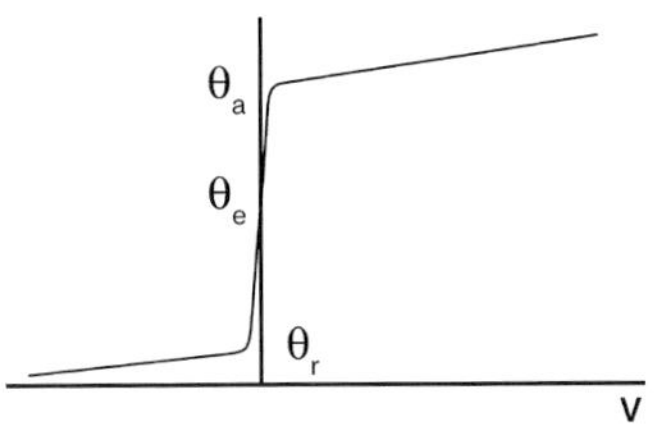

Fig. 2.5 At any deviation from the equilibrium contact angle θ_e, the liquid drop is in a state of slow "microscopic motion", which abruptly transforms into a state of "macroscopic motion" after a critical contact angle θ_a or θ_r is reached.

2.3.1 Line Tension

A number of attempts have been made to improve Neumann-Young's equation and to make it more theoretically justified. The most important of them is the introduction of line tension. It has been recognized for a long time that the drop profile cannot retain its spherical shape up to the three-phase contact line. It is shown below that the action of surface forces results in a substantial deviation of the drop shape in the vicinity of the three-phase contact line from a spherical shape. This results in the formation of a transition zone, where the influence of surface forces is important and cannot be ignored. However, if we still want to consider a spherical droplet, then the existence of the transition region can be effectively taken into account by replacing the whole transition region by an additional free energy located on the three-phase contact line. This consideration in a way is similar to the introduction of an interfacial tension. The additional free energy is referred to as line tension, τ. If the line tension is taken into account, then the excess free energy of the system, F, should be rewritten as

$$F = \gamma S + PV + \pi\Re^2(\gamma_{sl} - \gamma_{sv}) + 2\pi\Re\tau \tag{7}$$

instead of Eq. (2) and, using the expression for the droplet profile, $h(r)$, the latter equation can be rewritten as

$$F = 2\pi\int_0^{\Re} r\left[\gamma\sqrt{1 + (h'_{id})^2} + Ph + \gamma_{sl} - \gamma_{sv}\right]dr + 2\pi\Re\tau \tag{8}$$

where τ is the line tension. The first two conditions of the excess free energy minimum result in the identical equation for the drop profile, Eq. (4). However, the transversality condition takes a new form:

$$\left(\frac{\gamma}{\sqrt{1 + h'^2}} + \gamma_{sl} - \gamma_{sv}\right)_{r=\Re} + \tau''\Re + \tau = 0 \tag{9}$$

If we now introduce a new equilibrium contact angle, which takes into account line tension as θ_{LNY} and use the previous definition of the contact angle θ_{NY} according to Eq. (1), then Eq. (9) can be rewritten as

$$\gamma(\cos\theta_{LNY} - \cos\theta_{NY}) + \tau' + \frac{\tau}{\Re} = 0 \tag{10}$$

For a sufficiently large macroscopic droplet, the derivative $\tau' = d\tau/d\Re$ can be neglected, hence the latter equation becomes

$$\cos\theta_{LNY} = \cos\theta_{NY} - \frac{\tau}{\gamma\Re} \tag{11}$$

It has been shown that the line tension can change its sign from positive to negative with the salt concentration [4].

It appears that Eq. (10) is more justified than Eq. (1). Unfortunately, the same problems with equilibrium and hysteresis of the contact angle are still unresolved in the case of Eq. (10). To take hysteresis into account, a new quasi-equilibrium line tension should be introduced and investigated. No such attempts have yet been made. However, we believe that investigations of the dynamic line tension is a front line of science in the area of wetting phenomena.

2.4 Surface Forces

The above-mentioned problems show that the nature of wetting phenomena should be reconsidered.

It is necessary for this purpose to introduce a new class of phenomena, surface phenomena, which are determined by the special forces acting in thin liquid films or layers in the vicinity of the three-phase contact line.

Surface forces are well known and are widely used in colloid and interface science. They determine the stability and in general the behavior of colloidal suspensions and emulsions. In the case of emulsions/suspensions, their properties and behavior (stability, instability, rheology, interactions and so on) are completely determined by surface forces acting between colloidal particles or droplets. This theory is widely referred to as DLVO theory [1]. A number of important modifications of DLVO theory have been undertaken recently [5].

All colloidal particles have a rough surface and in a number of cases even chemically inhomogeneous surface (living cells, proteins). Roughness and inhomogeneity of colloidal particles can modify surface forces substantially: their nature, magnitude and range of action. However, roughness and inhomogeneity of the surface of colloidal particles do not influence the main phenomenon: all their interactions and properties are determined by the action of surface forces.

Surprisingly, it is a different story with wetting phenomena: it is widely (and erroneously) accepted that roughness and inhomogeneity of a solid substrate in contact with liquids can *by itself*, without consideration of surface forces acting in the vicinity of the three-phase contact line, explain the wetting properties of liquids on solid substrates. As a result, the influence of surface forces in the case of the kinetics of wetting and spreading is much less recognized than in the case of colloidal suspensions/emulsions, in spite of the same nature of surface forces.

It has been established that the range of action of surface forces is usually of the order of 0.1 μm [1]. In the vicinity of the three-phase contact line, $r=\Re$, where the liquid profile, $h(r)$, tends to zero thickness. The latter means that close to the three-phase contact line, surface forces come into play and their influence cannot be ignored because they have the same order of magnitude as the capillary forces and at smaller thickness even overcome capillary forces. The action of surface forces determines all equilibrium, quasi-equilibrium and dynamic properties of liquids on solid substrates [2, 6].

Let us ask a question: "Where is a region located that determines the contact angle and wetting properties?". It is obvious that this region cannot be located either in the bulk of the liquid or in the bulk of the solid substrate. The latter means that the region that determines all wetting properties is located in the vicinity of the three-phase contact line. However, this is just the region where the thickness of the droplet profile tends to zero and, hence, surface forces come into play.

A manifestation of the action of surface forces is the disjoining pressure. To explain the nature of the disjoining pressure, let us consider the interaction of

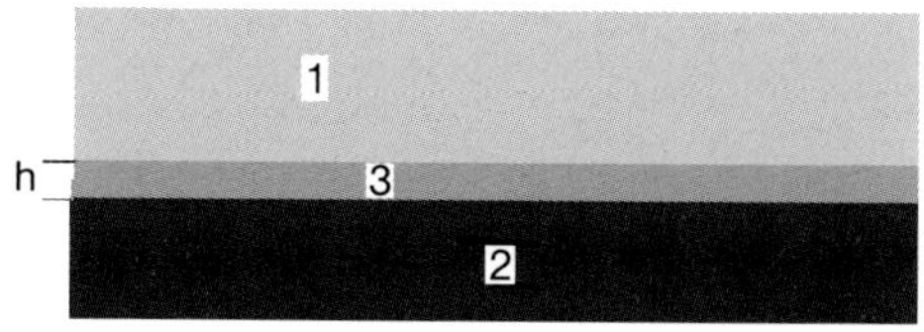

Fig. 2.6 Measurement of interaction between two thick plates 1 and 2, possibly made of different materials, with a thin layer, 3, in between.

two thick plain parallel plates (not necessarily of the same nature) divided by a thin liquid layer of thickness h (aqueous electrolyte solution, for example). There is a range of experimental methods to measure the interaction force between these two plates as a function of the thickness, h (gravity action is already taken into account) (Fig. 2.6) [1].

If h is larger than $\sim 10^{-5}$ cm $= 0.1$ μm, then the interaction force is equal to zero. However, if $h < 10^{-5}$ cm, then an interaction force appears. This force may depend on the thickness h in a very peculiar way. The interaction force divided by the surface area of the plate has the dimensions of pressure and is referred to as the disjoining pressure. Note that the term "disjoining pressure" is rather misleading, because the mentioned pressure can be both disjoining (repulsion between plates) and conjoining (attraction between plates). It is important to emphasize that we do not assume from the beginning that the natures of the two plates are identical: they can be of a different nature. The latter means that one plate can be a solid substrate and the second can be air, which corresponds to a thin liquid film on a solid substrate. The latter models the liquid layer in the vicinity of the three-phase contact line.

In the proximity of any interface there are boundary layers, where all properties differ from the corresponding bulk properties. The nature of surface forces (or disjoining pressure) is an overlapping of boundary layers in the vicinity of the three-phase contact line (Fig. 2.7). Let the thickness of the boundary layers be δ. If the thickness of droplet layer is thin enough, that is, $h \approx \delta$, these boundary layers overlap (Fig. 2.7), which results in an additional free energy of the system. The derivative of the excess free energy with respect to the thickness is the disjoining pressure.

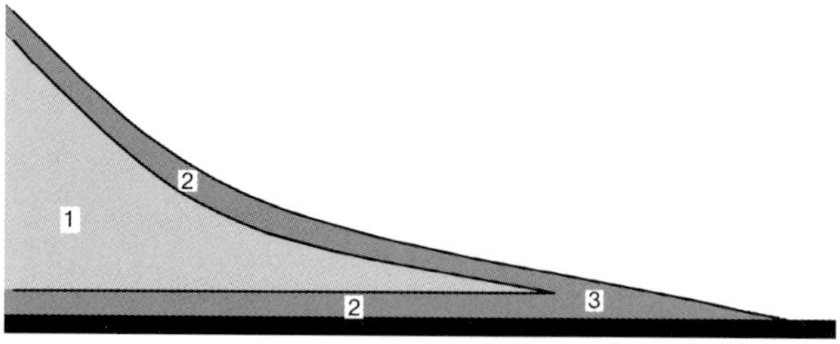

Fig. 2.7 (1) Bulk liquid and (2) boundary layers in the vicinity of liquid/air and liquid/solid interfaces and (3) region where boundary layers overlap. The latter is the region where surface forces act.

Below we review briefly the phenomena that result in the formation of the above-mentioned surface forces. The above-mentioned characteristic scale $\delta \approx 10^{-5}$ cm determines the characteristic thickness, where disjoining pressure acts.

2.5 Components of the Disjoining Pressure

Several physical phenomena have been identified for the appearance of the disjoining pressure. We consider below only three of them, which are important in thin layers of aqueous electrolyte solutions.

2.5.1 Molecular or Dispersion Component

Let us start with the one of the two the most investigated components of the surface forces: the molecular or dispersion component. It is well known that at large distances all neutral molecules interact with each other and the energy of this interaction is proportional to constant/r^6, where r is the intermolecular distance. Let us examine two plates made of different materials placed at a distance h from each other (Fig. 2.6) with a film of an aqueous electrolyte solution in between. After summing all interactions of all molecules in the system, we obtain the following expression for the molecular or dispersion components of the disjoining pressure:

$$\Pi_m = -\frac{A_H}{6\pi h^3}; \; A_H = A_{33} + A_{12} - A_{13} - A_{23} \tag{7}$$

where A_H is referred to as the Hamaker constant, named after Hamaker, who carried out these calculations around a half century ago [1]. The Hamaker constant, A_H, depends on the properties of the phases 1, 2 and 3 through the Hamaker constants, A_{ij}, of phases i and j. Expression (7) shows that the Hamaker constant can be either positive (attraction) or negative (repulsion).

In the case of oil droplets on a glass surface, when the dispersion component is the only component of the disjoining pressure acting in thin films, the dispersion interaction is repulsive, i.e. the Hamaker constant is negative. Below we consider only the latter case (thin liquid films on solid substrates) when the Hamaker constant is negative in the case of complete wetting (see below). For this purpose, we rewrite Eq. (7) as

$$\Pi_m = \frac{A}{h^3}; \; A = -\frac{A_H}{6\pi} \tag{8}$$

and note that constant A is referred to below as the Hamaker constant. Note that now a positive Hamaker constant A means a repulsion and a negative A means an attraction.

The characteristic value of the Hamaker constant is $A \approx 10^{-14}$ erg. The latter value of the Hamaker constant shows that at a thickness of the liquid layer $h \approx 10^{-7}$ cm, the dispersion component of the disjoining pressure is $\Pi_m \approx 10^{-14}/10^{-21} = 10^7$ dyn cm^{-2}. Let us consider a small water droplet of radius $\rho \approx 0.1$ cm on a solid substrate (Fig. 2.2) with a surface tension $\gamma \approx 75$ dyn cm^{-1}. The capillary pressure, P, inside the spherical oil droplet is

$$P = \frac{2\gamma}{\rho} \approx \frac{150}{0.1} = 1.5 \times 10^3 \text{ dyn cm}^{-2}$$

The latter is obviously much smaller than the disjoining pressure in the vicinity of the three-phase contact line. Let us assume for the moment that the droplet shape remains spherical until contact with the solid substrate. However, as we have already shown above, the capillary pressure is much lower than the disjoining pressure and cannot counterbalance it. The latter means that the action of surface forces distorts substantially the spherical shape of droplets in the vicinity of the three-phase contact line. That means that droplets cannot retain a spherical shape up to the contact line!

Before going further and discussing the next electrostatic component of disjoining pressure, we should say a few words about the double electrical layer (DEL).

2.5.2
Double Electrical Layers

Neutral molecules of many salts, acids and alkalis dissociate into ions (cations and anions) in water, forming aqueous electrolyte solutions. For example, NaCl dissociates with the formation of an Na^+ cation and a Cl^- anion.

In aqueous electrolyte solutions, the majority of solid surfaces acquire a charge. These charges are mostly rigidly fixed on the solid surface and can usually be moved only with the solids. There are two main mechanisms of formation of charge on a solid surface in aqueous electrolyte solutions: the dissociation of surface groups and unequal adsorption of different types of ions. According to these mechanisms, many solid surfaces in aqueous solutions acquire a negative surface charge.

Let us consider the distribution of ions in the close vicinity of a negatively charged surface in contact with aqueous electrolyte solution. The electroneutrality condition requires equal concentrations of cations and anions in the bulk solution far from the charged surface. However, close to the charged surface, according to Coulomb's law, the free cations are attracted by the negatively charged solid surface and the negatively charged ions are repulsed from the same surface. As a result, the concentration of cations is higher near the surface and the concentration of anions is lower than the corresponding concentration in the bulk solution. Diffusion will attempt to make exactly the opposite in comparison with the Coulomb interaction: to decrease the concentration of cations near the surface and to increase the concentration of anions. As a result of these two opposing trends near the negatively charged surface, a layer of a finite thickness is created, inside which the concentration of cations reaches its maximum near the surface and decreases monotonically into the depths of the solution to its bulk value, whereas the concentration of anions increases monotonically from its minimum value near the surface to its bulk value in the depth of the solution. This layer, where the concentrations of cations and anions differ from their bulk values, is referred to as the diffusive part of the double electrical layer (DEL). The characteristic thickness of the diffusive part of the DEL is the Debye length, R_d. The characteristic value of the Debye length is $R_d = 3\times10^{-8}/c^{1/2}$ cm, where the electrolyte concentration, c, is expressed in mol L^{-1}. The latter expression shows that the higher the electrolyte concentration, the thinner is the DEL. The DEL is formed by two parts: the charged surface (usually negatively charged) with immobile ions and the diffusive part. The electrical potential of the charged solid surface is the zeta-potential, ζ. The characteristic value of ζ is equal to $RT/F = 25$ mV, where R is the universal gas constant, T is the absolute temperature and F is the Faraday constant.

2.5.3
Electrostatic Component of the Disjoining Pressure

Now we can continue with the examination of the next component of the disjoining pressure: the electrostatic component.

Let us return to the examination of two charged plates (not necessarily of the same material) in aqueous electrolyte solution. The plates are assumed to be charged equally or oppositely, i.e. there is a DEL near each of them. The sign of the charge of the diffusive part of each DEL is opposite to that of the corresponding plate. If the clearance between the plates is $h >> R_d$, the DELs of the plates do not overlap and there is no electrostatic interaction of the plates. However, if the thickness of the clearance, h, is comparable to the thickness of the DEL, R_d, then the DELs overlap, resulting in an interaction between the plates. If plates are equally charged, their diffusive layers are also equally charged, i.e. the repulsion appears as a result of their overlapping (the electrostatic component of the disjoining pressure is positive in this case).

If the plates are oppositely charged, then as a result of the overlapping of opposite charges an attraction appears, and the electrostatic component of the disjoining pressure is negative in this case.

There are a number of approximate expressions for the electrostatic component of the disjoining pressure [1]. For example, in the case of low zeta potentials of both surfaces, the following relation is valid [1]:

$$\Pi(h) = \frac{\varepsilon\kappa^2}{8\pi} \frac{2\zeta_1\zeta_2 \cosh\kappa h - (\zeta_1^2 + \zeta_2^2)}{\sinh^2\kappa h} \tag{9}$$

where $1/\kappa = R_d$. Equation (9) shows that the disjoining pressure does not vanish even in the case when only one of the two surfaces is charged (e.g. $\zeta_1 = 0$). The physical reason for this phenomenon is the deformation of the DEL if the distance between the surfaces is smaller than the Debye radius.

The theory based on the calculation of the disjoining pressure based on the indicated two components, i.e. dispersion, Π_m and electrostatic, $\Pi_e(h)$, is referred to as the DLVO theory after the names of its authors (B.V. Derjaguin, L.D. Landau, E.J. Verwey and J.T.G. Overbeek). According to the DLVO theory, the total disjoining pressure is a sum of the two components, i.e., $\Pi(h) = \Pi_m(h) + \Pi_e(h)$. The DLVO theory made possible the explanation of a wide range of experimental data on the stability of colloidal suspensions/emulsions and the statics and kinetics of wetting.

However, it was understood later that these two components alone are insufficient for explaining the phenomena in thin liquid films and layers and in colloidal dispersions. The problem is that there is a third important component of disjoining pressure, which becomes equally important in a number of cases.

2.5.4
Structural Component of the Disjoining Pressure

This component of disjoining pressure is caused by the orientation of water molecules in the vicinity of aqueous solution/solid substrate or aqueous solution/air interfaces. Let us recall that all water molecules can be modeled as an electric dipole.

In the vicinity of a negatively charged interface, a positive part of water dipoles is attracted to the surface. That is, the negative part of dipoles is directed in the opposite direction and the next water dipoles are facing a negatively charged part of dipoles, which in turn results in the orientation of the next layer of dipoles. However, thermal fluctuations try to destroy this orientation (Fig. 2.8).

As a result of these two opposite trends, a finite layer forms where the structure of water dipoles differs from the completely random bulk structure. If now we have two interfaces with water dipoles oriented close to each of them (or even one of them), then at a close separation, comparable to the thickness of the hydration layer, these surfaces "feel each other", that is, the hydration layers overlap. The latter results either in attraction or repulsion of these two surfaces.

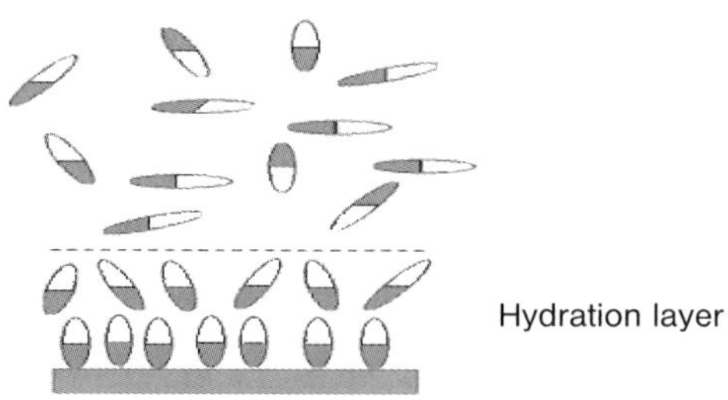

Fig. 2.8 Formation of a hydration layer of water dipoles in the vicinity of a negatively charged interface.

Unfortunately, so far there is no firm theoretical background of the structural component of disjoining pressure and we are unable to predict in which case the structure formation results in attraction and in which case a repulsion. As a consequence, only a semi-empirical equation exists, which gives the dependence of the structural component of disjoining pressure on the thickness of the liquid film [1]:

$$\Pi_S(h) = K e^{-\lambda h} \tag{10}$$

where K and λ are constants. There is a clear physical meaning of the parameter $1/\lambda$, which is the correlation length of water molecules in aqueous solutions. The latter gives $1/\lambda \approx 10$–15 Å, which is the characteristic thickness of the hydration layer.

However, we are still far from a complete understanding of the pre-exponential factor K, which at the current stage can be extracted only from experimental measurements of the disjoining pressure.

Currently it is assumed [1] that the disjoining pressure of thin aqueous films is equal to the sum of the three components:

$$\Pi(h) = \Pi_m(h) + \Pi_e(h) + \Pi_s(h) \tag{11}$$

In Fig. 2.9, the dependences of the disjoining pressure on the thickness of a flat liquid film are presented for the cases of complete wetting [curve 1, which corresponds to the dispersion or molecular component of disjoining pressure, $\Pi_m(h)$] and partial wetting [curve 2, which corresponds to the sum of all three components of the disjoining pressure, according to Eq. (11)] [2]. The disjoining pressure represented by curve 1 in Fig. 2.9 corresponds to the case of complete wetting, for example, oil droplets on glass substrate, and curve 2 corresponds to the case of partial wetting, for example, aqueous electrolyte solutions on glass substrates.

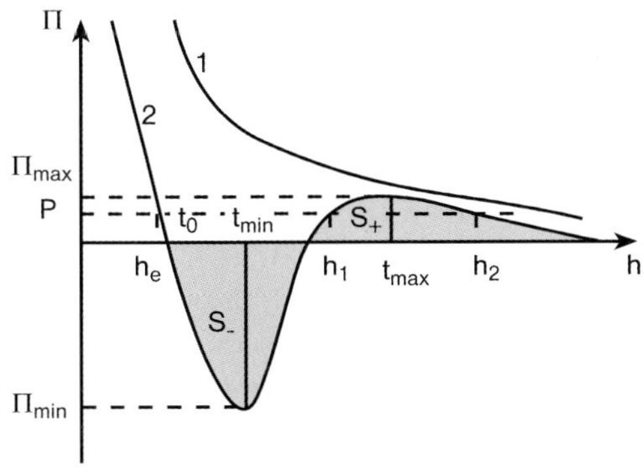

Fig. 2.9 Two types of isotherms of disjoining pressure: (1) complete wetting; (2) partial wetting.

2.6
Thin Liquid Films

The excess free energy of a flat liquid film of thickness h on a solid substrate is

$$F/S = \gamma + Ph + f_{\mathrm{D}}(h) + \text{constant}$$

where $f_{\mathrm{D}}(h)$ is the excess free energy caused by the presence of disjoining pressure. The first two conditions of the equilibrium require

$$P = \Pi(h) \tag{12}$$

and

$$\frac{\mathrm{d}\Pi(h)}{\mathrm{d}h} < 0 \tag{13}$$

where $\Pi(h) = -[\mathrm{d}f_{\mathrm{D}}(h)/\mathrm{d}h]$ is the disjoining pressure.

Condition (13) shows that the thin films in the case of complete wetting (curve 1, Fig. 2.9) are stable in the whole region of the disjoining pressure action. However, in the case of the partial wetting (curve 2) flat films are stable only in the intervals $0<h<t_{\mathrm{min}}$ (α-films) and $t_{\mathrm{max}}<h$ (β-films). However, in the interval $t_{\mathrm{min}}<h<t_{\mathrm{max}}$ flat films are unstable and cannot be measured. The latter shows that experimental measurement of the disjoining pressure on h is possible only in the intervals $0<h<t_{\mathrm{min}}$ (α-films) and $t_{\mathrm{max}}<h$ (β-films) and it is impossible to measure the disjoining pressure of flat films in the interval $t_{\mathrm{min}}<h<t_{\mathrm{max}}$. We show below that the situation is even worse than that: flat films with $t_0<h<t_{\mathrm{min}}$ also cannot be created experimentally. Unfortunately, so far it has proved possible to measure the disjoining pressure of flat films on flat solid substrates only.

Let $0<P<\Pi_{\mathrm{max}}$, then there are three solutions of Eq. (12). The first solution, h_{e}, corresponds to the absolutely stable α-film, the second solution, h_1, corresponds to the absolutely unstable film [see the stability condition (13)] and the third solution, h_2, corresponds to the metastable β-film (Fig. 2.9). It is possible to show that β-films possess higher excess free energy than α-films, that is, β-films are metastable, i.e. they can exist only over a limited time interval, whereas α-films are absolutely stable. Hence, in the case of the equilibrium of liquid in the capillary (see below), only the absolutely stable α-film of the thickness h_{e} can exist in front of the meniscus at equilibrium.

At equilibrium, flat liquid films must be in equilibrium with the vapor of the same liquid in the ambient air. The latter means equality of the chemical potentials of molecules in the thin liquid film and in the vapor. With the given value of the partial vapor pressure, p, in the ambient air, the thickness of a flat film in equilibrium with the vapor is found from the equality of the chemical potential of the molecules of water in vapor and a thin film, which gives

$$\Pi(h) = \frac{RT}{v_m} \ln\left(\frac{p_s}{p}\right) \qquad (14)$$

where v_m is the molar volume of liquid and p_s is the pressure of the saturated vapor at temperature T. The vapor is assumed to obey the ideal gas law.

Equality of chemical potentials of the bulk liquid and vapor results in

$$P = \frac{RT}{v_m} \ln\left(\frac{p_s}{p_e}\right) \qquad (15)$$

Equation (15) shows that $P>0$ corresponds to undersaturation, whereas $P<0$ corresponds to oversaturation. Equation (15) can also be considered from a different point of view. Let us consider a meniscus in a flat chamber (between two plates) in equilibrium with its own vapor of pressure p or a cylindrical drop also in equilibrium with its own vapor. Comparison of Eq. (15) and the definition of the excess pressure, P, shows that any meniscus can be in equilibrium with its own vapor only at undersaturation. The same consideration shows that droplets can be in equilibrium with its vapor only at oversaturation.

It is amazing how many researchers forgot about this plane and straightforward consideration: as soon as one sees an expression such as "let us consider an equilibrium droplet on a solid substrate", one should immediately ask if the author realized that it is possible only from a theoretical point of view, because there is no reliable way to keep oversaturation over a solid substrate for a long enough time to reach equilibrium, which can be achieved only over hours!

The impossibility of keeping oversaturation for a sufficiently long period predetermines also the impossibility of measurement of the disjoining pressure isotherm in the range of oversaturation, that is, of flat film thicknesses from t_0 to t_{min} (Fig. 2.9), in spite of thermodynamics, which allows us to carry out such experimental measurements. The measurement of disjoining pressure in this range really is a challenge. That is, after all, experimental measurements of the disjoining pressure are currently possible only in the following two ranges of flat film thickness: $h<t_0$ (α-films) and $h>t_{max}$ (β-films) [1].

Equation (12) shows that flat thin films in the range $h<t_0$ (α-films) and $h>t_{max}$ (β-films) can be in equilibrium only with menisci at undersaturation, whereas flat thin liquid films in the range $t_0<h<t_{min}$ (Fig. 2.9) can be in equilibrium only at oversaturation with drops of the same liquid.

At equilibrium, the vapor pressure is fixed by external conditions and the same is true for P [according to Eq. (15)] and cannot be arbitrary as stated in Eq. (1).

2.7 Disjoining Pressure and Equilibrium Contact Angles

Let us examine the equilibrium state of the meniscus of an aqueous solution of an electrolyte in a slit capillary i.e. between two solid planes, where H is the half-width of the clearance. Let us assume that the x-axis is directed parallel to the axis of the capillary (see Fig. 2.10 a, b). The profile of the meniscus is described by the dependence $h(x)$, the equation for which is now to be deduced.

A liquid is assumed to be volatile, that is at the equilibrium the walls of the capillary, are covered with equilibrium fluid film of thickness h_e. The expression for the excess free energy of the system, shown later in Fig. 2.10 a, b, should include the three terms described below. We can add any constant term to the expression of the free energy, that is, the term "excess free energy" below indicates "excess in comparison with the energy of the equilibrium thin film of thickness h_e".

These three terms are the following: excess energy connected with the curved shape of the liquid/air interface; excess energy appearing as a result of the additional volume of liquid in the system; and excess energy caused by the presence in the system of the overlapping of boundary layers near the "three-phase contact line" of the meniscus. Note that in reality at equilibrium there is no three-phase contact line because the meniscus profile meets the flat α-film but not the bare dry solid surface, because a bare solid surface is impossible under equilibrium conditions.

The expression for the excess energy, F, is the integral over the whole volume, including unknown dependence of the liquid profile, $h(x)$, and its first derivative (see [2] for details):

$$F = \int [\gamma(\sqrt{1 + h'^2} - 1) + P_e(h - h_e) + f_D(h) - f_D(h_e)]dx$$

According to the previous considerations, only liquid shapes can be materialized in nature, those corresponding to the minimum of excess free energy. Application of this principle, (A), to the expression for the excess free energy gives the following equation:

$$\frac{\gamma h''}{(1 + h'^2)^{\frac{3}{2}}} + \Pi(h) = P \tag{16}$$

where the prime indicates the derivative with respect to the variable x. Equation (16) is an ordinary differential equation of second order. The second requirement, (B), results in the same condition as before, $\gamma/(1 + h'^2)^{\frac{3}{2}} > 0$, which is always satisfied. However, as already mentioned, these two conditions are not sufficient in the case of curved layers (see below).

First, let us try to understand what precisely Eq. (16) describes. For this purpose, we consider an equilibrium meniscus in a flat capillary (slit) in the case of partial wetting.

Equation (16) describes three different zones presented in Fig. 2.10 a, b. If the thickness h is sufficiently large and lies outside the zone of action of surface forces (that is, roughly $h > 10^{-5}$ cm), then the disjoining pressure can be neglected in Eq. (16), $\Pi(h) = 0$, and instead of Eq. (16) we have Eq. (4). The latter equation is the well-known Laplace equation, which describes a spherical meniscus with the so far unknown contact angle θ_e (zone 1, Fig. 2.10 a, b). According to Eq. (4), in zone 1 (meniscus) only the capillary force acts. In zone 3, only flat liquid films are present, all derivatives vanish and only the disjoining pressure acts in this zone according to Eq. (16). However, according to the same equation, both capillary forces and disjoining pressure are important in the transition zone 2 (Fig. 2.10 a, b).

It is necessary to specify now what precisely we mean by saying "the equilibrium contact angle, θ_e". Usually one thinks of the angle at the three-phase contact line at equilibrium. However, there is *no* three-phase contact line at equilibrium!

Sufficiently far in from of the meniscus, a flat equilibrium α-film of constant thickness is situated, hence all derivatives in Eq. (16) vanish and Eq. (16) reduces to Eq. (12), for determination of the thickness of equilibrium film (zone 3, Fig. 2.10 a, b). That is, inside zone 3 only surface forces (disjoining pressure) act.

Finally, in the intermediate zone between the previous two zones, the spherical meniscus and the flat α-film, there is a transition zone [2], where a smooth transition from the meniscus to the flat film takes place. We refer to this zone as a transition zone (zone 2, Fig. 2.10 a, b). Inside the transition zone, both cap-

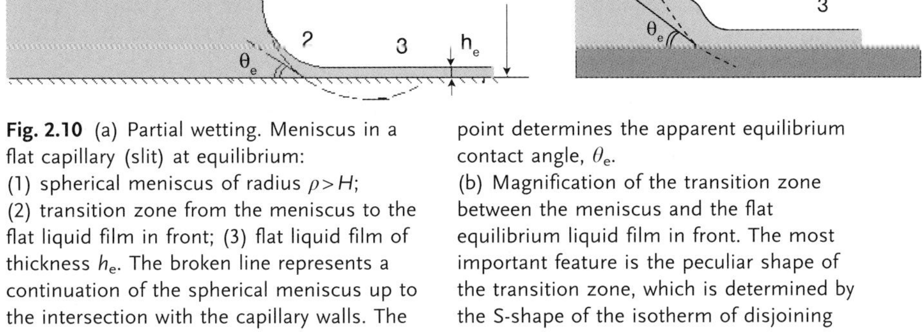

Fig. 2.10 (a) Partial wetting. Meniscus in a flat capillary (slit) at equilibrium:
(1) spherical meniscus of radius $\rho > H$;
(2) transition zone from the meniscus to the flat liquid film in front; (3) flat liquid film of thickness h_e. The broken line represents a continuation of the spherical meniscus up to the intersection with the capillary walls. The intersection determines the apparent "three-phase contact line" and the tangent at this point determines the apparent equilibrium contact angle, θ_e.
(b) Magnification of the transition zone between the meniscus and the flat equilibrium liquid film in front. The most important feature is the peculiar shape of the transition zone, which is determined by the S-shape of the isotherm of disjoining pressure in the case of partial wetting.

illary and surface forces are equally important. Hence the whole Eq. (16) describes the liquid profile inside this zone.

Inside the transition zone all thicknesses are passed:
1. starting from those outside the region of surface forces action (spherical meniscus) through thickness corresponding to β-films;
2. next thicknesses $t_{min} < h < t_{max}$, where flat liquid films are unstable;
3. then thicknesses from t_{min} to the equilibrium α-film, h_e.

The question arises of the stability of the transition zone in the interval of thicknesses $t_{min} < h < t_{max}$, where flat liquid films are unstable. The answer is as follows: the stability conditions of curved films is very different from those for flat liquid films. Roughly, the stability of curved layers can be explained as follows. The width of the "dangerous region" in the interval of thicknesses $t_{min} < h < t_{max}$, where flat liquid films are unstable, is sufficiently small that disturbances appearing in this "dangerous region" are "healed" from the outside by the adjacent stable regions. Thus, the curvature of the liquid profile in the transition zone provides stabilizing effect inside the "dangerous region" of thicknesses.

From the mathematical point of view, the fourth condition of equilibrium mentioned earlier should be used to check the stability of solutions of Eq. (16). This condition reads as follows: (D) solution of the Jacoby equation, $u(x)$, should not vanish at any position x inside the region under consideration. The equation

$$\left(\Phi_{hh} - \frac{d}{dx}\Phi_{hh'}\right)u - \frac{d}{dx}(\Phi_{h'h'}u') = 0, \; u(0) = 0 \tag{17}$$

where now

$$\Phi = \gamma\left(\sqrt{[1 + h'^2} - 1\right) + P_e(h - h_e) + f_D(h) - f_D(h_e)$$

is referred to as the Jacoby equation.

The condition (17) allows the determination of the stability of any solution of Eq. (16). Condition (D) is a generalization of Derjaguin's stability condition (13), which is valid only for flat liquid films. Condition (17) allows the stability condition of any non-flat liquid shape to be checked. The existence of non-flat equilibrium liquid shapes on flat substrates has been shown using the latter stability condition [7].

For solving the differential equation of second order, Eq. (16), two boundary conditions should be imposed. We see below that actually we have three boundary conditions instead. This looks like a contradiction. The two boundary conditions in the center of capillary are $h(0) = H$, $h'(0) = -\infty$, and the third boundary condition far from the meniscus, which corresponds to the smooth transition of the meniscus into the flat thin equilibrium film in front, is

$$h(x) \to h_e, \; x \to \infty \tag{18}$$

Hence we have three boundary conditions instead of the two required. It should be remembered that the unique vapor pressure inside the capillary, which corresponds to the equilibrium, is also unknown or P in Eq. (16) is also unknown. The third boundary condition (18) should be used for the determination of this unknown value of P. It is possible to show that the boundary condition (18) after some transformation can be rewritten in the following way [2]:

$$\gamma \cos \theta_e = \gamma + P_e h_e + \int_{h_e}^{\infty} \Pi(h) dh \qquad (19)$$

which allows the determination also of the unknown equilibrium contact angle θ_e. Equation (19) was obtained many years ago using a completely different thermodynamic consideration and named after the two scientists who deduced it: the Frumkin-Derjagin's equation [1]. Equation (19) was considered for many years as independent of the condition (16). Above we showed that the condition (19) can be directly deduced from Eq. (16).

We analyze briefly below the consequences of Eq. (19). First, Eq. (19) expresses the unique equilibrium contact angle, θ_e, via the isotherm of disjoining pressure. Equation (19) can be rewritten approximately as

$$\cos \theta_e \approx 1 + \frac{S_+ - S_-}{\gamma}$$

where S_+ and S_- (Fig. 2.9) are areas limited by the isotherm of the disjoining pressure, $\Pi(h)$, above and below the h axis, respectively. As $\cos \theta_e < 1$, the difference $\Delta S = S_+ - S_-$ must be negative and the presence of a loop in the region of negative values on the dependence $\Pi(h)$ is a necessary requirement for the existence of partial wetting.

In the case when ΔS is positive, as in the case of complete wetting (curve 1, Fig. 2.9), we obtain according to Eq. (19) $\cos \theta_e - 1$, which is impossible. Detailed examination of the meniscus and the transition zone profiles in this case shows [2] that the continuation of a spherical meniscus does not intersect the flat equilibrium film, as shown in Fig. 2.11, but passes above continuations of an equilibrium flat film.

From macroscopic point of view, the case of complete wetting corresponds to zero equilibrium contact angle; however, a microscopic picture is more sophisticated than that (Fig. 2.11).

An important consequence of the considerations on complete wetting is that the meniscus can be at equilibrium only inside capillaries. However, drops of the same liquid on the flat surface made of the same material as the capillary walls cannot be at equilibrium, but spread out completely.

Note that in the case of partial wetting, both the meniscus and drops on a flat surface can be at equilibrium but at different vapor pressures in the ambient air and with slightly different equilibrium contact angles.

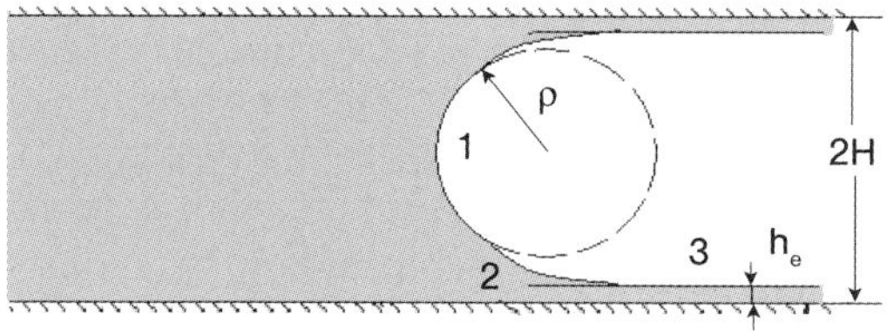

Fig. 2.11 Complete wetting. Meniscus in the flat capillary (slit) at equilibrium: (1) spherical meniscus of radius $\rho < H$; (2) transition zone from the meniscus to the flat liquid film in front; (3) flat liquid film of thickness h_e. The broken line represents a continuation of the spherical meniscus. In the case of complete wetting, the continuation does not intersect either the capillary walls or the flat film in front. The apparent contact angle, θ_e, cannot be introduced in this case.

The major problem with the equilibrium contact angles is that they are very difficult to measure. Hysteresis of contact angles made the measurement of the equilibrium contact enormously sophisticated if possible at all.

2.8
Hysteresis of Contact Angles from a Microscopic Point of View: Surface Forces

At this stage we are able to explain the nature of the hysteresis of contact angles via the S-shape of the isotherm of disjoining pressure (curve 2 in Fig. 2.9). First we recall what the hysteresis of contact angles in capillaries means. Let us consider a meniscus in the case of partial wetting in a flat capillary (Fig. 2.12 a, b) [2, 3].

If we increase the pressure under the meniscus then the meniscus does not move but changes its curvature to compensate for the excess pressure and, as a consequence, the contact angle increases. The meniscus does not move until some critical pressure and critical contact angle, θ_a, are reached. After further increase in pressure, the meniscus starts to advance. A similar phenomenon takes place if we decrease the pressure under the meniscus: it does not recede until a critical pressure and corresponding critical contact angle, θ_r, are reached. The latter means that in the whole range of contact angles, $\theta_r < \theta < \theta_a$, the meniscus does not move macroscopically. It is obvious that, on the smooth homogeneous solid substrate, only one contact angle corresponds to the equilibrium position and all the rest do not. Based on that idea in Fig. 2.5 we present the dependence of the contact angle on the velocity of motion, which shows that all contact angles θ in the range $\theta_r < \theta < \theta_a$ and $\theta \neq \theta_e$ correspond to slow microscopic advancing or receding of the meniscus. This microscopic motion changes abruptly to a macroscopic motion as soon as θ_r or θ_a is reached.

The explanation of the dependence presented in Fig. 2.5 is based on the S-shaped isotherm of the disjoining pressure in the case of partial wetting. This shape determines the special shape of the transition zone in the case of an equilibrium meniscus (Fig. 2.10 b). In the case of an increase in the pressure be-

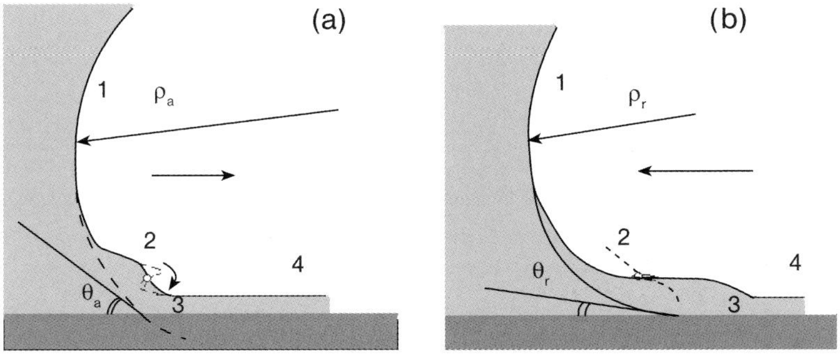

Fig. 2.12 Hysteresis of contact angles in flat capillaries in the case of partial wetting (S-shaped isotherm of disjoining pressure). (a) Advancing contact angle. (1) A spherical meniscus of radius ρ_a; (2) transition zone with a "dangerous" marked point (see explanation in the text); (3) zone of flow; (4) flat films. Close to the marked point a dashed line shows the profile of the transition zone just after the contact angle reaches the critical value θ_a, "caterpillar motion".
(b) Receding contact angle. (1) A spherical meniscus of radius $\rho_r < \rho_a$; (2) transition zone with a "dangerous" marked point (see explanation in the text); (3) zone of flow; (4) flat films. Close to the marked point a dashed line shows the profile of the transition zone just after the contact angle reaches the critical value θ_a.

hind the meniscus (Fig. 2.12 a), a detailed consideration [2, 3] of the transition zone shows that close to the "dangerous" point marked in Fig. 2.12 a the slope of the profile becomes steeper with increasing pressure. In the range of very thin films (region 3 in Fig. 2.12 a) there is a zone of flow. Viscous resistance in this region is very high, which is why the advance of the meniscus proceeds very slowly. After some critical pressure behind the meniscus has been reached, the slope at the "dangerous" point reaches $\pi/2$, after which the flow stepwisely occupies the region of thick films and the fast "caterpillar" motion starts as shown in Fig. 2.12 a.

In the case of a decrease in the pressure behind the meniscus, the events proceed according to Fig. 2.12 b. In this case, up to some critical pressure the slope in the transition zone close to the "dangerous" point becomes increasingly flat. In the range of very thin films (region 3 in Fig. 2.12 b) there is a zone of flow. Viscous resistance in this region is very high, which is why the receding of the meniscus proceeds very slowly. After some critical pressure behind the meniscus has been reached, the profile in the vicinity of the "dangerous" point shows discontinuous behavior, which is obviously impossible. This means that the meniscus will start to slide along the thick β-film, that is, the meniscus will move fast, leaving behind the thick β-film. The latter phenomenon (the presence of a thick β-film behind the receding meniscus of aqueous solutions in quartz capillaries) has been confirmed experimentally [1].

Now we should make some comments on the roughness of the substrate. In the case of a chemically homogeneous substrate, the presence of roughness does not influence the physical phenomena resulting in surface forces action (disjoining pressure). The only difference is that the profile of the transition zone becomes more sophisticated and there are more possibilities for hysteresis. Equilibrium and hysteresis contact angles on rough surfaces have never been considered from this point of view and will be the subject of future investigations.

Acknowledgments

This research was supported by the Royal Society and the Engineering and Physical Sciences Research Council. The author would like to express his gratitude to Manuel Velarde for valuable discussions.

References

1 Derjagin, B.V., Churaev, N.V., Muller, V.M. *Surface Forces*. Consultants Bureau, Plenum Press, New York (**1987**).
2 Starov, V.M. *Adv. Colloid Interface Sci.*, *39*, 147–173 (**1992**).
3 Martynov, G.A., Starov, V.M. and Churaev, N.V. Hysteresis of contact angle on smooth homogeneous solid substrates. *Colloid J.* (USSR Academy of Sciences, English Translation), *39*, 406–417 (**1977**).
4 Platikanov, D., Nedyalkov, M., Nasteva, V. *J. Colloid Interface Sci.*, *75*, 620 (**1980**).
5 Churaev, N., Sobolev, V. Predictions of contact angles on the basis of the Frumkin–Derjagin approach. *Adv. Colloid Interface Sci.*, *61*, 1–16 (**1995**).
6 Starov, V.M., Kalinin, V.V., Chen, J.-D. Spreading of liquid drops over solid substrata. *Adv. Colloid Interface Sci.*, *50*, 187–222 (**1994**).
7 Starov, V. Nonflat equilibrium liquid shapes on flat solid surfaces. *J. Colloid Interface Sci.*, *269*, 432–441 (**2003**).

3
Investigation of Plateau Border Profile Shape with Flow of Surfactant Solution Through Foam Under Constant Pressure Drop Using the FPDT Method

Pyotr M. Kruglyakov and Natalia G. Vilkova

3.1
Theoretical Background

Liquid foams have many industrial applications, including fire-fighting, froth flotation, surfactant accumulation and fractionation, oil recovery, cosmetics, food foams and foaming metal melts to produce metallic foams [1–4].

Foam drainage and stability are important problems in colloid chemistry. Nevertheless, many aspects of these problems are not completely clear because of process complications which occur in foams.

3.1.1
Foam Drainage

Foam drainage has for a long time been the subject of intensive investigations, both theoretically and experimentally [2–19]. Two different approaches have been used for the investigation of foam drainage: process description at the local scale and (or) at the macroscopic scale with solid-like surfaces or with mobile fluid-like surfaces.

Foam drainage is a complicated process. At the moment of formation, the liquid content in the foam is usually considerably larger than that at hydrostatic equilibrium. For this reason, already during foam generation (or immediately after its formation) the liquid starts to drain out of the foam. The "excess" liquid in the foam films drains into the Plateau borders, then flows through them down from the upper to the lower foam layers following gravity until the gradient of the capillary pressure equalizes the gravitational force ($\mathrm{d}P_\sigma/\mathrm{d}l = \rho g$, where l is a coordinate in the direction opposite to gravity).

Simultaneously with drainage from films into borders, the liquid begins to flow out from the foam when the pressure in the lower foam layers exceeds the external pressure.

The rate of foam drainage is determined not only by the hydrodynamic characteristics of the foam (border shape and size, liquid-phase viscosity, pressure gradi-

Colloids and Interface Science Series, Vol. 2
Colloid Stability: The Role of Surface Forces, Part II. Edited by Tharwat F. Tadros

ISBN: 978-3-527-31503-1

ent, mobility of the liquid/air interface, etc.) but also by the rate of internal foam (foam films and borders) collapse and the breakdown of the foam column. The decrease in the average foam dispersity leads to the liberation of excess liquid, which delays the establishment of hydrostatic equilibrium. However, the liquid drainage causes an increase in the capillary and disjoining pressure, both of which accelerate further bubble coalescence and foam column breakdown.

A number of differential equations have been obtained for the characterization of drainage processes in local foam layers, first derived by Kann and Druzhinin [5] and Krotov [6].

Krotov obtained foam drainage equations for the one-dimensional variant of the immobile dispersion phase of foam in gravitational and centrifugal fields when either films and borders are in hydrodynamic equilibrium or the equilibrium is absent [5, 8].

In particular, for the simplest one-dimensional case of foam drainage in a gravitational field, the differential equation is

$$\frac{\mathrm{d}V_\mathrm{L}}{\mathrm{d}\tau} = \beta\sqrt{V_\mathrm{L}}\frac{\partial^2 V_\mathrm{L}}{\partial l^2} + \frac{\beta}{2\sqrt{V_\mathrm{L}}}\left(\frac{\partial V_\mathrm{L}}{\partial l}\right)^2 + 2\rho g \times 2V_\mathrm{L}\frac{\partial V_\mathrm{L}}{\partial l} \tag{1}$$

where V_L is the volume fraction of the liquid in the foam, ρ is the density of the liquid phase, g is the acceleration due to gravity, l is the coordinate opposite to the direction of gravity, a and β are constants and τ is time.

Krotov [8] gave the most complete analysis of foam drainage equations. In particular, he noted that Eq. (1) and also other more general equations (accounting for the movement of the dispersed phase, absence of equilibrium between films and borders, etc.) are quasi-linear parabolic partial differential equations of second order and pointed out the existence of an analogy between these equations and equations of thermoconductivity [8].

Krotov also obtained the analytical solutions of these equations for some very simple foam states. For example, he obtained the equation of liquid distribution along the foam height in a gravitational field, for the equilibrium state in a centrifugal field and in steady-state flow, the equation for limited liquid flow under the influence of capillary sucking at a defined foam height, etc. He also investigated the problem of the description of slight (with regard to amplitude) wave disturbances of the dispersion medium of a draining foam.

In the late 1980s and during the last 10–15 years, foam drainage and especially the analysis of foam drainage theory have become a very active field again [4, 10–19].

The most important findings of these publications should be noted:

1. The solitary-wave solution of foam drainage equation [4, 12], which is used in a process called "forced drainage" and has been the topic of numerous experiments [12–15, 17, 18]. In these experiments, a target was set to account for viscous dissipation in both the Plateau border and the nodes (vertices) of the liquid network.

2. Analysis of the influence of different physicochemical factors (surface viscosity, adsorption, Marangoni effect and others) on the solution of the foam drainage equation [14].

3. Attempts to divide the hydrodynamic resistance in forced drainage experiments into contributions of nodes and Plateau borders [14, 15, 17].

3.1.2
Foam Pressure Drop Technique

This technique, which consists in creating a reduced pressure in the Plateau borders of a foam using a porous plate and measuring this pressure difference with the help of a micromanometer [2, 3, 20], provides great advantages for investigations of different properties of foams. The technique was first developed about 30 years ago by Khristov et al. [20]. Since then, it has been further improved and is now used as the Foam Pressure Drop Technique (FPDT) [21–30].

The main idea of the FPDT is that the pressure above the foam that is formed in a vessel with a well-wettable porous plate bottom is increased or the pressure under the plate is decreased.

The pressure difference should obviously not exceed the capillary pressure of the porous plate, which is equal to $2\sigma \cos\theta / r_p$, where r_p is the radius of the pores and θ is the contact angle of wetting the glass. Under this condition only the liquid (and not gas) would pass through the plate's pores.

During the process of draining the liquid from a foam, the pressure difference (P'_L) balances with the increasing capillary pressure.

At equilibrium, the pressure difference in the liquid phase becomes equal to

$$P'_L = \Delta P_0 + \rho g l \qquad (2)$$

where ΔP_0 is the applied reduced pressure under the porous plate, ρ is the density of the liquid phase, g is the acceleration due to gravity and l is the height from the porous plate.

At a pressure difference considerably higher than $\rho g l$, the foam becomes identical as regards the pressure in the liquid phase. In such foams, the radii of curvature of Plateau borders also become identical. This is the main difference between the FPDT and traditional experiments where the foam properties are a function of the foam column height.

The excess pressure in the foam liquid phase can be measured with a micromanometer. There are different kinds of micromanometers with open and closed capillary tubes [2, 3]. Its main units are a porous plate and a capillary tube, part of which is filled to a certain level with the solution used for foam formation.

The porous plate brings the foam liquid and the solution in the capillary tube into contact. The other part of the capillary tube is filled with gas, which changes its volume according to the pressure.

It is possible to measure and investigate the evolution of the expansion ratio of foam, foam dispersity and its evolution, the optical properties of foam, foam drainage kinetics and kinetics of establishing pressure in the foam liquid phase, kinetics of coarsening of foam (internal foam collapse) and kinetics and establishment of the equilibrium profile of Plateau borders using the FPDT [2, 3].

The FPDT can also be used to develop a number of foam applications for the accumulation of surfactants in foam and for foam fractionation and foam separation [3].

Combination of the FPDT and thin-film pressure balance (TFPB) is a very promising way to find quantitative correlations between film and foam properties, especially in investigations of foam stability and foam drainage [3, 29].

At high pressure, foam drainage has various peculiarities that, compared with gravitational drainage, make it easier to model and compute this hydrodynamic process.

It is possible to study foam drainage and the flow-rate of liquid through the foam at constant or variable Plateau border radius using the FPDT.

3.1.3 Hydroconductivity

The volumetric flow-rate over the unit of foam cross-section, which is perpendicular to the direction of flow, can be expressed as

$$Q = Q_f + Q_B = (H_B + H_f)(\rho g - \operatorname{grad} P) \tag{3}$$

where Q_f is the volumetric flow-rate over the film, Q_B is the volumetric flow-rate over Plateau borders, H_B, H_f are the local hydroconductivities of borders and films, respectively, P is the hydrostatic pressure and ρ is the density of the liquid.

It was shown in Krotov's work [7] that H_B may be expressed for three-dimensional foams as

$$H_B = \frac{w_B V_B}{3} \tag{4}$$

where V_B is the volume part of the border liquid,

$$w_B = \frac{v_B}{\rho g \cos\alpha} \tag{5}$$

and v_B is the average linear rate in the single border.

The film hydroconductivity is expressed by

$$H_f = \frac{w_f V_f}{1.5} \tag{6}$$

where

$$w_f = \frac{v_f}{\rho g \cos \alpha'} \tag{7}$$

and v_f is the average linear velocity of liquid in the films.

This means that the hydroconductivity of a large number of disorderly oriented borders is equal to one-third of the same independent borders oriented in the direction of flow (and two-thirds for films).

The average velocity of liquid flow in a single Plateau border ($\bar{v}_B$) was determined by a numerical method using the Navier-Stokes equation in the work of Leonard and Lemlich [31]:

$$\bar{v}_B = \frac{r^2}{306\eta} \rho g \cos \alpha' \tag{8}$$

where α is the angle between the border and the direction of the shortest path of flow for the case of the gas/liquid interface being considered as a solid wall. For the flow velocity in a film with the same assumption the equation obtained was [7, 32]

$$v_f = \frac{h^2}{12\eta} \rho g \cos \alpha' \tag{9}$$

where η is the dynamic (bulk) viscosity and h is the thickness of the film.

The relation V_B with the geometric (structural) parameters of foam bubbles is given as

$$V_B = C_B \frac{r^2}{a^2} \tag{10}$$

where C_B is a dimensionless coefficient dependent on the bubble shape.

For the dodecahedron model [3] C_B=0.209 and for the Kelvin cell (regular minimal tetrakaidecahedron) [12, 13, 15] $C_B^K = 0.171$.

Substituting the values v_B, V_B and v_f, V_f in Eqs. (4) and (6) and then in Eq. (3), we obtain the dependence of the volumetric velocity on the geometric parameters of foam. These equations are valid for the local cross-sections of foams.

For the macroscopic description of foam drainage, it is necessary to have additional data on the parameters V and H and their evolution in the process of drainage.

It is necessary to know the Plateau border shape and profile, the degree of mobility of gas/liquid interfaces and the contributions to viscous losses of both vertices and Plateau borders in order to obtain a realistic macroscopic equation for liquid flow through foam and the foam drainage equation.

The flow of dispersion medium from foams is one of the main stages of their destruction. In the hydrodynamics of a dispersion system with elastic surfaces, the mobility of these surfaces, which greatly influences the total flow velocity, is very important.

The aim of this work was to investigate (with the help of the FPDT method) the Plateau border profile shape during liquid flow through the foam and the influence of the surface viscosity on the flow-rate according to the Leonard-Lemlich, Desai, Kumar and Nguyen equations. The FPDT method is also used in our work for accounting for the viscous dissipation in both the Plateau border and the nodes.

3.2
Experimental Investigation of the Liquid Flow Through the Foam

Foam from a foam generator was introduced into a cell (1) with two porous plates (3, 4) which were in contact with the liquid (dispersion medium) under the same (by the liquid flow through the foam constant along the Plateau border radius) or under the different pressure drops $P_L^{max} = P_0 - \Delta P_{min}$ and $P_L^{min} = P_0 - \Delta P_{max}$ at the upper and lower porous plates, respectively (Fig. 3.1). The electrical conductivity was determined using the electrodes (5, 6) and the conductometer (7); the volumetric flow-rates were measured using the graduated glass tubes (8, 9).

The volumetric flow-rate Q_{exp} of the surfactant solution through a single Plateau border with the length L is equal to

$$Q_{exp} = Q/N_{el} \tag{11}$$

where Q is the total experimental volumetric flow-rate and N_{el} is the effective number of the Plateau borders (which was determined by the electro-hydrody-

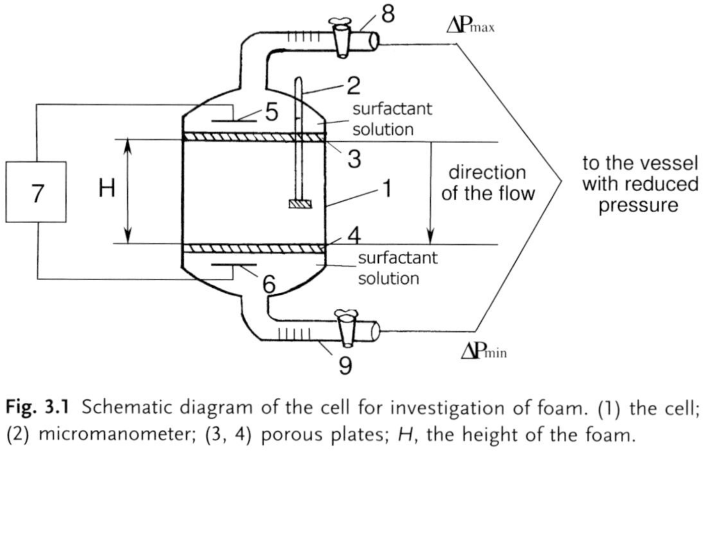

Fig. 3.1 Schematic diagram of the cell for investigation of foam. (1) the cell; (2) micromanometer; (3, 4) porous plates; *H*, the height of the foam.

namic analogy between current intensity and liquid flow-rate through the border [3, 33]).

The surface viscosity was measured using the disk rotation method [30]. The Plateau border radius r was measured with a capillary micromanometer.

3.3 Results and Discussion

3.3.1 Liquid Flow Through the Foam with Constant Plateau Border Radius

Liquid flow through the foam with constant (with respect to the foam height) Plateau border radius (r) was investigated in previous work [23, 30]. It was established [23] that the experimental drainage rate (ν_{exp}) in the foam from sodium dodecyl sulfate (SDS) with common black films and from NP-20 was substantially (factor of 2–9) larger than that calculated using the Leonard-Lemlich equation (ν_{th}) (at Plateau border radii from 50 to 120 μm).

We investigated [30] the liquid flow through foam stabilized with SDS, Triton X-100 with electrolyte, gelatine, glycerol added and variable surface viscosity. The Plateau border radii varied in our experiments from 30 to 100 μm (including the range of radii where surface immobility was observed).

We compared the experimental results obtained with the FPDT method with values calculated using the Desai and Kumar and Nguyen equations [34–37], which take into account the surface viscosity. (Note that Desai and Kumar [34–36] estimated the surface mobility from the movement of dye through the foam. They pointed out that dyes may influence the foam properties.)

For the estimation of the border surface mobility, Desai and Kumar [34–36] used a foam bubble model of a pentagonal dodecahedron, the border shape of which is idealized as a pipe whose cross-section is an equilateral triangle and its vertex has zero velocity. A solution of the Navier-Stokes equation for this model was derived by the method of successive approximations involving surface viscosity and assuming that border edges are immobile (at the border–film transition line):

$$\beta = \frac{\nu_{exp}}{\nu_{th}} = f(a) \tag{12}$$

where ν_{exp} is the volumetric flow velocity through the border with a mobile surface, ν_{th} is the theoretical flow velocity through a border with an immobile surface and a is the inverse value of the surface viscosity η_s:

$$a = 0.176\, r\eta/\eta_s \tag{13}$$

Graphical and algebraic expressions for the $\beta(a)$ function were given by Desai and Kumar [34–36].

We also compared the experimental drainage rate with the rate obtained by the Nguyen equation [37], which also takes into account surface mobility:

$$\nu_N = \frac{0.16\, r^2 \rho g}{\eta}\left[\frac{a}{\sqrt{B_0}(C + B_0^{0.63})} + 0.02\right] \tag{14}$$

where a=0.0655, C=0.209 and $B_0 = \eta_s/\eta r$.

First we studied the liquid flow through the foam with constant (with respect to foam height) Plateau border radius (assuming that the viscous dissipation is in the Plateau borders).

Comparison of our experimental drainage rates with the values calculated using the Desai and Kumar and Nguyen equations is presented in Table 3.1. As can be seen, the immobile surface of the Plateau border was observed in the foam from SDS with lauryl alcohol and gelatine added (solutions 1 and 9). Note that the surface immobility of the Plateau border is predicted by the Desai and Kumar and Nguyen equations (parameter $\beta = \nu_{exp}/\nu_L = \nu_N/\nu_L = 1$ for the same solutions).

Surface mobility was observed experimentally in the foam from Triton X-100 and in the foam from SDS (with common black films) (solutions 7 and 8). In solutions 7 and 8 the ratios ν_{exp}/ν_L were equal to 4.4 and 2.6, respectively, and were more than those calculated using the Desai and Kumar and Nguyen equations (β=1.4, 1.2; ν_N/ν_L=2.44, 1.7). The Boussinesq numbers for these solutions were 2 and 10, respectively. We assume that such surface mobility may exist owing to influence of the surface tension gradient along the Plateau border. Note that when B_0=0.06 (in the foam from Triton X-100 with glycerol added, solution 6), ν_{exp}/ν_L=10 and ν_N/ν_L=33. We assume that the viscous dissipation in the nodes can influence the liquid flow through the foam at large Plateau border radii and volume viscosity. It is necessary to study this case more in detail.

Table 3.1 Estimation of the Plateau border surface mobility[a)].

No.	Surfactant solution	$r\cdot(10^6$ m)	η_s (N s m^{-1})	β	ν_N/ν_L	ν_{exp}/ν_L
1	SDS+0.5×10^{-4}% LOH+0.1 mol dm^{-3} NaCl	30	4.5×10^{-6}	1	1.0	1
2	SDS+0.334 mol dm^{-3} NaCl	30	2.8×10^{-7}	1	1.25	1
3	SDS+0.1 mol dm^{-3} NaCl	30	2.8×10^{-7}	1	1.25	1
4	SDS+0.4 mol dm^{-3} NaCl	68	2.8×10^{-7}	1.2	1.7	1.2
5	SDS+0.1 mol dm^{-3} NaCl	90	3.2×10^{-7}	1.2	1.68	3.2
6	Triton X-100+0.1 mol dm^{-3} NaCl+ 72% glycerol	90	1.2×10^{-7}	11	33	10
7	Triton X-100+0.1 mol dm^{-3} NaCl	60	1.2×10^{-7}	1.4	2.44	4.4
8	SDS+0.1 mol dm^{-3} NaCl	32	3.2×10^{-7}	1.2	1.7	2.6
9	10^{-2} mol dm^{-3} DDS +0.2% gelatine+ 0.1 mol dm^{-3} NaCl	30	(5–8)×10^{-5}	1	1	1

a) ν_N was calculated with the Nguyen equation, v_L with the Leonard-Lemlich equation and β with the Desai and Kumar equation.

3.3.2
Comparison of Experimental Plateau Border Profile with that Calculated on the Assumption of a Mobile Border

For the investigation of the liquid flow through the foam under different pressure drops (P_L^{max} and P_L^{min}, Fig. 3.1) it is necessary to study the Plateau border profile (the change in the Plateau border curvature with height) and the volumetric flow-rate.

We investigated the Plateau border profile in the foam from SDS and Triton X-100 with additions of different compounds (electrolyte, glycerol and gelatine) and compared the experimental Plateau border profile and the calculated profile assuming that border surfaces are mobile. For this purpose we used the cell shown in Fig. 3.1.

With the assumption that the border surfaces are immobile and when $dP_L/dl \gg \rho g$ [where P_L is the liquid pressure in the Plateau border, ρ is the liquid density, g is the acceleration due to gravity and l is the coordinate in the direction opposite to the liquid flow (Fig. 3.1)] the theoretical volumetric flow-rate Q is determined by the Leonard-Lemlich equation:

$$Q = 0.16\frac{f}{\eta}\left(\frac{dP}{dl}\right)r^4 \qquad (15)$$

where r is the Plateau border radius, f is the geometric dimensionless coefficient, 3.3×10^{-3} [31], η is the dynamic viscosity of the solution and dP/dl is the pressure gradient in the liquid phase.

The dependence $r(l)$ was calculated by [2, 3]

$$r^3 = r_{min}^3 + \frac{(r_{max}^3 - r_{min}^3)l}{L} \qquad (16)$$

where l is the distance from the cell bottom, $L=1.35H$, H is the straight-line distance from the plane of the porous plate to the other plate, r_{min} and r_{max} are the minimum and maximum Plateau border radii at the lower and upper porous plates which were under different pressure drops $P_{L_{max}}$ and $P_{L_{min}}$ respectively, and 1.35 is the winding coefficient for the foam bubble as a pentagonal dodecahedron.

The Plateau border radius was calculated using

$$r = \frac{\sigma}{P_\sigma} \qquad (17)$$

where σ is the surface tension and P_σ is the capillary pressure (the difference between the pressure in the foam bubbles P_b and the pressure in the liquid phase P_L).

In a polyhedral foam (content of the liquid phase approximately 0.5%), the absolute value of the excess pressure in the liquid phase ($\Delta P_L = P_L - P_0$) is much lower than the excess bubble gas pressure $\Delta P_b = P_b - P_0$, therefore $P_\sigma \approx -\Delta P_L$ (where P_0 is the pressure in the surrounding medium).

As mentioned above, the border surface mobility was estimated by the parameter β; the dependence $\beta = f(\alpha)$ was plotted as a straight line in $\log\beta$ vs. $\log\alpha$ coordinates [33].

With the help of this straight line, we obtained the approximate equation

$$\beta = 1 + 5.4\,\alpha^{0.5} \tag{18}$$

From Eqs. (12), (15) and (18) we obtained [33] Eq. (19) for calculating the volumetric flow-rate of the solution taking into account the surface mobility:

$$Q'_{\text{th}} = \frac{0.16\,f}{\eta}\left(\frac{\mathrm{d}P}{\mathrm{d}l}\right) r^4 (1 + 5.4\,\alpha^{0.5}) \tag{19}$$

and Eq. (20) for calculating the Plateau border curvature profile:

$$\frac{l}{L}\left[r^3_{\text{max}} - r^3_{\text{min}} + 3 \times 1.98\left(\frac{\eta}{\eta_s}\right)^{0.5}(r^{3.5}_{\text{max}} - r^{3.5}_{\text{min}})\right] + r^3_{\text{min}} + 1.98\,r^{3.5}_{\text{min}}\left(\frac{\eta}{\eta_s}\right)^{0.5} = r^3 + 1.98\,r^{3.5}\left(\frac{\eta}{\eta_s}\right)^{0.5} \tag{20}$$

We calculated [33, 38] also the Plateau border profile (with the help of the Nguyen equation and the surface mobility taken into account):

$$\frac{K\sigma a(\eta/\eta_s)^{0.5} r^{4.13}}{\eta(\eta_s/\eta)^{0.63} \times 4.13} + \frac{5.2 \times 10^{-4}\sigma r^3}{3\eta} = \frac{K\sigma a(\eta/\eta_s)^{0.5}(r^{4.13}_{\text{max}} - r^{4.13}_{\text{min}})l}{4.13\,\eta(\eta_s/\eta)^{0.63} L} + \frac{5.2 \times 10^{-4}\sigma(r^3_{\text{max}} - r^3_{\text{min}})l}{3\,\eta L} + \frac{K\sigma a(\eta/\eta_s)^{0.5} \cdot r^{4.13}_{\text{min}}}{4.13\,\eta(\eta_s/\eta)} + \frac{5.2 \times 10^{-4}\sigma r^3_{\text{min}}}{3\,\eta} \tag{21}$$

and the volumetric flow-rate Q''_{th} of the solution through the Plateau border with minimum and maximum radii r_{min} and r_{max}, respectively:

$$Q''_{\text{th}} = \frac{K\sigma a\left(\frac{\eta r_{\text{max}}}{\eta_s}\right)^{1.13} r^3_{\text{max}} - K\sigma a\left(\frac{\eta r_{\text{min}}}{\eta_s}\right)^{1.13} r^3_{\text{min}}}{4.13\,\eta L} + \frac{5.2 \times 10^{-4}\sigma(r^3_{\text{max}} - r^3_{\text{min}})}{3\,\eta L} \tag{22}$$

The Plateau border radii r_3 and r_2 (Table 3.2) were calculated using Eqs. (20) and (21).

Although the volumetric flow-rates calculated by Eqs. (19) and (22) (at $r_{\text{min}} = 8$ μm and $r_{\text{max}} = 32$ μm) are higher than those calculated using the Leonard-

Table 3.2 The Plateau border radii[a] in foams from SDS and Triton X-100.

No.	Surfactant solution	ΔP_{max} (kPa)	ΔP_{min} (kPa)	L (m)	r_{exp} (10^{-6} m)	r_1 (10^{-6} m)	r_2 (10^{-6} m)	r_3 (10^{-6} m)
1	10^{-3} mol dm^{-3} SDS + 0.4 mol dm^{-3} NaCl	8	3	0.02	6.6	8.64	8.65	8.71
2	10^{-3} mol dm^{-3} SDS + 0.1 mol dm^{-3} NaCl	8	3	0.02	8	8.07	8.1	8.1
3	10^{-3} mol dm^{-3} Triton X-100 + 0.4 mol dm^{-3} NaCl	8	3	0.02	8.6	8.64	8.7	8.73
4	10^{-2} mol dm^{-3} SDS + 0.1 mol dm^{-3} NaCl + 0.2% gelatine	4	1	0.02	17	23.9	23.9	23.9
5	10^{-3} mol dm^{-3} Triton X-100 + 0.4 mol dm^{-3} NaCl	4	1	0.02	24	25.5	26	26.5
6	10^{-3} mol dm^{-3} SDS + 0.1 mol dm^{-3} NaCl	4	1	0.02	24	23.9	24.1	24

a) The Plateau border radii are given at $l = 1$ cm.

Lemlich equation (by a factor of 1.2) in the foam from SDS with common black films and in the foam from Triton X-100, the Plateau border radii calculated using Eqs. (20) and (21) were larger only by 0.5–1% (in the middle part of the Plateau border) than the Plateau border radii when assuming that border "walls" are immobile.

The experimental Plateau radii in the foam from SDS + 0.1 mol dm^{-3} NaCl and in the foam from Triton X-100 + 0.4 mol dm^{-3} NaCl were the same as calculated with Eqs. (20) and (21) at high pressure gradients ($\Delta P_{max} = 8$ kPa and $\Delta P_{min} = 3$ kPa) and were equal to 8 and 8.6 μm, respectively.

The Plateau border profiles were also studied at lower pressure gradients, $\Delta P = 3$ kPa ($\Delta P_{max} = 4$ kPa and $\Delta P_{min} = 1$ kPa), and minimum radius 8 μm in the foam from SDS and Triton X-100 with addition of electrolyte, gelatine and glycerol.

The experimental Plateau border radius in the foam from SDS + 0.1 mol dm^{-3} NaCl was equal to 24 μm (at $l = 1$ cm) and was identical with the experimental Plateau border radius in the foam from Triton X-100 + 0.4 mol dm^{-3} NaCl and with that calculated by Eq. (20) (Table 3.2, solutions 5 and 6).

In the foam from SDS and gelatine with Newton black films, the Plateau border radii differed from those calculated with Eq. (21).

In the foam from SDS with Newton black films, the Plateau border radius was less by 20% than that calculated with Eq. (16) with the surface immobility taken into account and was equal to 6.6 μm ($\Delta P_{max} = 8$ kPa, $\Delta P_{min} = 3$ kPa and $l = 1$ cm).

A similar decrease in the Plateau border radii was observed in the foam from SDS and 0.2% gelatine. The experimental Plateau border radius in the foam from SDS + 0.1 mol dm^{-3} NaCl + 0.2% gelatine was 17×10^{-6} m (at $l = 1$ cm) and differed by 28% from that calculated with Eq. (16). We assume that in this case the surface tension gradient may influence greatly the Plateau border profile.

It was shown that in the solutions of SDS and gelatine, formation of surface-active complexes is possible. This process results in a decrease in surface tension [39].

A decrease in the surface tension of protein solutions with time was also observed [40]. It was established that the surface tension of 0.1% gelatine solution decreased by 6 mN m^{-1} within 78 min. Possibly this process leads to a decrease in Plateau border radius during the flow of surfactant solution through the foam.

3.3.3 Influence of Surface Tension Decrease on the Plateau Border Profile

In our experiments [30, 33], we observed that the Plateau border radii (in foam from SDS with Newton black films and gelatine added) were different by 17–28% from those calculated using the Leonard-Lemlich equation (Table 3.2). With assumption of a linear decrease in surface tension along the surface of the Plateau border [$\sigma=\sigma_0-Kl$, where $K=0.5$ (we assume that the surface tension in our experiments is changed from the value of 32×10^{-3} N m^{-1} for the initial surfactant solution to 28×10^{-3} N m^{-1} after the adsorption of lauryl alcohol) and σ_0 is the surface tension of the surfactant solution], taking into account the capillary pressure gradient:

$$\frac{dP_\sigma}{dl}=\frac{r\frac{d\sigma}{dl}-\frac{\sigma dr}{dl}}{r^2} \tag{23}$$

and the Leonard–Lemlich dependence, Eq. (15), we obtained

$$A+r^3-=\frac{r^2}{K}(\sigma_0-Kl)\frac{dr}{dl} \tag{23}$$

where $A=Q\eta/0.16fK$ and

$$\frac{r^2dr}{A+r^3}=-\frac{Kdl}{\sigma-Kl} \tag{24}$$

Integrating Eq. (23) written in the form

$$\int_{r_{max}}^{r}\frac{d(A+r^3)}{3(A+r^3)}=\int_0^l\frac{d(\sigma_0-Kl)}{\sigma_0-Kl} \tag{25}$$

we obtain Eq. (26) for calculating the Plateau border profile with the surface tension gradient taken into account:

$$A+r^3=\left(\frac{\sigma_0-Kl}{\sigma_0}\right)(A+r^3_{max}) \tag{26}$$

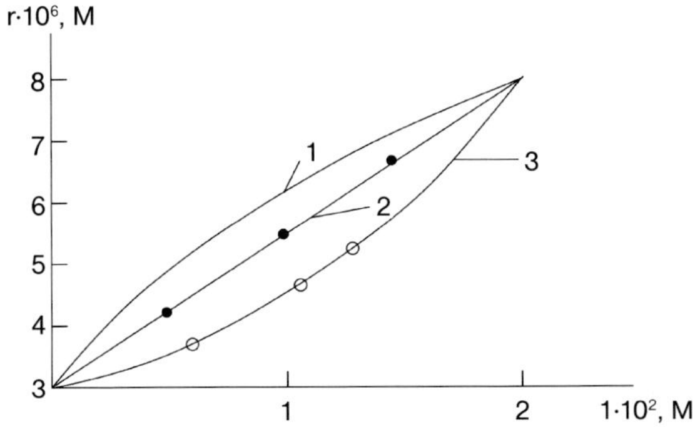

Fig. 3.2 The dependence $r(l)$ in the foam from 10^{-3} mol dm^{-3} SDS + 0.4 mol dm^{-3} NaCl: $\Delta P_{max}=8$ kPa and $\Delta P_{min}=3$ kPa. 1, Calculation with the Leonard-Lemlich equation; 2, calculation with surface tension gradient taken into account (Eq. 26); 3, in the foam from SDS + 0.4 mol dm^{-3} NaCl (experimental data).

where the parameter A was calculated by

$$\frac{A + r_{max}^3}{A + r_{min}^3} = \left(\frac{\sigma_0}{\sigma_0 K l}\right)^3 \tag{27}$$

[This equation was obtained by integration of the left-hand side of Eq. (25) from r_{min} to r_{max} and its right-hand side from 0 to l].

With the help of Eq. (26), we calculated the Plateau border radii (at $\Delta P_{max}=8$ kPa and $\Delta P_{min}=3$ kPa) in the foam from SDS + 0.4 mol dm^{-3} NaCl with the surface tension gradient taken into account (Fig. 3.2). It was shown that in this case the experimental Plateau border profile (in its middle part) was less by only 6.3% (Fig. 3.2, curve 2).

Hence the surface tension gradient may be one of the causes of the discrepancy between the experimental Plateau border profile and then calculated one.

3.3.4
Comparison of the Experimental and Calculated Volume Flow-rates

We also investigated the volume flow-rate through the foam with the help of our cell (Fig. 3.1). The direction of the liquid flow is shown in Fig. 3.1.

From Eqs. (15) and (17), we determined the volume flow-rate through the Plateau border with the definite radii r_{min} and r_{max}:

$$Q_{th} = \frac{0.16 f (r_{max}^3 - r_{min}^3)}{3\eta L} \tag{28}$$

We also compared our experimental volumetric flow-rate with that calculated using Eqs. (19) and (22), which were obtained with the help of the Desai and

Kumar and Nguyen equations taking into account the Plateau border surface mobility [33].

The experimental volume rates and the results of calculation are presented in Table 3.3. It can be seen that in the foam from the solution of Triton X-100 + 0.4 mol dm^{-3} NaCl the ratio Q_{exp}/Q_{th} is equal to 4.3, 5.2 and 5.9 at maximum (ΔP_{max}) and minimum (ΔP_{min}) pressure drops equal to 3 and 1 kPa, 4 and 1 kPa and 5 and 1 kPa, respectively, assuming immobility of the surfaces.

We calculated the ratios Q_{exp}/Q'_{th} and Q_{exp}/Q''_{th} with the surface mobility taken into account. It was established that the ratio Q_{exp}/Q'_{th} (Table 3.1) was 1.6, 2.1 and 2.3 at maximum (ΔP_{max}) and minimum (ΔP_{min}) pressure drops equal to 3 and 1 kPa, 4 and 1 kPa and 5 and 1 kPa, respectively, with the cross-section of the Plateau border in the form of the "spherical" triangle taken into account. The Boussinesq numbers in this case were changed from 12 to 20 and Q_{exp}/Q''_{th} was equal to 2.7, 3.5 and 3.9 for the same pressure gradients (Table 3.3).

As can be seen from Table 3.3, the experimental volume flow-rates (at different pressure drops) are somewhat higher than those calculated with the help of the Desai and Kumar and Nguyen equations. We assume that the surface tension gradient along the Plateau border (which increases with the pressure gradient) greatly influences the surface mobility.

An insignificant deviation of the experimental volume flow-rate from the theoretical value was observed for the foam from Triton X-100 with glycerol added. The ratios Q_{exp}/Q'_{th} were equal to 0.93 and 1.1 at maximum (ΔP_{max}) and minimum (ΔP_{min}) pressure drops equal to 2 and 1 kPa and 3 and 1 kPa, respectively.

Table 3.3 Comparison of the experimental volume flow-rates Q_{exp} through the Plateau border with theoretical values [with the assumption that the border surfaces are immobile (Q_{th}) or mobile (Q'_{th} and Q''_{th})].

Surfactant solution	ΔP_{max} (kPa)	Q_{th} (10^{17} m^3 s^{-1})	Q_{exp}/Q_{th} *(Leonard-Lemlich)*	Q_{exp}/Q'_{th} Desai and Kumar)		Q_{exp}/Q''_{th} (Nguyen)	B_0 [b)]
				A [a)]	B [a)]		
1×10^{-3} mol dm^{-3} Triton X-100	3	888	4.3	1.98	1.6	2.7	12
+0.4 mol dm^{-3} NaCl	4	909	5.2	2.6	2.1	3.5	15
	5	915	5.9	2.87	2.3	3.9	20
2×10^{-3} mol dm^{-3} Triton X-100	2	289	6.5	1.12	0.93	2.55	0.6
+0.4 mol dm^{-3} NaCl	3	317	7.7	1.36	1.1	2.28	0.8
+35% glycerol							
3×10^{-3} mol dm^{-3} SDS+	3	874	3.8	2.9	2.4	2.47	5×10^3
0.2% gelatine+0.1 mol dm^{-3} NaCl	4	702	2.9	2.8	2.3	1.5	6×10^3

a) A, cross-section of the Plateau border in the shape of a triangle (these shape formed at the direct contact of three the same cylinders), B, cross-section of the Plateau border in the shape of a spherical triangle. The Plateau border radius was calculated from the equality of the areas of the triangle and the spherical triangle.

b) B_0 was calculated with r_{min} taken into account.

At the same pressure gradients B_0 was equal to 0.6 and 0.8 and the ratios Q_{exp}/Q''_{th} were equal to 2.55 and 2.28.

In the foam from SDS with gelatine added (Table 3.3), the ratio of the experimental volumetric flow-rate to that calculated with the Leonard-Lemlich equation was equal to 3.8 at $\Delta P_{max}=3$ kPa and $\Delta P_{min}=1$ kPa. The ratios of Q_{exp} to Q'_{th} and Q''_{th} (taking into account the surface viscosity) were 2.4 and 2.47, respectively, at the same ΔP_{max} and ΔP_{min}.

The excess of the experimental volume flow-rate above the calculated Q'_{th} and Q''_{th} [B_0 in this case is equal to $(5\text{–}6)\times10^3$] may be due to the formation of a Plateau border profile which is different from that calculated by Eq. (1) (this problem is mentioned above).

We also assume that the surface tension decrease in due time leads to a decrease in the Plateau border radius during the flow of surfactant solution through the foam. That is why a decrease in the Plateau border radius (r_{max} and r_{min}) is possible during the foam drainage and the ratios Q_{exp}/Q'_{th} and Q_{exp}/Q''_{th} will also decrease, becoming equal to 2 and 1 at a pressure gradient of 3 kPa ($\Delta P_{max}=4$ kPa and $\Delta P_{min}=1$ kPa).

3.4 Foam Drainage Investigations Using the Pressure Established When Pressure Drop Is Created in the Liquid Phase of Foam

The foam drainage velocity can also be determined on investigation of the pressure established when a pressure drop is created in the liquid phase of foam.

In polyhedral foam (dodecahedral cell), when all liquid is situated in the Plateau border (border foam), the expansion ratio (n) can be expressed by

$$n = \frac{V_F}{V_L} = \frac{4.8\,a^2}{r^2} = \frac{4.8\,a^2}{\sigma^2}\,P_\sigma^2 \tag{29}$$

where V_F is the foam volume, V_L is the volume of the liquid in the foam, a is the length of the Plateau border and r is the radius of curvature. Then the drainage velocity will be equal to

$$v_B = \frac{1}{KP_\sigma^3}\cdot\frac{dP_\sigma}{d\tau} \tag{30}$$

where $K=3.2/l_0\sigma^2$, l_0 is the distance from the measuring point to the foam top, σ is the surface tension and F is the cross-section of the vessel.

Such a method permits one to compare the experimental results with the calculations (for example, with data obtained using the Leonard-Lemlich or Nguyen equation) in the presence of information about the border profile during foam drainage.

Experimental results of foam drainage investigations and simultaneously measured border profiles showed that the drainage velocity depends on the surfactant type and at the beginning of the drainage process (when the border radii are large) it exceeds 10–20 times the calculated value and then decreases [2, 3, 41]. This reflects the change in the influence of tangential mobility of border surfaces on the drainage velocity.

3.5 Conclusions

Recent theoretical and experimental studies were focused on the process called "forced drainage" in order to estimate the hydrodynamic resistance of the Plateau borders and nodes. However, the interpretation of these results is contradictory.

The FPDT method gives great possibilities for the study of foam drainage and division of the hydrodynamic resistance into the parts of nodes and borders.

It was established that the experimental volume flow-rate Q_{exp} was substantially (by a factor of 3–8) larger than that calculated with the Leonard-Lemlich equation at all pressure gradients.

The experimental volume flow-rates in the foam from Triton X-100 with glycerol added were the same as those calculated using the Desai and Kumar equation with surface mobility and "spherical" triangle in the cross-section of the Plateau border taken into account.

In the foam from SDS with gelatine added, Q_{exp} was somewhat higher than the calculated Q'_{th} and Q''_{th} values, which may be due to (a) the formation of a Plateau border profile that is different from the calculated one, (b) a decrease in the surface tension in protein solutions or (c) a change in the effective number of Plateau borders due to gel formation.

References

1 J.J. Bikerman, *Foams*, Vol. 3. Springer, Berlin, **1973**.
2 P.M. Kruglyakov, D. Exerowa, *Pena i Pennye Plenki*. Khimiya, Moscow, **1990**.
3 D. Exerowa, P.M. Kruglyakov, *Foam and Foam Films*. Elsevier, Amsterdam, **1998**.
4 D. Weaire, S. Hutzler, *The Physics of Foams*. Oxford University Press, Oxford, **1999**.
5 K.B. Kann, S.A. Druzhinin, *Kolloid. Zh.*, **1979**, *41*, 667.
6 V.V. Krotov, *Dokl. Akad. Nauk SSSR*, **1980**, *254*, 402.
7 V.V. Krotov, *Kolloid. Zh.*, **1980**, *42*, 1092.
8 V.V. Krotov, *Kolloid. Zh.*, **1981**, *43*, 286.
9 I.I. Goldfarb, K.B. Kann, Schreiber, *Izv. Akad. Nauk SSSR*, **1987**, *2*, 103.
10 V. Goldstein, I.I. Goldfarb, Schreiber, *Ind. Int. J. Multiphase Flow*, **1996**, *22*, 991.
11 G. Verbist, D. Weaire, A. Kraynik, *J. Phys.: Condens. Matter*, **1996**, *8*, 3715.
12 S.A. Koehler, S. Hilgenfeldt, H.A. Stone, *Phys. Rev. Lett.*, **1999**, *82*, 4232.
13 S.A. Koehler, S. Hilgenfeldt, H.A. Stone, *Langmuir*, **2000**, *16*, 6327.
14 M. Durand, D. Langevin, *Eur. Phys. J.*, **2000**, *E7*, 35.
15 S.J. Neethling, H.T. Lee, J.J. Cillers, *J. Phys.: Condens. Matter*, **2002**, *14*, 331.

16 S. A. Magrabi, B. Z. Dlugogorski, G. L. Jameson, *AIChE J.*, **2001**, *47*, 314.
17 A. Saint-Jalmes, Y. Zhang, D. Langevin, *Eur. Phys. J.*, **2004**, *15*, 53.
18 S. A. Koehler, S. Hilgenfeldt, E. R. Weeks, H. A. Stone, *J. Colloid Interface Sci.*, **2005**, *276*, 439.
19 O. Pitois, C. Fritz, M. Vignes-Adler, *J. Colloid Interface Sci.*, **2005**, *282*, 458.
20 Khr. Khristov, P. M. Kruglyakov, D. Exerowa, *Colloid Polym. Sci.*, **1979**, *257*, 506.
21 L. L. Kuznetsova, P. M. Kruglyakov, A. F. Koretski, *Zh. Prikl. Khim.*, **1979**, *52*, 1294.
22 Khr. Khristov, D. Exerowa, P. M. Kruglyakov, *Kolloid. Zh.*, **1981**, *43*, 101.
23 L. L. Kuznetsova, P. M. Kruglyakov, *Dokl. Akad. Nauk SSSR*, **1981**, *260*, 928.
24 Khr. Khristov, D. Exerowa, P. M. Kruglyakov, *Colloid Polym. Sci.*, **1983**, *261*, 265.
25 P. M. Kruglyakov, D. Exerowa, Khr. Khristov, *Adv. Colloid Interface Sci.*, **1992**, *40*, 257.
26 P. M. Kruglyakov, D. Exerowa, Khr. Khristov, *Langmuir*, **1991**, *7*, 1846.
27 N. G. Fokina, P. M. Kruglyakov, *Trudy Vsesoyuznogo Seminara po Kolloidnoi Khimii i Phisico-Khimicheskoi Mekhanike i Bioaktivnykh Dispersnykh System.* Nauka, Moscow, **1991**, p. 71.
28 C. Stubenrauch, A. V. Makievski, Khr. Khristov, D. Exerowa, R. Miller, *Tenside Surfact. Deterg.*, **2003**, *40*, 196.
29 C. Stubenrauch, Khr. Khristov, *J. Colloid Interface Sci.*, **2005**, *286*, 710.
30 N. G. Vilkova, PhD Thesis, Moscow State University, **1992**.
31 R. A. Leonard, R. Lemlich, *Chem. Eng. Sci.*, **1965**, *20*, 790.
32 L. D. Landau, E. M. Lifshitz, *Mekhanika Sploshnykh Sred.* GITTL, Moscow, **1954**.
33 N. G. Vilkova, P. M. Kruglyakov, *Mendeleev Commun.*, **2004**, *1*, 22.
34 D. Desai, R. Kumar, *Chem. Eng. Sci.*, **1982**, *37*, 1361.
35 D. Desai, R. Kumar, *Chem. Eng. Sci.*, **1983**, *39*, 1525.
36 D. Desai, R. Kumar, *Chem. Eng. Sci.*, **1984**, *39*, 1559.
37 A. Nguyen, *J. Colloid Interface Sci.*, **2002**, *249*, 194.
38 N. G. Vilkova, P. M. Kruglyakov, *Colloids Surf.*, **2005**, *263*, 205.
39 V. N. Izmailova, S. P. Derkach, K. V. Zotova, R. G. Danilova, *Kolloid. Zh.*, **1993**, *55*, 54.
40 A. A. Trapeznikov, V. G. Vins, T. U. Shirokova, *Kolloid. Zh.*, **1981**, *43*, 322.
41 L. L. Kuznetsova, P. M. Kruglyakov, *Kolloid Zh.*, **1983**, *45*, 1076.

4
Physical Chemistry of Wetting Phenomena

Nicolay V. Churaev and Vladimir D. Sobolev

4.1
The State of the Theory of Wetting

Wetting is a widespread phenomenon that plays an important role both in natural science and in chemical technology. Wetting phenomena control processes such as suction, spreading and mass transfer in porous bodies in their application to flotation, evaporation of liquids from porous bodies, kinetics of drying, capillary condensation and suction and mutual displacement of non-miscible liquids.

As a macroscopic characteristic of wetting, the equilibrium contact angle θ_0 formed by a bulk liquid on the substrate was used. Complete wetting corresponds to θ_0 values near to 0 and $\theta_0 > 90^\circ$ corresponds to the region of non-wetting. For droplets and capillary menisci of spherical shape, contact angles, neglecting gravity, may be calculated using the well-known Young equation:

$$\gamma_{\rm lv} \cos \theta_0 = \gamma_{\rm sv} - \gamma_{\rm sl} \tag{1}$$

where $\gamma_{\rm sv}$ is the interface free energy of the solid surface in contact with vapor and correspondingly $\gamma_{\rm sl}$ in contact with liquid and $\gamma_{\rm lv}$ is the surface tension of the liquid.

As shown in Fig. 4.1, this equation determines also the condition of equilibrium of forces, when the normal force component $f = \gamma_{\rm lv} \sin \theta_0$ is equilibrated by elastic reaction of the substrate. In some cases such deformation may be not reversible. For substrates with a high modulus of elasticity, a small local projection arises along the contact line of a droplet. Its height depends on the mechanical properties of the substrates and for rubber, for instance, equals about 0.1 μm [1]. For lightly deforming gels, sometimes used as a substrate, surface deflections may even be observed visually [2, 3].

The theory of wetting of the elastic deforming bodies was developed by Rusanov [4]. In this case, expansion of the wetting perimeter is associated with performance of the work of deformation. This effect may be also expressed as a

Colloids and Interface Science Series, Vol. 2
Colloid Stability: The Role of Surface Forces, Part II. Edited by Tharwat F. Tadros

ISBN: 978-3-527-31503-1

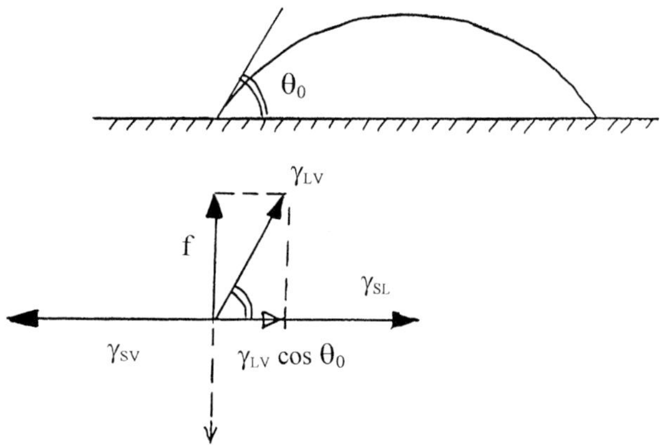

Fig. 4.1 Equilibrium contact angle of a drop θ_0 and interfacial tension in a zone of three-phase contact.

"special" positive line tension [5]. Transfer of the meniscus in that case is accompanied by the migration of a wave of surface deformation, which may influence the rate of establishment of the state of equilibrium between the meniscus and substrate.

The formation of a transition zone (fluid or gaseous) between meniscus and the substrate results in the conclusion that the normal force f (Fig. 4.1) is not applied to some point on the solid surface, but is redistributed over some transition zone, which may be several microns long [6, 7]. In the case of rough substrates, the mean contact angle may be calculated using the Wenzel-Derjaguin equation:

$$\theta_n = K \cos \theta_0 \tag{2}$$

where K is the roughness coefficient, which is equal to the ratio of the actual surface to its projection on the horizontal plane.

For contact angles $\theta_0 < 90^\circ$, an increase in K values leads to a decrease in the observable contact angle, as compared with a flat surface. At high values of K, contact angle values tend to zero. Incidentally, it should be noted that for practical purposes, rough surfaces are often used for the attainment the conditions of complete wetting.

On the basis of Eq. (1), it becomes possible to consider different conditions of wetting. For instance, complete wetting corresponds to the condition when $\gamma_{lv} = \gamma_{sv} - \gamma_{sl}$. The difference is large enough at small values of γ_{sl}, which correspond to a strong interaction of the liquid with a solid substrate, that is, to a lyophilic surface. The second limiting case corresponds to a lyophobic surface, $\gamma_{sv} < \gamma_{sl}$, when the contact angle becomes $> 90^\circ$. Between these two limiting cases, the region of contact angles $0 < \theta_0 < 90^\circ$ is related, which corresponds to cases of non-complete wetting. The main part of practically important cases is situated just in this region of contact angles.

The effects will be considered for, as one example, the case of menisci formed in a flat slit having a thickness $2H$, which is small enough to neglect gravity effects (Fig. 4.2). In the case of non-complete wetting, extension of the central, non-disturbed part of the meniscus forms, in the state of equilibrium, a non-zero contact angle θ_0 with the pore walls. The latter are covered with an adsorption or wetting film, the thickness h_0 of which is determined by the isotherm of disjoining pressure $\Pi(h)$. Between the bulk part of the meniscus (zone 1, Fig. 4.2), having constant curvature r, and the flat film, with a thickness h_0, the transition zone 2 is formed. The state of equilibrium is determined by constancy of pressure in all parts of the system under consideration [8, 9]:

$$\gamma_{\mathrm{lv}} K(h) + \Pi(h) = P_{\mathrm{c}} = \mathrm{constant} \tag{3}$$

where γ_{lv} is the surface tension of the liquid and h and $K(h)$ are the local thickness and local surface curvature of the transition zone, respectively. Here $P_{\mathrm{c}} = \gamma_{\mathrm{lv}}/r$ is the meniscus capillary pressure, where r is the radius of its non-disturbed part (curve 4, Fig. 4.2).

Equation (3) is applicable for gentle profiles of a transition zone, $\mathrm{d}h\,\mathrm{d}x \ll 1$, when within the zone it is possible to use the local values of disjoining pressure $\Pi(h, x)$. Outside the transition zone, at $K=0$, the capillary pressure is equal to the disjoining pressure of a flat wetting film, that is, $P_{\mathrm{c}} = \Pi(h)$. Extension l of the transition zone may be estimated according an equation $l \approx (Hh_0)^{1/2}$, derived in [10].

The case of complete wetting, when meniscus profiles do not intersect the pore walls (curve 3, Fig. 4.2) and a contact angle is not formed, demands special consideration in more detail, as described earlier [6]. Finally, in the absence of surface forces, Eq. (3) transforms into the equation of capillary pressure with constant radius of curvature, $r = H$ (curve 4, Fig. 4.2).

Let us now consider the modern state of methods of theoretical predictions of contact angle values, based on the theory of surface forces [8, 9]. First, let us consider the simple case of non-polar liquids, where only one type of surface forces (dispersion force) is acting.

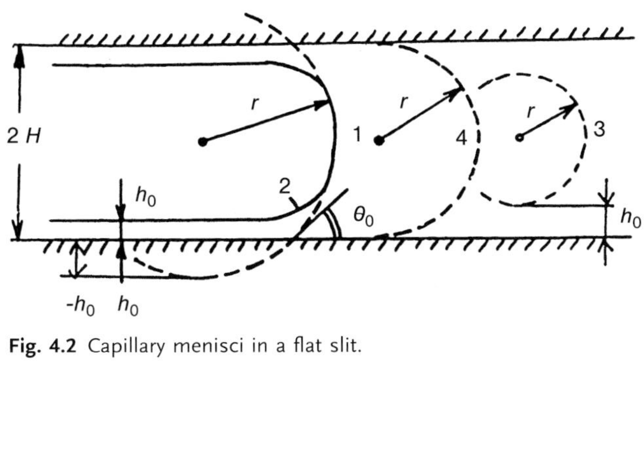

Fig. 4.2 Capillary menisci in a flat slit.

Calculations of contact angles are based in this case on the Gibbs adsorption equation $\Gamma = -(d\gamma_{sv}/d\mu)_T$, which includes the chemical potential of the liquid μ and interfacial free energy of the solid substrate in contact with vapor, γ_{sv}. The theoretical approach, developed by Frumkin and Derjaguin, was based on integration of the Gibbs equation of adsorption $\Gamma = h/v_m$ and $\mu = -v_m \Pi$, where v_m is the molar volume of the liquid, h is the wetting film thickness and $\Pi(h)$ is the disjoining pressure acting in the film. $\Pi(h)$ is equal to the difference between the pressure acting on the surface of the flat film and the pressure acting in the bulk liquid with which the film is in a state of equilibrium.

Integration of the transformed Gibbs equation results in following expression:

$$\gamma_{sv} - \gamma_{sl} = \gamma_{lv} + \int_{h_0}^{\infty} \Pi(h) dh \tag{4}$$

The Frumkin-Derjaguin method, subsequently developed further [8, 9, 11], gives an equation for calculating the value of the equilibrium contact angle:

$$\cos\theta_0 = 1 + (1/\gamma_{lv}) \int_{h_0}^{\infty} \Pi(h) dh = 1 + G_0(h)/\gamma_{lv} \tag{5}$$

where G_0 is the excess free energy of wetting film with equilibrium thickness $h = h_0$, γ_{lv} is surface tension and $\Pi(h)$ is the isotherm of disjoining pressure, which additively includes three basic components of forces: dispersion forces Π_m, electrostatic forces Π_e and structural forces Π_s.

The equilibrium film thickness h_0 may be determined as the lowest root of the equation $\Pi_m(h) + \Pi_e(h) + \Pi_s(h) = 0$. The calculated films of greater thickness are metastable and their stability may be disturbed.

Let us consider now the application of the theory of wetting to the simple case of non-polar liquids on a non-deformable solid substrate.

4.2
Non-polar Liquids

In this case, for the analysis of conditions of wetting, numerical methods of computer simulation have often been used. For instance, using the method of molecular dynamics, the state of a Lennard-Jones fluid in a flat slit formed between two crystalline surfaces was investigated [12, 13]. The parameter $\alpha = \alpha_{sl}/\alpha_{ll}$, which characterizes the relation of the potential of pair interaction of substrate molecules (s) with liquid (l), α_{sl}, to the potential α_{ll} which characterizes the interaction between liquid molecules itself, was used. It was shown that an increase in α leads to transition from non-wetting at $\alpha = 0.2$ to complete wetting at $a=0.7$. In agreement with general principles, strengthening of the interaction between the liquid and the surface improves the wetting.

Contact angles calculated on the basis of meniscus profiles in thin slits [11, 12] include the contribution of dispersion forces and also (in a hidden form) the influence of structural forces as a result of a non-uniform distribution of liquid density over the slit thickness.

In the case of non-polar liquids, the calculations may be restricted by action on dispersion forces only. The macroscopic theory developed by Dzyaloshinskii et al. [14] gives in this case the following expression for the isotherm of disjoining pressure $\Pi(h)$ in thin wetting films:

$$\Pi_{\mathrm{m}}(h) = \frac{\hbar}{8\pi h^3} \int_0^\infty \frac{(\varepsilon_{\mathrm{s}} - \varepsilon_{\mathrm{l}})(\varepsilon_{\mathrm{l}} - 1)}{(\varepsilon_{\mathrm{s}} + \varepsilon_{\mathrm{l}})(\varepsilon_{\mathrm{s}} + 1)} d\chi = \frac{A_{\mathrm{slv}}}{6\pi h^3} > 0 \tag{6}$$

where $\hbar$ is Planck's constant, $\varepsilon_{\mathrm{s}}(\xi)$ and $\varepsilon_{\mathrm{l}}(\xi)$ are functions which express the frequency dependence of the dielectric permeability of the solid substrate and liquid, respectively, and h is the film thickness. The parameter A_{slv} for historical reasons is called the Hamaker constant, because an equation similar in form, but not nature, was proposed earlier on the basis of a microscopic approach by Hamaker. Because for condensed matter $\varepsilon>1$, the sign of the dispersion forces depends on the sign of the difference between the functions $\varepsilon_{\mathrm{s}}(\xi)$ and $\varepsilon_{\mathrm{l}}(\xi)$ in the overall interval of frequencies. The main contribution gives the range of frequencies of the order of 10^{16} rad s^{-1}.

For liquids on low-polarity substrates, when $\varepsilon_{\mathrm{s}}<\varepsilon_{\mathrm{l}}$, the disjoining pressure is negative. In this case, wetting films become unstable, which leads to non-wetting. In contrast to this, on solid substrates more polar than the liquid, wetting films are stable. For example, non-polar liquids always effectively wet the surface of metals.

Equation (6) allows one, in particular, to trace the transition from wetting to non-wetting when the temperature is raised. Because values of the dielectric permeability depend on density, taking into account a higher coefficient of thermal expansion of liquids than solids, a rise in temperature may by accompanied by a transition from $\varepsilon_{\mathrm{l}}>\varepsilon_{\mathrm{s}}$ to $\varepsilon_{\mathrm{l}}<\varepsilon_{\mathrm{s}}$. The temperature of this transition, T_{w}, is located in the region of the approach of the dielectric permeability of the liquid to that of the substrate. Quantitative evaluations were obtained on the basis of statistical mechanics taking into account not only the dispersion forces, but also the structural change of the liquid in the field of surface forces of the substrate [15, 16].

It should be noted that Eq. (5) includes, in the general case, several components of the disjoining pressure, such as dispersion, electrostatic and structural forces. In the special case when only dispersion forces are acting, Eq. (5) adopts the following simple form:

$$\cos\theta_0 = 1 + A_{\mathrm{slv}}/12\,\pi h_0^2\,\gamma_{\mathrm{lv}} \tag{7}$$

where $A_{\mathrm{slv}}<0$ is the Hamaker constant, the value of which is determined by Eq. (6).

4.3
Low-energy Surfaces

Some progress in the theoretical prediction of contact angles was achieved in the case of low-energy surfaces, when consideration may be limited on taking into account the dispersion forces only.

In this case, contact angles are formed at $A_{slv} < 0$, when the whole isotherm is located in the region of negative $\Pi(h)$ values, as shown by curve 3 in Fig. 4.3. Here h_0 is the equilibrium thickness of the film, formed as a result of vapor adsorption. The film is stabilized by the action of short-range van der Waals potential. Much thicker films lose stability owing to the realization in this region of the condition $d\Pi/dh > 0$.

The influence of the adsorption film on the contact angle is considered in the Young equation:

$$\gamma_{lv} \cos\theta_0 = \gamma_s + \gamma_{sl} - \pi_s \tag{8}$$

where π_s is surface pressure of an adsorption film and γ_s is the free energy of the solid surface in contact with its own vapor. Adsorption of vapor decreases the surface free energy up to $\gamma_{sv} = \gamma_s - \pi_s$, where π_s is given by

$$\pi_s = RT \int_0^{p_s} \Gamma \, d\ln p \tag{9}$$

where Γ is the isotherm of adsorption, R the gas constant and T the absolute temperature.

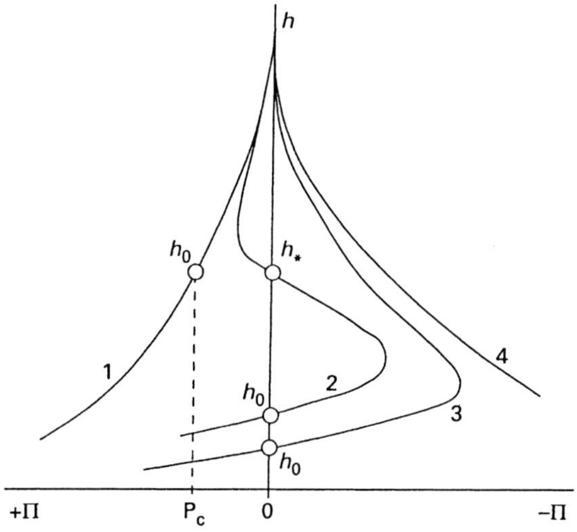

Fig. 4.3 Various isotherms of disjoining pressure of wetting films on a flat substrate.

With poor wetting, that is, with a weak interaction between the liquid and solid substrate, it is possible to neglect the effect of surface pressure. However, in general, the influence of vapor adsorption may be significant. For calculating the values of contact angles by means of Eqs. (8) and (9), besides the isotherm $\Gamma(p)$, the values of γ_s and γ_{sl} must be known.

In the case when the wetting film thickness is small, it is preferable to use, instead of the isotherm of disjoining pressure, the isotherm of vapor adsorption. In this case, on the basis of the Frumkin-Derjaguin approach, the following equation has been derived for the contact angle:

$$\cos\theta_0 = 1 - \left(\frac{RT}{\gamma_{lv}}\right) \int_{\Gamma_0}^{\infty} \ln\left(\frac{p}{p_s}\right) d\Gamma \tag{10}$$

where Γ is the adsorption at relative vapor pressure p/p_s at which the film is in equilibrium with a droplet. When the bubble size is not very small, Γ may be related to $p/p_0=1$. It should be noted that Eq. (10) may be obtained directly from Eq. (5) using the condition of equality of chemical potentials in the vapor and liquid phases.

All the theoretical expressions include equilibrium values of contact angles θ_0. Experimentally, advancing θ_a and receding θ_r contact angles are measured. Hysteresis of contact angles, observed also in the case of flat homogeneous surfaces, might be caused by friction on contact lines or by the formation of projections [3]. As a result, many states of mechanical equilibrium can be formed in the region of contact angles between θ_a and θ_r. In this case, transition to the state of real equilibrium may demand a long time, as reported in the literature [17, 18]. In the case of an easily deformed substrate the effect may be connected with the action of the normal component of force, acting on the line of three-phase contact, $f=\gamma_{lv}\sin\theta_0$.

When comparing measured values with theoretical evaluations of equilibrium contact angles, one needs to take into account that equilibrium values lie between θ_a and θ_r. Often the recommended determination of the equilibrium contact angle as the mean/arithmetic mean is a simplification, which is not sufficiently grounded.

Let us discuss the known results of measurements of advancing contact angles θ_a (Fig. 4.4), conducted using the unity method for different liquids on low energy surfaces [19, 20]. The results obtained for various liquids, from non-polar pentane ($\gamma_{lv}=15.65\times10^{-4}$ N m^{-1}) to water ($\gamma_{lv}=72.75\times10^{-4}$ N m^{-1}), on the hydrophobic surface of Fluorad FC 721 are shown by curve 1. Figure 4.4 shows the dependences of $\cos\theta-1$ and the surface free energy of wetting films $G(h_0)=\gamma_{lv}(\cos\theta_0-1)$ on the surface tension of liquids. The region of non-polar liquids (from pentane to *cis*-decalin) is separated in the Fig. 4.4 by the horizontal broken line. The position of the line corresponds to $G(h_0)=-35\times10^{-3}$ J m^{-2}, which relates to limiting possible values of the Hamaker constant. An increase in the absolute value of $G(h_0)$ in this region corresponds to a corresponding rise

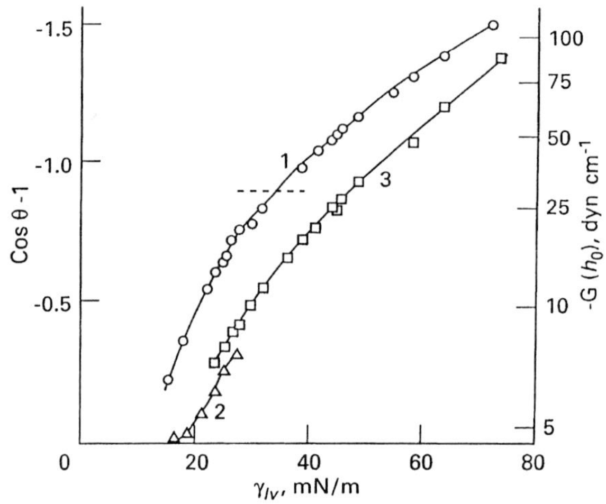

Fig. 4.4 Dependences of equilibrium contact angle values and free energy of wetting film on the surface tension of liquids [19, 20].

in the refractive index of the liquid, from 1.349 to 1.481. It should be noted that data obtained for alkanes (the part of curve 1 below the broken line) differs from that obtained earlier, shown by the curve 2. This means that Teflon possesses a smaller surface energy than the surface, covered with FC 721. The part of curve 1 above the broken line corresponds to more polar liquids [from dimethylformamide (DMF) to water]. In this case, calculations cannot be limited by the theory of dispersion forces only. Here one needs to take into account also polar forces, such as electron-donor (ED) and electron-acceptor (EA) interactions, including hydrogen bonding, which makes a supplementary contribution to the interfacial energy.

Polarization of the interface surfaces as a result of electron transfer leads to a change in orientation of the structure of the liquid near the surface. The structural changes that occur dampen exponentially with departure from the surface with correlation length λ of the order of the correlation length in the bulk of the liquid. Overlapping of boundary layers with changed structure in thin films or interlayers leads to a polar component of the disjoining pressure $\Pi_s(h)$ or structural forces [9]. Both the value and the sign of these forces depend on the value of the dipole moment of the surface dipoles, their surface orientation and density [21]. This approach permits polar forces to be represented as a special case of structural forces and the use for the isotherm of structural forces of the well-known equation

$$\Pi_s = K \exp(-h/\lambda) \qquad (11)$$

where K and λ are parameters that characterize the value and long-range action of polar forces, respectively.

Let us continue the analysis of the values of the contact angles of polar liquids related to the part of curve 1 in Fig. 4.4 above the broken line. In the case of non-polar isolator surfaces, when no inter-phase polarization occurs, the appearance of polar forces is not excluded. Numerical modeling of the liquid structure near neutral surfaces has shown that molecular dipoles in the liquid phase may orient parallel to the surface [22]. A similar structure of boundary layers is formed by dipole liquids, including water, near the boundary with the vapor phase [23–25]. It was shown that a preferred normal orientation of dipoles near surfaces results from overlapping of boundary layers with repulsion forces between surfaces. It may be proposed that parallel orientation of dipoles will result in an inverse effect, namely in attractive forces, as in the case of hydrophobic surfaces in water.

When in thin liquid films, formed on low-energy surfaces, forces of structural attraction arise, Eq. (7) should be supplemented by a polar term:

$$\cos\theta_0 = 1 + (1/\gamma)\,[(A_{slv}/12\,\pi h^2) + K\lambda\,\exp\,(-h_0/\lambda)] = 1 + G(h_0)/\lambda \tag{12}$$

where γ is the surface tension of the liquid. For low-energy surfaces $K<0$ and, as follows from Eq. (11), $\Pi_s<0$.

The increase in negative values of the surface free energy, $G(h_0)$, for polar liquids (corresponding to the top part of curve 1 in Fig. 4.4) might be caused by the appearance of two components of repulsion forces in liquid films, namely dispersion and structural forces. The contribution of the latter component increases with the polarity of a liquid and becomes determining in liquids with hydrogen bonds, e.g. glycerine and water (the two last points on the top of curve 1). Extracting the contribution of dispersion forces to the free energy, $G^d(h_0) = -5\times10^{-11}$ J m^{-2} and using of the Hamaker constant $A_{slv} = -7\times10^{-21}$ J and $h_0 = 2\times10^{-7}$ m, we obtain the polar component of the free energy, $G^p(h_0)$, which is equal to -1.02×10^{-9} J m^{-2}.

Assuming that in the expression for polar component, $G^p(h_0) = K\lambda\,\exp\,(-h_0/\lambda)$, $\lambda = 1$ nm, we obtain $K = -0.08$ Pa, which is close to known values for hydrophobic attractive forces in water. Here was used a λ value equal to the length of correlation in bulk water.

Using the dependence of $G(h_0)$ on surface tension (curve 1) shown in Fig. 4.4, it is possible to assess the $K\lambda$ values also for other polar liquids. The product $K\lambda$ characterizes the polar component of the work of adhesion of liquids to the low-energy FC 721 substrate. The values of $K\lambda$ decrease with the surface tension of liquids, which is proportional to the energy of intermolecular bonds. The increase in energy of intermolecular bonds in liquids results in increased forces of structural attraction and, consequently, increased contact angles.

For non-polar polytetrafluoroethylene (PTFE) with a higher surface energy (curve 3, Fig. 4.4), the absolute value of $G(h_0)$ decreases, demonstrating decreasing structural forces due to the reduced lyophobic nature of the substrate. Therefore, the results shown in Fig. 4.4 might be explained by the joint action of attractive dispersion and structural forces, destabilizing wetting films.

4.4
High-energy Surfaces

The high-energy surfaces of dielectrics and metals, in accordance with the condition discussed above ($a_{sl} > a_{ll}$), are wetted well by non-polar liquids and several polar liquids. In this case, the isotherm of disjoining pressure is disposed in the region of positive disjoining pressure Π (Fig. 4.3, curve 1), which correspond to complete wetting. The isotherm is used in this case for calculations of the wetting film thickness on flat surfaces and profiles of a transition zone between film and meniscus [26–28].

In a slit pore (Fig. 4.2), the radius of curvature of the non-disturbed meniscus 3 is smaller than the half pore thickness $H/2$ on the value of film thickness h_s, which equals

$$h_s = H - r = h_0 + \frac{r}{\gamma_{lv}} \int_{h_0}^{\infty} \Pi(h) \mathrm{d}h \tag{13}$$

where $\Pi(h)$ is the isotherm of the disjoining pressure of a wetting film with thickness h_0 given by the capillary pressure of the non-disturbed meniscus $P_c = \gamma_{lv}/r = \Pi(h_0)$, being in equilibrium with the film.

Calculations have shown that for an isotherm (11) with parameters $K = 10^7$ N m^{-2} and $\lambda = 1$ nm, the ratio H/r depends on relative vapor pressure and equals 1.1 at $p/p_s = 0.99$, 1.5 at $p/p_s = 0.98$, and 1.8 at $p/p_s = 0.97$.

The disjoining pressure and film thickness of non-polar liquids on high-energy surfaces are determined mainly by dispersion forces and depend on the Hamaker constant value $A_{slv} > 0$:

$$\Pi = A_{slv}/6\pi h^3 \tag{14}$$

Calculated on the basis of Eq. (6), the values of the Hamaker constants for wetting films on a quartz surface are equal to 1.7×10^{-20} J for octane, 1.5×10^{-20} J for decane, 1.3×10^{-20} J for dodecane and tetradecane, 0.9×10^{-20} J for benzene, and 0.56×10^{-20} J for carbon tetrachloride. On much more highly energetic metal surfaces values of A rise to 2×10^{-19} J, with corresponding increases also in the film thickness. Thus, for instance, in capillaries with radii of the order of 1 μm the wetting film thickness in equilibrium with the meniscus may reach about 2.5 nm.

However, one needs to take into account that the wetting film thickness on curved surfaces depends not only on surface forces, but also on the curvature of the surface. The film thickness h on a solid surface with constant radius of curvature r is determined by the following equation:

$$[m\gamma_{lv}/(r - h)] + \Pi(h) = (RT/v_m) \ln(p/p_s) \tag{15}$$

where $m = 1$ refer to cylindrical surfaces and $m = 2$ to a film on a spherical surface; R is the gas constant, T is the temperature, v_m is the molar volume of the liquid and γ is its surface tension. The stability of the curved film is determined by the condition $d\Pi/dh < m\gamma_{lv}/(r-h)^2$.

For non-polar liquids, where only dispersion forces are acting and $\Pi = A/6\pi h^3$, capillary condensation occurs with increase in vapor pressure, when the film approaches the critical thickness, which is equal to $h_c = (Ar^2/2\pi\gamma_{lv})^{1/4}$.

On the surface of a hard sphere of radius r, films may preserve their stability in equilibrium with saturated vapor, when their thickness equals $h_0 = (Ar/6\pi\gamma_{lv})^{1/3}$. The films may preserve their stability in supersaturated vapor, but rupture, forming bubbles on the surface, after attaining some mean critical thickness:

$$h_c = c + [c(c + 2r)]^{1/2} \qquad (16)$$

where $c = A/16\pi\gamma_{lv}$.

All the evaluations relate to non-polar liquids, when dispersion forces are acting, depend only on the value of the Hamaker constant.

For water films with a mean thickness of about 10 nm on a wavy glass surface with a mean height of the glass roughness of about 0.75 nm (as was shown in a study [29] using X-ray diffraction analysis), the thickness of water films is greater in holes and smaller in projections. The difference was equal to about 0.42 nm.

In the framework of macroscopic theory [14], surfaces of dielectrics cannot be wetted by liquids which are more polar than the surface. A typical example is non-wetting by liquid mercury of the surfaces of glass and quartz. Advancing contact angles of mercury on quartz lie between 131° and 139°.

By means of Eq. (7), it is possible to assess the gap between mercury and the glass surface. Assuming $\theta_0 \approx 130°$, $\gamma_{lv} = 470$ mN m^{-1} and a Hamaker constant equal to $A_{slv} = -3 \times 10^{-19}$ J, calculated on the basis of spectral data [30], we obtain $h_0 = 0.15$ nm. This result is close to that found for mercury in porous glass from comparison of the glass porosity with the measured amount of mercury imbibed [31].

4.5 Polar Liquids

The prediction of the conditions of wetting and contact angle values for aqueous solutions is more difficult, because in this case, in addition to dispersion, structural and electrostatic forces are also acting. The latter depend on the values of the charges of the film surface and the substrates, regulated by the electrolyte concentration and adsorption of ionic surfactants. Structural forces arise when particles or surfaces covered by boundary layers with a modified water structure come into contact. Overlapping of boundary layers of water results in repulsive forces and these change the free energy of the system. The thickness of boundary layers and the degree of structural changes in the modified zone reflect the intensity of manifestation of structural forces depending on the nature of contact with the substrate.

Different types of isotherms of disjoining pressure are shown in Fig. 4.3. In the region $\Pi > 0$ the isotherms were calculated theoretically and confirmed experimentally. At negative Π values, when films are unstable, which corresponds to the condition $d\Pi/dh > 0$, the isotherms might be calculated only on the basis of well-known equations of the theory of surface forces. Isotherms, represented in analytical form, might be used directly by using Eq. (5), including known expressions for different components of disjoining pressure [7–9, 11].

4.6
Hydrophobic Surfaces

Let us first discuss the influence of dispersion forces on wetting conditions. For this purpose, were performed calculations of the Hamaker constants A for aqueous films on dielectrics and metal surfaces. For water films they were found to be positive and equal, for instance, $A = 10^{-20}$ J on a quartz surface; 2×10^{-20} J on calcite, 3.8×10^{-20} J on sapphire and of the order of 10^{-19} J for wetting films on metal substrates [38, 39]. However, in spite of high A values for metals, which, it seemed, may secure complete wetting, water on a gold substrate formed contact angles up to 30–40° [37]. It was shown [37, 40, 41] that non-complete wetting of a gold surface was connected with hydrophobic impurities. Purification of the gold surface in high vacuum restored complete wetting of gold in water. In contrast to water, the same impurities do not affect complete wetting of metals by non-polar liquids, which wet completely both hydrophilic and hydrophobic parts of a heterogeneous surface.

The high sensitivity of contact angles of aqueous solutions to hydrophobic impurities indicates the important role of structural forces in wetting phenomena. So, for instance methylation of a quartz surface, while not changing electrostatic and dispersion forces, sharply changes the wetting conditions. The contact angles increase up to $\theta_a = 88°$ and $\theta_r = 72°$ when coverage by methyl groups increases to 0.72 [29]. At a low degree of coverage the contact angles decrease, up to complete wetting by water in the absence of methyl groups on the surface. Like methylation, thermal treatment of a quartz surface also diminishes the number of surface hydroxyl groups [30].

The influence of the number of hydrophilic and hydrophobic sites has been investigated in detail by means of surfaces composed of mixed monolayers oriented normal with respect to the substrate, with different proportions of molecules having different numbers of methyl and hydroxyl end-groups [31–33]. At a molar concentration of hydroxyl end-groups of >45%, the external surface of the monolayer start to wet, resulting in the formation on the surface of a compact water monolayer. At a lower molar concentration a rapid increase in contact angles takes place, up to 95–100°. The molecular-dynamics simulation of mixed layers results in analogous conclusions [34]. Analysis of the form of microdroplets, sitting on the outer surface of the monolayer formed, has shown that the contact angles decrease from 135° to 5–17° depending on the molar concentrations of surface methyl and hydroxyl groups. The remarkable action of the hydrophobic sites or inclusions is connected with manifestation of hydrophobic attractive forces. Corresponding to this case the isotherm of surface forces is shown in Fig. 4.3 by curve 3. The effect of hydrophobic attraction is strengthened by simultaneous action of the hydrophobic water–air interface. The hydrophobicity of the water surface in contact with air was confirmed by physical methods using vibrational spectroscopy [35]. It may be supposed that hydrophobicity of the water surface may be caused, as in the case of solid hydrophobic substrates, by a decrease in water density in thin films and orientation of dipoles of water molecules parallel to the surface [23, 25].

The decrease in water density near hydrophobic surfaces results in the well-known effect of slippage of water near the hydrophobic walls, which changes the profile of the flowing liquid. The equation for water flow rate V in a cylindrical capillary, taking into account the slippage effect, takes the following form

$$V = (r^2 \Delta P / 8\eta l)(1 + 4\eta\chi/r) \qquad (17)$$

where r is the capillary radius, ΔP is pressure drop for a capillary length l, η is the viscosity and $\chi = 2V_0 l / r\Delta P$ is the coefficient of slippage.

The effect of liquid slippage was first discovered for mercury in quartz capillaries [36] and the coefficient of slippage was assessed as $(5 \pm 3) \times 10^{-4}$ cm^2 dyn^{-1} s. For water in methylated (hydrophobic) capillaries with radius $r \approx 1$ μm, on the basis of comparative measurements of flow rates of water and non-polar perchloromethane in the same capillary, the value of slippage coefficient χ determined using the above equation (17) was equal to $(3 \pm 1) \times 10^{-4}$ cm^2 dyn^{-1} s [37], which is close to the value obtained earlier for mercury. The values of coefficients of slippage obtained for water were subsequently used for calculations of charges and potentials of hydrophobic thin capillaries ($r \approx 1$ μm) using the modified Helmholtz-Smoluchowski equation, taking into account the effect of slippage [38].

Using molecular dynamics, structural changes of water were investigated using one of the water models (ST2) in a thin (2.5 nm) slit between two hydrophobic walls, the walls containing no sites which are able to interact with water molecules [39]. It was shown that near to such surfaces the density of water decreases and dipoles of water molecules are oriented preferably parallel to the surface. More than 25% of intermolecular bonds in boundary layers are destroyed.

Later, using the Monte Carlo method, comparative calculations of density profiles in an interlayer of 3.6 nm thickness between two hydrophilic and two hydrophobic surfaces showed [40] that in the first case density oscillations were observed, which damped on the surface, and in the second case the water density decreased slowly approaching the wall.

Oscillations of density in boundary layers of water near hydrophilic surfaces have received direct experimental confirmation using new apparatus for direct measurements of forces between two crossed glass cylinders covered with mica [41]. In addition, the applicability of Eq. (11) was shown for averaging repulsive oscillating forces and the incoming parameters were determined as $K = 3 \times 10^6$ N m^{-2} and $\lambda = 1$ nm.

These results were subsequently substantiated theoretically, by the calculations of oscillating forces of hydrophilic repulsion using statistical mechanics with the Percus-Jewic approximation for correlation function [42]. In the calculations performed, water molecules were modeled as dipole hard spheres and electrolyte solutions were presented as point ions.

Also taken into account are forces caused by the excluded volume and dipole orientation. This results in oscillating density profiles in boundary layers of water having a thickness of about 2 nm and a period of oscillation of the order of the molecular size (0.28 nm). It was shown that the results obtained are in agreement with those obtained earlier [41]. In the same work were measured the forces of attraction of mica surfaces, hydrophobized by adsorption of cetyltrimethylammonium bromide (CTAB) in 10^{-3} M aqueous solutions of NaCl and KBr. The forces decrease exponentially with correlation length $\lambda = 1$ nm and $K \approx 10^7$ N m^{-2} in Eq. (7). Therefore, forces of hydrophobic attraction are described in the same way by Eq. (11) with a similar coefficient K, but with the opposite sign.

Adsorption of non-anionic surfactants on solid hydrophilic surfaces leads to worsening of wetting because of binding of surfactant hydrophilic groups with the surface and orientation of molecule hydrophobic moieties in the aqueous phase. Further, adsorption of molecules also proceeds on the outer film surface in contact with air. This leads to hydrophilization, which may decrease hydrophobic attraction.

In Table 4.1 are collected parameters extracted from several experimental studies. They may be useful at last for preliminary discussion.

The degree of hydrophobicity is characterized here by the values of contact angles. The $-K\lambda$ values give the work of adhesion W_a of water to a hydrophobic surface. The isotherm of disjoining pressure, expressed by Eqs. (11) and (12),

Table 4.1 Values of parameters.

θ_a (°)	$-K \times 10^2$ (Pa)	λ (nm)	$-K\lambda \times 10^{-7}$ (N m^{-1})	Surface
65	2	1	26	mica
94	4.8	1.25	60	mica
95	4	1.4	56	mica
97	3.6	2.25	80	mica
100	3	3	90	glass
113	4.8	2.5	120	mica

leads to the following equation for the value of the equilibrium contact angle formed by water with a hydrophobic surface:

$$\cos\theta_0 = 1 + G_s(h_0)/\gamma_{lv} = 1 + K\lambda/\gamma_{lv} \tag{18}$$

where $G_s(h_0) = K\lambda \exp(-h_0/\lambda)$ is the excess free energy of wetting of a hydrophobic surface. Because of a wetting equilibrium, the film thickness is very small, so a simplification $\exp(h_0/\lambda) = 1$ has been used for deriving Eq. (18). This makes Eq. (18) similar to some known relations, including that for the work of adhesion [1].

However, later, more long-range forces between hydrophobized surfaces with characteristic length from 13 to 62 nm were discovered. They are changed by the usual short-range forces for the case of approaching hydrophobized surfaces in water.

Figure 4.5 shows as an example the results of direct measurements of forces of hydrophobic attraction in water between two crossed mica cylinders, the surfaces of which were hydrophobized using adsorption layers of a fluorocarbon surfactant [43]. The ordinate shows measurements of attraction force F, related to the cylinder radius R, on a logarithmic scale as a function of the distance between the surfaces.

The dependence obtained can be approximated by two linear parts with different slopes, corresponding to a characteristic distance $\lambda_d \approx 16$ nm for distances more than 20 nm and more short-range forces and $\lambda = 2$–3 nm for small distances. The second region is shown in the inset of Fig. 4.5, demonstrating that the dependence is really exponential. Similar results were obtained much later [43–45], where it was shown that the structure of the boundary layers is changed

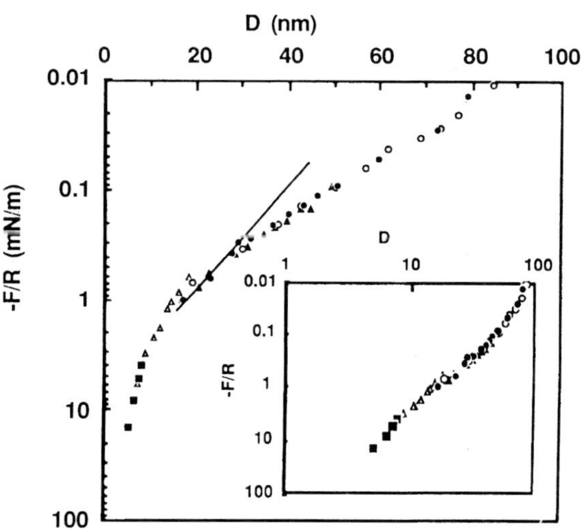

Fig. 4.5 Results of direct measurements of interaction forces for hydrophobic surfaces of crossed cylinders versus distance between them [43].

at distances up to 10–15 nm from the surface. These and other results led to the conclusion that these short-range forces acting between two hydrophobic surfaces are reversible and correspond to structural changes in water boundary layers near to hydrophobic surfaces.

The nature of the long-range forces of hydrophobic attraction deserves separate discussion.

Recently, it has been confirmed that linkage of hydrophobic surfaces in water is caused by merging of air nano-bubbles, existing on them before contact. When surface contact occurs, blisters merge with rupture of capillary bridges and overcoming the originating capillary forces is necessary for the subsequent separation of surfaces.

It has been shown that in the case of hydrophobic surfaces in degased water, a capillary bridge filled with vapor appears in the region of contact of approaching cylinders or the spheres applied as modeling bodies in devices for direct measurement of surface forces. On approaching of surfaces they jump into a state of contact [46–48]. For separation of hydrophobic surfaces from each other in this case, overcoming of the capillary pressure of meniscuses restricting bridges is needed. The reality of this effect is demonstrated by the apparent decrease in long-range hydrophobic forces after degassing of water [49, 50].

It has been shown [51] that a correction for the effect of slippage of water relative to a surface for correct measurement of the attractive force of hydrophobic surfaces in water is necessary.

4.7 Hydrophilic Surfaces

Let us now consider the forces acting between hydrophilic surfaces in water and aqueous solutions. The action of several components of surface forces can be manifested simultaneously, which makes the situation more difficult and varied.

First, the effect of most long-range electrostatic forces caused by interaction of charged surfaces of the substrate and the wetting film through an aqueous layer must be considered. With good wetting of surfaces, the action of structural forces of the hydrophilic repulsion originating from overlapping of approaching boundary layers of water with the changed structure become appreciable on thinning of an aqueous layer. These changes depend on the number, activity and way of arrangement of the surface sites interacting with molecules of water. Further, the action of dispersion forces must be considered in all cases. In aqueous wetting films on emulsion solid carriers less polar than water, dispersion forces are attractive ones that degrade the stability of films.

In the case of hydrophilic surfaces, structural forces are described by the same exponential law (Eq. 11), but values of the parameter K, unlike hydrophobic surfaces, are positive. The characteristic length λ is close to the length of correlation in bulk water and is 1–2 nm. The disjoining pressure of films can be calculated using the known equation following from the condition equality of the chemical potentials of water in vapor and liquid phases: $\Pi(h) = -(RT/v_m) \ln (p/p_s)$.

Adsorption isotherms of water vapor on surfaces of quartz with various extents of wetting ability have been obtained [52]. At full coverage of the surface by hydroxyl groups, the isotherm $\Pi(h)$ is well described by Eq. (11) with parameters $K=0.2$ Pa and $\lambda=1.5$ nm. With a smaller extent of a hydroxylation of surfaces the contribution of electrostatic and dispersion forces to the isotherm is also manifested. As these forces, unlike structural forces, can be attractive forces in this case, partial passing of the $\Pi(h)$ isotherm negative values of the disjoining pressure become possible, which results in finite contact angles. Hence the isotherm obtained in [52] at a contact angle $\theta=40°$ is well described by DLVO theory, i.e. with action only of dispersion and electrostatic forces.

In addition to adsorbed films, studies of the isotherm of a disjoining pressure of wetting films formed in a Teflon cell on a solid plane substrate by the method of drawing off a liquid from a cell [53] have also been carried out. The diagram of the cell is shown in Fig. 4.6. The film is formed on arrival of a meniscus of a liquid in a cylindrical tube 1 to a polished plate 2 as a result of a pressure pumped down in a cell 3 by the liquid level decrease in the mobile vessel 4 connected by a flexible tube with the cell. The liquid is sucked out from the cell through the thin slot between the plate and butt-end tubes.

The disjoining pressure $\Pi=\rho g H$ is determined by the difference in fluid levels H in a vessel and in a film. In this expression, ρ is the density of the liquid and g is the acceleration due to gravity. The thickness of the equilibrium plane wetting film is measured by methods of optical interference or ellipsometry.

In Fig. 4.7, results of measurements of the thickness dependence of films on the disjoining pressure Π by ellipsometry are shown for the simplest case of films of non-polar tetradecane on a mica surface where the basic contribution to film stability is made by dispersion forces. As in this case, according to the

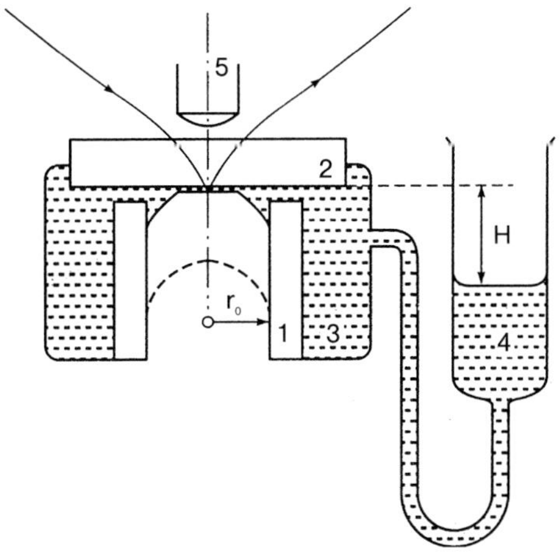

Fig. 4.6 Scheme of cell to obtain the isotherms of disjoining pressure of wetting films.

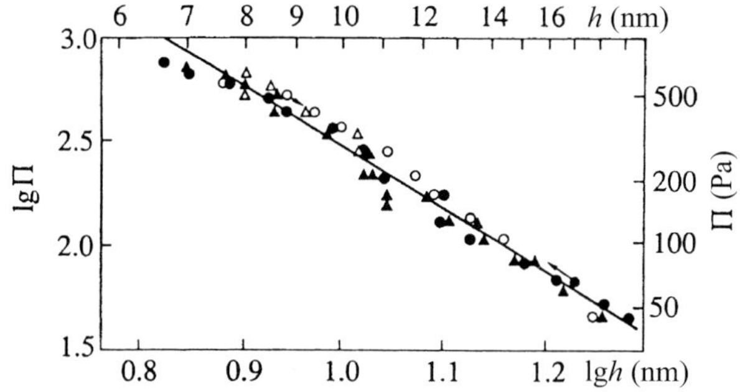

Fig. 4.7 The isotherm of disjoining pressure of wetting films of *n*-tetradecane is shown in logarithmic coordinates.

theory, the isotherm follows the exponential law $\Pi \sim h^n$, the obtained dependence of film thickness h on disjoining pressure Π is plotted in logarithmic coordinates. As the thickness of films does not exceed 20 nm, Eq. (6) of the theory of non-retarded forces $\Pi = A/6ph^3$ can be used for interpretation of the data obtained. The slope of the linear relation obtained in Fig. 4.7 leads to a value $n = 2.95 \pm 0.07$, which is close to the theoretical value of $n = 3$. The corresponding value of the Hamaker constant is $A = 5.5 \times 10^{-21}$ J, which coincides with the theoretical calculation [7].

For thicker wetting films of tetradecane on a steel surface, the good agreement of isotherms with the theory of retarded dispersion forces, when the disjoining pressure diminishes more sharply with increase in film thickness, has been achieved and the isotherm of a disjoining pressure has the form $\Pi = B/h^4$. In this experiment, value of the constant $B = 1.6 \times 10^{-24}$ J m, close to the theoretical value (1.75×10^{-24} J m) has been obtained. Thus, for films of a non-polar liquid on surfaces of dielectrics and metals the isotherms of disjoining pressure are in good agreement with the theory of dispersion forces.

The situation looks more difficult in the case of wetting films of aqueous solutions. In this case, in addition to dispersion forces, the influence of the electrostatic forces corresponding to the charge of the film surfaces and substrate and also the structural repulsive force arising on approaching and overlapping of boundary layers of water with the modified structure must be considered.

The electrostatic component of the disjoining pressure of wetting films under the condition of constancy of electrical potentials ψ_1 and ψ_2 of surfaces of the substrate and film, respectively, can be calculated using the equation

$$\Pi_\psi = \varepsilon\varepsilon_0(2\psi_1\psi_2 \cos \kappa h - \psi_1^2 - \psi_2^2)/2 \sinh^2 \kappa h \qquad (19)$$

where ε_0 is the dielectric constant, ε is the dielectric permittivity, κ is the inverse Debye length and h is the thickness of the wetting film. The same expression can also be used for the case of constancy of surface charge σ of a film by re-

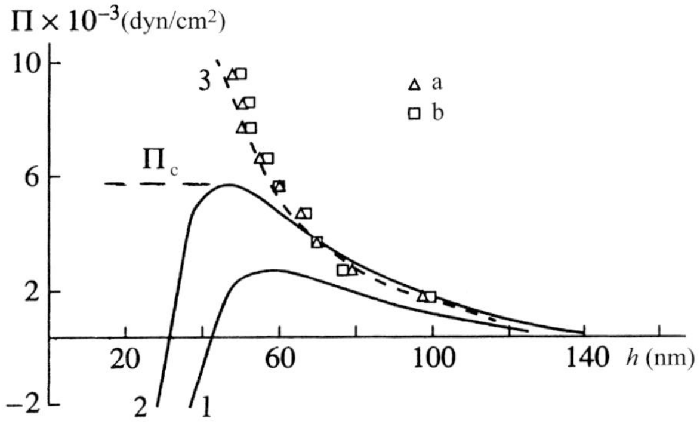

Fig. 4.8 An isotherm of disjoining pressure, $\Pi(h)$, of wetting films for aqueous solution of 10^{-5} M NaCl on the surface of quartz (points). Solid curves are results of calculations of electrostatic forces under conditions of constancy of potentials $\psi_1 = -150$ mV and $\psi_2 = -25$ mV (curve 1) and −35 mV (curve 2). Curve 3 was calculated under the condition of constancy of charges.

placing the ψ_1 potential of the substrate by $\psi_{1\infty}$, the potential of the surface of the substrate at $h \to \infty$.

Figures 4.8 and 4.9 show the dependences of the thickness of wetting films of aqueous solution of NaCl on their disjoining pressure. The points are the experimental data and the solid curves are the theoretical isotherms obtained under the condition of constancy of the potential of the substrate, $\psi_1 = -150$ mV, and the potentials of the film surface, $\psi_2 = -25$ mV (curve 1), and also $\psi_1 = -125$ mV and $\psi_2 = -35$ mV (curve 2). The dotted curve 3 in Fig. 4.8 was obtained under the condition of constancy of the charge of the film surfaces. The potentials ψ_1 were measured independently using the electrokinetic method in quartz capillaries [54]. The accepted range of values of ψ_2 corresponds to known experimental data for foam films of the same composition. Results of calculations under the condition of constancy of surface charge are shown in Fig. 4.9 by dotted curves. The experimental points measured for a concentration of 10^{-4} M in Fig. 4.9 are in accordance with this condition for thickness $h > 60$ nm.

The potential of ion adsorption was high enough that the electric field of a film free surface could not change the value of the charge of a substrate, i.e., the charge of the substrate surface was constant. For the dissociation mechanism of the formation of a constant charge, for example in the case of a quartz, a high potential of dissociation of surface OH groups is necessary.

As shown in Figs. 4.8 and 4.9, at large thicknesses of solution films obtained under conditions of constancy of potential and of charge coincide. Deviations of the experimental data from theoretical values (with ψ=constant) begin to be manifested on approaching the critical disjoining pressure $\Pi_c = (\varepsilon\varepsilon_0\kappa^2/2)\psi_2^2$ corresponding to loss of stability electrolyte film due to a change of sign of electrostatic forces [7]. If the condition of constancy of potentials is fulfilled, the film

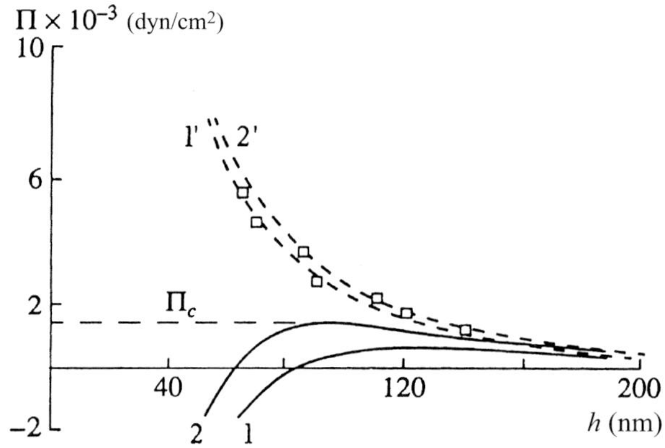

Fig. 4.9 Isotherms of disjoining pressure, $\Pi(h)$, of wetting films for aqueous solutions of 10^{-4} M NaCl (1) and 10^{-4} M KCl (2) on the surface of quartz. Solid curves are results of calculations of electrostatic forces under the condition of constancy of potentials $\psi_1 = -150$ mV and $\psi_2 = -25$ mV (curve 1), $\psi_1 = -125$ mV and $\psi_2 = -35$ mV (curve 2). Results of calculations under condition of a constancy of charges are shown by curves 1′ and 2′.

would be ruptured on approaching the critical pressure Π_c. Conservation of stability of films is caused by the implementation, in this case, of another boundary condition, namely constancy of charges of film surfaces. In accordance with the theory, films are stable and their disjoining pressure increases with decrease in film thickness.

At higher concentrations of an electrolyte (Fig. 4.9), the divergence between the dotted experimental curves and theoretical calculations shown by solid curves 1 and 2 increases. The experimental data, shown by points, are in better agreement with the results of calculations made for the condition of constancy of charge (dotted curves). Employment of two different potentials of a film surface ψ_2 (−25 and −35 mV) affected slightly the results of calculations.

4.8 Methods of Control of Surface Wetting

Most popular methods are chemical and adsorption modifications of a surface. In the first case, formation of chemical bonds between modifying agents and surface centers or molecules of the surface layer of a solid substrate (for example, methylation) is used. This method is usually applied for hydrophobization of surfaces and contact angles depend on the extent of methylation. The contact angles increase with extent of methylation m of the surface of quartz, approaching 30–40° at $m = 20\%$, 50–60° at $m = 60\%$, and approximately 80–90° at full methylation of a surface.

For the adsorption modification of a surface with the purpose of changing the conditions of wetting, surfactants are used. With increasing concentration of non-ionic Triton X-100 in aqueous solution up to 10^{-4} M there was an increase in contact angle of the surface of quartz from an initial 4° up to 33°, and to 5° with further concentration increase up to 1.6×10^{-3} M. The same effect has been detected for the longer chain Triton X-305. A similar result was observed for particles of quartz in aqueous solutions of non-ionic TN 101 containing 10 oxyethyl (EO) groups in a molecule. The initial increase in contact angle is explained by adsorption of the hydrophilic part of the molecules of a surfactant on the hydrophilic sites of the surface. Thus, the adsorbed molecules are oriented by their hydrophobic fragments in a solution. At saturation of the hydrophilic sites, further adsorption occurs owing to the formation of a second adsorbed layer on the hydrophobic tails of the first layer. For aqueous solutions of $C_{12}EO_8$, the contact angles on the surface of quartz from 15° up to 20° passing through a poorly expressed maximum near the critical micellar concentration (CMC).

Thus, solutions of non-ionic surfactants of low concentration can be used for some reduction in the wetting ability of the surface of quartz.

However, with a higher molecular weight surfactant, an increase in its concentration leads to only a gradual decrease in contact angles to 20° on approaching the CMC in the case of the hydrophobic surface of quartz having an initial contact angle of 90°. In this case the formation of a second adsorbed layer did not occur. The same result was also obtained for non-ionic Sintamid-5. The values of the static contact angle measured by the method of a sessile drop on a plane quartz surface hydrophobized with trimethylchlorosilane decreased from 103° to 42° on approaching the CMC and subsequently remained constant.

An ionogenic surfactant added to water has a much stronger influence on the conditions of wetting. The widest application for wetting control is provided by cationic surfactants since the surface of the majority of inorganic natural and construction materials are usually negatively charged. Electrostatic interaction results in surface hydrophobization due to the orientation of molecules of a cationic surfactant by hydrophobic groups inside the solution. In contrast, adsorption of an anionic surfactant from aqueous solutions, as shown by experiments [55], increases the thickness of wetting films with increasing negative charge of the surface of quartz, which improves wetting.

As an example illustrating the influence of cationic surfactants, the concentration dependences of the potentials of surfaces of wetting films ψ_1 (curve 1) and ψ_2 (curve 3) on the surface of quartz are shown in Fig. 4.10. Curve 2 presents for comparison results of measurements of ψ_1 for a surface of paraffin. Figure 4.11 presents results of measurements of the advancing angle θ_a for aqueous solutions of CTAB in the presence of 10^{-4} KCl [56]. The ψ_1 potentials of quartz surfaces have been measured by the method of capillary electrokinetics [56]. The ψ_2 potentials of the free surface of a film were calculated using the results of measurements of the thickness of solutions of free films [57].

Advancing contact angles were measured in quartz capillaries by extrapolation of the linear section of the dependence of the displacement rate of a solution

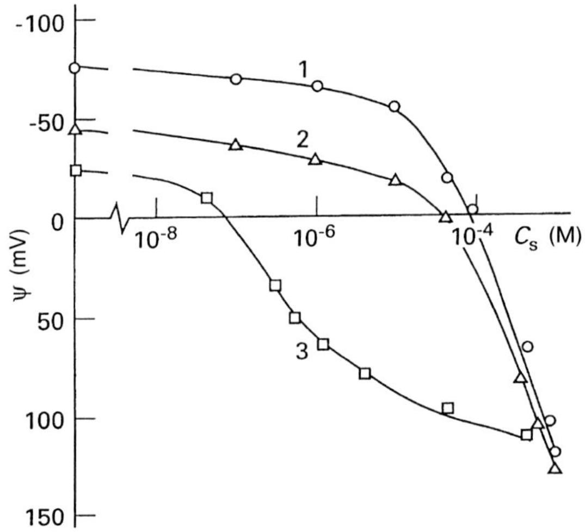

Fig. 4.10 Dependences of ψ_1 potentials for quartz (curve 1) and paraffin (curve 2) surfaces and ψ_2 potential for solution–air interface versus concentration C_s of CTAB solutions.

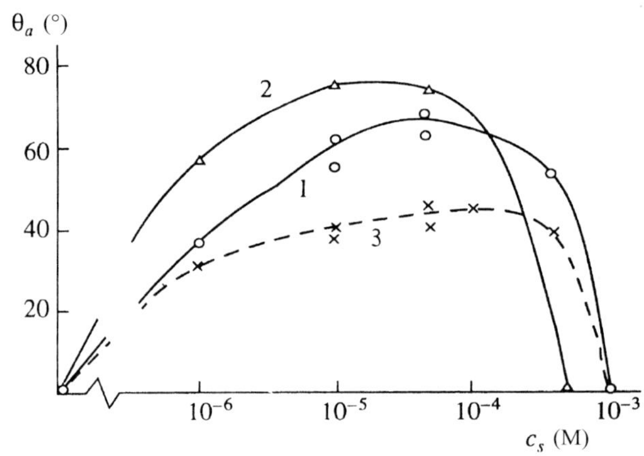

Fig. 4.11 Dependences of advancing contact angle θ_a on concentration of CTAB solutions. Advancing angles are obtained by pressure measurements at zero displacement velocity of the meniscus of a solution (curve 1) in quartz capillaries and at beginning of motion of a gas bubble in the same capillary (curve 2). For comparison, results of measurements of static values of contact angle for gas bubbles on a flat quartz surface are shown by curve 3.

meniscus in a capillary on pressure drop on a zero rate (curve 1) [58] and by measurement of the pressure drop corresponded to the beginning of motion of a gas bubble in the same capillary (curve 2) [59]. For comparison, curve 3 presents results of measurements of static values of contact angles for gas bubbles on a flat surface of quartz. The dependences of the contact angles on the concentration of surfactant on mica and on quartz of 10^{-6} M KCl show the same shape.

For a background solution (10^{-4} M KCl) in the absence of surfactant and at a surfactant concentration of 10^{-3} M, close to the CMC, complete wetting occurs. This indicates that in both the first and second cases the film surfaces bounded by quartz and air have high potentials of the same sign. In a background electrolyte solution they are both negative ($\psi < 0$) and at the CMC they are both positive ($\psi > 0$). The values of the contact angles are maximum at a surfactant concentration of about 10^{-5} M [56]. In this region of concentration the potentials of the surface of the wetting film and the substrate have opposite signs, which leads to the appearance of forces of electrostatic attraction, destabilizing the film. At the same time, an increase in contact angles occurs.

The maximum value of the equilibrium contact angle, θ_0, is estimated theoretically to be 50°. This value is higher than that of the advancing contact angle (curves 1 and 2, Fig. 4.11), but is close to values of the static contact angle shown in Fig. 4.11 (curve 3). Such dependences of contact angles on the concentration of CTAB solutions having a maximum have been obtained in many studies. A correlation of the dependences of contact angle values on concentration of CTAB solutions with floatability of particles of quartz, calcite, kaolinite, synthetic lime carbonate [60] and pyrite [61] has been detected.

Consider now the opportunities for surface wetting control by changes in electrolyte composition. For example, an increase in 1:1 electrolyte concentration and also a decrease in solution pH result in a decrease in contact angles on a quartz surface mainly owing to a decrease in the values of the ψ_1 potentials and a corresponding decrease in the force of electrostatic repulsion of the surfaces of the wetting films [52, 62].

For the prediction of the conditions of surface wetting, Eq. (12) with a term added to account for influence of electrostatic forces on the value of the contact angle can be used. As a first approximation, sufficient for choosing ways of changing the contact angles, the following expression for the equilibrium contact angle, including three basic surface forces, dispersion, electrostatic and structural, can be used:

$$\cos\theta = 1 + (1/\gamma)\,\{A_{\text{slv}}/12\,\pi h^2 - [\varepsilon\varepsilon_0(\psi_1 - \psi_2)^2/2h] + K\lambda\,\exp(-h/\lambda)\} \qquad (20)$$

where γ is the surface tension of the liquid, A_{slv} is the Hamaker constant for a solid–liquid film–gas system calculated using the theory of dispersion forces, ε_0 is the dielectric constant, ε is the dielectric permittivity, κ is the inverse Debye length and h is the thickness of the wetting film; ψ_1 and ψ_2 are the potentials of the surface of the substrate and the film, respectively, and K and λ are parameters of the isotherm of structural forces.

This equation indicates that values of contact angles can be controlled mainly by the modification of the potentials of the surfaces of the substrate and the film, for example, using adsorbed surfactant. Thus, for speeding up the spontaneous water spreading on oxidized silicon, the commercial non-ionic surfactant Trisiloxane was used [63]. At a relative humidity of air $\varphi < 30\%$, normal spreading occurred when the radius of the base of a drop depended linearly on $\tau^{0,1}$, where τ is time. With increase in φ the spreading is speeded up, but the front of wetting becomes unstable owing to the Marangoni effect. The highest rate of spreading of Trisiloxane surfactant on the hydrophobic surface of silicon is attained on approaching the boundary of miscibility when solutions of the surfactant are dispersions containing vesicles [64]. The maximum rate of spreading of Trisiloxane surfactant is observed for non-polar or low-polarity solid surfaces [65]. The initial velocities of spreading of the Trisiloxane surfactant correlate with the temperature of phase transition [66]. Aqueous solutions of TMS with small molecular hydrophilic ring do not spread on hydrophobic silicon surfaces. With increasing number of ethyl groups up to $n=4$ the rate of spreading increases. Micellar solutions ($n=6$) spread out slowly or stop spreading out in a few seconds [67].

Hence there is a good opportunity for the control of wetting of surfaces on the basis of Eq. (20) by changing the composition of the solution and adsorption modification of the surfaces, changing the contribution of various components of surface forces. Choosing the best way for control the wetting of surfaces depends, naturally, on the conditions and properties of the initial system.

Acknowledgment

This work was supported by the Russian Foundation for Basic Research, Project 05-03-32979.

References

1 Summ B. D., Goryunov Yu., *Fizikokhimicheskie osnovy smachivaniya i rastekaniya (Physicochemical Bases of Wetting and Spreading)*. Khimiya, Moscow, **1976**.
2 Rusanov A. I., *Kolloid. Zh.*, **1975**, *37*, 678.
3 Shanahan, M. E. R., de Gennes P. G., in *Adhesion, Vol. 11. Papers Presented at the 24th Annual Conference on Adhesion and Adhesives*, London, **1986**, p. 11.
4 Rusanov A. I., *Mendeleev. Commun.*, **1996**, 30.
5 Berenshtein G. V., D'jachenko A. M., Rusanov A. I., *Kolloid. Zh.*, **1985**, *47*, 9.
6 Derjaguin B. V., Starov V.M., Churaev N. V., *Kolloid. Zh.*, **1982**, *44*, 871.
7 Derjaguin B. V., Starov V. M., Churaev N. V., *J. Colloid Interface Sci.*, **1982**, *89*, 16.
8 Derjaguin B. V., Churaev N. V., *Smachivayushchie Plenki (Wetting Films)*. Nauka, Moscow, **1984**.
9 Derjaguin B. V., Churaev N. V., Muller V. M., *Surface Forces*. Consultants Bureau/Plenum Press, New York, **1987**.
10 De Feijter J. A., Vrij A., *Electroanal. Chem. Interfacial Electrochem.*, **1972**, *37*, 9.

11 Churaev N.V., *Liquid and Vapor Flows in Porous Bodies Surface Phenomena, Surface Phenomena, Topics in Chemical Engineering*, Vol. 13. Gordon and Breach, New York, **2000**.
12 Sikkenk J.N., Indeken J.O., van Leeuwen J.M.J., Vossneck E.O., Bakker A.F., *J. Stat. Phys.*, **1988**, *52*, 23.
13 Nijmeijer M.J.P., Bruin C., Bakker A.F., *J. Phys.: Condens. Matter*, **1992**, *4*, 15.
14 Dzyaloshinskii I.E., Lifshits E.M., Pitaevskii L.P., *Zh. Eksp. Teor. Fiz.*, **1959**, *37*, 229.
15 de Gennes P.G., *Rev. Mod. Phys.*, **1985**, *57*, 827.
16 Percus J.K., *J. Stat. Phys.*, **1987**, *47*, 801.
17 Martynov G.A., Starov V.M., Churaev N.V., *Kolloid Zh.*, **1977**, *39*, 472.
18 Bayramli E., van de Ven T.G.M., Mason S.G., *Colloids Surf.*, **1981**, *3*, 279.
19 Li D., Neumann A.W., *J. Colloid Interface Sci.*, **1992**, *148*, 190.
20 Li D., Xie M., Neumann A.W., *Colloid Polym. Sci.*, **1993**, *271*, 573.
21 Belaya M.L., Feigelman M.V., Levadny V.G., *Langmuir*, **1986**, *3*, 648.
22 Eggebrecht J., Tompson S.M., Gubbins K.E., *J. Chem. Phys.*, **1987**, *86*, 2299.
23 Brodskaya E.N., Rusanov A.I., *Kolloid. Zh.*, **1986**, *48*, 3.
24 Yang D., Sullivan D.E., Tjipto-Margo B., Gray C.G., *J. Phys.: Condens. Matter.*, **1991**, *3*, 109.
25 Il'in V.V., Khryapa V.M., Churaev N.V., *Fiz. Mnogochastichnykh Sistem*, **1991**, *18*, 50.
26 Zorin Z., Platikanov D., Kolarov T., *Colloids Surf.*, **1987**, *22*, 147.
27 Kolarov T., Zorin Z., Platikanov D., *Colloids Surf.*, **1990**, *51*, 37.
28 Churaev N.V., *Rev. Phys. Appl.*, **1988**, *23*, 975.
29 Crawford R., Koopal L.K., Ralston J., *Colloids Surf.*, **1987**, *27*, 57.
30 Zhuravlev L.T., *Colloids Surf.*, **1993**, *74*, 71.
31 Hato M., Minamikawa H., Okamoto K., *Chem. Lett.*, **1991**, *6*, 1049.
32 Ulman A., Evans S.D., Shnidman Y., *J. Am. Chem. Soc.*, **1991**, *113*, 1499.
33 Ulman A., Evans S.D., Sharma R., *Thin Solid Films*, **1992**, *211*, 810.
34 Hautman J., Klein M.L., *Phys. Rev. Lett.*, **1991**, *67*, 1763.
35 Du O., Freysz E., Shen Y.R., *Science*, **1994**, *264*, 826.
36 Tolstoj D.M., *Dokl. Akad. Nauk SSSR*, **1952**, *85*, 1329.
37 Churaev N.V., Sobolev V.D., Somov A.N., *J. Colloid Interface Sci.*, **1989**, *97*, 1089.
38 Muller V.M., Sergeeva I.P., Sobolev V.D., Churaev N.V., *Kolloid. Zh.*, **1986**, *48*, 606.
39 Sonnenschein P., Heinzinger K., *Chem. Phys. Lett.*, **1983**, *102*, 550.
40 Forsman J., Woodward C.E., Jonsson B., *Langmuir*, **1997**, *13*, 5159.
41 Israelachvili J.N., Pashley R.M., *Nature*, **1983**, *306*, 249.
42 Trokhimchuk F., Henderson D., Wasan D.T., *J. Colloid Interface Sci.*, **1999**, *210*, 320.
43 Hato M., *J. Phys. Chem.*, **1996**, *100*, 18530.
44 Claesson P.M., Herder P.C., Blom C.E., Ninham B.W., *J. Colloid Interface Sci.*, **1987**, *118*, 68.
45 Kekicheff P., Spala O., *Langmuir*, **1994**, *10*, 1584.
46 Ishida N., Inone T., Miyahara M., *Langmuir*, **2000**, *16*, 6377.
47 Ishida N., Sakamoto M., Miyahara M., Higashitani K., *Langmuir*, **2000**, *16*, 568.
48 Yakubov G.E., Butt H.J., Vinogradova O.I., *J. Phys. Chem. B*, **2000**, *104*, 3407.
49 Meager L., Craig V.S., *Langmuir*, **1994**, *10*, 2731.
50 Craig V.S., Ninham B.W., Pashley R.M., *Langmuir*, **1999**, *15*, 1562.
51 Vinogradova O.I., Horn R.G., *Langmuir*, **2001**, *17*, 1604.
52 Gee M.L., Healy T.W., White L.R., *J. Colloid Interface Sci.*, **1990**, *140*, 450.
53 Churaev N.V., *Kolloid. Zh.*, **2000**, *62*, 581.
54 Sergeeva I.P., Sobolev V.D., Churaev N.V., *Kolloid. Zh.*, **1981**, *43*, 744.
55 Ershov A.P., Zorin Z.M., Churaev N.V., *Kolloid. Zh.*, **1995**, *57*, 329.
56 Sergeeva I.P., Sobolev V.D., Churaev N.V., *Kolloid. Zh.*, **1990**, *52*, 972.
57 Exerova D., Churaev N.V., Kolarov T., *Adv. Colloid Interface Sci.*, **2003**, *104*, 1.
58 Churaev N.V., Ershov A.P., Zorin Z.M., *J. Colloid Interface Sci.*, **1996**, *177*, 589.
59 Churaev N.V., *Kolloid. Zh.*, **1994**, *56*, 707.

60 Kydros K., Matis K., Stalidis G., *J. Colloid Interface Sci.*, **1993**, *155*, 409.

61 Drzymala J., *Adv. Colloid Interface Sci.*, **1994**, *50*, 143.

62 Sergeeva I. P., Sobolev V. D., Madzharova E. A., Churaev N. V., *Kolloid. Zh.*, **1995**, *57*, 805.

63 Cachile M., Cazabat A. M., *Langmuir*, **1999**, *15*, 1515.

64 Wagner R., Wu Y., Czichocki G., Berlepsch H., *Appl. Organomet. Chem.*, **1999**, *13*, 201.

65 Wagner R., Wu Y., Berlepsch H., *Appl. Organomet. Chem.*, **2000**, *14*, 177.

66 Wagner R., Wu Y., Berlepsch H., *Monatsh. Chem.*, **1999**, *130*, 237.

67 Wagner R., Wu Y., Berlepsch H., *Appl. Organomet. Chem.*, **1999**, *13*, 845.

5
The Intrinsic Charge at the Hydrophobe/Water Interface

James K. Beattie

5.1
Introduction

There is accumulating evidence that water at hydrophobic surfaces acquires a negative charge at neutral pH by the preferential adsorption of hydroxide ions. This occurs whatever the hydrophobic surface, be it gas (air), liquid (oil) or solid (e.g. Teflon). Qualitative evidence for this effect has been reported for decades, mainly from the pH dependence of the zeta potential. In 1969, Exerowa used measurement of the diffuse double layer potential of surfactant-free films to estimate that their isoelectric point was at pH 4.5 and hence that the difference in the adsorption energies of the hydroxide and hydronium ions is of the order 7 kcal mol^{-1} [1].

Since then, many other measurements with different techniques on different systems have confirmed an isoelectric point for various hydrophobe/water interfaces around pH 3–4. In the following, we summarize the evidence from these reports, offer an explanatory hypothesis to account for the observations and indicate some of the outstanding questions that remain.

5.2
Oil Droplets

Already in 1941, Dickinson wrote, "It has long been known that droplets of oil, dispersed in water, carry an electrical charge which is usually negative. This charge is thought to be due to the adsorption of hydroxyl ions and gives rise to an electrical double layer across the interface" [2]. Three years earlier, his Liverpool colleague Carruthers had shown that the electrophoretic mobilities of solid octadecane particles and liquid octadecane droplets were almost identical, negatively charged above an isoelectric point of about pH 2.5 in 0.01 M NaCl [3]. The pH dependence of the electrophoretic mobility, which increases from the isoelectric point to a limiting value above pH 8–10, suggests that hydroxide ion is responsible for the negative charge.

Colloids and Interface Science Series, Vol. 2
Colloid Stability: The Role of Surface Forces, Part II. Edited by Tharwat F. Tadros

ISBN: 978-3-527-31503-1

The matter was almost neglected over the next half-century [4] until the Sofia group revived interest in the phenomenon with a seminal paper in 1996 by Marinova et al. titled "Charging of Oil–Water Interfaces Due to Spontaneous Adsorption of Hydroxyl Ions" [5]. These authors made a number of important observations. They found that the zeta potential was independent of the nature of the oil for hexadecane, dodecane, xylene and perfluoromethyldecalin, concluding that "the magnitude of the surface charge depends mostly on the composition of the aqueous phase, while the nature of the oil phase is of secondary importance." The zeta potential decreased (became less negative) as the aqueous NaCl electrolyte concentration was increased from 0.1 to 10 mM, consistent with typical double-layer compression and not with adsorption of chloride ions. They excluded the adsorption of hydrogen carbonate and carbonate ions by making measurements in 10^{-3} M Na_2CO_3 solution at pH 9.8 and finding no difference in the zeta potential from an emulsion of xylene droplets at the same pH and ionic strength but in the absence of carbonate. Finally, they took precautions to exclude ionic and surface-active impurities. The consistent results obtained by different methods of preparing the emulsion samples, with different oils, and the general agreement with previous observations made with different materials and different techniques indicate that adventitious contamination is not the cause of this effect.

Measurements of the electrophoretic mobility allow calculation of the zeta potential, but this of course gives only the charge of the diffuse part of the double layer. The total surface charge was ultimately measured by us in a pH-stat experiment [6]. As an oil-in-water (O/W) emulsion is homogenized, hydroxide ions are adsorbed on the newly created surface, releasing protons. The pH of the suspension would drop if base were not added to maintain a constant pH. The amount of base required gives the total surface charge. If the droplet size is measured at the same time, to give the total surface area of the O/W interface, the surface charge density can be calculated. We measured the droplet size by electroacoustics, but other methods could have been used. The advantage of the electroacoustic technique is that concentrated emulsions can be measured directly without dilution. With 2–5 vol.% emulsions an easily measurable pH effect is obtained.

The results are consistent with previous observations: (1) the surface charge density is nearly independent of the nature of the oil, for hexadecane, perfluoromethyldecalin and squalene (Fig. 5.1 a); (2) the pH dependence of the zeta potential shows a decrease below pH 7 leading to an isoelectric point at pH 3–4 (Fig. 5.1 b); and (3) the zeta potential is independent of the identity of the co-anion of the electrolyte salt, for NaCl, NaI and $NaClO_4$, indicating that these anions do not compete with hydroxide in determining the surface charge (Fig. 5.1 b).

Two new features emerge from these results. First, the surface charge at pH 9 of $-4.9\ \mu C\ cm^{-2}$ ($-49\ mC\ m^{-2}$) for hexadecane is much larger than the diffuse layer charge calculated from the zeta potential of $0.63\ \mu C\ cm^{-2}$, implying that much of the surface charge is compensated by counterion (Na^+) condensation in the

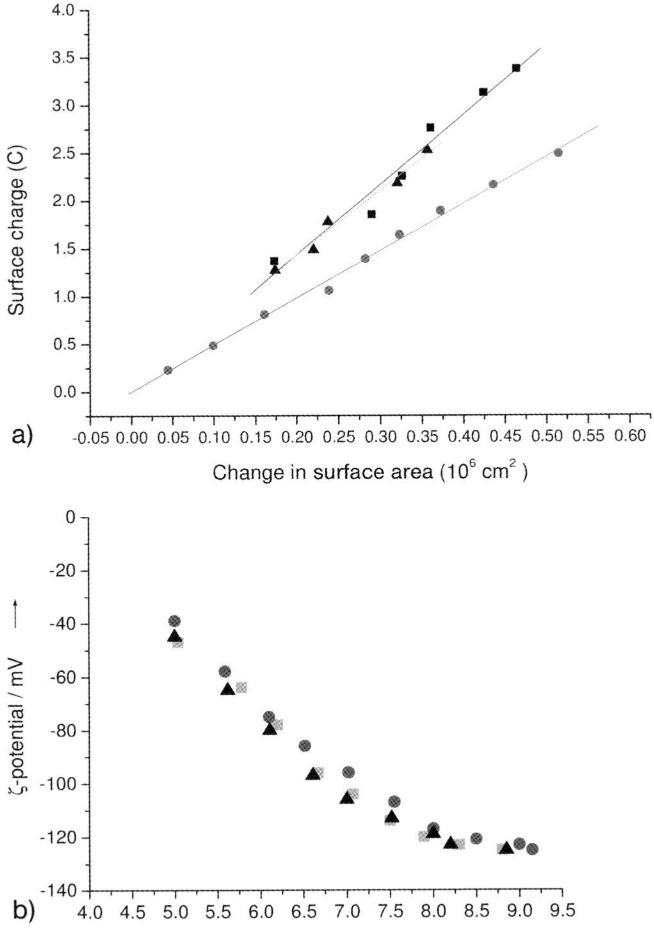

Fig. 5.1 (a) The surface charge at pH 9 in 240 mL of 0.2 mM NaCl for 2 vol.% emulsions of hexadecane (●) (–4.9 μC cm^{-2}), perfluoromethyldecalin (■) (–7.3 μC cm^{-2}) and squalene (▲) (–6.7 μC cm^{-2});
(b) the pH dependence of the zeta potential of hexadecane emulsions in 0.1 mM $NaClO_4$ (■), NaI (●) and NaCl (▲).

Stern or stagnant layer. Second, the quantity of hydroxide formed and adsorbed at pH 7 is much larger than the equilibrium concentration of 10^{-7} M at this pH. In the pH-stat experiment, this hydroxide is added externally to maintain pH 7. However, if base is not added, the pH decreases and an emulsion is still formed, with a lower zeta potential (see Fig. 5.1 b). This requires that the hydroxide is formed by autolysis of water as the new surface is created. This latter feature is not apparent when small surface area experiments are performed, but becomes evident with the use of the relatively concentrated 2 vol.% emulsions.

5.3 Gas Bubbles

Gases present a low dielectric surface to water, as does oil. The differences between these two hydrophobic surfaces are that the density of gases is low, so that dispersion forces will be insignificant, and that the interface is presumably more diffuse than the oil/water divide.

The negative charge on air bubbles has been investigated since the 19th century. In 1914, McTaggart extended the 1861 observations of Quincke using a cylindrical rotating cell to counteract the bubble buoyancy, a technique that is still used today. The subject has been well reviewed by Graciaa et al., so only the principal conclusions will be described here [7, 8].

The negative charge is independent of the identity of the gas, for air, hydrogen, oxygen or nitrogen, as would be expected. The zeta potential decreases with increasing electrolyte concentration, in the manner typical of double-layer compression. At low electrolyte concentrations the measured zeta potentials differ among early reports from –60 to –100 mV. In a recent study a value of –35 mV was measured at pH 5.8, rising to –110 mV at pH 10 [9]. The effect of the addition of NaCl is to decrease the zeta potential, consistent with double layer compression but not with adsorption of the chloride anion in competition with hydroxide. Most significantly, the pH dependence of the zeta potential is very similar to that observed for the O/W interface, with an isoelectric point around pH 3 from earlier studies and just above pH 4 in the later work [9].

5.4 Thin Films

Another method of investigating the air/water (or the oil/water) interface is to measure the disjoining pressure of thin films. This pressure arises from the interaction between the two film surfaces and is consequently sensitive to the charges on the surfaces. In pioneering work, Exerowa measured the diffuse double layer potential of the air/water interface with this technique to be –30 mV at pH 6 in 0.1 mM KCl and –80 mV at this pH in 10^{-5} M KCl. Isoelectric points between pH 4 and 5 are estimated by a long extrapolation, as the films become unstable as the potential decreases [1]. The technique and results have recently been comprehensively reviewed [10].

5.5 Solid Hydrophobic Surfaces

The interface between water and a solid hydrophobic material without any functional, dissociable groups develops a similar substantial negative charge. In a comprehensive study the Dresden group measured the zeta potential and the surface conductivity of Teflon AF films in different electrolytes [11]. Zeta potentials of –120 mV were observed at pH 9 in 0.1 mM KCl. The pH dependence of the zeta potentials gave a common isoelectric point of pH 4, independent of the KCl concentration, indicating that chloride ion was not responsible for the negative charge. Similar isoelectric points have been observed with other inert hydrophobic polymers [12–15].

By measuring the surface conductivity of the water/Teflon interface, Zimmermann et al. [11] calculated the surface charge in the stagnant layer. They found, for example, that in 10^{-5} M KOH, the compensated charge was somewhat less than half the diffuse layer charge. This is in contrast to the results from the electroacoustic study described above, in which the stagnant layer charge was measured to be about eight times greater than the diffuse layer charge. The streaming potential measurements, however, could not measure the stagnant layer charge directly; it had to be calculated assuming that the ions retained their bulk conductivity values even in the stagnant layer. This means that the calculated stagnant layer charge is a lower limit and is particularly sensitive to the high bulk value for the conductivity of the hydroxide ion. The electroacoustics results indicate that the stagnant layer conduction of the hydroxide ion at the oil/water interface is low, suggesting perhaps that the mobility of the hydroxide ion in the stagnant layer could be much lower than in the bulk, accounting for the discrepancy between the two values.

Two other significant points emerge from this important paper of Zimmermann et al. One is that as acid is added the zeta potential approaches zero and passes smoothly to positive values; it does not approach zero asymptotically (this has also been reported recently for the air/water interface of microbubbles, which also becomes positively charged below pH 4 [9]). The other is that in 10^{-5} M HCl the calculated stagnant layer charge is greater than the diffuse layer charge. Both observations suggest that protons are not neutralizing hydroxide ions to give an uncharged surface without ions, but are adsorbing to compensate the negative charge of the hydroxide ions, while some at least remain ionized. This would be consistent with the autolysis hypothesis described below. Regrettably, the surface conductivity at the isoelectric point of pH 4 was not reported; such a measurement would serve to resolve this question.

5.6 Self-assembled Monolayers

Alkyl chains tethered on solid supports are another means of generating a hydrophobic surface. Thus a C_{18}-trimethoxysilane on silica also shows an isoelectric point of pH 4. This was ascribed to neutralization of the underlying silanol groups but the results can be reinterpreted as the generation of adsorbed hydroxide ion above this pH [15]. Similarly, C_{18}-thiol on gold shows an isoelectric point at pH 4, with the zeta potential obtained from streaming currents decreasing with KCl concentration, indicating that the chloride ion is not adsorbed [16].

Perhaps unexpectedly, impedance spectroscopy measurements showed no pH dependence of the surface layer capacitance between pH 4 and 9.5, despite a large change in the zeta potential over this pH range [17]. This may be related to the contribution of the gold substrate [16].

Atomic force microscopy measurements with a C_{16}-alkanethiol tip made on C_{16}-alkanethiol films prepared on gold-coated silicon wafers gave puzzling results. At pH 9.5 the force is repulsive, as expected, and decreases with ionic strength of KCl solutions between 0.1 and 10 mM, consistent with electrostatic double layers. However, at neutral pH the force is attractive, which is not expected if both probe and film remain negatively charged [18, 19]. Furthermore, the attraction is independent of ionic strength from 0.1 mM to 0.1 M KNO_3, so the attraction is not due to screening of the electrostatic repulsion [20]. In the same set of experiments, the forces between the C_{16}-alkanethiol tip and methoxytri(ethylene glycol)-terminated alkanethiol monolayers were repulsive over the pH range 9.7–4.4 in 1 mM KCl and became attractive at pH 3.6; they also showed the expected decrease with ionic strength at pH 6 in KNO_3 solutions [19]. This is the behavior expected of negatively charged interfaces and suggests that there was something unusual about the alkanethiol film.

5.7 Surface Tension

It is well known that addition of salts to water results in an increase of its surface tension. According to the Gibbs adsorption equation, this requires that the surface is depleted in ions relative to their bulk concentration [21], and would appear to contradict the idea that the surface is charged by adsorption of hydroxide ions. Early, but controversial, work by Jones and Ray [22, 23], however, revealed a decrease in surface tension by a few hundredths of a percent as salt was added up to 1–2 mM concentrations, followed by the expected increase. Recently second harmonic generation experiments have confirmed the enhanced adsorption of iodide ions at the interface, which saturates at about 1 mM NaI concentration, consistent with the Jones-Ray effect for this salt [24]. These observations could be seen to be inconsistent with the effects of added salts on the zeta potential of oil droplets, which showed no specific anion effects up to 10 mM salt concentrations [25]. The optical experiment was limited to the iodide ion, however, and does not exclude the adsorption of other anions as well. Moreover, the Jones-Ray effect appears to be general among various salts they employed, all showing a minimum in the surface tension at concentrations of a few mM. There appears to have been no study of the pH dependence of the effect, which is needed to clarify the relative contributions of hydroxide ion and other anions.

5.8
Theory

There have been many theoretical studies of the air/water interface and some of the oil/water interface, but most have ignored the presence of the intrinsic negative charge and hence have failed to capture the essential physics of the problem. A general review of theory and experiment appeared in 2002 [26]. Of those that include electrolytes, most used concentrations greater than 10 mM, above which dispersion forces begin to have effects [27, 28], or used single ions [29, 30] or a single ion-pair [31]. The problem appears to be the size of the computational resources required. To keep the ionic concentration below 5 mM and yet have, say, 10 ion pairs in the system to simulate the electrostatic interactions would require 10–100 times the sizes of the molecular dynamics systems currently being used [32].

It is generally agreed, of course, that the structure of water alters as the interface is approached and that there are some preferred orientations of the water dipoles at the surface. In agreement with experimental surface spectroscopy results [33], the most common model has the hydrogen atoms of the water preferentially oriented to the low dielectric oil or air phase. A simple physical rationalization of this orientation is that it minimizes the image charge repulsion [34]. Since the image charge repulsion increases as the square of the charge, it is lower for the sum of the two charges (δ+) on the hydrogens than the single larger charge (2δ–) on the oxygen. Two molecular dynamics calculations agree that the orientation of the water dipoles results in a potential drop of $\sim$0.5 V over the 0.5 nm thickness of the surface layer, a result that will be part of the working hypothesis presented below [30, 34].

5.9
The Autolysis Hypothesis

Almost all of the experimental evidence is consistent with the idea that the interface of water of low ionic strength with low dielectric constant surfaces acquires a negative charge above pH 3–4 by the adsorption of hydroxide ions. The surface charge behaves as a typical double layer: the zeta potential decreases with increasing electrolyte concentration with no evidence for competitive adsorption of chloride co-ions up to at least 10 mM concentration. In the absence of salts, protons can act as the counterions.

The structure of the interfacial water that leads to this effect appears to be more or less independent of the nature of the low dielectric surface that it encounters. Extending the useful approach of Stubenrauch and von Klitzing [10], the diffuse layer charge can be calculated from the measured zeta potentials at various ionic strengths for different surfaces, according to the equation for water at 298 K [21]:

$$\sigma_d = -11.74\, c^{1/2} \sinh(19.46\, \xi)$$

where σ_d is the diffuse layer charge, c the electrolyte concentration and ξ the zeta potential. The results are plotted in Fig. 5.2 for air, oil and Teflon. Despite the disparate systems included, the results are remarkably similar: the zeta potentials are essentially equal at the same salt concentrations, independent of the identity of the hydrophobic surfaces. This justifies the conclusion that they reflect the properties of water at these low dielectric constant surfaces and are not very dependent on the properties of that material, be it gas, liquid or solid.

The same conclusion can be drawn from the observation of similar isoelectric points of pH 3–4 for the various systems, again independent of the properties of the different materials. This indicates that the autodissociation constant of the interfacial water has increased by six orders of magnitude over that of bulk water, with pK_w at the surface, $pK_w^s \approx 8$, since $[H^+]^s = [OH^-]^s = 10^{-4}$ M.

The origin of this enhanced autolysis could be the large electric field gradient at the interface. It is known that weak electrolytes undergo enhanced dissociation in an electric field – the second Wien effect. The calculated potential drop of 0.5 V at the interface over about 0.5 nm corresponds to a field of 10^9 V m^{-1}. Indeed, there is experimental evidence of enhanced autolysis of water by factors of 10^7–10^9 in the strong electric fields of bipolar ion-exchange membranes, the so-called "water-splitting" effect [35, 36].

Molecular dynamics simulations of the autolysis of bulk water ascribe the separation of hydroxide and hydronium ions to a solvent fluctuation that produces an electric field [37]. At the surface, this electric field could arise from the orientation of the interfacial water molecules. Indeed, among the 30 026 distinct ways to arrange 20 water molecules with different hydrogen bond networks, a simulation showed that a few spontaneous self-dissociated to give separated protons and hydroxyl ions [38].

It remains to account for the preferential adsorption of hydroxide over hydronium ions at the surface. Adsorption need not require specific adsorption sites.

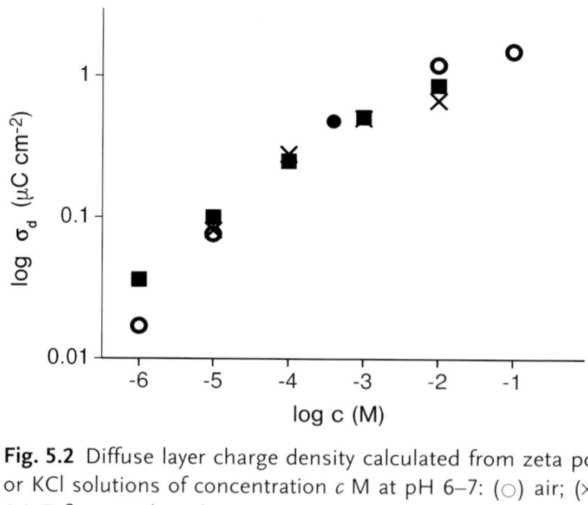

Fig. 5.2 Diffuse layer charge density calculated from zeta potentials in NaCl or KCl solutions of concentration c M at pH 6–7: (○) air; (×) xylene; (■) Teflon; (●) hexadecane.

The IUPAC definition is "an increase in the concentration of a dissolved substance at the interface ... due to the operation of surface forces". If water is oriented at the interface so that its positive hydrogen atoms are pointed away from the water phase, then all anions should be repelled by the negative oxygen end of the surface water. Since this is not observed for hydroxide ions, other, presumably hydrogen-bonding, forces must dominate. There have been many models of the structures of the hydronium and hydroxide ions. One recent image of the hydronium ion is of a pyramidal $H_9O_4^+$ structure with some lone pair density on the central oxygen that would be repelled by the surface water [31]. In contrast, models of the hydroxide ion include a square-pyramidal structure with the hydroxide hydrogen pointing upwards [39], where it could hydrogen bond to the surface water oxygen. New models are required that incorporate the hydrogen bonding hydration of the surface hydroxide ions into the overall surface water structure to ascertain why hydroxide ion is preferentially adsorbed. The surface charge density of 5 $\mu C\,cm^{-2}$ equals a charged ion every 3 nm^2, the hydration of which comprises a significant perturbation of the water structure.

5.10 Excluded Explanations

A number of alternative explanations for the negative charge at the oil/water or air/water interface have been advanced over the years that can be excluded by the evidence available now. These will be described briefly, without attribution!

- *Oriented dipoles.* Net orientation of water molecule dipoles can lead to an electric field, as described above, but cannot by itself lead to the charge separation required for the observed electrokinetic effects.
- *Dipole–dipole attraction.* The hydroxide ion is dipolar, whereas the proton nominally is not. Hence the hydroxide ion could have a favorable interaction with the oriented water dipoles, whereas a tetrahedral proton would not. The effect would have to be unique to the hydroxide ion, however, as other dipolar anions do not behave differently to non-dipolar ones in affecting the zeta potential of the hexadecane/water interface [40]. Furthermore, most model structures of the hydronium ion also have a dipole moment, so the distinction between the hydroxide and hydronium ions is not as rigorous as initially described.
- *Dispersion effects.* The hydroxide ion is probably more polarizable than the hydronium ion and so would be more favorably attracted to the oil/water interface. The difference between the ions should disappear, however, at the air/water interface, where dispersion forces would be negligible. The evidence suggests, on the contrary, that both interfaces behave similarly.
  Differences in dispersion forces between the oriented water dipoles and the ions would also favor the hydroxide ion, but the effect should also apply to other polarizable ions.

- *Born hydration energies.* It has been suggested that the lower hydration energy of the larger hydroxide ion would favor its accumulation in the interfacial region of lower dielectric constant, compared with the larger hydration energy of the smaller hydronium ion. If this were the case, however, even larger anions with even lower hydration energies would be even more preferred; this is not observed experimentally [24]. Hence an adequate explanation must include properties specific to the hydroxide and hydronium ions.
- *Hydronium ion repulsion.* Rather than hydroxide ion adsorption, the alternative description of the negative surface charge arising from hydronium ion repulsion has been dismissed on concentration grounds. It was argued that at pH 9 there are too few hydronium ions (10^{-9} M) to account for the observed negative charge. The converse argument also applies, however, at pH 5: there are too few hydroxide ions at pH 5 (10^{-9} M) to account for the observed charge. Both arguments ignore the enhanced autolysis of water at the surface. An adequate explanation must account for both the adsorption of hydroxide ions and the concomitant exclusion of hydronium ions.

5.11
Conclusions and Outstanding Questions

Almost all of the experimental evidence cited is consistent with the idea that water at extended hydrophobic surfaces acquires a negative charge above pH 3–4 from the preferential adsorption of hydroxide ions. For most of the experiments this could arise from the adsorption of the hydroxide naturally present in water. (A surface charge of 5 $\mu C\,cm^{-2}$ on a surface of 10 cm^2 with a water depth of 1 cm requires an adsorbed hydroxide concentration of only 5×10^{-8} M.) However, our experiments with more concentrated emulsions demand that additional hydroxide ions are created by the enhanced autolysis of water at the surface. The common isoelectric point of about pH 4 observed for the various surfaces suggests that the autolysis constant at the surface is of the order 10^{-8}, significantly increased from the pK_w value of 14 for bulk water.

Enhanced autolysis of water at the surface could account for the absence of a pH dependence of the surface impedance, if there are both hydronium ions and hydroxide ions present at the isoelectric point. Measurements of the pH dependence of the surface conductivity are required to resolve this matter.

Similarly, the pH dependence of the surface tension with various electrolytes needs investigation to elucidate the various factors that contribute to the Jones-Ray effect. It may emerge that the hydroxide adsorption is the underlying cause of the decrease in surface tension at low electrolyte concentrations [41].

The actual surface charge has been measured only for some oil droplets. This needs to be done for air bubbles and for solid hydrophobic surfaces, to determine if the surface charge is similarly much higher than the diffuse layer charge calculated from the zeta potential. This information is needed to resolve issues regarding the contribution of the hydroxide surface charge on the Jones-Ray effect.

Finally, theoretical studies must begin afresh. Because electrostatic forces are both powerful and long range, simulations which ignore the intrinsic surface charge are irrelevant. Models are required which incorporate the charge and the structures of the hydroxide and hydronium ions into the nature of the surface water, both its structure and its thermodynamics.

Acknowledgments

The author acknowledges continued helpful discussions and assistance from Professor Robert J. Hunter and Dr. Alex M. Djerdjev, with financial support from the Australian Research Council.

References

1 D. Exerowa, *Kolloid-Zeitschrift und Zeitschrift für Polymer* **1969**, *232*, 703.

2 W. Dickinson, *Transactions of the Faraday Society* **1941**, *37*, 140.

3 J.C. Carruthers, *Transactions of the Faraday Society* **1938**, *34*, 300.

4 A.J. Taylor, F.W. Wood, *Transactions of the Faraday Society* **1957**, *53*, 523.

5 K.G. Marinova, R.G. Alargova, N.D. Denkov, O.D. Velev, D.N. Petsev, I.B. Ivanov, R.P. Borwankar, *Langmuir* **1996**, *12*, 2045.

6 J.K. Beattie, A.M. Djerdjev, *Angewandte Chemie International Edition* **2004**, *43*, 3568.

7 A. Graciaa, P. Creux, J. Lachaise, *Surfactant Science Series* **2002**, *106*, 825.

8 A. Graciaa, P. Creux, J. Lachaise, in *Encyclopedia of Surface and Colloid Science*, Vol. 2 (ed. A.T. Hubbard), Marcel Dekker, New York, **2002**, p. 1876.

9 M. Takahashi, *Journal of Physical Chemistry B* **2005**, *109*, 21858.

10 C. Stubenrauch, R. von Klitzing, *Journal of Physics: Condensed Matter* **2003**, *15*, R1197.

11 R. Zimmermann, S. Dukhin, C. Werner, *Journal of Physical Chemistry B* **2001**, *105*, 8544.

12 A. Bismark, M.E. Kumra, J. Springer, *Journal of Colloid and Interface Science* **1999**, *217*, 377.

13 D. Möckel, E. Staude, M. Dal-Cin, K. Darcovich, M. Guiver, *Journal of Membrane Science* **1998**, *145*, 211.

14 C. Werner, U. König, A. Augsburg, C. Arnhold, H. Körber, R. Zimmermann, H.-J. Jacobasch, *Colloids and Surfaces A: Physicochemical and Engineering Aspects* **1999**, *159*, 519.

15 A. Hozumi, H. Sugimura, Y. Yokoygawa, T. Kameyama, O. Takai, *Colloids and Surfaces A: Physicochemical and Engineering Aspects* **2001**, *182*, 257.

16 R. Schweiss, P.B. Welzel, C. Werner, W. Knoll, *Langmuir* **2001**, *17*, 4304.

17 R. Schweiss, C. Werner, W. Knoll, *Journal of Electroanalytical Chemistry* **2003**, *540*, 145.

18 C. Dicke, G. Hähner, *Journal of Physical Chemistry B* **2002**, *106*, 4450.

19 C. Dicke, G. Hähner, *Journal of the American Chemical Society* **2002**, *124*, 12619.

20 H.J. Kreuzer, R.L.C. Wang, M. Grunze, *Journal of the American Chemical Society* **2003**, *125*, 8384.

21 R.J. Hunter, *Foundations of Colloid Science*, 2nd edn., Oxford University Press, Oxford, **2001**, pp. 65, 323.

22 G. Jones, W.A. Ray, *Journal of the American Chemical Society* **1937**, *59*, 187.

23 G. Jones, W.A. Ray, *Journal of the American Chemical Society* **1941**, *63*, 288.

24 P.B. Petersen, J.C. Johnson, K.P. Knutsen, R.J. Saykally, *Chemical Physics Letters* **2004**, *397*, 46.
25 G.V. Franks, A.M. Djerdjev, J.K. Beattie, *Langmuir* **2005**, *21*, 8670.
26 L.R. Pratt, A. Pohorille, *Chemical Reviews* **2002**, *102*, 2671.
27 P. Jungwirth, D.J. Tobias, *Journal of Physical Chemistry B* **2002**, *106*, 6361.
28 M. Mucha, T. Frigato, L.M. Levering, H.C. Allen, D.J. Tobias, L.X. Dang, P. Jungwirth, *Journal of Physical Chemistry B* **2005**, *109*, 7617.
29 L.X. Dang, *Journal of Chemical Physics* **1999**, *110*, 1526.
30 L.X. Dang, T.-M. Chang, *Journal of Physical Chemistry B* **2002**, *106*, 235.
31 M.K. Petersen, S.S. Iyengar, T.J.F. Day, G.A. Voth, *Journal of Physical Chemistry B* **2004**, *108*, 14804.
32 I.-F. W. Kuo, C.J. Mundy, *Science* **2004**, *303*, 658.
33 M.G. Brown, D.S. Walker, E.A. Raymond, G.L. Richmond, *Journal of Physical Chemistry B* **2003**, *107*, 237.
34 S.I. Mamatkulov, P.K. Khabibullaev, R.R. Netz, *Langmuir* **2004**, *20*, 4756.
35 A. Alcaraz, P. Ramírez, J.A. Manzanares, S. Mafé, *Journal of Physical Chemistry B* **2001**, *105*, 11669.
36 R. Simons, *Electrochimica Acta* **1984**, *29*, 151.
37 P.L. Geissler, C. Dellago, D. Chandler, J. Hutter, M. Parrinello, *Science* **2001**, *291*, 2121.
38 J.-L. Kuo, C.V. Ciobanu, L. Ojamäe, I. Shavitt, S.J. Singer, *Journal of Chemical Physics* **2003**, *118*, 3583.
39 B. Chen, I. Ivanov, J.M. Park, M. Parrinello, M.L. Klein, *Journal of Physical Chemistry B* **2002**, *106*, 12006.
40 J.K. Beattie, A.M. Djerdjev, G.V. Franks, G.W. Warr, *Journal of Physical Chemistry B* **2005**, *109*, 15675.
41 M. Manciu, E. Ruckenstein, *Advances in Colloid and Interface Science* **2003**, *105*, 63.

6
Surface Forces in Wetting Phenomena in Fluid Systems

Hiroki Matsubara and Makoto Aratono

6.1
Overview of Wetting Transition of Alkanes on a Water Surface

Wetting phenomena have attracted significant attention since the early stages in the history of colloid chemistry. These phenomena have been investigated from both theoretical and practical perspectives because of their substantial importance in industrial operations such as coating, painting and lubrication. Many of the studies on these phenomena have been conducted on systems with solid surfaces; however, the heterogeneity of the solid surfaces made the experiments less reproducible and prevented a quantitative understanding of the phenomena. From such a viewpoint, experiments in three-fluid systems are considerably useful for understanding the principles of wetting phenomena.

The wetting of a liquid on another material has been conventionally classified by means of the initial spreading coefficient (S_i) into the complete and partial wetting states, as follows:

$$S_i = \sigma^{AW\text{-}0} - \sigma^{OW} - \sigma^{AO} \tag{1}$$

where $\sigma^{AW\text{-}0}$, σ^{OW} and σ^{AO} represent the air–water, oil–water and air–oil interfacial tensions, respectively. The superscript 0 indicates that the air–water interfacial tension is measured in the absence of oil. When $S_i > 0$, a liquid drop placed on a surface spreads out to form a film of uniform thickness; on the other hand, when $S_i < 0$, it exists in the form of a lens. A consensus was recently reached regarding the third wetting state; in this state, called pseudo-partial wetting, a liquid drop spreads as a microscopically thick layer on the surface at equilibrium and the excess oil is in the form of one or more lenses.

Brochard-Wyart et al. [1] first explained the wetting phenomena by considering the free energy of a wetting film of unit surface area as a function of the film thickness d and by using the van der Waals interaction between the two bulk bodies on both sides of the film, as follows:

Colloids and Interface Science Series, Vol. 2
Colloid Stability: The Role of Surface Forces, Part II. Edited by Tharwat F. Tadros

ISBN: 978-3-527-31503-1

$$F(d) = \sigma^{OW} + \sigma^{AW} - \frac{A}{12\,\pi d^2} \tag{2}$$

Then, they discussed the four possible outcomes of this theory, as shown in Fig. 6.1 [1]. From this theory, the origin of pseudo-partial wetting can be attributed to the competition between the positive initial spreading coefficient, which favors spreading, and the positive Hamaker constant, which favors a thinner film. These give rise to the minimum in $F(d)$ at a finite thickness, as shown in (c). When the initial spreading coefficient is positive but the Hamaker constant is negative, the free energy decreases monotonically as d increases, as shown in (d). Hence the minimum free energy is achieved when a uniform thickness of oil exists on the surface (complete wetting). If the amount of oil is not sufficient to cover the entire surface, a pancake-like complete wetting region can be observed to coexist with a thinner film. When the spreading coefficient has a negative value as shown in Fig. 6.1a and b, the minimum in $F(d)$ occurs at (or very close to) $d=0$ and partial wetting is expected irrespective of the sign of the Hamaker constant.

Ragil and co-workers experimentally verified this theory by demonstrating a sequence of two transitions for the wetting of pentane on water with increasing temperature [2–4]. The first transition occurs from partial to pseudo-partial wetting at 25 °C. If the temperature is increased further, a second transition from pseudo-partial to complete wetting occurs at 53 °C. Ellipsometry revealed that

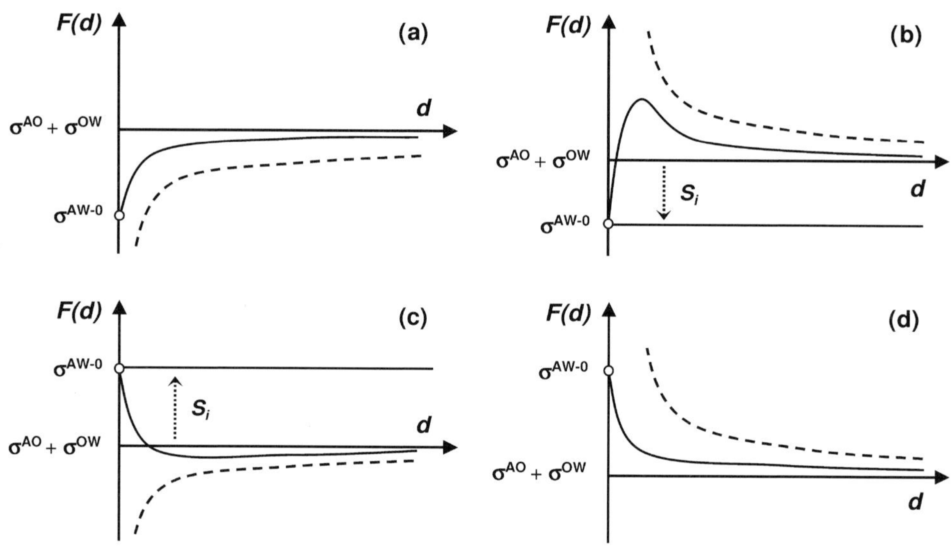

Fig. 6.1 Schematic representation of the free energy $F(d)$ of the air–water interface as a function of the thickness d of the oil film. The long-range potential is shown as a dotted line. (a) $S_i<0$ and $A>0$ (partial wetting); (b) $S_i<0$ and $A<0$ (partial wetting); (c) $S_i>0$ and $A>0$ (pseudo-partial wetting); (d) $S_i>0$ and $A<0$ (complete wetting) [15].

the film thickness increases abruptly from almost zero to several molecules at the first transition and diverges continuously to a macroscopically thick film at the second transition.

They explained that the first-order transition is caused by the change in the sign of the initial spreading coefficient from negative to positive at 25 °C, where the Hamaker constant of pentane is positive. On the other hand, regarding the second transition, a necessary condition for the continuous growth of the wetting layer is that the sign of the Hamaker constant should change. This constant has two contributions that have opposite signs, namely the dispersion interaction term and zero-frequency term. The latter contributes negatively but its absolute value is smaller than the positive contribution from the dispersion term at room temperature. The dispersion term

$$A = \frac{3h\nu}{8\sqrt{2}} \frac{(n_1^2 - n_3^2)(n_2^2 - n_3^2)}{(n_1^2 + n_3^2)^{\frac{1}{2}}(n_2^2 + n_3^2)^{\frac{1}{2}}\left\{(n_1^2 + n_3^2)^{\frac{1}{2}} + (n_2^2 + n_3^2)^{\frac{1}{2}}\right\}} \tag{3}$$

decreases with increasing temperature; hence, the sign of the Hamaker constant changes [5]. The authors theoretically estimated that this occurs at 52 °C for pentane on water. A sequence of two transitions has also been observed experimentally with varying temperature for hexane on brine [6] and with varying salinity for heptane on brine [7]. A continuous transition from pseudo-partial to complete wetting of octane on glucose solutions was induced by varying the glucose concentration [8].

However, the spreading coefficient increases monotonically from the partial wetting regime (shown in Fig. 6.1 a) to the pseudo-partial wetting regime (shown in Fig. 6.1 c); this would lead to a continuous phase transition, which is in contrast to the experimental results. Bonn and Ross pointed out that the order of the wetting transition is determined by the short-range forces in the free energy [9]. Unfortunately, very little is currently known about the precise shape of $F(d)$ for a small thickness. Although Eq. (2) does not include the short-range forces, it avoids this problem by connecting the long-range force to $F(0) = \sigma^{\text{AW-0}}$ in molecular dimensions. It should be mentioned that Ragil and co-workers were able to predict the spreading coefficients and first-order wetting transition temperatures by applying the Cahn theory of wetting to the adsorption of n-alkanes at the air–water interface [3, 10–12].

In this chapter, we summarize our recent studies on the transition of alkanes on surfactant solutions from the partial to pseudo-partial wetting regime. In Section 6.2, experimental results are used to describe that in an adsorbed surfactant film, the phase transition from the two-dimensional (2D) gas to liquid expanded (LE) phase at the air–water interface induces the spreading of alkane molecules on the film. In Section 6.3, a theoretical model for the transition from partial to pseudo-partial wetting is presented after confirming the generality of the wetting transition by using different combinations of surfactants and oils. From the quantitative agreement of the theoretical predictions with the observed wetting transitions, it is concluded that the origin of the short-range forces for the wetting of alkanes on water can be explained by the changes in the enthalpy and entropy when oil molecules leave the bulk and enter the film. In Section 6.4, it is shown that the free-energy profile calculated on the basis of our theory also explains that for hexadecane on dodecyltrimethylammonium bromide (DTAB) solutions, the line tension changes sign from positive to negative across the transition from partial to pseudo-partial wetting. The chapter concludes with a summary in Section 6.5.

6.2 Transition from Partial to Pseudo-partial Wetting Induced by Surfactant Adsorption at the Air–Water Interface

The initial spreading coefficient can be tuned widely by adding a surfactant. Therefore, an investigation of the effects of surfactant adsorption on the wetting transition of an alkane lens can provide useful information on the origin of the first-order wetting transition. In this section, in order to provide direct evidence of surfactant-induced wetting transition, we present the experimental results obtained from interfacial tensiometry and ellipsometry for quaternary systems comprising air, hexadecane and an aqueous solution of DTAB [13–15].

In general, ionic surfactant molecules adsorbed at the air–water interface are in the 2D gas phase of low surface coverage in a dilute region. The air–water interfacial tension decreases only slightly with increasing concentration in this region and then decreases rapidly in the LE phase. The transition concentration between the 2D gas and LE phases is roughly in the range 0.5–1.0 mmol kg^{-1} for the aqueous solution of DTAB. By considering the facts that the oil–water interfacial tension decreases with increasing concentration without any surface transitions and the air–oil interfacial tension is essentially independent of the surfactant concentration, a change in the sign of the initial spreading coefficient is expected to occur during the transition from the 2D gas to LE phase.

In Fig. 6.2, all the data obtained from surface tension measurements are plotted as a function of the surfactant concentration m. σ^{AW} represents the air–water interfacial tension in the presence of hexadecane. This is measured by a drop-shape analysis of a drop of water that is in equilibrium with bulk hexadecane in the vapor phase and is hanging from a cylindrical glass tip. Below $m = 0.5$ mmol kg^{-1}, $\sigma^{AW\text{-}0}$ and σ^{AW} are almost the same within the experimental error, confirming that the alkane is in the partial wetting regime. Above this concentration, $\sigma^{AW} < \sigma^{AW\text{-}0}$, which indicates that the surfactant at the air–water interface is mixed with oil (pseudo-partial wetting).

We have previously derived equations to calculate the interfacial density of surfactant molecules by using the interfacial tensions for quaternary three-phase systems in which the additional degree of freedom arises from the curvature of the oil lens. The interfacial density of each interface is deduced simply from the slope of the corresponding interfacial tension curve although rigorous thermo-

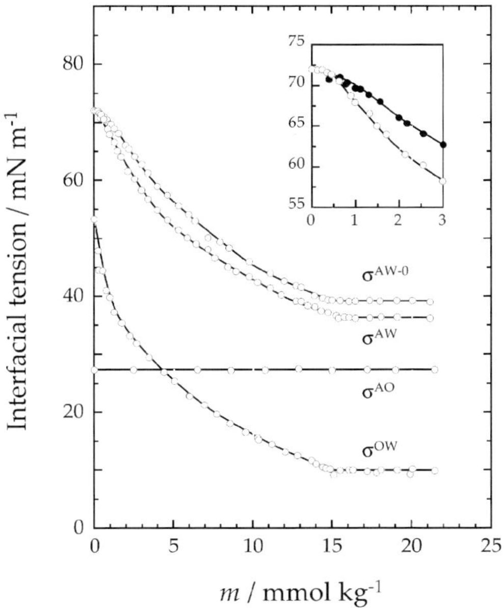

Fig. 6.2 Interfacial tension versus surfactant concentration m for hexadecane on DTAB solutions: (1) $\sigma^{\text{AW-0}}$, (2) σ^{AW}, (3) σ^{AO}, (4) σ^{OW}. All data were obtained at 25 °C. The inset shows $\sigma^{\text{AW-0}}$ and σ^{AW} around the wetting transition [14].

dynamic treatment is complicated [16]. For this purpose, a proper set of independent thermodynamic variables is chosen, namely, temperature T, pressure p, pressure within the oil lens, p^{O}, and concentration in the aqueous phase, m:

$$\Gamma^{\alpha\beta} = -(m/2RT)(\partial\gamma^{\alpha\beta}/\partial m)_{\text{T,p,p}^{\text{O}}} \tag{4}$$

The air–water interfacial density in the absence of hexadecane increases linearly with m up to the break point at 0.8 mmol kg^{-1}, where it jumps from ca. 0.2 to 0.6 μmol m^{-2} (Fig. 6.3). At higher concentrations, the interfacial density increases smoothly with m and approaches a saturation value near the critical micelle concentration (cmc). A similar behavior is observed for the air–water interfacial density in the presence of hexadecane; however, the jump in the interfacial density occurs at 0.5 mmol kg^{-1} and is more pronounced. A comparison of these interfacial densities reveals that the presence of hexadecane enhances the adsorption of surfactant molecules at low bulk concentrations and reduces it at high concentrations. On the other hand, the fact that the air–water interfacial density in the presence of hexadecane also differs from the oil–water interfacial density indicates that the hexadecane film at the air–water interface is so thin that it does not behave like bulk hexadecane.

When an oil molecule is transferred from the bulk (oil lens) to the film, there is an enthalpic cost because the interactions within the film are weaker than those within the bulk drop. However, the existence of surfactant molecules in the film can reduce this cost; this is because they augment the lateral van der Waals interactions between the hydrophobic chains and increase the mixing en-

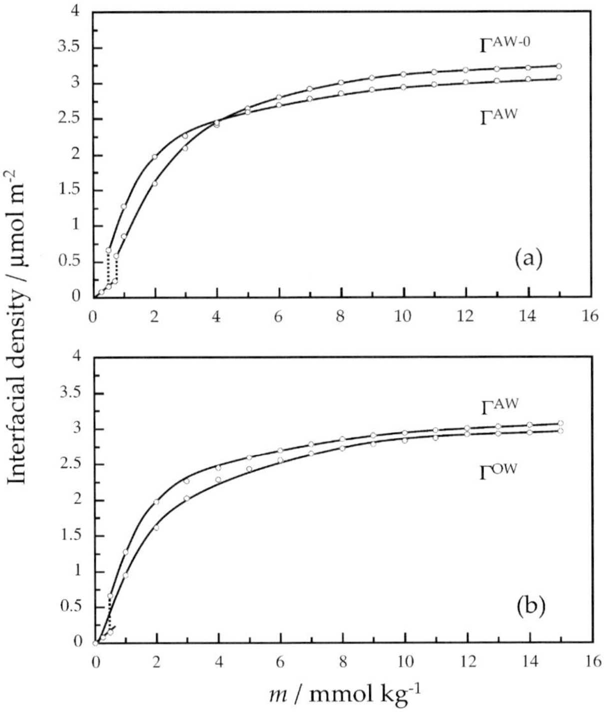

Fig. 6.3 Interfacial density versus *m* at 25 °C. Comparison of interfacial densities between (a) air–water interfaces and (b) air–water and oil–water interfaces [14].

tropy of the surfactant and oil molecules within the film. Therefore, the presence of hexadecane drives the phase transition from the 2D gas to the LE phase at the air–water interface; further, the abrupt increase in the interfacial density of DTAB molecules that happens simultaneously drives the first-order wetting transition from partial to pseudo-partial wetting. At higher concentrations, the interfacial density is determined by the balance of the van der Waals attraction between the chains and the electrostatic and steric repulsions between the head groups. The presence of hexadecane molecules permits a greater distance between the head groups without compromising the van der Waals interactions in the chain region. Consequently, at a given concentration of DTAB, the interfacial density of DTAB molecules in the LE phase reduces due to the presence of hexadecane.

Ellipsometric measurements provide convincing evidence that the film thickness is in fact of the order of molecules. The coefficient of ellipticity, $\bar{\rho}$, is defined as the imaginary part of r_p/r_s at the Brewster angle, where r_p and r_s are the complex Fresnel reflection coefficients for p- and s-polarized light, respectively. The coefficient of ellipticity at the air–water interface in the absence of a hexadecane lens, $\bar{\rho}^{AW\text{-}0}$ and that in its presence, $\bar{\rho}^{AW}$, are plotted against the surfactant concentration together with the initial spreading coefficient calculated from the surface tension data in Fig. 6.2. This plot is shown in Fig. 6.4. At low concentrations, both of these coefficients are almost constant and are nearly equal to the coefficient of ellipticity of pure water; this is in agreement with the tensiometry results. In the presence of an oil lens, the ellipticity changes discontinuously at 0.8 mmol kg^{-1}; this happens slightly above the break of the surface tension curve. From a simple model calculation in which DTAB and hexadecane adsorbed at the air–water interface are treated as an isotropic monolayer with dielectric constant ε, $\bar{\rho}$ can be used to determine the film thickness through the following equations [17, 18].

$$\bar{\rho} = \frac{\pi}{\lambda}\frac{\sqrt{\varepsilon_1 + \varepsilon_2}}{\varepsilon_1 - \varepsilon_2}\eta \tag{5}$$

$$\eta = \frac{(\varepsilon - \varepsilon_1)(\varepsilon - \varepsilon_2)}{\varepsilon}d \tag{6}$$

where ε_1 and ε_2 are the dielectric constants of air and water, respectively. For the air–water interfacial film comprising simple cationic surfactants in the LE phase, it has been found that $\bar{\rho}$ can be adequately explained by a model in which the hydrocarbon region is described as an oil film whose dielectric constant is equal to that of a liquid hydrocarbon with a similar chain length [19, 20].

From these equations, it is found that after the wetting transition, the thickness of the hydrocarbon layer is almost constant at a value of 0.7 nm corresponding to the interfacial density of ca. 3.1 μmol m^{-2}. The decrease in $\bar{\rho}$ above the wetting transition arises principally from the contribution of the additional

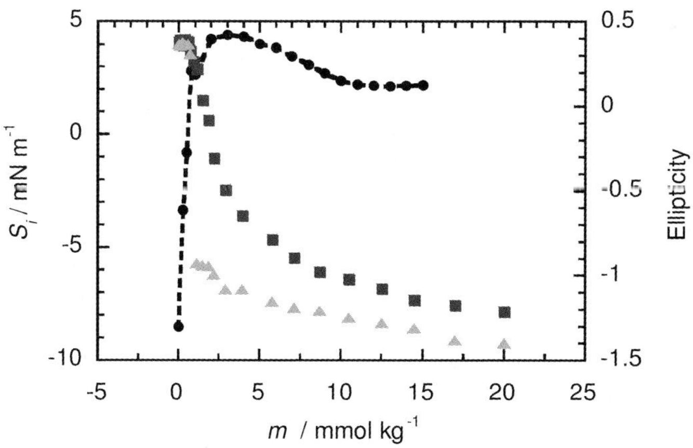

Fig. 6.4 Coefficient of ellipticity, $\bar{\rho}$, versus m for the air–DTAB solution interface without oil (■) and with hexadecane (▲). The y-axis on the left shows the initial spreading coefficient S_i (dashed line).

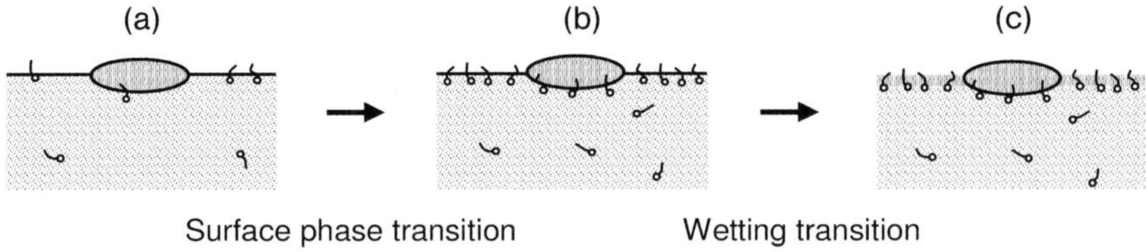

Fig. 6.5 Schematic illustration of the wetting scenario of hexadecane on DTAB solutions: (a) partial wetting is observed in the 2D gas phase, (b) surface phase transition from the 2D gas to LE phase occurs and (c) entropy gain accompanied by the phase transition induces the wetting transition from partial to pseudo-partial wetting.

polar head groups and bromide counterions of the surfactant molecules, which replace oil molecules in the mixed monolayer. The interfacial density of DTAB is ca. 0.7 μmol m^{-2} from the interfacial tension measurement just after the phase transition; therefore, approximately 80% of the molecules in the monolayer are hexadecane. As the concentration increases, the total interfacial density remains approximately constant but hexadecane molecules are replaced by DTAB molecules in the monolayer. The interfacial film at the cmc is composed of approximately 80% DTAB and 20% hexadecane, which is exactly the reverse of the composition just after the cmc.

The observations for hexadecane on DTAB solutions are summarized schematically in Fig. 6.5. At a bulk concentration of 0.5–0.8 mmol kg^{-1}, the adsorbed molecules undergo a phase transition from dilute 2D gas comprising surfactant molecules to an LE film. The first-order wetting transition from partial to pseudo-partial wetting accompanies this surface phase transition. Just above the transition concentration, the mixed monolayer is composed principally of oil molecules. As m increases, the thickness of the mixed monolayer remains approximately constant, but oil molecules are gradually replaced by surfactant molecules. Taking these factors into consideration, the enthalpic cost and entropic gain accompanied by the transfer of oil molecules from the bulk to the film appear to be important for the transition from partial to pseudo-partial wetting. In the following section, we will verify our concept of the origin of this wetting transition by comparing the transition concentrations of systems comprising different combinations of surfactants and oils. Then, we will propose a theoretical approach based on a simple statistical model to explain the first-order wetting transition.

6.3
Generality of Surfactant-induced Wetting Transition and Theoretical Prediction of the Wetting Transition Using a 2D Lattice Model

In order to confirm the generality of the surfactant-induced wetting transition, we used ellipsometry to study the spreading of a short linear alkane, dodecane, and a long branched alkane, squalane, on DTAB solutions. The results are qualitatively the same as those obtained for hexadecane on DTAB solutions. At low surfactant concentrations, the ellipticity has almost the same value as that of pure water; at the transition concentration, it drops abruptly to the negative value corresponding to the mixed monolayer of oil and surfactant molecules. However, from a quantitative viewpoint, the wetting transition of dodecane occurs at a lower surfactant concentration than that of hexadecane and that of squalane occurs at a higher concentration. By considering that the reported surface tensions are 23.35 [21], 27.05 [22] and 30.7 mN m^{-1} [23, 24] for dodecane, hexadecane and squalane at 25, 25 and 20 °C, respectively, it can be concluded that the transition concentration increases with the air–oil interfacial tension for the same surfactant.

As mentioned in the previous section, the occurrence of the transition from partial to pseudo-partial wetting is considered to be governed by the reduction in the enthalpic cost and entropic gain, which accompanies the transfer of oil molecules from the oil lens to the monolayer in the presence of surfactant molecules at the air–water interface. The increase in the air–oil interfacial tension in the order of dodecane, hexadecane and squalane implies that the cohesive energy between the oil molecules increases in the same order. Since the transition occurs when the chemical potential of oil in the monolayer becomes the same as that in the oil lens, the entropic gain must be increased to compensate for the stronger cohesive energy of the longer alkane. When the monolayer is dilute 2D gas, this increase is achieved simply by increasing the surface density of the surfactant molecules. Therefore, the longer the alkane, the higher is the value of the transition concentration at which the chemical potential of the surfactant in the monolayer and that in the bulk solution are the same.

Transitions from partial to pseudo-partial wetting were also observed for dodecane and hexadecane on tetradecyltrimethylammonium bromide (TTAB) solutions and for hexadecane on dibucaine hydrochloride (DC HCl) solutions. The transition concentrations of wetting on TTAB solutions decrease by a factor of four, reflecting that TTAB molecules are more surface active than DTAB molecules. By considering the existence of wetting transitions of squalane on DTAB solutions and hexadecane on DC HCl solutions, it can be concluded that the surfactant-induced wetting transition has some generality irrespective of the molecular structure of the oil and surfactant molecules. However, in our previous study, it was demonstrated that hexadecane on AOT solutions does not exhibit an analogous wetting transition. Therefore, more experiments are required in order to understand the relation between the wetting transition and molecular structure in greater detail.

Taking into consideration (1) the fact that the wetting behavior of alkanes on surfactant solutions can be reasonably explained by the van der Waals interaction between the hydrophobic chains and (2) the mixing entropy of the surfactant and oil molecules within the film, it appears logical to regard the mixed monolayer as a 2D regular solution. Based on this concept, we recently proposed a theoretical model that provides the transition concentrations and the variation in the composition of the wetting film; this model is in good agreement with experimental observations [25].

We regarded the monolayer as a lattice for considering the short-range forces; in this lattice, sites can be occupied by surfactant molecules or oil molecules or they can be empty. With the Bragg-Williams approximation, the canonical partition function of the monolayer is defined as

$$Q(N_+, N_-, N_0, M, T) = \frac{M!}{N_+!N_0!(M - N_+ - N_0)!}\frac{M!}{N!(M - N_-)!}q_0^{N_0}q_+^{N_+}q_-^{N_-}$$
$$\exp[-(zN_+^2 u_{++} + zN_0^2 u_{00} + 2zN_+N_0u_{+0} + zN_-^2u_- + 2\gamma N_+N_-u_{+-})/2MkT] \quad (7)$$

where the molecular partition functions q_i are chosen to have the same zero of energy; the pair potential between species i and j is denoted u_{ij}. Apart from the oil and surfactant molecules in the wetting layer, the distribution of the counterions of the surfactant in the electrical double layer is also taken into account in Eq. (7). The contributions of each component are represented by the subscripts 0, + and –; z denotes the number of nearest neighbors within each layer and γ denotes those between layers. This equation can be simplified by using the electroneutrality of the adsorbed film, i.e. $N_+ = N_-$, as follows:

$$Q(N_0, N_s, M, T) = \frac{(M!)^2}{(N_s!)^2N_0!(M - N_0 - N_s)!(M - N_s)!}q_0^{N_0}q_s^{N_s}$$
$$\exp[-(N_0^2\beta_{00} + 2N_0N_s\beta_{0s} + N_s^2\beta_{ss})/2MkT] \quad (8)$$

where $\beta_{ss}=zu_{++} + zu_{--} + z\gamma u_{+-}$, $\beta_{00}=zu_{00}$, $\beta_{0s}=zu_{0+}$ and $q_s=q_+q_-$. From Eq. (8), the Helmholtz free energy of the monolayer is given by

$$F^\sigma(N_0, N_s, M, T) = MkT\begin{bmatrix} x_0 \ln x_0 + 2x_s \ln x_s + (1 - x_0 - x_s)\ln(1 - x_0 - x_s) \\ +(1 - x_s)\ln(1 - x_s) - x_0 \ln q_0 - x_s \ln q_s \\ +(x_0^2\beta_{00} + 2x_0x_s\beta_{0s} + x_s^2\beta_{ss})/2kT \end{bmatrix} \quad (9)$$

Then, the chemical potentials of the oil and surfactant molecules in the monolayer can be calculated as

$$\mu_0^\sigma = -kT\ln q_0 + kT\ln[x_0/(1 - x_0 - x_s)] + (x_0\beta_{00} + x_s\beta_{0s}) = \mu_0^\theta \quad (10)$$

and

$$\mu_s^\sigma = -kT\ln q_s + kT\ln[x_s/(1 - x_0 - x_s)] + kT\ln[x_s/(1 - x_s)] + x_s\beta_{ss} + x_0\beta_{0s}$$
$$= \mu_s^\theta + 2kT\ln\left(\frac{m}{m^\theta}\right) \quad (11)$$

where m is the molality of the surfactant in solution, m^θ is the standard molality and $\mu_i^\ominus$ is the standard chemical potential of the oil and surfactant. For given values of m, Eqs. (9) and (10) allow us to calculate x_S and x_O. Hence, from the equation

$$\sigma^{AW} = \sigma^{AW-0} + \frac{\Gamma_m}{M}(F^\sigma - N_S\mu_S - N_O\mu_O^\theta) \tag{12}$$

the total free energy change (per unit area) can be calculated. The appropriate values of the molecular partition functions and interaction parameters can be determined by fitting the surface tension of pure surfactant solutions. For details of the fitting procedure, see [25].

Figure 6.6 shows the calculated σ^{AW} for a given m as a function of the oil composition in the monolayer and x_O of hexadecane on DTAB solutions. It should be noted that the free-energy curve has a double minimum for lower concentrations and the thermodynamically stable state changes from the primary minimum to the secondary minimum at m=0.8–0.9 mmol kg^{-1}. This is very close to the wetting transition concentration determined by tensiometry and ellipsometry. At higher concentrations, there is only a single minimum corresponding to the mixed monolayer of oil and surfactant.

In order to combine these calculations and the concept of the initial spreading coefficient, the short-range interaction described by the lattice model should be matched with the non-retarded van der Waals interaction described by the equation

$$U(d) = -A/12\,\pi d^2 \tag{13}$$

where $d=e'+\delta$; δ is the thickness of the monolayer and e' is the thickness of the oil film above the monolayer. Ellipsometry revealed that the thickness of the

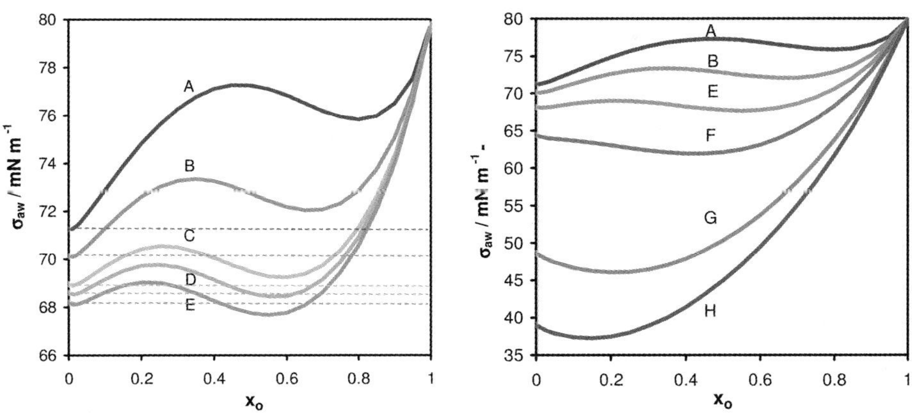

Fig. 6.6 Plots of the surface free energy against the mole fraction of oil in the mixed monolayer for bulk concentrations of (A) 0.2, (B) 0.5, (C) 0.8, (D) 0.9, (E) 1.0, (F) 2.0, (G) 8.0 and (H) 15.0 mmol kg^{-1}. The dotted lines indicate the energy of the minimum at low x_O [25].

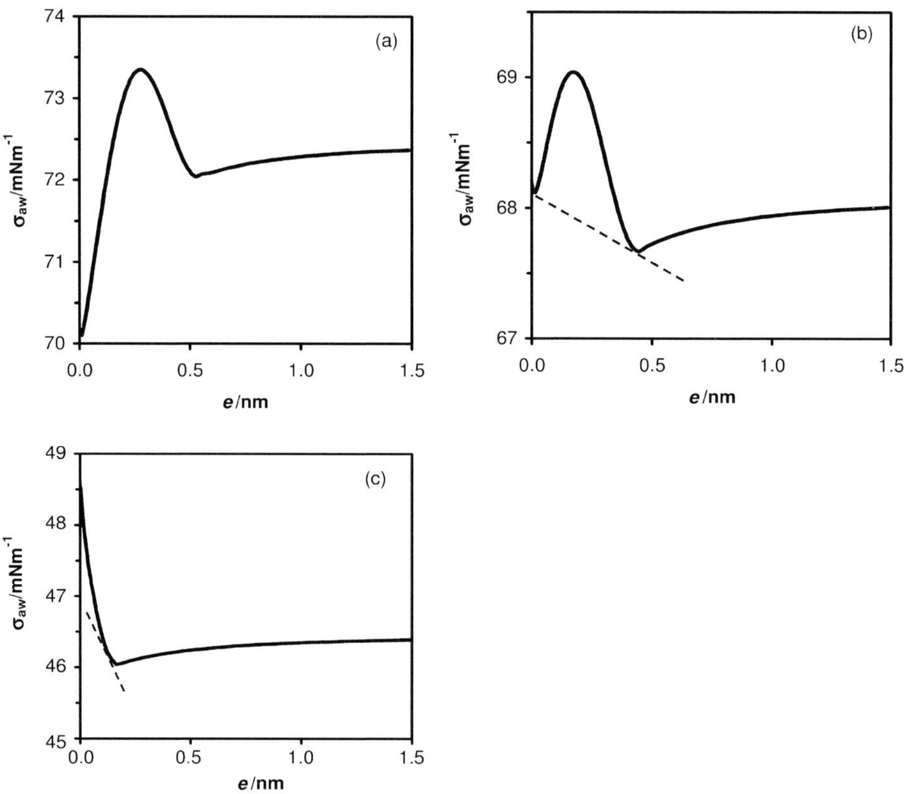

Fig. 6.7 Surface-free-energy diagrams for DTAB + hexadecane: (a) 0.5, (b) 1.0 and (c) 8.0 mmol kg^{-1}. The dotted lines are tangents to the free-energy curves [25].

mixed oil–surfactant monolayer after the wetting transition is roughly constant at ~0.7 nm. The long-range potential thus has a value of $U(e'=0) = -0.4$ mJ m^{-2}. In this model, the equilibrium spreading coefficient of hexadecane on DTAB solutions is

$$S_{eq} = \sigma^{AW} - \sigma^{OW} - \sigma^{AO} = U(e' = 0)$$

which is in good agreement with the experimental value of S_{eq} obtained by tensiometry. The results are shown in Fig. 6.7 in the form of surface free-energy diagrams as a function of the equivalent oil film thickness e, which is given by $x_0\delta$ for the short-range contribution and by $x_0\delta + e'$ for the long-range contribution. The asymptotic value at large e should yield the experimental value of $\sigma^{AO}+\sigma^{OW}$.

For an oil lens on a surfactant solution of 0.5 mmol kg^{-1}, the minimum free energy is clearly at $e=0$, i.e. partial wetting. For 1.0 mmol kg^{-1}, the initial spreading coefficient becomes positive and the free-energy minimum appears at a small but finite e. Hence the stable state becomes a mixed monolayer of oil and surfactant molecules rather than a dry film. Above the wetting transition, the monolayer is nearly complete and comprises 60% oil and 40% surfactant. Further, for 8.0 mmol kg^{-1}, the minimum is at a small x_0, but the composition of oil in the mixed monolayer decreases to 20%. Although these values depend to some extent on the parameter used, the fact that the 2D regular solution model is in accordance with real experiments suggests that the entropy of mixing and the lateral van der Waals interaction between the oil and surfactant molecules are extremely important for the surfactant-induced wetting transition. This agreement in the transition concentration is also obtained for the two other choices of surfactant and alkane chain length (dodecane on DTAB solutions and hexadecane on TTAB solutions).

6.4
Line-tension Behavior Near the Transition from Partial to Pseudo-partial Wetting

The magnitude of the line tension τ is theoretically estimated to be of the order of 10^{-11}–10^{-12} N [26]. This considerably small tension is normally negligible compared with other forces acting in the system, such as the surface tension and gravitational force. However, since the line tension contributes to the free energy in the form of τ/r, where r is the radius of the lens, it plays an increasingly important role as the lens becomes smaller. One of the most important aspects of the line tension of a lens system is that it takes either positive or negative values depending on the situation; however, the surface tension never takes negative values. This is why the line tension is regarded as an important factor for nucleation [27, 28], cell fusion [29] and emulsion coalescence [30, 31].

In 1972, White concluded theoretically that the microscopic contact angle is different from the macroscopic contact angle because of the modification of the drop shape near the three-phase contact [32]. Later, this was linked with the concept of line tension proposed by Indekeu as the interface displacement model [33–35]. For an oil lens placed on a water surface, such modification is caused by the surface forces between the air and water phases across the oil phase in the close vicinity of the three-phase contact region, where the air–oil and oil–water interfaces are microscopically close to each other. This can be understood from the following equation:

$$\tau = \int_{-\infty}^{\infty} \left[\frac{1}{2} \gamma^{\mathrm{AO}} \left(\frac{\mathrm{d}l}{\mathrm{d}x} \right)^2 + V(l) \right] \mathrm{d}x + \text{constant} \tag{14}$$

where l denotes the interface displacement $l(x)$; the first term in the integrand accounts for the increase in the air–water interfacial area and the second term accounts for the surface forces acting through the liquid drop. $V(l)$ is termed the interface potential and is identical with the surface free energy $F(d)$ when its reference state is taken as the sum of air–oil and oil–water interfacial tensions. After some transformation, the equilibrium line tension can be calculated by integrating the square root of the free-energy profile from the thickness of a film in equilibrium with a lens, l_1, to the microscopically large thickness l_2 where the modification of the drop profile by surface forces is negligible, i.e.

$$\tau = \xi \int_{l_1}^{l_2} \mathrm{d}l [2\gamma^{\mathrm{AW}} V(l)]^{\frac{1}{2}} \tag{15}$$

where ξ is the bulk correlation length.

For the first-order wetting transition to the complete wetting regime, two minima exist at a finite and an infinite film thickness in the free-energy profile. A negative contribution J_1 and a positive contribution J_2 occur before the wetting transition where the negative contribution from J_1 exceeds J_2 (Fig. 6.8 a). The magnitude of J_1 reduces as the transition point is approached and, finally, only a positive contribution remains in the integral. Consequently, the line tension has a positive sign at the complete-wetting transition point. If the long-range surface force decays slowly as a function of the thickness, the line tension would be allowed to diverge at the transition point. On the other hand, when the wetting transition occurs between two microscopic films of different thicknesses, the line tension can change its sign from positive to negative as expected from Fig. 6.7 b.

The interface displacement model has been verified experimentally only a few times because of the experimental difficulties involved in measuring the line tension. Law and co-workers measured the line tension of *n*-octane and *n*-octene droplets on a silicon wafer coated with hexadecyltrichlorosilane; then, they demonstrated that the line tension changes sign from negative to positive with a diverging slope near the first-order wetting transition [36, 37]. Pompe measured the line tension of hexaethylene glycol on a silicon wafer and obtained positive values below a contact angle of 6° and negative values above that [38]. In the following paragraphs, we will present the first experimental proof of the line tension behavior at the first-order transition from partial to pseudo-partial wetting for hexadecane on DTAB solutions [39]. For measuring the line tension, we used an interferometer originally developed by Aveyard et al. [40].

The images of hexadecane lenses obtained for two specific concentrations are shown in Fig. 6.9. Before the wetting transition concentration (partial wetting regime), the small lenses formed by stirring merge spontaneously into a large lens as the equilibrium is approached. The behavior of the oil lens after the wetting transition (pseudo-partial wetting regime) leads to a remarkable contrast: large lenses break into small lenses without stirring. We analyze the dihedral angle δ and the lens radius r from these interference fringes; then, we calculate the line tension from the Neumann-Young equation as follows:

$$\sigma^{AO}\cos\left(\frac{\sigma^{OW}\delta}{\sigma^{OW}+\sigma^{AO}}\right)+\sigma^{OW}\cos\left(\frac{\sigma^{AO}\delta}{\sigma^{OW}+\sigma^{AO}}\right)=\sigma^{AW}-\frac{\tau}{r} \qquad (16)$$

Before the wetting transition, the plot of the left-hand side of Eq. (16), y, versus r^{-1} has a negative slope corresponding to a line tension of $+8.04\times10^{-12}$ N, as shown in Fig. 6.10. In contrast, the plot has a positive slope after the wetting transition and the line tension decreases to a negative value of -19.7×10^{-12} N.

The free-energy profile $F(d)$ can be regarded as the interface potential $V(l)$ when the free energy is assumed to originate from the sum of air–oil and oil–water interfacial tensions. From the fact that the line tension takes a positive value in the partial wetting regime, it is considered that the maximum in the free-energy curve in Fig. 6.7 a, which is caused by the mixing of the oil and surfactant with an unfavorable ratio, overcomes the negative contributions on both the sides of the maximum. The line tension takes less positive values as the wetting transition is approached. In the pseudo-partial wetting regime, $V(l)$

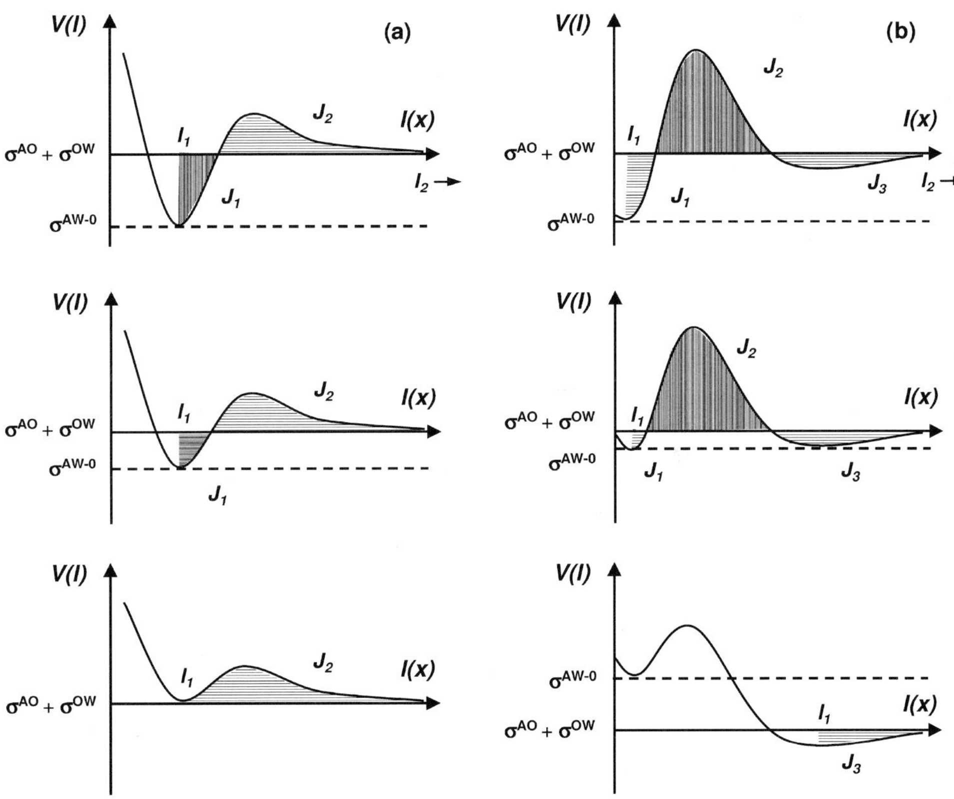

Fig. 6.8 Interface displacement profile for the first-order wetting transition; transition from (a) partial to complete wetting and (b) partial to pseudo-partial wetting.

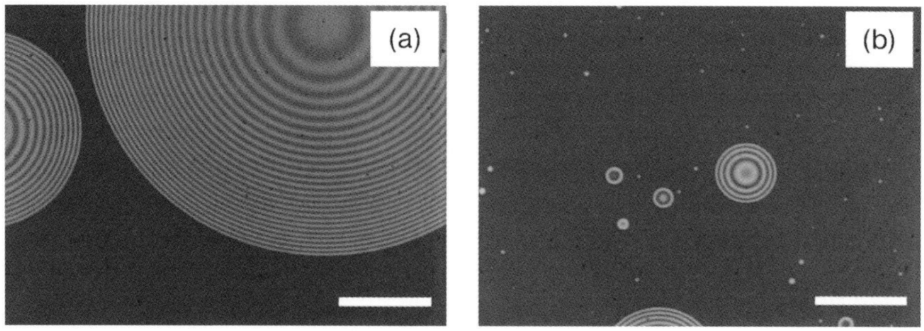

Fig. 6.9 Image of hexadecane lenses left for 3 h on an aqueous solution, obtained before and after the wetting transition: (a) 0.505 and (b) 1.013 mmol kg^{-1}. The white bar indicates a length of 50 μm [39].

should be integrated from the thickness of the mixed monolayer so that the integration along the van der Waals potential to an infinite thickness leads to a negative line tension.

The manner in which the negative line tension promotes the spontaneous rise of the three-phase contact line is somewhat analogous to the stabilization of emulsions by the addition of surfactant molecules. From such a viewpoint,

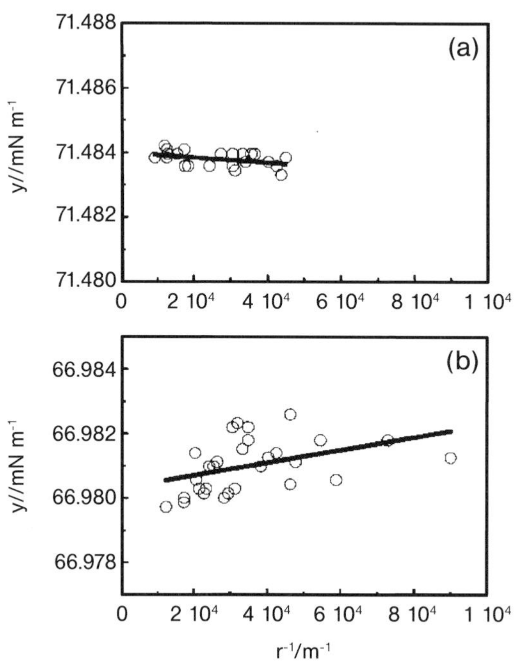

Fig. 6.10 Plots of the left-hand side of Eq. (16) versus the reciprocal of the lens radius: (a) 0.505 and (b) 1.013 mmol kg^{-1} [31].

the surfactants that induce the wetting transition from partial to pseudo-partial wetting can be considered to be a type of "linactant" (line active agent), which reduces the free energy associated with the three-phase contact line. Systematic studies on the role of surfactants in the pseudo-partial wetting transition can facilitate industrial applications to a significant extent.

6.5 Conclusion

We employed surface tensiometry and ellipsometry to demonstrate that the increase in the surface density of surfactant molecules at the air–water interface drives the wetting transition of the oil lens from partial to pseudo-partial wetting. A quantitative analysis of the wetting transition by means of a 2D lattice model suggested that the origin of the wetting transition is the increase in the mixing entropy and the lateral van der Waals interaction within the mixed oil–surfactant monolayer. Moreover, reliable line-tension measurements revealed that the line tension changes sign from positive to negative at the wetting transition. Small lenses merge into a large lens in the partial wetting regime and a large lens spontaneously splits into small lenses in the pseudo-partial wetting regime. The 2D lattice model also provided a useful link between the macroscopic wetting transition and the line tension.

Acknowledgment

We are grateful to Prof. Colin Bain for providing advice on ellipsometric measurements and the development of the 2D lattice model.

References

1 Brochard-Wyart, F., Di Meglio, J. M., Quere, D., De Gennes, P. G. *Langmuir,* **1991**, *7*, 335.
2 Ragil, K., Meunier, J., Broseta, D., Indekeu, J. O., Bonn, D. *Phys. Rev. Lett.* **1996**, *77*, 1532.
3 Ragil, K., Bonn, D., Broseta, D., Meunier, J. *J. Chem. Phys.* **1996**, *105*, 5160.
4 Ragil, K., Bonn, D., Broseta, D., Indekeu, J. O., Kalaydjian, F., Meunier, J. *J. Pet. Sci. Eng.* **1998**, *20*, 177.
5 Israelachvili, J. N. *Intermolecular and Surface Forces.* Academic Press, New York, **1991**.
6 Bonn, D., Pauchard, L., Shahidzadeh, N., Meunier, J. *Physica* **1999**, *263*, 78.
7 Bertrand, E., Dobbs, H., Broseta, D., Indekeu, J., Bonn, D., Meunier, J. *Phys. Rev. Lett.* **2000**, *85*, 1282.
8 Pfohl, T., Riegler, H. *Phys. Rev. Lett.* **1999**, *82*, 783.
9 Bonn, D., Ross, D. *Rep. Prog. Phys.* **2001**, *64*, 1085.
10 Cahn, J. W. *J. Chem. Phys.* **1977**, *66*, 3667.
11 Dobbs, H. *J. Chem. Phys.* **2001**, *114*, 468.
12 Dobbs, H., Bonn, D. *Langmuir,* **2001**, *17*, 4674.

13 Aratono, M., Kawagoe, H., Toyomasu, T., Ikeda, N., Takiue, T., Matsubara, H. *Langmuir* **2001**, *17*, 7344.

14 Matsubara, H., Ikeda, N., Takiue, T., Aratono, M., Bain, C. D. *Langmuir* **2003**, *19*, 2249.

15 Wilkinson, K., Bain, C. D., Matsubara, H., Aratono, M. *Chem. Phys. Chem.* **2005**, *6*, 547.

16 Aratono, M., Toyomasu, T., Ikeda, N., Takiue, T. *J. Colloid Interface Sci.* **1999**, *218*, 412.

17 Drude, P. *The Theory of Optics*. Dover, New York, 1959.

18 Meunier, J. In *Light Scattering by Liquid Surfaces and Complementary Techniques*, Langevin, D., ed. Marcel Dekker, New York, 1992.

19 Casson, B. D., Bain, C. D. *J. Phys. Chem. B* **1998**, *102*.

20 McKenna, C. E., Knock, M. M., Bain, C. D. *Langmuir* **2000**, *16*, 5853.

21 *CRC Handbook of Chemistry and Physics*, 74th edn., Ed. David R. Lide, CRC Press, Boca Raton, Fl, **1993**.

22 Jasper, J. J. *J. Phys. Chem. Ref. Data* **1972**, *1*, 841.

23 Fletcher, P. D. I., Nicholls, R. J. *Phys. Chem. Chem. Phys.* **2000**, *2*, 361.

24 Semal, S., Bautheir, C., Voue, M., Vanden Eynde, J. J., Gouttebaron, R., De Coninck, J. *J. Phys. Chem. B* **2000**, *104*, 6225.

25 Matsubara, H., Aratono, M., Wilkinson, K., Bain, C. D. *Langmuir*, **2006**, *22*, 982.

26 Rowlinson, J. S., Widom, B. *Molecular Theory of Capillarity*. Oxford University Press, Oxford, 1982.

27 Scheludko, A., Chakarov, V., Toshev, B. *J. Colloid Interface Sci.* **1981**, *82*, 83.

28 Toshev, B. V., Platikanov, D., Scheludko, A. *Langmuir* **1988**, *4*, 489.

29 Karatekin, E., Sandre, O., Brochard-Wyart, F. *Polym. Int.* **2003**, *52*, 486.

30 Churaev, N. V., Starov, V. M. *J. Colloid Interface Sci.* **1985**, *103*, 301.

31 Denkov, N. D., Petsev, D. N., Danov, K. D. *J. Colloid Interface Sci.* **1995**, *176*, 189.

32 White, L. R. *J. Chem. Soc., Faraday Trans. 1* **1977**, *73*, 390.

33 Indekeu, J. O. *Physica (Amsterdam)* **1992**, *183A*, 439.

34 Dobbs, H. T., Indekeu, J. O. *Physica (Amsterdam)* **1993**, *201A*, 457.

35 Indekeu, J. O. *Int. J. Mod. Phys. B* **1994**, *8*, 309.

36 Wang, J. Y., Betelu, S., Law, B. M. *Phys. Rev. Lett.* **1999**, *83*, 3677.

37 Wang, J. Y., Betelu, S., Law, B. M. *Phys. Rev. E* **2001**, *63*, 31601.

38 Pompe, T. *Phys. Rev. Lett.* **2002**, *89*, 76102.

39 Takata, Y., Kikuchi, Y., Matsubara, H., Takiue, T., Aratono, M. *Langmuir*, **2005**, *21*, 8594.

40 Aveyard, R., Clint, J. H., Nees, D., Paunov, V. *Colloids Surf. A* **1999**, *146*, 95.

7
Aggregation of Microgel Particles

Brian Vincent and Brian Saunders

7.1
Introduction to Microgel Particles

This chapter is essentially a review of the literature concerning the stability and aggregation of microgel particles. Comprehensive reviews of other features of these types of particles can be found in previous articles [1–3].

Microgel particles are cross-linked polymer (latex) particles, which change their volume (that is, swell or de-swell) according to (1) the solvency conditions of the medium in which they are dispersed and (2) the density of the cross-linking moieties within the particles. One may write an osmotic balance between these two terms, as follows [4],

$$\Pi_{\text{osm}} + \Pi_{\text{el}} = 0 \tag{1}$$

The first term in Eq. (1) is the osmotic mixing term, which relates to the polymer–solvent interaction parameter (χ) through the expression

$$\Pi_{\text{osm}} = -\frac{N_A kT}{\nu_s[\phi + \ln(1-\phi) + \chi\phi^2]} \tag{2}$$

where ϕ is the (average) polymer volume fraction in the microgel particles, N_A is Avogadro's number, k is the Boltzmann constant, T is the absolute temperature and ν_s is the molecular volume of the solvent molecules. Equation (2) is valid for *uncharged* microgel particles. For *charged* microgel particles, there is an additional osmotic contribution (to the basic χ-containing term) from the charge-balancing counterions within the microgel particles [4]. In the simplest case, where the counterions may be treated ideally, one may write [4] this additional (ionic) contribution to the osmotic term as

$$\Pi_{\text{ion}} = kT\frac{f}{V} \tag{3}$$

Colloids and Interface Science Series, Vol. 2
Colloid Stability: The Role of Surface Forces, Part II. Edited by Tharwat F. Tadros

ISBN: 978-3-527-31503-1

where f is the number of *mobile* (that is, non-*condensed*) counterions per particle; f clearly relates to the bulk particle charge, Q_b.

The second term in Eq. (1) refers to the elastic term, which restricts swelling and is given by

$$\Pi_{el} = \frac{2N_A XkT}{V_0}\left[\left(\frac{\phi}{2\phi_0}\right) - \left(\frac{\phi}{\phi_0}\right)^{\frac{1}{3}}\right] \tag{4}$$

where X is the number of cross-link sites per particle and V_0 and ϕ_0 are, respectively, the particle volume and volume fraction of polymer in the "unswollen" microgel particles. Note that ϕ_0 is not actually equal to unity, since there is considerable evidence that, once swollen with solvent, microgel particles always retain some solvent, even when de-swollen at high temperatures, provided that they remain in dispersion. For example, Rasmusson et al. [5] estimated the value of ϕ for poly(*N*-isopropylacrylamide) (PNIPAM) particles, containing 9% cross-linking monomer, as a function of temperature, in the absence of electrolyte. For this particular system, they showed that $\phi \approx 0.1$ at 25 °C, increasing to ~ 0.2 in the region of the LCST for PNIPAM solutions ($\sim$32 °C) and to ~ 0.5 at 40 °C.

By substituting Eqs. (2) and (4) into Eq. (1) (and realizing that $\phi/\phi_0 = V_0/V$, where V is the particle volume in the swollen state), one may obtain, for neutral microgel particles, a relationship between V/V_0 (the volume swelling ratio, R, which is always >1) and χ, for a given cross-link density (X/V_0), although this relationship has to be solved numerically.

Because of the uncertainty in the value for V_0, discussed above, it is more usual to express experimental values for the volume swelling ratio as $S = V/V_{max}$ (always <1), where V_{max} is the *maximum* value of V, for a given cross-link density. One may also use (equivalently) the *diameter* swelling ratio, $S_d = d/d_{max}$, where d and d_{max} are the corresponding particle diameters (as obtained, for example, using dynamic light scattering); clearly, $S = S_d^3$.

The more common ways in which χ may be varied systematically are either by a change in the solvent (composition) or by a change in the temperature. PNIPAM microgel particles are perhaps the most commonly studied systems of this sort, which show both a temperature response [1–3] and a response to solvency changes, for example, adding short-chain alcohols to the aqueous phase [6]. With respect to the temperature response, PNIPAM microgel particles deswell on heating (aqueous solutions of corresponding high molecular weight, linear PNIPAM chains have a lower critical solution temperature around 32 °C). Bouillot and Vincent were able to make microgel particles which demonstrated the *inverse* temperature effect, that is, they swelled on heating [7]. These particles were based on interpenetrating networks of polyacrylamide (PAM) and poly(acrylic acid) (PAA).

For weakly charged microgels, the solution pH may be used to vary Q_b. For example, polyvinylpyridine (PVP) microgel particles have no bulk charge at neutral or higher pH but become progressively more positively charged as the pH is lowered below 4.5, through protonation of the N atoms in the (weakly basic) pyridine moieties [8]. Alternatively, one may alter the ionic balance, inside and outside the particles, by adding electrolyte to the system. One may think of this, equivalently, as a "screening effect" on the charge repulsion within the particles.

An alternative way of using osmotic balancing, inside and outside the microgel particles, to control the particle size is to add a non-penetrating polymer. Saunders and Vincent demonstrated this effect for PNIPAM microgel particles in aqueous solution, to which poly(ethylene oxide) (PEO) chains were added [10] and also for polystyrene (PS) microgel particles in ethyl benzene, to which free PS chains were added [11]. It was recently shown by Bradley et al. [12] that, if one used a microgel based on copolymers of NIPAM and AA, then, at low pH values ($\sim$3), PEO chains of MW $\sim$3000 (depending on the cross-link density) actually entered the microgel particles, up to some maximum absorption limit, before the concentration of free PEO chains started to build up. The entry of the PEO chains into the particles in this case is driven by the strong H-bonding interaction between the H atoms of the –COOH moieties and the ether O atoms of the PEO chains. For these systems, therefore, the size of the microgel particle passes through a maximum with increasing PEO concentration. It is interesting that the particle volume in this case can actually exceed the value of V_{max} in the absence added polymer!

Since microgel particles are usually prepared in the de-swollen state, by a dispersion or polymerization route, there are normally charged groups present at the periphery of the particles, in addition to any bulk charge that may be present. These arise from the initiator residues at the end of the constituent polymer chains in the particles. For PVP particles, Fernández-Nieves et al. [9] were able to deconvolute the separate contributions of the bulk charge, Q_b (protonated pyridine groups), and the surface charge, Q_s (amidinium groups), to the total particle charge, Q, using conductometric and potentiometric titrations.

7.2 Stability and Aggregation of Microgel Particles: Theoretical Background

7.2.1 Interparticle Forces

As with hard (non-swelling) latex particles, one can deconvolute the total, pairwise interaction free energy, $G(r)$, where r is the center–center separation between any two microgel particles, into several contributions. The primary, *longer range* interactions (i.e. for $r>d$ or $h>0$, where $h=r-d$) are the van der Waals attraction (G_A) and an electrostatic repulsion (G_E) if the particles have surface and/or bulk charged groups.

In addition, since microgel particles are soft, they may possibly interpenetrate when they touch, that is, it is possible for r to have values $<d$ (i.e. $h<0$). One way of thinking about this is to consider such an interaction as being a "steric" interaction, similar in nature to that for hard particles carrying adsorbed or grafted polymer chains. Thus, in principle, microgel particles ought to show incipient flocculation, when the solvency of the medium is adjusted such that $\chi > 0.5$. However, the authors are not aware of any experiments where this effect has been studied with microgel particles, *per se*.

In addition, microgel particles may show other interactions when polymers are added to the dispersion. First, bridging interactions may be observed if the added polymer associates with (but does not significantly penetrate into) the polymer chains comprising the microgel particles. Again, this effect has not been observed directly with microgel particles, as far as the authors are aware, but has been observed with sterically stabilized particles by Cawdery and Vincent [13], who investigated the stability of PS particles carrying grafted PEO chains, to which poly(acrylic acid) (PAAc) was added. At low pH values ($<\sim 4$), a strong H-bonding interaction occurs between the ether O atoms of the PEO chains and the –OH groups of the carboxylic acid moieties, leading to bridging flocculation of the PS-g-PEO particles by the PAAc chains. Second, depletion interactions may occur if the added polymer chains do *not* associate with the polymer chains comprising the particles. Some examples of depletion flocculation of this type for microgel particles will be described later (Section 7.3.2). Note (1) that the depletion interactions in these cases are of the "soft-wall" type, analyzed by Jones and Vincent [14, 15], and (2) Cawdery and Vincent [13] also observed depletion flocculation in their systems, described above, at *high* pH values, when the H-bonding interaction disappears, through ionization of the –OH moieties.

For the majority of microgel dispersions, whose aggregation behavior has been studied in the literature, the only significant interparticle interactions that need to be considered are G_E and G_A. Although, in principle, one could consider using the standard equations in the literature for $G_E(h)$ for hard-sphere particles, this term is not easily calculated, because of difficulties in establishing the relevant ("surface") potential term to ascribe, and, to a lesser extent, knowing exactly the ionic strength in the bulk solution if ions penetrate into the microgel particles (there will be a Donnan partitioning of ions). The latter effect is only significant for relatively concentrated microgel dispersions. However, the difficulty in ascribing a value to the ("surface") potential is a serious one. One needs to know the potential distribution *from the center* of the particle out into the bulk solution, not just the potential distribution from the surface, as with hard particles (such as PS latex particles). There is a directly related problem in ascribing the concept of a "zeta potential" (ζ) to microgel particles. Where is the plane of shear? More importantly, how does one calculate the equivalent of a "zeta potential" from an experimental electrophoretic mobility (u) value? Oshima [16], Fernández-Nieves et al. [17] and others have considered these problems and we will not discuss them further here, as they are rather complex and detailed and the issues involved are still not fully resolved. Suffice to say, one should be very wary indeed of simply using classical equations relating ζ to u, derived for charged, *hard sphere* particles, and then using the related classical equations for calculating $G_E(h)$ between microgel particles. It is safer simply to offer qualitative explanations of, for example, the effect of adding electrolytes on the aggregation behavior of charged microgel particles.

One is on somewhat firmer ground in discussing the long-range van der Waals interactions, $G_A(h)$, between microgel particles. For the purposes of calculating this function, one may use the classical equations for hard spheres, but remembering that microgel particles consist of a *mixture* of polymer and solvent molecules, at a given (average) polymer volume fraction (ϕ). Vincent [18] has given the following equation for the Hamaker constant (A) for two-component mixtures (1 + 2):

$$A = \left[\phi A_2^{\frac{1}{2}} + (1-\phi)A_1^{\frac{1}{2}}\right]^2 \tag{5}$$

Hence, if the Hamaker constant of the solvent (A_1) and that of the polymer (A_2) are known and assuming that there is a uniform distribution of polymer segments throughout the particles, then the Hamaker constant of the particles may be calculated using Eq. (5). Of course, one needs to know ϕ and this is not straightforward, since one needs to know both V and V_0 (recall that $\phi = V/V_0$), and, as we have discussed earlier, V_0 is not easily obtained, unless *all* the solvent is removed from the particles. This problem has been discussed further by Rasmusson et al. [5].

One interesting feature, which occurs for microgel particles but is not pertinent to hard-sphere dispersions, is that one is able to tune $G_A(h)$ by varying ϕ, for example, for PNIPAM particles, by varying the temperature. Indeed, around room temperature, where PNIPAM particles are swollen (and ϕ becomes very small, typically less than a few percent), the Hamaker constant of the particle approaches that of the medium and the interparticle van der Waals attraction becomes negligible.

7.2.2 Criteria for Dispersion Stability

The classical approach which has evolved for describing the stability/aggregation behavior of (dilute) *hard-sphere* colloidal dispersions is based on the original ideas of Derjaguin and Landau [19] and Verwey and Overbeek [20] (the "DLVO" theory), in which one considers the magnitude of any potential energy barriers (maxima) and of any potential energy minima, in the total pairwise interaction $G(h)$, that is the sum of all the separate interactions described in the previous section. The reader is referred to any of the major textbooks on colloid science for a full description of these concepts. The question is, how far can these ideas be applied to *soft* microgel particles? There is no intrinsic problem, provided one remembers the points, raised in the last section, concerning the nature of and the difficulties in calculating, the various component interactions.

As was described in the previous section, $G_A(h)$ for swollen microgel particles is negligible (and $\chi < 0.5$), and, hence (in the absence of added polymers which could introduce bridging or depletion interactions), one cannot induce aggregation by simply reducing $G_E(h)$, that is, by reducing Q close the zero (e.g. by a pH change) or adding sufficient inert electrolyte, as with conventional hard-sphere particles. There is no driving force to induce aggregation. Of course, one is able to "tune in" a $G_A(h)$ contribution, of controlled magnitude, by increasing ϕ, for example, in the case of PNIPAM particles, by increasing the temperature. This concept was investigated by Rasmusson et al. [5] and their work will be described in a later section. Suffice to say here, that one may describe this effect in terms of the depth of the free energy minimum (G_{min}) introduced into the form of $G(h)$, where

$$G(h) = G_A(h) + G_E(h) + G_s(h) \tag{6}$$

where $G_S(h)$ is the steric interaction, associated with any interpenetration of the microgel particles ($h<0$). For *hard* spheres this is a step (delta) function (i.e. $G_S \to \infty$, for $h<0$), but for soft, microgel particles the magnitude of $G_S(h)$, for $h<0$, will depend on the polymer segment density distribution near the periphery of the particles (as determined, for example, by small-angle neutron scattering [21]). For hard sphere particles, $G_A(h)$ can be approximated by the following equation:

$$G_A(h) = -\frac{Ad}{24h} \tag{7}$$

Many authors have discussed the depth of G_{min} required to achieve flocculation. For example, Napper [22] suggested that weak flocculation would occur if G_{min} were somewhat greater than the thermal energy ($\sim 3/2\,kT$) available to the particles. Scales and co-workers [23, 24] suggested that reversible flocculation occurs when $G_{min} < 5\,kT$ and irreversible flocculation occurs when $G_{min} > 10\,kT$. Vincent et al. [25], however, claimed that, for cases of weak, *reversible* flocculation, such arguments are spurious and that a thermodynamic approach should be used, in line with the analogy between weak particle flocculation and molecular condensation processes. They postulated that the free energy of flocculation, ΔG_{floc}, of particles may be split into two contributions:

$$\Delta G_{floc} = \Delta G_i + \Delta G_{hs} \tag{8}$$

where ΔG_{hs} (= $-T\Delta S_{hs}$) is the entropic contribution associated with the flocculation of hard spheres in the absence of interparticle interactions. ΔG_{hs} per particle is positive and therefore opposes the flocculation process. ΔS_{hs} depends on the particle volume fraction, ϕ_p; its magnitude decreases with increasing ϕ_p. Feigin et al. [26] calculated ΔG_{hs} based on a simple lattice model. They estimated that, at 20 °C, ΔG_{hs} decreases from 12.2 to 3.0 kT, over the ϕ_p range 1×10^{-5}–1×10^{-1}.

ΔG_i is the interaction free energy term associated with the (non-hard sphere) interactions between the particles. It is a function of the floc structure and the depth of the relevant free energy minimum (G_{min}) in the interaction free energy–particle separation curve for two interacting particles. ΔG_i may be related to G_{min} at low ϕ_p through $\Delta G_i = zG_{min}$, where z is the average coordination number of a particle in a floc. It may be assumed, at low ϕ_p values, that ΔG_i is independent of ϕ_p. For strongly aggregating (coagulating) dispersions, ΔG_i (negative) $\gg \Delta G_{hs}$ (positive) in magnitude and the aggregation process is spontaneous and irreversible. However, for weakly interacting particles ($G_{min} < \sim 10\,kT$), ΔG_i and ΔG_{hs} are of the same order of magnitude. If $\Delta G_i < \Delta G_{hs}$ in magnitude, then the dispersion is thermodynamically stable, since ΔG_{floc} is then positive – see Eq. (8). On the other hand, if $\Delta G_i > \Delta G_{hs}$ in magnitude, then weak, reversible flocculation occurs, leading to colloidal phase separation. Such colloidal phase separation has been widely reported for dispersions of weakly flocculating particles, since the first systematic study by Long et al. [27].

7.3 Experimental Studies of Microgel Aggregation

7.3.1 Temperature- and Electrolyte-induced Homoaggregation

In this section we will discuss selected examples from the microgel literature which are illustrative of the theory presented in Section 7.2.

We consider first PNIPAM "homopolymer" microgels. Note that the term "homopolymer" is not strictly correct because all microgel particles contain a cross-linking co-monomer. However, the concentration of the cross-linking co-monomer is usually much lower than that of the other monomers present. PNIPAM dispersions were first reported by Pelton and Chibante [28], who studied their critical flocculation temperature (CFT) as a function of $CaCl_2$ concentration. Snowden and Vincent [29] made similar studies as a function of NaCl concentration. Both sets of authors showed that the CFT decreased with increasing electrolyte concentration. However, it was Daly and Saunders [30] who conducted the first comprehensive study of the effect of added electrolyte on the particle swelling and stability of PNIPAM particles. Figure 7.1 shows the variation of the hydrodynamic diameter of PNIPAM microgel particles, as a function of temperature, in the presence of various concentrations of added NaCl. Added NaCl causes a substantial decrease in diameter, in particular at concentrations in excess of 0.1 mol dm^{-3}. This is due to the poorer solvency conditions established for the PNIPAM chains (i.e. χ is increased). Na^+ and Cl^- ions have a high charge density and promote the formation of extensive hydration spheres owing to their relatively strong ion–dipole interactions with water molecules. Hence, at high concentrations of these ions, some of the water molecules available for hydrating the PNIPAM chains are no longer available.

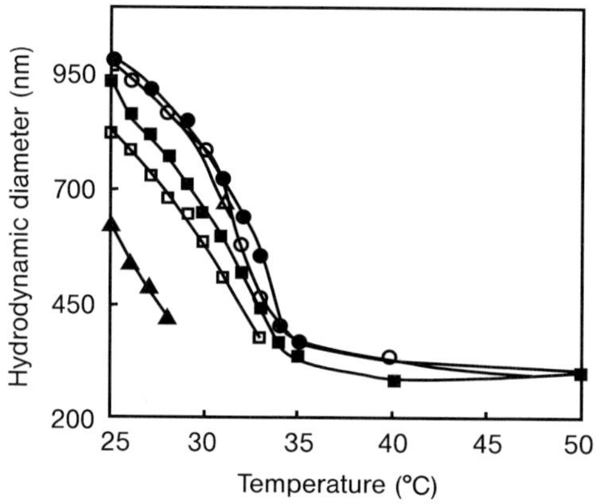

Fig. 7.1 Variation of the hydrodynamic diameter, as a function of temperature, for PNIPAM microgel particle dispersions, at different NaCl concentrations: 0 (●), 10^{-4} (○), 10^{-3} (△), 10^{-2} (■), 10^{-1} (□) and 0.5 mol dm^{-3} (▲).

A second important consequence of salt addition is the suppression of electrostatic interactions between neighboring PNIPAM particles. This results in temperature-induced flocculation. Flocculation may be followed using the wavelength exponent method [27]. The wavelength exponent (n) is the magnitude of the gradient obtained from a plot of log(optical density) versus log(wavelength) for light passing through a given dispersion. This parameter is very sensitive to particle aggregation [27]. Figure 7.2, taken from Daly and Saunders' work [30], shows the variation of the wavelength exponent with temperature for PNIPAM microgel particle dispersions. The onset of aggregation is apparent from an abrupt decrease in the value of n. This implies again that flocculation occurs at a critical temperature, namely the CFT. The CFT moves to lower temperatures with increasing NaCl concentration, as observed earlier by Snowden and Vincent [29].

Daly and Saunders [30] also compared the dependence on NaCl concentration of the CFT values of the PNIPAM particle dispersions with the corresponding dependence of the LCST values for homopolymer PNIPAM solutions in water;

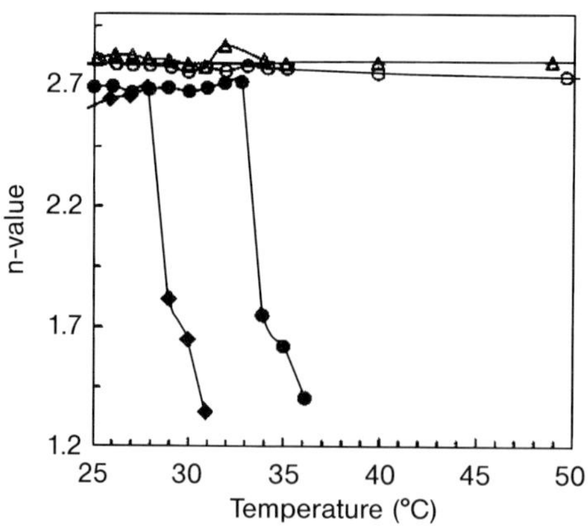

Fig. 7.2 Wavelength exponent (n), as a function of temperature, for PNIPAM particle dispersions, at different NaCl concentrations: 10^{-2} (△), 10^{-1} (●) and 0.5 mol dm^{-3} (◆) and in the absence of electrolyte (○).

these results are shown in Fig. 7.3. These data indicate that flocculation of the microgel particles occurs at about the same temperature as the LCST for PNIPAM (except perhaps for the points corresponding to pure water). Napper [22] suggested that the CFT value for sterically stabilized particles, with terminally grafted polymer chains, should correlate with the theta temperature of the corresponding linear polymer chains in solution, under corresponding conditions (the theta temperature is the extrapolated value of the LCST in the limit of infinite molecular weight). In Daly and Saunders' experiments, the PNIPAM particles may be flocculating under better than theta (i.e. $\chi < 0.5$) solution conditions owing to aggregation arising from the longer range van der Waals attraction between the particles (cf. incipient flocculation discussed in Section 7.2.1).

Daly and Saunders [30] also investigated the *nature* of the added electrolyte on the CFT of PNIPAM dispersions. There is a long history [31] of the so-called "Hoffmeister" or lyotropic series regarding different ions and their effects on particle aggregation. Figure 7.4 shows the effect of anion type on the hydrodynamic diameter of PNIPAM microgel particles, measured at 25 °C. The anions have the following order in terms of their ability to "salt-out" hydrophilic colloids [31]: $Cit^{3-} > Cl^- > SCN^-$. The data shown in Fig. 7.4 are consistent with this order. Figure 7.5 shows the CFT values of these PNIPAM dispersions in the presence of a 0.5 mol dm^{-3} concentration of different electrolytes containing a range of anions. The order of CFTs also generally follows the lyotropic series, i.e. the CFTs increase in the order: $Cit^{3-} < Cl^- < Br^- < I^- < SCN^-$. The more strongly structure-breaking anions (e.g. Cit^{3-}) have a higher charge density and compete more effectively for water of hydration. The effect of different cations on microgel particle size and CFT values, at equivalent concentrations, has also been investigated [30]. However, the results did not follow any particular trend. This is commonly observed for cations and may be a consequence of factors other than direct ion hydration by water molecules.

Duracher et al. [32] investigated the flocculation behavior of poly(*N*-isopropylmethacrylamide) (PNIPMAM) microgel particles. Linear PNIPMAM has an

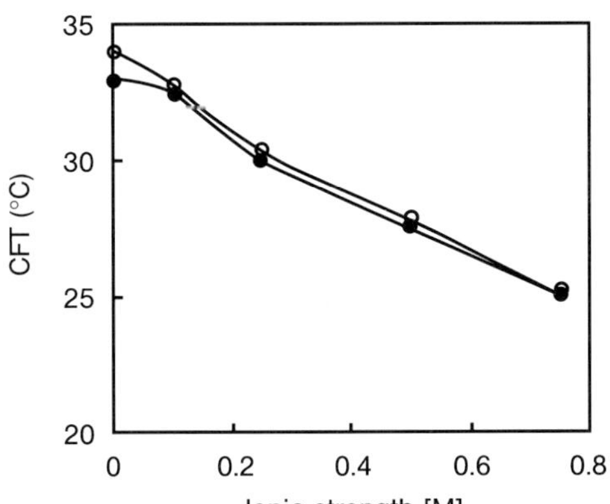

Fig. 7.3 Variation of CFT for PNIPAM microgel particle dispersions with ionic strength (NaCl concentration). The CFT data from this work are shown (○), in addition to LCST values obtained for homopolymer PNIPAM (●); from ref. [53].

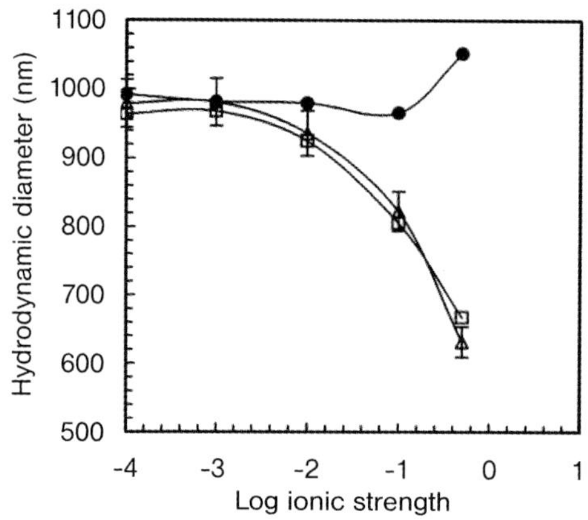

Fig. 7.4 Variation of the hydrodynamic diameter of PNIPAM microgel particles, as a function of ionic strength, for NaCl (△), NaSCN (●) and Na_3Cit (□), at 25 °C.

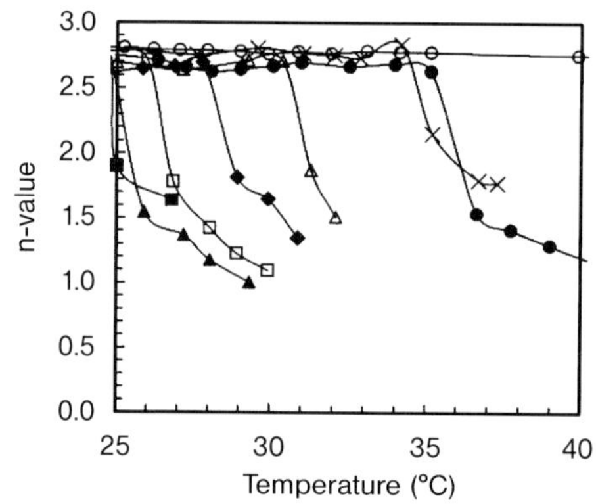

Fig. 7.5 Wavelength exponent (n) as a function of temperature for PNIPAM particle dispersions in the presence of NaCl (◆), NaSCN (●), Na_3Cit (□), NaI (×), NaBr (△), Na_2CO_3 (▲), $NaCH_3COO^-$ (■) and no electrolyte (○). The ionic strengths were all 0.5 mol dm^{-3}.

LCST of ~44 °C (the corresponding value for PNIPAM is ~32 °C). They showed, like Daly and Saunders [30], that there is some correlation between the CFT of the microgel particles and the LCST of the homopolymer in solution, as a function of added salt concentration. They also showed that the CFT of the microgel particles depended on their degree of internal cross-linking.

Cross-linked poly(2-vinylpyridine) (P2VP) microgel particles swell at pH values below ~4 owing to protonation of the N atoms in the pyridine rings; they also have a (positive) surface charge arising from the initiator used [33]. Fernández-Nieves et al. [34, 35] investigated the aggregation of dispersions of P2VP microgel particles as a function of NaCl concentration at pH 9 (i.e. in the *unswollen* state). They showed, from small-angle static light scattering studies, that the fractal dimension (d_f) of the aggregates decreased, from 2.13 at 0.25 mol dm^{-3} NaCl, to 1.82 at 3.0 mol dm^{-3} NaCl. The explanation offered was that the depth of the free energy minimum (G_{min}) in the pair interaction was increasing, such that the aggregation became less reversible with increasing salt concentration. At high salt concentrations the value of d_f reached the limiting

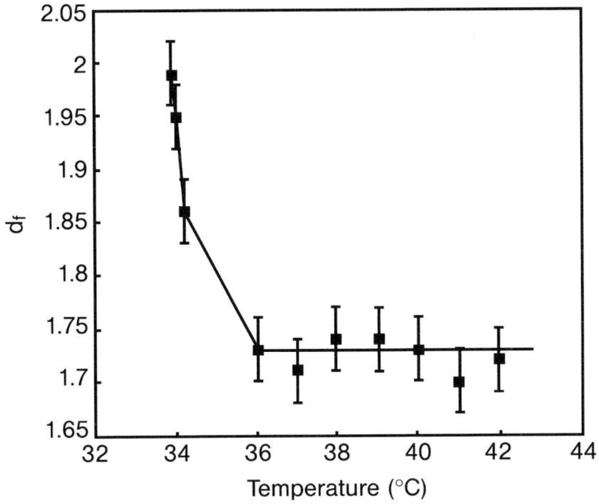

Fig. 7.6 Fractal dimension of PNIPAM flocs, in 1 mol dm^{-3} NaCl, as a function of temperature.

value generally found for irreversible, diffusion-controlled aggregation ($d_f \approx 1.7$–1.8). Indeed, based on Eqs. (5) and (7), the authors [34, 35] calculated that the Hamaker constant of the PNIPAM microgel particles changed from 6.0×10^{-20} J at 0.25 mol dm^{-3} NaCl to 6.6×10^{-20} J at 3 mol dm^{-3} and the value of G_{min} increased from 2.5 kT to 12.5 kT over the same NaCl concentration range. Results on similar systems by the same group [34] from *dynamic* light scattering studies were less easy to explain, however, especially at very high salt concentrations (>2 mol dm^{-3} NaCl).

Routh and Vincent [36] carried out small-angle static light scattering studies on the fractal dimensions of PNIPAM aggregates as a function of temperature, at a fixed NaCl concentration (1 mol dm^{-3}) and pH 9. Their results are shown in Fig. 7.6. The PNIPAM particles contained a small amount of acrylic acid as co-monomer. However, although these microgel particles would, therefore, be strongly charged at pH 9, the electrostatic repulsion between the particles would be negligible at 1 mol dm^{-3} salt concentration. The corresponding CFT value was 29 °C. It is clear from Fig. 7.6 that, as the temperature is increased, the value of d_f decreases, reaching the diffusion-controlled limiting value around 36 °C. The most likely explanation again is that, as the particles contract with increasing temperature, the van der Waals attraction between the particles becomes stronger, i.e. G_{min} becomes deeper.

Rasmusson et al. [5] also studied the aggregation of PNIPAM microgel particles as a function of temperature and NaCl concentration. Figure 7.7 shows a similar plot to that of Daly and Saunders (Fig. 1 in [30]), but this time the hydrodynamic size is plotted as a function of NaCl concentration, at different temperatures. This figure clearly illustrates the strong decrease in solvency of water for PNIPAM with increasing NaCl concentration, particularly at 25 °C. The CFT of the PNIPAM microgel dispersions, as a function of NaCl concentration, is shown in Fig. 7.8 (cf. Fig. 3 in [30]). There are three distinct regions. Below

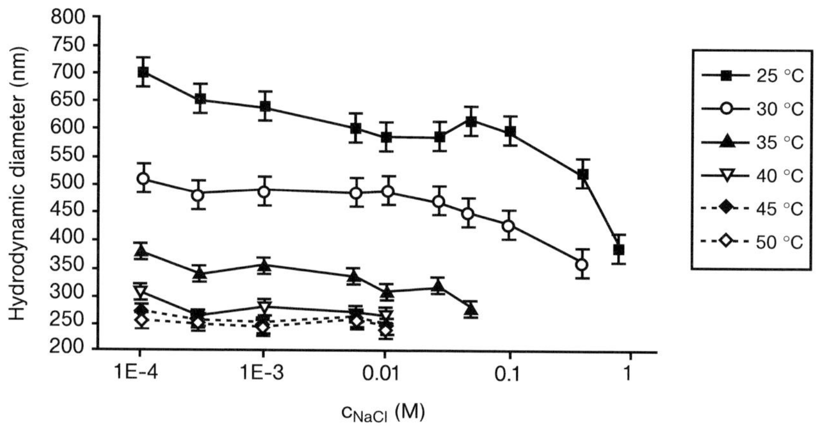

Fig. 7.7 Hydrodynamic diameter of PNIPAM particles, from dynamic light scattering measurements, as a function of log[NaCl], at different temperatures.

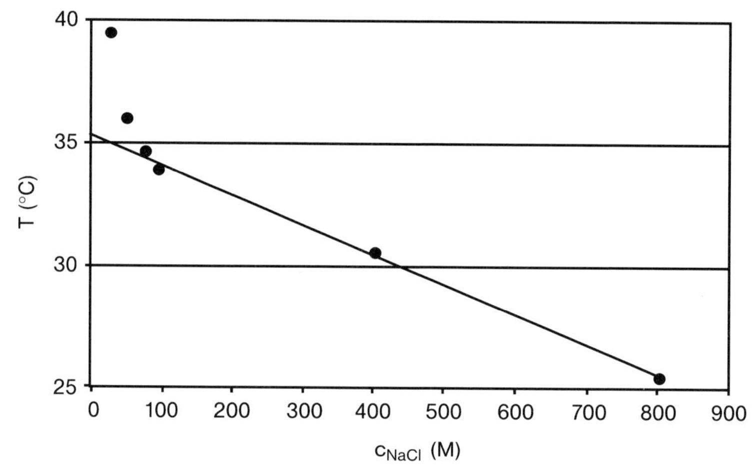

Fig. 7.8 Critical flocculation temperature of PNIPAM particle dispersions as a function of NaCl concentration.

$\sim$10 mmol dm^{-3} NaCl, no flocculation occurred at any temperature (over the range studied). At these very low NaCl concentrations, the electrostatic repulsion is sufficiently strong to prevent any aggregation. At NaCl concentrations $> \sim$100 mmol dm^{-3} the electrostatic repulsion is eliminated and the decrease in CFT values simply reflects the decrease in solvency of water for poly(NIPAM) at these high salt concentrations and the corresponding increased Hamaker constant of the particles, referred to earlier. Between 10 and 100 mol dm^{-3} NaCl concentration, there is an interplay between the electrostatic and van der Waals interactions controlling the CFT. Rasmusson et al. analyzed their data in terms of the theory presented in Section 7.2.2. They showed that, at the particle concentration used, the critical value of G_{min} for the onset of flocculation (i.e. where $\Delta G_{floc}=0$) was $\sim 8kT$, in line with the calculations reported above by Fernando-Nieves et al.

Wu et al. [37] studied (experimentally and theoretically) the equilibrium phase-coexistence behavior of PNIPAM particles in water as a function of temperature. The effective pair potential between the particles was represented as a Sutherland-like potential, where the size and energy parameters are correlated with particle radius and the solution osmotic second virial coefficients, obtained from static and dynamic light scattering experiments. They modeled the fluid–solid coexistence boundary, using a first-order perturbation model for the fluid and a cell model for the solid (crystal). They predict that a solid phase may appear (i.e. freezing) at both low and high temperatures. The low-temperature case is the classical hard-sphere one which occurs at high ϕ_p and which is entropically driven. The high-temperature case, however, is associated with the flocculation of the particles when they de-swell and is energetically driven.

There have been a number of investigations of the effect of electrolyte and temperature on the stability of microgel particles containing NIPAM plus other co-monomers. Benee et al. [38] investigated P(NIPAM/VL)] microgel particles (VL=vinyl laurate), using a range of techniques. It was found that the critical aggregation concentrations of NaCl for those microgel particles were lower than for PNIPAM microgel particles. This indicates that the Hamaker constant is greater for these copolymer microgel particles at a given value of ϕ, presumably reflecting the greater hydrophobicity of VL compared with NIPAM.

Garciá-Salinas et al. [39] investigated the effects of added electrolyte and temperature on the aggregation of P(NIPAM/AMPS) microgel particles (AMPS=2-acrylamido-2-methylpropanesulfonic acid). AMPS is a more hydrophilic monomer than NIPAM. They reported that the temperature-induced aggregation of the particles was not always reversible; on reducing the temperature again below the CFT value, in the presence of 0.5 mol dm^{-3} NaCl, in particular if the system was left in the flocculated state for any length of time. Aggregation was only always reversible if the dispersion temperature was reduced to below 20 °C, that is, well below the CFT value. They suggested that this irreversibility, on standing, arises from entanglements between polymer chains from contacting particles in an aggregate. This concept had been proposed previously by Zhu and Napper [40] for PNIPAM chains grafted to the surface of polystyrene particles after they have flocculated.

Kratz et al. [41] investigated P(NIPAM/AA) (AA=acrylic acid) microgel particles. Interestingly, they reported a two-stage swelling transition for the microgel particles containing the highest AA contents, i.e. those prepared using 12.5 wt.% AA. A low-temperature swelling transition at around 30 °C was found at low pH values, which is expected for this type of system. The higher temperature swelling transition of about 50 °C is intriguing. It may be due to a second-stage collapse associated with the microgel periphery. This idea was suggested at about the same time by Daly and Saunders [42]. The P(NIPAM/AA) dispersions of Kratz et al. exhibited reversible flocculation when heated in the presence of 0.2 mol dm^{-3} NaCl. They found that the CFT increased with increasing AA incorporated into their microgels. This reflects the increased electrostatic repulsion between particles with higher AA incorporation.

Zha et al. [43] investigated P(NIPAM/DMAEMA) (DMAEMA = dimethylaminoethyl methacrylate) microgel particles. However, less than 2.5 mol% DMAEMA was incorporated into the microgel particles. They found similar trends to those of Daly and Saunders [30]. Structure-making anions caused greater particle de-swelling and lower CFT values than structure-breaking anions. The CFT values for the P(NIPAM/DMAEMA) microgels were similar in magnitude to those for PNIPAM microgels reported by Daly and Saunders [30], reflecting the low level of DMAEMA incorporated into the microgel particles.

An interesting microgel system would be one that shows the *reverse* electrolyte effect, that is, it swelled, rather than de-swelled, on addition of electrolyte. Polyampholyte-based microgel particles, which contain an equal number of moles of a strong cationic monomer and a strong anionic monomer, would be expected to show this effect. Such a system has been investigated by Neyret and Vincent [44]. They synthesized, using an inverse microemulsion polymerization route, microgel particles containing equal numbers of moles of AMPS and MADQUAT [2-(methacryloyloxy)ethyltrimethylammonium chloride)]. These particles were strongly aggregated in water, but re-dispersed on adding sufficient electrolyte. This led to swelling of the microgel particles, as counterions entered the microgel particles to reduce the mutual electrostatic attraction between the positive and negative segments. Hence, a reduction in the particle Hamaker constant now occurs, leading to weaker interparticle attraction.

The internal structure of PNIPAM microgel particles has recently been studied using small-angle neutron scattering (SANS) [45]; the reader is directed to that publication for more information about the technique and its usefulness for studying the structure of microgel particles. Here, we simply present a brief discussion of the use of SANS to obtain qualitative insights about the nature of the flocculated particles. Figure 7.9 shows the effect of NaCl concentration on the neutron scattering profiles for PNIPAM dispersions in D_2O, measured at 25 °C. Data, plotted as the product of the scattered intensity [$I(q)$] and the square of the scattering vector (q), versus q, are shown for two NaCl concentrations: 0.60 mol dm^{-3}, where the dispersion is stable, and 0.71 mol dm^{-3}, where the dispersions had been observed to flocculate at 25 °C. Corresponding data for the microgel particles dispersed in pure D_2O, measured at 32 and 50 °C, are also shown for comparison. The profiles at 32 and 50 °C, in the absence of NaCl, correspond to the partially expanded and collapsed states, respectively. At high q values the gradient is approximately zero at 32 °C, indicating $I(q) \sim q^{-2}$, corresponding to Lorentzian scattering, suggesting that the PNIPAM chains within the microgel particles are in a solution-like environment. On the other hand, at 50 °C, $I(q) \sim q^{-4}$, over the whole q range, corresponding to Porod scattering, which one expects for harder particles. At 25 °C, in the presence of NaCl at both 0.60 and 0.71 mol dm^{-3}, it can be seen that both SANS profiles show a

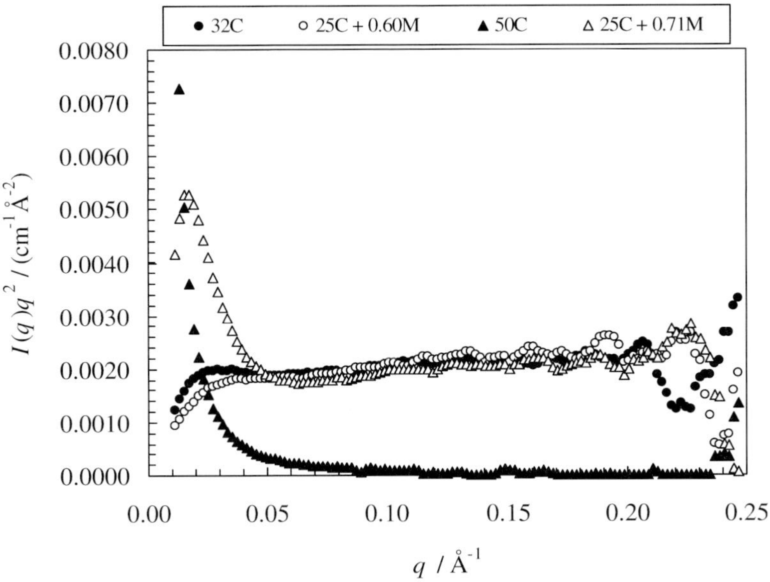

Fig. 7.9 Kratky plots for PNIPAM microgel dispersions in D_2O, measured in the presence of NaCl at 25 °C and in the absence of NaCl at 32 and 50 °C.

strong scattering at high q, again indicative of solution-like polymer chains within the microgel particles. This is despite the fact that the dispersion containing 0.71 mol dm^{-3} was aggregated (also evidenced by the high scattering at low q in this case). These data show that aggregation occurs for PNIPAM microgel particles whilst the particles are only partially collapsed. This implies that, in the presence of sufficient salt, at 25 °C, the interparticle van der Waals interactions become strong enough to induce flocculation of the microgel particles, even though the Hamaker constant of the particles, A, has not reached its maximum value of A_2 (see Eq. 5). These results are consistent with the conclusions from the work of Routh and Vincent [36] and Rasmusson et al. [5] discussed above, where it was shown that microgel particles retain a considerable volume fraction of water molecules under conditions where flocculation is induced, either by increasing temperature of by increasing electrolyte concentration.

Another interesting result from SANS experiments on PNIPAM particle dispersions is that the segment volume fraction is not uniform across these microgel particles and that ϕ, therefore, is an *average* value [21, 45]. This, of course, has implications for calculation of the van der Waals attraction between the particles.

7.3.2
Depletion-induced Aggregation

The first reported study of inducing flocculation of microgel particles by adding non-adsorbing polymer of which we are aware is that of Clarke and Vincent [46]. These authors prepared a series of cross-linked polystyrene (PS) particles in water, containing different percentages of divinylbenzene as the cross-linking monomer. These particles were then re-dispersed in ethylbenzene. In this solvent, the average value of ϕ for the swollen microgel particles varied from 0.10 to 0.63, with increasing cross-linker concentration. The effect of adding homopolymer PS was then investigated. It was found that weak, reversible flocculation occurred above a critical free polymer concentration ($\phi_{2,f}$), which was strongly dependent on the microgel particle concentration; $\phi_{2,f}$ decreased with increasing particle concentration. This is consistent with the theory of weak flocculation discussed in Section 7.2.2, in terms of the interplay of ΔG_i (determined by G_{min}) and ΔG_{hs} (determined by the particle concentration). In this case G_{min} is controlled by $\phi_{2,f}$ through the depletion interaction, rather than the van der Waals interaction, which remains negligible in these systems.

Depletion flocculation of PNIPAM microgel particles in aqueous media was first investigated by Snowden and co-workers [29, 47, 48]. The showed that such microgel particles could not be flocculated at 25 °C by adding sodium poly(styrene sulfonate) (NaPSS) of MW 50 000, at least up to concentrations of NaPSS of 0.8 wt.%, but did so at 40 °C and that the minimum amount of NaPSS required to do so decreased if NaCl was also present.

Fernández-Nieves et al. [49] have shown, using the wavelength exponent turbidity method described earlier, that P2VP microgel particles de-swell and eventually reversibly flocculate on the addition of dextran (MW 70 000) to the dispersion.

Rasmusson et al. [5] extended the studies of Snowden and co-workers [29, 47, 48]. They investigated the minimum concentration of NaPSS (MW 100 000) required to induce flocculation of PNIPAM microgel particles as a function of temperature in the absence of added NaCl. Their results are shown in Fig. 7.10. Over the relatively small temperature range studied (23–33 °C) the particles do shrink (by about a factor of 2), but this is still not sufficient for the van der Waals attraction to play a dominant role. Hence depletion attraction must be the more significant interaction occurring here. The question then arises of why less NaPSS is required to induce flocculation at higher temperatures. As the particles become smaller with increasing temperature, so the depletion interaction should be smaller at a given NaPSS concentration, and more NaPSS would seemingly be required to reach the condition for G_{min} to be large enough for ΔG_{floc} to be zero (the boundary condition for flocculation). One possibility has to do with the variation in "softness" of periphery of the PNIPAM particles with temperature. As mentioned in Section 7.2.1, Jones and co-workers [14, 15] have discussed the depletion interaction for "soft" particles, in particular, for particles carrying terminally grafted polymer chains. They demonstrated that the depletion interaction becomes weaker, the softer the grafted layer, for a given

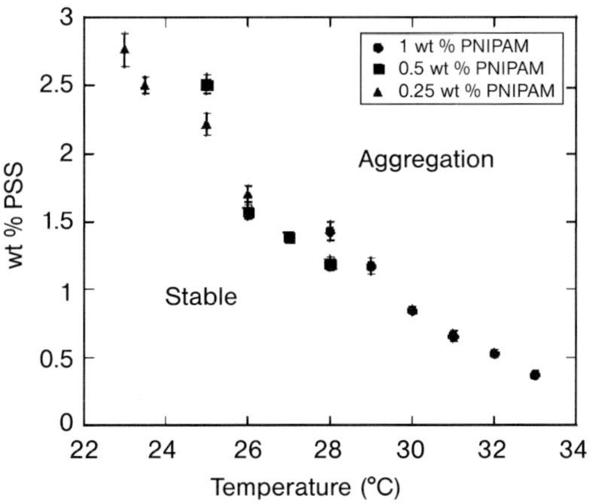

Fig. 7.10 Minimum wt.% of added PSS (MW 100000 g mol^{-1}) required to induce flocculation of the PNIPAM particles as a function of temperature, for three PNIPAM concentrations.

core size. This is due to (partial) interpenetration of the free polymer chains into the periphery of the grafted layer. A similar situation could arise for swollen microgel particles. Any interpenetration of the free chains into the soft particle reduces the effective depletion layer thickness of the free chains. It may well be that with increasing temperature the microgel particles become harder, so that it is less easy for free NaPSS chains to penetrate the periphery of the microgel particles. Hence, the depletion interaction becomes stronger and less NaPSS is required to reach the $\Delta G_{floc}=0$.

These results beg the question of just how far the free polymer chains can actually penetrate into the microgel particles. Clearly, this depends on the relative sizes of the "mesh" within the interior of the microgel particle network and of the polymer chains in solution. The former may vary from the periphery to the interior of the microgel particles, since, as discussed earlier, the cross-link density is rarely uniform. Moreover, there really has to be an energy gain (attraction) for the free polymer chains to enter the microgel particle network, in order to overcome the translational and configurational entropy losses suffered by a free chain entering the network. As mentioned in Section 7.1, Bradley et al. [12] have recently considered this in terms of PEO chains entering P(NIPAM/AAc) particles in water. The energy gain in this case comes largely from the strong H-bonding interaction (at low pH values) between the PEO chains and the AAc moieties inside the particles. The extent (and rate) of interpenetration, for a given microgel particle, depends strongly on the PEO molecular weight.

7.3.3
Heteroaggregation

Fernández-Barbero et al. [50, 51] studied the heteroaggregation of (equal numbers of) hard, negatively charged PS particles and soft, positively charged P2VP microgel particles, as a function of ionic strength, at room temperature. They used static light scattering to obtain the heteroaggregate fractal dimensions and dynamic light scattering to follow the kinetics. At low electrolyte concentrations the heteroaggregates were observed to be highly branched ($d_f \approx 1.6$), since both repulsive and attractive forces are present in the structures. The kinetics was essentially diffusion controlled. However, at very high electrolyte concentrations ($\sim 1\ \text{mol dm}^{-3}$) very compact clusters were obtained ($d_f \approx 2.3$), reflecting the weakness of the attractive forces under these conditions. This behavior is similar to the homoaggregation of microgel particles in this respect, as discussed earlier.

Snoswell et al. also recently [52] studied the heteroaggregation of cationic P2VP microgel particles and anionic PS particles. The resulting heteroaggregated particles were then concentrated by vacuum filtration, freeze-dried and characterized by mercury porosimetry and electron microscopy. Heteroaggregation, at a constant KCl concentration of $0.01\ \text{mmol dm}^{-3}$, was "arrested" at various time intervals by the addition of anionic silica nanoparticles, thereby limiting the size of the flocs. The pore volume increased from 58 to 63 vol.% as the aggregation time prior to "arrest" was increased from 15 to 120 s, a result of changes in packing efficiency within the filter cake. In a second set of experiments, the aggregation time prior to arrest was maintained at 120 s, while the KCl concentration was varied between 0.01 and $10\ \text{mmol dm}^{-3}$. The pore volume of the aggregates decreased from 63 to 54 vol.% as the electrolyte concentration increased, in accord with the changing Debye length. The inclusion of soft deformable microgels resulted in aggregates with higher mechanical strength and porosity than obtained with heteroaggregates of anionic and cationic hard latex particles. Furthermore, incorporation of swellable microgels within a porous structure offers potential for creating novel structures suitable for controlled release applications.

Acknowledgments

The authors would like to thank all their former students and co-workers who have worked with them in this field over the years, together with all the companies and governmental funding agencies who have funded this work. They would also like to congratulate Professors Platikanov and Exerowa, to whom this chapter is dedicated, as they celebrate their 70th birthdays, and B.V. in particular would like to thank them both for their great friendship over many years.

References

1 M.J. Murray, M.J. Snowden, *Adv. Colloid Interface Sci.*, **1995**, *54*, 73–98.
2 B.R. Saunders, B. Vincent, *Adv. Colloid Interface Sci.*, **1999**, *80*, 1–25.
3 B.R. Saunders, B. Vincent, *Encyclopedia of Surface and Colloid Science*, Marcel Dekker, New York, **2002**, 4544–4559.
4 A. Fernández-Nieves, A. Fernández-Barbero, B. Vincent, *J. Chem. Phys.*, **2003**, *119*, 10383–10388.
5 M. Rasmusson, A. Routh, B. Vincent, *Langmuir*, **2004**, *20*, 3536–3542.
6 H.M. Crowther, B. Vincent, *Colloid Polym. Sci.*, **1998**, *276*, 46–51.
7 P. Bouillot, B. Vincent, *Colloid Polym. Sci.*, **2000**, *278*, 74–79.
8 A. Loxley, B. Vincent, *Colloid Polym. Sci.*, **1997**, *38*, 6129–6134.
9 A. Fernández-Nieves, A. Fernández-Barbero, B. Vincent, F.J. de las Nieves, *Macromolecules*, **2000**, *33*, 2114–2118.
10 B.R. Saunders, B. Vincent, *J. Chem. Soc., Faraday Trans.*, **1996**, *92*, 3385–3389.
11 B.R. Saunders, B. Vincent, *Colloid Polym. Sci.*, **1997**, *275*, 9–17.
12 M. Bradley, J. Ramos, B. Vincent, *Langmuir*, **2005**, *21*, 1209–1215.
13 N. Cawdery, B. Vincent, in *Colloidal Polymer Particles*, ed. J.W. Goodwin, R. Buscall, Academic Press, New York, **1995**, 245–275.
14 A. Jones, B. Vincent, *Colloids Surf.*, **1989**, *42*, 113–138.
15 A. Milling, B. Vincent, S. Emmett, A. Jones, *Colloids Surf.*, **1991**, *57*, 185–195.
16 H. Oshima, *Adv. Colloid Interface Sci.*, **1995**, *62*, 189–195.
17 A. Fernández-Nieves, A. Fernández-Barbero, F.J. de las Nieves, B. Vincent, *J. Phys.: Condens. Matter*, **2000**, *12*, 3605–3614.
18 B. Vincent, *J. Colloid Interface Sci.*, **1973**, *42*, 270–285.
19 B.V. Derjaguin, L.D. Landau, *Acta Physiochim.*, **1944**, *14*, 1073–1085.
20 E.J.W. Verwey, J.Th. Overbeek, *Theory of the Stability of Lyophobic Colloids*, Elsevier, Amsterdam, **1948**.
21 H.M. Crowther, B.R. Saunders, S.J. Meares, T. Cosgrove, B. Vincent, S.M. King, G.E. Yu, *Colloids Surf.*, **1999**, *152*, 327–333.
22 D.H. Napper, *Polymeric Stabilization of Colloidal Dispersions*, Academic Press, London, **1983**.
23 M. Bevan, P. Scales, *Langmuir* **2002**, *18*, 1474–1484.
24 P. Scales, F. Grieser, D. Furlong, T.W. Healy, *Colloids Surf.*, **1986**, *21*, 55–68.
25 B. Vincent, P.F. Luckham, F.A. Waite, *J. Colloid Interface Sci.*, **1980**, *73*, 508–521.
26 R. Feigin, J. Dodd, D.H. Napper, *Colloid Polym. Sci.*, **1981**, *259*, 1027–1030.
27 J.A. Long, D.W.J. Osmond, B.Vincent, *J. Colloid Interface Sci.*, **1973**, *42*, 545–553.
28 R.H. Pelton, P. Chibante, *Colloids Surf. A*, **1986**, *20*, 247–257.
29 M.J. Snowden, B. Vincent, *J. Chem. Soc., Chem. Commun.*, **1992**, 1103–1105.
30 E. Daly, B.R. Saunders, *Langmuir*, **2000**, *16*, 5546–5552.
31 D.J. Shaw, *Introduction to Colloid and Surface Chemistry*, 4th edn., Butterworth, London, **1993**, 235.
32 D. Duracher, A. Elaïssar, C. Pichot, *Colloid Polym. Sci.*, **1999**, *277*, 905–910.
33 A. Loxley, B. Vincent, *Colloid Polym. Sci.*, **1997**, *275*, 1108–1114.
34 A. Fernández-Nieves, A. Fernández-Barbero, B. Vincent, F.J. de las Nieves, *Langmuir*, **2001**, *17*, 1841–1846.
35 A. Fernández-Nieves, J.S. van Duijneveldt, A. Fernández-Barbero, B. Vincent, F.J. de las Nieves, *Phys. Rev. E*, **2001**, *64*, 1–10.
36 A. Routh, B. Vincent, *Langmuir*, **2002**, *18*, 5366–5369.
37 J. Wu, S. Huang, Z. Hu, *Macromolecules*, **2003**, *36*, 440–448.
38 L.S. Benee, M.J. Snowden, B.Z. Chowdhry, *Langmuir*, **2002**, *18*, 6025–6030.
39 M.J. Garciá-Salinas, M.S. Romaro-Cano, F.J. de las Nieves, *J. Colloid Interface Sci.*, **2002**, *248*, 54–61.
40 P.W. Zhu and D.H. Napper, *J. Colloid Interface Sci.*, **2004**, *268*, 380–386.
41 K. Kratz, T. Hellweg, W. Eimer, *Colloids Surf. A*, **2000**, *170*, 137–149.
42 E. Daly, B.R. Saunders, *Phys. Chem. Chem. Phys.*, **2000**, *2*, 3187–3193.

43 L. Zha, J. Hu, Z. Wang, S. Fu, M. Luo, *Colloid Polym. Sci.*, **2002**, *280*, 1116–1121.

44 S. Neyret, B. Vincent, *Polymer*, **1997**, *38*, 6129–6134.

45 B. R. Saunders, *Langmuir*, **2004**, *20*, 3925–3932.

46 J. Clarke, B. Vincent, *J. Chem. Soc., Faraday Trans.*, **1981**, *77*, 1831–1843.

47 M. J. Snowden, B. Vincent, *Am. Chem. Soc. Symp. Ser.*, **1993**, *532*, 153–160.

48 M. J. Snowden, N. Marston, B. Vincent, *Colloids Polym.*, **1994**, *272*, 1273–1280.

49 A. Fernández-Nieves, A. Fernández-Barbero, B. Vincent, F. J. de las Nieves, *Prog. Colloid Polym. Sci.*, **2000**, *115*, 134–136.

50 A. Fernández-Barbero, A. Loxley, B. Vincent, *Prog. Colloid Polym. Sci.*, **2000**, *115*, 84–87.

51 A. Fernández-Barbero, B. Vincent, *Phys. Rev. E*, **2000**, *63*, 1–7.

52 D. R. E. Snoswell, T. J. Rogers, A. M. Howe, B. Vincent, *Langmuir*, **2005**, *21*, 11439–11445.

53 T. G. Park, A. S. Hoffman, *Macromolecules*, **1993**, *26*, 5045–5050.

8 Progress in Structural Transformation in Lyotropic Liquid Crystals

Idit Amar-Yuli and Nissim Garti

Abstract

Polar lipids and certain surfactants are known to form thermodynamically stable lyotropic liquid crystals (LLC) when mixed with water. The major phases are lamellar (L_α), normal and reverse hexagonal (H_I and H_{II}) and cubic bicontinuous and discontinuous structures (V_I, V_{II} and I_I, I_{II}, respectively). Theoretically, the transformation sequence of the phases with increasing water content is $I_{II} \rightarrow H_{II} \rightarrow V_{II} \rightarrow L_\alpha \rightarrow V_I \rightarrow H_I \rightarrow I_I$. Lyotropic liquid crystal transformation has been extensively studied and was found to take place also upon surfactant modifications (head or tail), addition of a guest molecule (hydrophilic or hydrophobic), co-surfactant or electrolyte and varying the temperature. The phases are of growing scientific and industrial interest because of their structural resemblance to human membranes through which drug passage of lipophilic compounds (vitamins, fats, oil and cholesterol) occurs and because of their high surface area and solubilization capacities. The variations in the phase formations, the phase behavior and phase transitions are of significant importance when designing a potential application for these systems. This chapter presents studies related to phase behavior and phase transitions as a function of different physical and chemical conditions. The relationship between surfactant geometry that includes tail volume, tail length and area per head group and the corresponding phase formation is stressed. Research demonstrating dependence of the phase behavior on the addition of a third component such as hydrophobic, hydrophilic guest molecule or co-surfactant is summarized. A brief overview of the instruments used in the above is also presented, illustrating their functionality in detecting the phase transitions and the unique information potentially extracted from each instrument. The main findings show successful control of lyotropic liquid crystal structure by altering the surfactant geometry in various respects: unsaturation site or degree, head or tail chain length and by an additional component such as linear or branched oils or alcohols.

Colloids and Interface Science Series, Vol. 2
Colloid Stability: The Role of Surface Forces, Part II. Edited by Tharwat F. Tadros

ISBN: 978-3-527-31503-1

8.1 Introduction

Scientists believe that phase transitions between lamellar and non-lamellar mesophases (hexagonal and cubic mesophases) can be of significant importance to the understanding of the mode of action of biological membrane lipid–water systems since similar transitions may be induced by a number of biologically relevant factors [1, 2] and because similar phase transitions can occur in many processes associated with biological membranes, during which the lipid bilayer structure is transiently disrupted (e.g. fusion and exo- and endocytosis) [2].

The ability to control the phase behavior and transformation between the mesophases is also of great importance for designing a potential application as a delivery vehicle.

Binary and ternary phase diagrams were constructed when studying the phase behavior of lyotropic liquid crystals (LLC). Surfactants, co-surfactants and additional guest molecules (hydrophilic or hydrophobic) were utilized to form a variety of LLC phases and transformations. In most cases, with increasing water content the phase transformations occur in the order cubic to hexagonal to lamellar in liquid crystals of type I (oil in water) and vice versa in type II. Some unique mesostructures, sometimes identified as ill-defined LLC such as sponge (L_3) or ribbon (R_I) phases, have also been detected [3].

Different instruments have been used to detect the phases and the phase transitions, with each technique providing unique information. The techniques can be divided into two categories: *direct techniques* – measurements which provide the phase identification (phase symmetry) such as small-angle X-ray scattering and polarized microscope – and *indirect techniques* – where the data allow characterization of the phases, such as nuclear magnetic resonance and rheological techniques.

8.2 Liquid Crystal Mesophases

8.2.1 Lamellar Mesophases

Lamellar mesophases (ideal mesostructure) consist of planar, parallel stacks of amphiphilic bilayers, forming a 1D lattice (Fig. 8.1). Lamellar mesophases are identified by equally spaced peaks, corresponding to a, $2a$, $3a$, etc., where a is the spacing between adjacent bilayers. Conventionally, one calls a mesophase "lamellar" when it is optically anisotropic and exhibits equally spaced peaks by small-angle X-ray diffraction [4].

The lamellar mesophase (L_α) is of intermediate viscosity to the more freely flowing micellar and hexagonal mesophases. The hydrogen-bond system, in the lamellar sheets, formed by the polar head group is strong compared with the

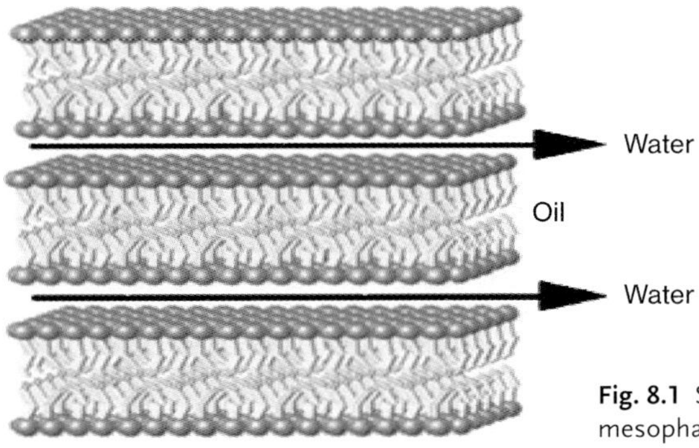

Fig. 8.1 Schematic presentation of lamellar mesophase.

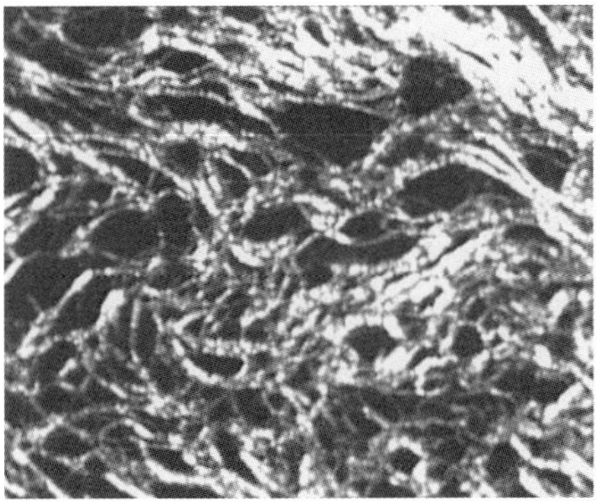

Fig. 8.2 Polarizing microscopy image of the lamellar mesophase at ca. 25 °C.

weak van der Waals interactions between the hydrocarbon chains, hence the lipids' chains can "melt", although the polar groups are associated into sheets [5].

L X-ray spectra are characterized by scattering peaks in the ratio 1:2:3, etc., characteristic of the inter-bilayer spacing and a diffuse 4.5 $Å^{-1}$ X-ray scattering band.

Like all anisotropic phases, lamellar mesophases exhibit distinct optical textures when viewed through a cross-polarized microscope. Typically the texture is "streaky" or mosaic-like, as can be seen in Fig. 8.2.

8.2.2 Hexagonal Mesophases (H_I, H_{II})

The anisotropic hexagonal phase consists of a dense packing of infinite cylindrical micelles, arranged on 2D hexagonal lattice. It is often identified by a characteristic "fan" texture under a cross-polarized microscope due to focal conic domains of columns (Fig. 8.3). This mesophase is of intermediate viscosity to lamellar and bicontinuous cubic mesophases [6].

In contrast to lamellar mesophases, which are equally curved towards both hydrophilic and hydrophobic regions, hexagonal mesophases exhibit two types.

Fig. 8.3 Polarizing microscopy image of the reverse hexagonal mesophase at ca. 25 °C.

The normal hexagonal (type I, termed H_I) consists of lipid cylinders arranged in a continuous aqueous matrix, and the reverse hexagonal (type II, termed H_{II}, shown in Fig. 8.4) consists of water-filled cylinders which are arranged in a continuous lipid matrix. In both cases, X-ray scattering has revealed that the chains are molten and the small-angle spectrum contains Bragg peaks in the ratio $1 : \sqrt{3} : \sqrt{4} : \sqrt{7}$, etc., corresponding to the reflections from the 2D hexagonal symmetry.

The hexagonal unit cell (the smallest periodic group from which the liquid crystal can be built) is of a primitive (P) type, which represents one infinite rod per unit cell corner. The hexagonal unit cell is described in terms of three sides of the cell, a, b and c, where $a = b$ represents the space between each corner of the cylinder micelles (termed the lattice parameter, a) and c is the infinite length of each rod. As a result, the angles between a and b are 60° and 120° and between a (or b) and c they are 90°.

Fig. 8.4 Schematic presentation of H_{II} mesophase including the structural parameters derived from SAXS: a, lattice parameter; d, distance between planes. The shaded void volumes between the cylinders in the hydrophobic regions are derived from the bending of the lipid monolayer and are discussed in the text.

8.2.3
Cubic Mesophases

Cubic mesophases exhibit the most complex spatial organization of all known LLC. The mesophases do not display an optical texture. So far seven different cubic structures have been discovered (Q^{212}, Q^{223}, Q^{224}, Q^{225}, Q^{227}, Q^{229} and Q^{230}). They can have either a micellar or a bicontinuous structure and their topology can be either normal (type I, oil-in-water) or inverse (type II, water-in oil). Of these, only three have an inverse and bicontinuous structure: the primitive (P) type [body-centered lattice (*Im*3*m*, denoted Q^{224})], the diamond (D) type (primitive lattice *Pn*3*m*, denoted Q^{229}) and, most frequently, the gyroid (G) type (*Ia*3*d*, denoted Q^{230}). The type of crystallographic space group determines the type of the "infinite periodic minimal surface" (IPMS) that is located at the bilayer misplane, which is the interface of two monolayer membranes [7]. The IPMS is infinite array of connected saddle surfaces with negative Gaussian curvature and zero mean curvature at all points. The IPMS divides the space into separated sub-spaces and can therefore be used to describe the interface between the incommensurable parts of the structure. Structure P contains two aqueous channel systems that are separated by a bilayer (Fig. 8.5). The unit cell has three mutually perpendicular aqueous channels which are connected to contiguous unit cells forming a cubic array. The structure D also contains a bilayer that separates two interpenetrating aqueous channel systems that form a diamond lattice (Fig. 8.5). Here, four aqueous channels of the D-surface meet at a tetrahedral angle of 109.5°. The G-surface lacks the twofold axis and mirror plane symmetries that are present both in the D- and P-phases. The aqueous compartments of the G-surface consist of two separate, left- and right-handed helical channels. The aqueous channels can extend through the matrix, as in the P-surface, but the centers of the water channels never intersect. Rather, they are connected to give rise to a helical arrangement. The *Ia*3*d* structure is the least hydrated of the three and has helical channels possessing opposing chiralities [4, 8].

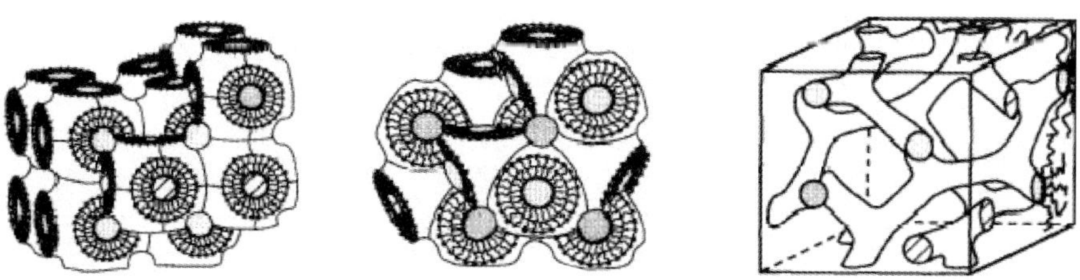

Fig. 8.5 From left to right: schematic presentations of Q_P primitive, Q_D diamond, and Q_G gyroid cubic lyotropic liquid crystals.

8.3
Mesophase Transformations

Most binary systems studied to date conform to a scheme (Fig. 8.6) where the phases are arranged in their "natural" sequence (although, in general, a given system will not necessarily display all phases shown in the scheme).

There are several major models that elucidate the tendency of the surfactant self-assembly in lyotropic liquid crystal transformations [9–11].

It was concluded by Israelachvili et al. [9] that the effective geometry of the amphiphilic molecules in the system determines the lipid aggregate shape. The geometric effect can be envisaged through the critical packing parameter, *CPP*:

$$CCP = \frac{V_S}{a_0 l} \tag{1}$$

where V_s is the hydrophobic chain volume, a_0 is the cross polar head-group area and l is the chain length of the molecule in its molten state [9]. It is known that reversed-type mesophases are formed from amphiphiles with $CPP > 1$.

Kirk et al. [10] and Gruner et al. [7, 11] used a spontaneous monolayer curvature to explain mesomorphic transitions, such as the L_α–H_{II} transition, in which the shape of the monolayer abruptly changes. The lipid monolayer, which forms a bilayer or rolls into tubes that pack as H_{II} mesophases, may be characterized by the spontaneous radius of curvature, R_0, which represents the minimum

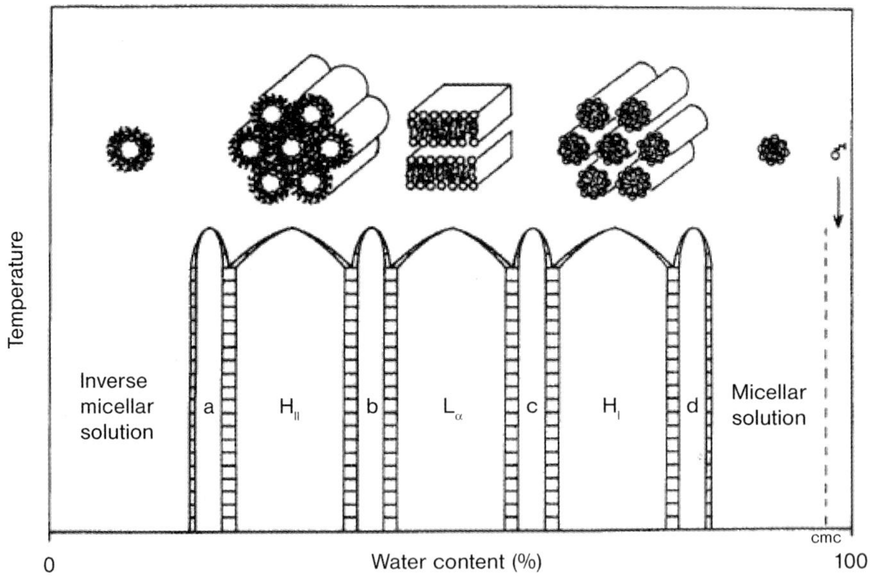

Fig. 8.6 Hypothetical lyotropic binary phase diagram when the phase transitions are induced by varying the water content.

elastic free energy state of the layer with respect to its bending degree. A small deviation from R_0 of a layer (in an H_{II} mesophase) changes the free energy per unit area by

$$dFeF = \frac{K}{2}\left(\frac{1}{R_I} - \frac{1}{R_0}\right)^2 \tag{2}$$

where R_I is the principal radius of curvature of the layer and K is a constant of the material. In general, R_0 is proposed to be independent of any particular phase structure and only to be a property of the monolayer. This theory predicts that as a result of a small R_0 in the L_α mesophase, $R_I \approx \infty$ and, according to Eq. (2), a large drop in the free energy may be accomplished by curling, such that $R_I \approx R_0$, which then leads to the formation of an H_{II} mesophase. If R_0 is large, then Eq. (2) shows that the driving force is diminished, causing L_α to be the preferable structure. In addition, for large R_0, large positive hydrocarbon packing free energies must be considered in the transition into the H_{II} mesophase.

According to Siegel [12], the initial step in the L_α–H_{II} or L_α–Q transitions is the formation of very short-lived micelles between apposed bilayers. These micelles may then transform into either an H_{II} phase or to new intermediate, interlamellar attachments (ILA), which are very long-lived. The ILA were suggested to have a structure similar to the unit cell in some bicontinuous cubic phases. The probability of the ILA formation largely depends upon the experimental parameter Z:

$$Z = \frac{a_L}{a_H} \tag{3}$$

where a_L is the average polar head group area in the L_α phase, just below the L_α–H_{II} phase transition temperature, and a_H is the average area per molecule in the H_{II} phase just above the transition temperature. The parameter Z is closely related to R_0 in Gruner's theory and to the critical packing parameter in Israelachvili's theory. For large values of Z the H_{II} phase is formed, and for Z values decreasing to the approximate value 1.2, the system will form isotropic phases.

Mesophases transformations in binary systems may be induced by varying the composition (water content) or the temperature.

Phase transitions which are driven primarily by changes in water content can be found in many ionic lipid systems. The phase diagram for lyso-phospholipids may fall under this category, although they are uncharged. Figure 8.7 shows the binary phase diagram of the 1-lyso-oleoylphosphatidylcholine–water system [13]. At ~7 wt.% of water and a wide range of temperatures (25–100 °C), a lamellar liquid crystal (L_α) was formed. Further hydration of the surfactant head group, by increasing water content, influenced the L_α flat curvature. As a result of increasing the curvature towards the lipids region, the mesophase transformed into cubic (I_I), normal hexagonal (H_I) and eventually to normal micellar solu-

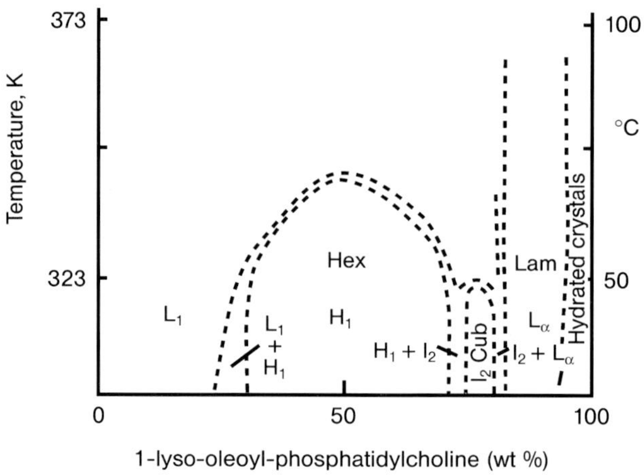

Fig. 8.7 Binary phase diagram of 1-lyso-oleoylphosphatidylcholine–water.

tion (L_1) with the highest curvature. One should note that the mesophases transition sequences depicted in Fig. 8.7 are similar to those portrayed in Fig. 8.6.

Transitions that are driven predominantly by temperature are typical for systems containing diacyl zwitterionic phospholipids. Figure 8.8 depicts the phase diagram of a more hydrophobic surfactant, diarachinoylphosphatidylethanolamine, and water [14]. It can be seen that the lamellar liquid crystal was formed at a relatively high temperature and constricted to a narrow temperature range, 80–90 °C. Temperature increase up to 120 °C can increase the volume of the hydrophobic chains by increasing their mobility and consequently caused a transformation into a more curved interface reversed hexagonal liquid crystal. One

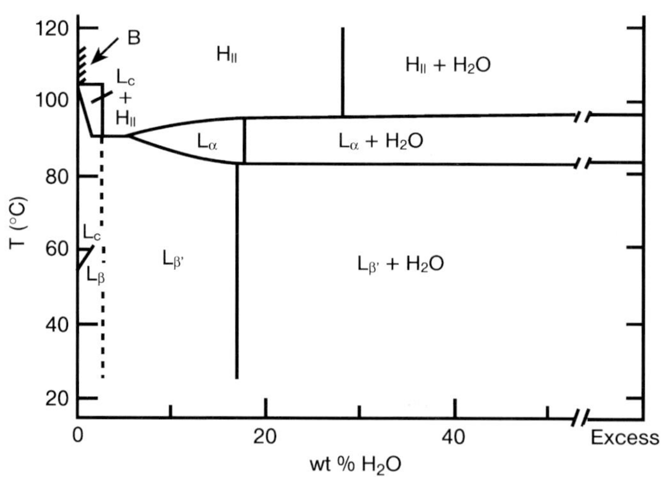

Fig. 8.8 Binary phase diagram of diarachinoylphosphatidylethanolamine–water.

should note that the phases were formed at all water contents excluding the excess water region [15].

Transitions which are affected from both water content and temperature can be shown in the extensively studied phase diagram of glycerol monooleate (GMO)–water (Fig. 8.9) [5, 16–18, 19]. When 5–20 wt.% of water was added to GMO, the polar heads of the lipid were hydrated and the water penetrated between the GMO molecules to form a lamellar (L_α) mesophase. If an additional amount of water of 20–40 wt.% was incorporated or sample heating occurred, the hydrocarbon chain disorder was enhanced, resulting in a transition from a lamellar mesophase to cubic mesophases (Q_G or Q_D) depending on the amount of water. Eventually, at $\sim 85\,^\circ$C, the cubic phase was transformed into a reversed hexagonal mesophase (H_{II}).

Larsson et al. [16] clarified the lamellar–cubic–hexagonal transitions which occur in the GMO–water system by an increasing tendency for chain divergence. They hypothetically assumed that lamellar, cubic and hexagonal phases have a negligible bilayer thickness, then, after elaborate calculations, showed that a reduction in the phases' polar head cross-sectional area corresponds to $1:3/4:1/\sqrt{3}$ $(1:0.75:0.58)$. In the lamellar phases there is a pronounced decrease in spacing with increase in temperature, indicating an increase in the cross-sectional area per molecule, along the chain (from the polar head to the methyl end-group). Near the transition temperature there is an increase in cross-sectional area per molecule of 0.5–1 Å^2 per $1\,^\circ$C [16]. When the temperature of the lamellar phase is raised, the increased molecular mobility of the hydrocarbon chains results in a demand for an increased molecular cross-sectional area, resulting in a strain on the stability of the phase. This strain may be appeased by a transition to a

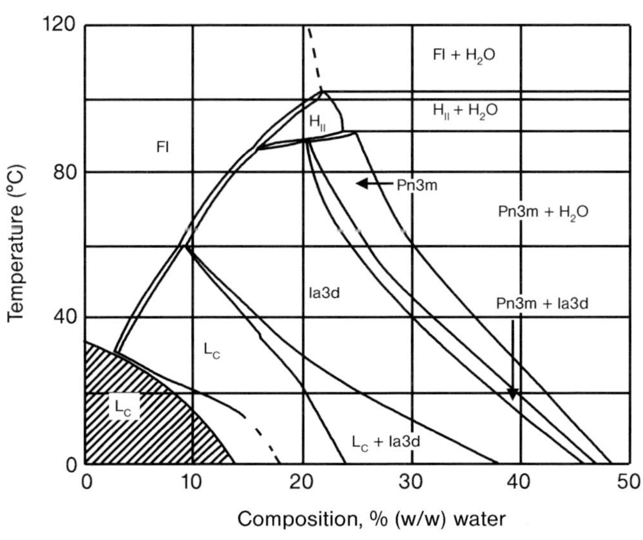

Fig. 8.9 Binary phase diagram of monoolein–water.

cubic structure, so that a close molecular packing at the polar–non-polar interfaces can persist while the relative volume of the non-polar hydrocarbon region is increased. At higher temperatures the same mechanism might result in the transition from cubic to a reversed hexagonal.

For the GMO–water system, at 25 wt.% water, in the cubic phase, Larsson et al. [16] estimated the molecular divergence as an increment of about 25% in the molecular cross-sectional area (head to tail's end). In the hexagonal phase, the corresponding divergence is estimated at ~50%.

8.3.1
Correlation Between Molecular Structure and Phase Behavior

Many studies have examined the effect of modifications to the surfactant structure (hydrophilic or hydrophobic group) and the effect of an additional component on the behavior of mesophases. From the results, one can (1) study the relationship between surfactant geometry and the corresponding mesophase formation, (2) acquire a formation methodology for a broad range of mesophases and phase transition and (3) manipulate in a controlled manner the mesophases for a required design system.

It has been suggested that the interfacial curvature which determines the phase behavior depends on the relationship between three elements: tail volume, tail length and area per head group. Therefore, we will present studies related to phase behavior and mesophase transitions as a function of these three elements.

8.3.2
The Tail Volume and/or Length (Binary System)

Misquitta and Caffrey [18] characterized the water and temperature dependence of several monoacylglycerol chain homologs. The objective of their study was to determine the quality of phase behavior prediction for a specific monoacylglycerols. Their analysis was based on existing phase diagrams for related chain homologs. Each given lipid was assigned a coordinate in N–T space as dictated by the length of the chain on either side of the olefin bond. The region between the double bond and the glycerol head group is referred to as the neck, denoted N, whereas the tail, denoted T, represents the region from the double bond to the methyl end. N and T values represent the number of the carbon atoms in the N and T regions, respectively, and were assumed to be linearly related to the transition temperatures. One series includes monoolein (18:1c9, $N=9$, $T=9$), monoeicosenoin (20:1c11, $N=11$, $T=9$) and monoeruccin (22:1c13, $N=13$, $T=9$), as seen in Fig. 8.10a. The second homologous series (Fig. 8.10b) has N fixed at a value of 10 whereas T varies from 5 (monopentadecenoin, 15:1c10) to 7 (monoheptadecenoin, 17:1c10).

In Fig. 8.10a, most of the transition temperatures increase linearly with increasing N values. For example, the transformation temperatures from H_{II} to FI

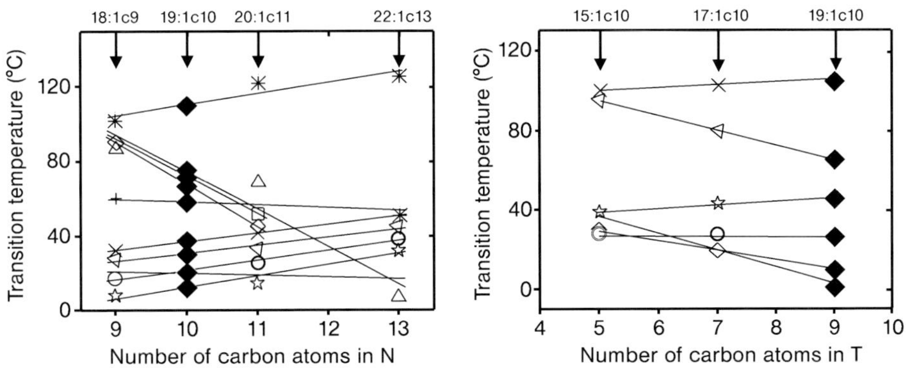

Fig. 8.10 (a) Phase transition temperature as a function of *N*, the number of carbon atoms in the neck portion of the monoacylglycerol acyl chain with *T*, the number of carbon atoms in the tail portion of the acyl chain, at a constant value of 9. The transitions are as follows: ☆ L_c-to-liquid crystal; ○ (L_c+*Ia*3*d*)-to-(L_α+ *Ia*3*d*); ○ (L_c+*Pn*3*m*)-to-(*Ia*3*d*+*Pn*3*m*); ◁ L_c-to-(L_α+FI); × dry lipid melting; + (L_α+FI)-to-(*Ia*3*d*+FI); △ (*Ia*3*d*+FI)-to-(*Pn*3*m*+H_{II}); ◇ (*Ia*3*d*+FI)-to-(H_{II}+FI); □ *Pn*3*m*-to-H_{II}; * H_{II}-to-FI. (b) Phase transition temperature as a function of *T* and *N*=10. The transitions are as follows: ○ dry lipid melting; ◇ L_c-to-(L_α+FI); + (L_α+FI)-to-(*Ia*3*d*+FI); ☆ (L_α+*Pn*3*m*)-to-(*Ia*3*d*+*Pn*3*m*); ◁ (L_α+water)-to-(*Pn*3*m*+water); × (*Pn*3*m*+water)-to-(FI+water).

are at *ca.* 100, 110, 120 and 125 °C for *N* values of 9, 10, 11 and 13, respectively. However transformation temperatures, from *Pn*3*m* to H_{II}, decrease with increasing *N* values.

Additional research by Qiu and Caffrey [20] compared the phase properties of monovaccenin and monoolein, two monoacylglycerols that differ structurally only in the position of the cis double bond. In the case of monoolein, the double bond is at the C9 position and in monovaccenin at C11. Since the molecules are positional isomers with similar effective lipid chain lengths, they exhibit similar phases in the same order with respect to temperature (–15 to 55 °C) and hydration as expected. However, there are slight differences in phase boundaries, which are shown in their phase diagrams (Figs. 8.9 and 8.11). For example, the *Pn*3*m* phase hydration boundary at 40 °C is at 46 wt.% in the monovaccenin system, as opposed to only 36 wt.% water in the monoolein system. Similarly, at 80 °C, the hydration boundaries are at 30 and 26 wt.% water, in the monovaccenin and monoolein system, respectively. For both molecules the *CPP* value is greater than unity. However, for the entire temperature range, the *CPP* value of monoolein is greater than that of monovaccenin, indicating that monoolein has a pronounced wedge shape at a fixed temperature. Bearing in mind that both molecules have identical theoretical values of a_0 and l, one can conclude that a larger *CPP* value indicates a highly curved lipid/water interface; hence less water is accommodated [17].

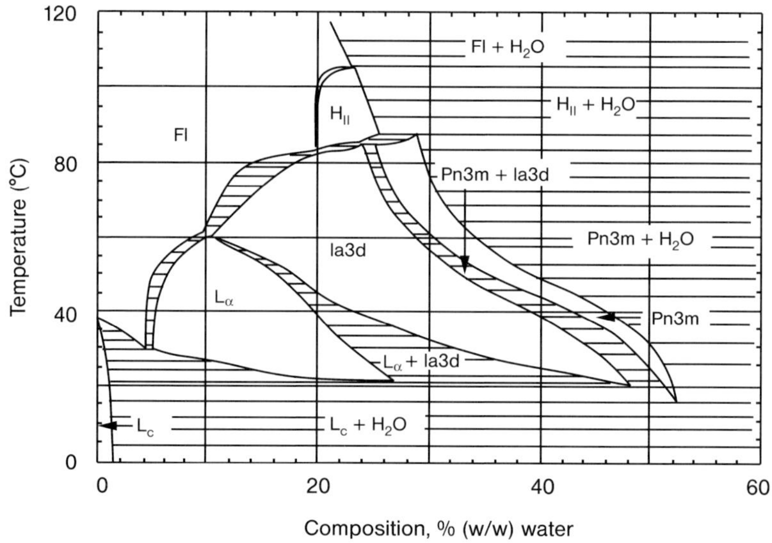

Fig. 8.11 Binary phase diagram of monovaccenin–water.

8.3.3
The Area per Head Group (Binary System)

The polymorphic phase behavior of dioleoylphosphatidylethanolamine (DOPE) and its *N*-methylated (in the head group) analogs, DOPE-Me, DOPE-Me_2 and DOPC (equal to PE-Me_3), has been investigated by Gruner et al. [7]. Phosphatidylcholines (PCs) and unsaturated phosphatidylethanolamines (PEs) are lipids that form lamellar and non-lamellar phases, respectively. Both head groups are identical except in the degree of methylation of the terminal quaternary nitrogen: in PE the nitrogen is bound to three hydrogens, whereas in PC each hydrogen has been replaced by a methyl group.

Upon methylation the L_α–H_{II} transition temperature increased. For the DOPE–water system the transition temperature was 5–10 °C. For DOPE-Me, the hexagonal phase was observed above 65 °C (on cooling from H_{II}, in copious water, the disordered state transformed very slowly into cubic phase) and for DOPE-Me_2 and DOPC at ~85 °C. Note that the largest change in the transition temperature occurs after a single methylation. Gruner et al. [7] showed that the relative magnitudes of R_0, which were inferred from the H_{II} lattice spacings versus temperature, increased with increasing methylation. They suggested that small R_0 lipids system (DOPE) exhibited facile, low-temperature L_α–H_{II} transitions, intermediate R_0 system (DOPE-Me) exhibited complex non-equilibrium transition behavior and were likely to form cubic phases and large R_0 systems with a very small curvature (DOPE-Me_2 and DOPC) were stable as L_α phases.

These results can be analyzed also using *CPP* theory. In this approach, increasing the degree of methylation of the head group increased only the parameter a_0, therefore decreased the *CPP* value and the preferred structure was lamellar.

Modifications of the surfactant head group based on varying the polyoxyethylene or polyglycerol units affected its hydrophilic–lipophilic properties more drastically and have been studied by many researchers.

The influence of the polyoxyethylene chain length on the phase behavior of a series of phytosterol ethoxylates with water was studied by Folmer et al. [21]. Sterols consist of four condensed rings with a secondary hydroxyl group on the first ring and a branched alkyl chain on the fourth. The secondary hydroxyl group has been ethoxylated to give sterol ethoxylates with much higher interface activity. The phase behaviors of phytosterols surfactants with 5, 10, 20 and 30 oxyethylene units are illustrated by the binary phase diagrams in Fig. 8.12 a–d [21]. The surfactant with five oxyethylene units is insoluble in water and did not form a single phase until the concentration of surfactant reached ~30 wt.% (at room temperature), where a lamellar phase was formed. At high water content, since the hydrophilic head group was too small for the surfactant to pack into discrete aggregates in water, a lamellar phase was formed with an excess of water. For the surfactant with 10 oxyethylene units, a micellar solution was formed at low surfactant concentrations (up to 15 wt.%) at temperatures below 60 °C. When the surfactant concentration was increased, an extended normal hexagonal phase (20–60 wt.% of surfactant) appeared. At higher temperatures, above 40 °C, mostly lamellar phase was formed. The mesophase sequences with increasing water content or decreasing surfactant concentration conform to the scheme that shown in Fig. 8.6. However, the temperature effect can be explained by the reorganization of the ethylene oxide chain to a more hydrophobic structure (dehydration of the head group). As a result, a transition from L_1 or H_I to L_α is observed and followed by a transition to L_2. The more hydrophilic surfactants with 20 and 30 oxyethylene units transformed from a micellar solution into a micellar cubic phase (at room temperature, at 20 wt.% for both systems) and before a hexagonal phase formed at higher concentration (25–~100 and 50–100 wt.% of surfactant, respectively). The general appearance of the phase diagrams is in agreement with a *CPP* of around unity for the 10 oxyethylene surfactant and with progressively lower *CPP* values for the longer polyoxyethylene chain surfactants. One should note that only the value of a_0 (in the *CPP* equation) increases as the polar head group increases from 10 to 20 up to 30 oxyethylene units. When the *CPP* value is close to unity, phytosterols with 5–10 oxyethylene units, mostly lamellar LLC was formed at the composition–temperature-dependent phase diagrams. However, with increasing oxyethylene units (20, 30), the *CPP* values are smaller than unity and hence the curvature increased to form non-lamellar phases, mostly cubic and hexagonal LLC.

Recently, Hossain et al. [22] studied the phase behavior of a similar system, polyoxyethylene cholesteryl ether and water [22]. The effect of the ethylene oxide chain length was examined using 10 and 15 ethylene oxide units. Before presenting the results, it is worth making a comparison between the phase behavior of cholesterol ethoxylated (10) and phytosterol ethoxylated (10), at room tem-

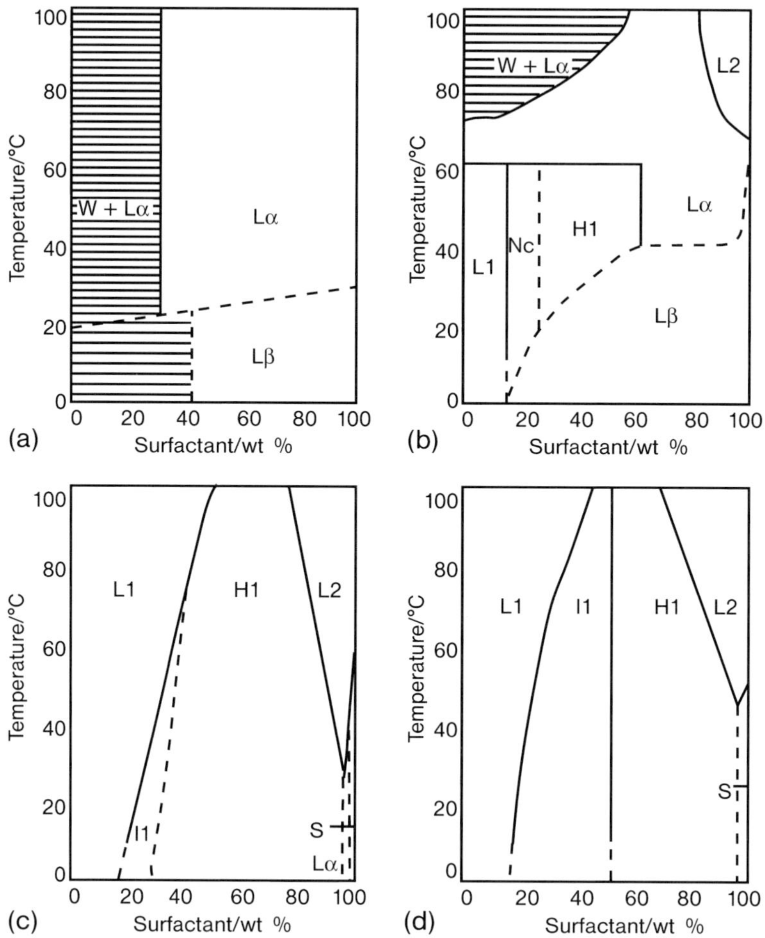

Fig. 8.12 Phase diagram of phytosterol with (a) 5, (b) 10, (c) 20 and (d) 30 oxyethylene units.

perature. As shown in Fig. 8.12 b, phytosterol ethoxylated (10) and water formed L_1 (up to 20 wt.% of surfactant), H_I (25–30 wt.%) and L_β phases (at 30–$\sim$100 wt.%) with increasing surfactant concentration. Replacing the surfactant's tail from phytosterol (extra methyl group on C24) to cholesterol affected the phase behavior. At high surfactant content (70–$\sim$100 wt.%), more surfactant interactions occurred, which caused a denser structure, hence L_α and "ill-defined" L_α LLC were formed (liquid crystal instead of gel, L_β). In the water-rich region L_1 was formed, similarly to the phytosterol ethoxylated system. However, when cholesterol ethoxylated was used, L_1 was formed even at higher surfactant concentrations (up to 40 wt.%). This result may be attributed to the fact that the cholesterol moiety is less bulky and more hydrophilic than the phytosterol, hence it tends to form highly curved oil-in-water micelles. This explanation may explain the formation of R_I (ribbon LC phase which is similar to H_I) at 50–70 wt.% of surfactant in the cholesterol ethoxylated system, instead of H_I at 25–30 wt.%, when phytosterol ethoxylated was used.

Increasing the number of ethylene oxide units, from 10 to 15 permits the formation of additional phases. Cholesterol ethoxylated (15) and water formed L_1 (up to 30 wt.% of surfactant), I_I (30–50 wt.%), H_I (50–60 wt.%), R_I (60–80 wt.%) and L_α phases (at 80–$\sim$100 wt.%) [22]. The H_I–R_I–L_α phase transitions are attributed to the increase in the packing constraint in the lipophilic core and consequently decrease in interfacial curvature of the aggregate. The cholesterol ethoxylated (15)–water showed a phase diagram of similar general appearance to the cholesterol ethoxylated (10)–water system, except for additional phases, H_I and I_I, similarly to the observation in the phytosterol-based systems. In addition, owing to decrease in lipophilicity of the cholesteric surfactant (from 10 to 15 ethylene oxide units), the domains of the mesophases with positive curvature shift towards higher surfactant concentration.

Kunieda and co-workers [23–26] also studied the phase behavior of mostly normal phases of LLC in binary systems that consist of hydrophilic surfactants such as polyoxyethylene oleyl or dodecyl ether, polyglyceryl didodecanoate and sucrose esters. Their major research included studies of the phase transitions upon modifications in the surfactant's head group in order to control the phase formation on changing the number of ethylene oxide or glycerol units. They achieved the transformation $L_\alpha \rightarrow H_I \rightarrow L_1$ like the previous researchers, but with a polyglycerol-based surfactant. In addition they were able to transform from the reverse type of hexagonal to normal upon increasing the polyoxyethylene units.

In the polyoxyethylene oleyl ether-based system with $\sim$2–20 ethylene oxide units, H_{II}, H_I, L_α, V_I, V_{II} (two kinds) and I_I mesophases were formed, as shown in the phase diagram for polyoxyethylene oleyl ether–water system in Fig. 8.13. To stress the correlation between the number or volume fraction of the ethylene oxide units and the phase behavior, Kunieda et al. [24] presented their results in a unique way. The phase diagram of the polyoxyethylene oleyl ether–water system is shown as a function of the volume fraction and number of ethylene oxide (EO) units in the surfactant molecule and weight fraction of polyoxyethylene oleyl ether. The weight fraction of polyoxyethylene oleyl ether in the system, denoted W_S, is plotted horizontally. The volume fraction of the hydrophilic part in the total surfactant molecule, denoted $\Phi_{EO}\Phi_S$, is plotted vertically, where Φ_S is the volume fraction of each surfactant in the system and Φ_{EO} is the volume fraction of the hydrophilic chain of the surfactant in the system. The number of the ethylene oxide units is denoted n.

One can note that H_{II} mesophase was formed in a short EO chain length and in a very narrow range of units (n=2.1–4.5) and weight fraction of surfactant (W_S=0.63–0.9). Upon increasing the EO units the hydration repulsion between the hydrophilic groups increased drastically in a closed space (water-in-oil structure). Inside the core of the H_{II} cylinder, the movement of the hydrophilic groups is confined, hence even a slight difference in the hydrophilic chain

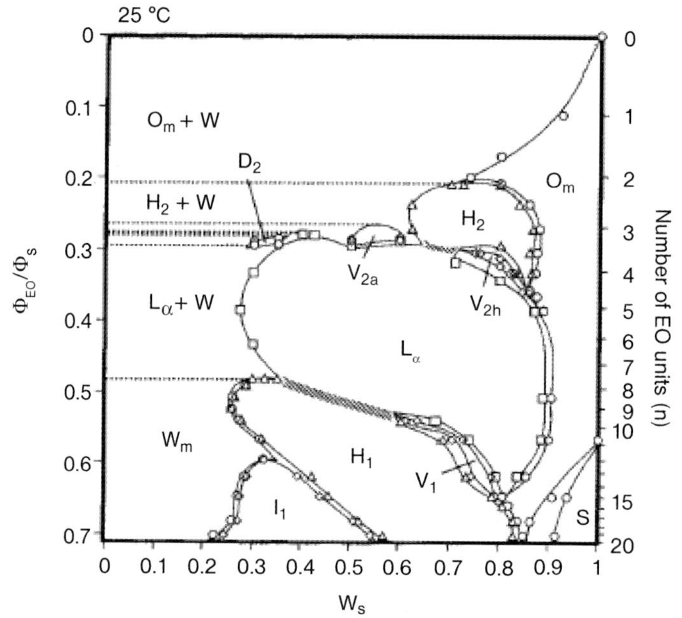

Fig. 8.13 Phase diagram of poly(oxyethylene) oleyl ether (POIE)–water system as a function of the volume fraction of EO chain in the surfactant molecule and weight fraction of POIE at 25 °C.

length affected significantly the interactions between them. Consequently, an $H_{II} \rightarrow L_\alpha$ transformation occurred. Between these two phases, very viscous bicontinuous cubic phases appeared between 8 and 15 EO units. The L_α phase was observed at the widest range of EO units (n=4.5–15, at fixed surfactant concentration). As a result of further increase in EO chain length, a very hydrophilic surfactant was formed and therefore an $L_\alpha \rightarrow H_I$ transformation was observed. It should be mentioned that between 11 and 20 ethylene oxide units, at lower surfactant concentrations, I_I mesophase was formed.

In an additional study, Kunieda et al. [25] showed the influence of polyglycerol chain length on the phase behavior of the polyglyceryl didodecanoate $[(C_{11})_2G_n]$–water system. In contrast to the polyoxyethylene-based system, in the $(C_{11})_2G_n$–water system only several phases were formed [23, 25]. With an increase in the head chain length, the surfactant changed from lipophilic to hydrophilic and the self-organized structure also changed from lamellar LLC to aqueous micellar solution phase via normal hexagonal LLC. $(C_{11})_{1.6}G_{4.5}$ formed solely lamellar LLC in a relatively high weight fraction of surfactant. With increase in the polyglycerol chain length, the surfactant layer became positive and a normal hexagonal phase was produced at a similar weight fraction of surfactant. However, in the long polyglycerol-chain surfactant-based system $[(C_{11})_{1.9}G_{21.3}]$, a W_m region exists almost over all the composition range. One should note that no cubic phases were observed even with long polyglycerol-chain surfactant systems, in comparison with the polyoxyethylene-based system. The low stability of the cubic phase could be attributed to the nature of the surfactant molecule. Since polyglycerol didodecanoate has two hydrophobic chains in addition to its relatively long head repeating chain, it is difficult to form spherical micelles. The repulsion between its hydrophilic chains may not be as strong as the polyoxyethylene chains, hence the interfacial curvature is limited.

From these results, which illustrate the relationship between the tail volume, tail length, area per head group and the corresponding phase behavior, the following conclusions can be derived. In systems containing monoglycerides with one double bond, one can control the stabilization of the reverse hexagonal phase upon changing the hydrophobic chain length, which can increase linearly its melting temperature and decrease the transformation temperature from the cubic to the hexagonal phase. It was demonstrated that the hydration boundaries of the cubic phase might also be controlled by changing the double bond site, which affected the *CPP* value owing to the formation of a surfactant with a more wedge shape. The formation of highly curved interface structures such as a normal hexagonal or ribbon form was indicated by determination of the tail's volume, which is related to the structure density.

The effect of the surfactant's area per head group on the structure obtained has been extensively studied [22–26]. The ability to form L_α or H_{II} phases can be shown by the degree of methylation of dioleoylphosphatidylethanolamine. It was shown that without methylation the preferred structure was H_{II} and upon *N*-methylation the temperature transition to the L_α phase decreased. Further methylation assists in stabilization of the L_α structure due to the increase in R_0 and decrease in the *CPP* value. In normal LLC it was demonstrated that in systems which contained surfactants with a polyoxyethylene chain various phases were formed. However, the phase formation may be controlled by changing the head chain length. If the required structure is lamellar, a short chain length should be utilized and if the formation of more highly curved LLC (such as hexagonal or cubic phases) is needed one should increase the head length. In addition, it was shown that in order to obtain various LLC structures one can use polyoxyethylene-based surfactant with one tail ($C_{18:1}$), but if a polyglycerol-based surfactant with two tails (C_{11}) is used the formation of the phases would be relatively restricted.

8.3.4 Guest Molecule Effect (Ternary System)

8.3.4.1 Hydrophilic Guest Molecule

The addition of a hydrophilic guest molecule to relatively hydrophobic surfactant and water mixtures affected the stability of specific phases. Mezzenga et al. [27] investigated the phase behavior of monoglyceride (~62 wt.% monolinolein and ~25 wt.% monoolein)–water mixtures, in the presence of mono-, oligo- and polysaccharides dissolved in the water phase. Both the concentration and the type of polysaccharide were found to have an effect on the structure and sta-

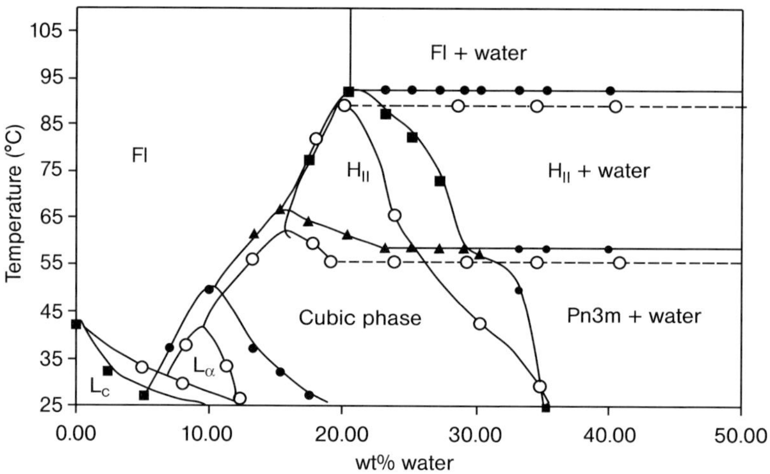

Fig. 8.14 Phase diagram of the monoglyceride and water solution of 5 wt.% glucose (dashed line, open symbols) compared with that of monoglyceride.

bility of the various liquid crystalline phases. Figure 8.14 shows the phase diagram of both binary (continuous lines, full symbols) and ternary (monoglyceride–glucose–water mixture designated by dashed lines and open symbols) systems [27].

One can see that the shape of the phase diagram remained similar and analogous liquid crystalline phases were detected; however, major changes can be noted on the boundaries of the liquid crystalline regions as a result of the hydrophilic guest molecule penetration. The lamellar and cubic regions shrank, whereas the transition from cubic to reverse hexagonal phase was observed at lower temperatures (by a maximum reduction of 5 °C). In addition, the lines corresponding to the onset of coexistence of H_{II} and water as well as the diamond cubic phase (*Pn*3*m*) and water were shifted towards lower water content. These results are consistent with a systematic increase in the *CPP* values, leading to the favorable $L_{\alpha} \rightarrow Q \rightarrow H_{II}$ shifts. Since glucose is a very hydrophilic compound with nearly zero partitioning in the lipid tail, its presence had no effect on the lipid domains; that is, both V_s and l remain constant. On the other hand, a_0 was reduced in the presence of hydrophilic guest molecules dissolved in water, which decreased the number of water molecules hydrating the lipid polar head, by introducing stronger hydrogen bond acceptors. This effect has already been described in the literature and is known as the Hofmeister effect [28–30]. In such a way, the curvature of L_{α} lipid/water interfaces decreased in the presence of glucose to yield the cubic phase or, similarly, that of the cubic phases was further reduced to yield H_{II}.

It should be stressed that according to the observations of Mezzenga et al. [27] and molecular dynamics simulations, the polysaccharide had to be small enough to fit within the water channels of liquid crystalline phases in order to shift the phase transition. However, as soon as the chain end-to-end distance of the polysaccharide approached the diameter of the water channels, the presence of polymeric sugars induced phase transitions toward structures with larger water channels ($Ia3d \rightarrow Pn3m$ transitions).

8.3.4.2 **Lipophilic Guest Molecule**

The structural transitions that occurred in LLC upon addition of a lipophilic guest molecule were extensively examined in different surfactant-based systems. The formation of non-lamellar liquid crystalline phases upon addition of alkanes to PC molecules was observed by Sjölund et al. [31] and was found to be dependent on the degree of acyl chain unsaturation. The chosen PC molecules were dipalmitoyl-PC (DPPC), 1-palmitoyl-2-oleoyl-PC (POPC), dioleoyl-PC (DOPC, also tested by Gruner et al. [7]) and dilinoleoyl-PC (DLiPC), lipids that with water form only a lamellar LLC mesophase, up to at least 90 °C. The addition of *n*-dodecane (C_{12}) induced the formation of reverse hexagonal and cubic phases in all kinds of PC-based systems. Figure 8.15 shows the ability of different PC–*n*-dodecane–water mixtures, with a dodecane : PC molar ratio of 2 : 1, to form an H_{II} or Q phase at different water concentrations at 45 °C. The amount of dodecane required to induce the formation of non-lamellar phases in PC–*n*-dodecane–water mixtures with a fixed water content increased in the order DLiPC ≈ DOPC < POPC < DPPC. The concentrations of water needed to form these phases, as in the case of *n*-dodecane, were dependent on the degree of acyl chain unsaturation of the PC molecules with the same order. The degree of unsaturation of the PC facilitated the formation of non-lamellar mesophases;

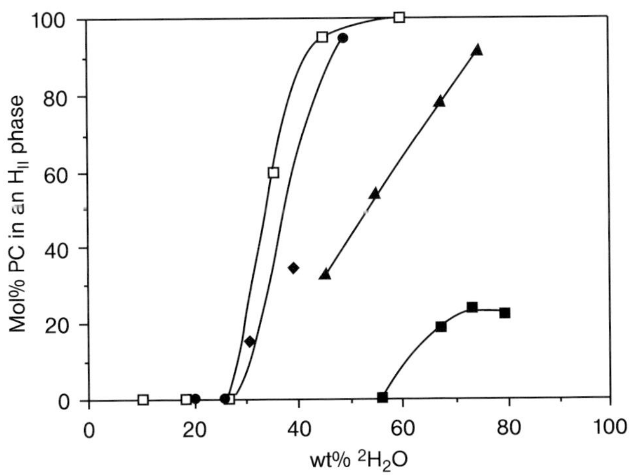

Fig. 8.15 Phase equilibria at 45 °C in different PC–*n*-dodecane–^2H_2O systems as a function of the water content. The dodecane : PC molar ratio is 2 : 1. (◆) DLiPC; (□) DOPC; (▲) POPC; (■) DPPC.

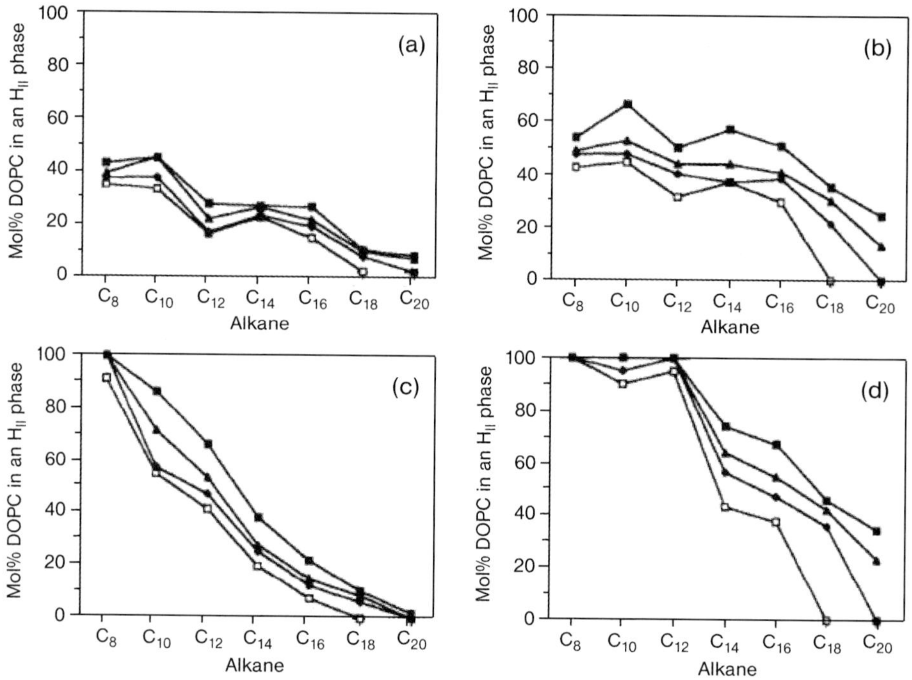

Fig. 8.16 Phase equilibria in DOPC–*n*-alkane–^2H_2O systems as a function of the alkane chain length at four temperatures: (□) 25, (◆) 35, (▲) 45 and (■) 55 °C. The water concentration was 39 wt.% (31 mol mol^{-1} DOPC) in (a), 49 wt.% (46 mol mol^{-1} DOPC) in (b), 35 wt.% (31 mol mol^{-1} DOPC) in (c) and 49 wt.% (54 mol mol^{-1} DOPC) in (d). The alkane concentration corresponds to 12 carbon atoms per lipid in (c) and (d).

however, there was no significant difference between mono- or diunsaturated molecules (dioleoyl-PC or dilinoleoyl-PC, respectively).

The effect of the hydrophobic guest molecule chain length (*n*-alkane with C_8–C_{20}) on the phase behavior of the DOPC–water system was also studied by Sjölund et al. [7]. In Fig. 8.16 a–d it can be seen that the chain length of the alkane plays an important role in the formation of the H_{II} mesophase and a certain number of CH_2 groups is not sufficient for the phase transition to occur (since the amount of alkane added was adjusted to give the same number of carbon atoms per DOPC molecule for the different alkanes). At 31 mol of water per mole of DOPC and 25 °C the ability of an alkane to induce an H_{II} phase was abolished when the alkane chain length was >16 (Fig. 8.16 a and c). The longest alkanes, C_{18} and C_{20}, are solid at this temperature, so the reason for their inability to induce an H_{II} phase was most probably a question of solubility. The chain length dependence observed in this work may be due to the fact that the location of the alkanes varies with the chain length so that the shorter ones are, at least partly, extended between the lipid acyl chains. As shown in Fig. 8.16c, an increase in alkane concentration from 12 to 24 alkane carbon atoms per mole of DOPC displaced the phase equilibria towards the H_{II} phase when the alkane chain length was <16. An increase in water concentration had a large effect on the DOPC–*n*-alkane–water systems. As shown in Fig. 8.16b and d at 49 wt.% water and 12 and 24 alkane carbon atoms per mole of DOPC, respectively, the formation of H_{II} was induced comparable to those at lower water concentrations. However, at 49 wt.% water the ability to promote H_{II} formation was less dependent on chain length at higher temperatures and even longer alkanes facilitate the formation of an H_{II} phase. The larger water content probably increased the lipid–water cylinder radius, and therefore also the volumes between the lipids. This effect could be the reason for the reduced chain length dependence with increasing water content.

The effect of bulkier guest molecules such as triacylglycerols (TAGs) on the phase transition in GMO–water mixtures was examined by Amar-Yuli and Garti [32]. In comparison with the PC-based systems, the GMO molecule tends to form lamellar and cubic phases at room temperature. In order to study the relationship between the guest molecule structure and the self-assembly of the GMO, TAGs with different fatty chain lengths (from C_2 to C_{18}) were used and divided into four categories of activities; TAG of very short chain length (triacetin), TAG of short chain length (tributyrin), TAG of medium chain length (tricaprylin) and TAGs of long fatty chain lengths (trilaurin, trimyristin and tristearin).

From the ternary phase diagram (Fig. 8.17) for GMO, triacetin and water, it can be seen that solely lamellar and cubic mesostructures were observed when the levels of triacetin did not exceed 10 wt.%. On comparing this phase diagram with the binary GMO–water system (at room temperature, Fig. 8.9) that was discussed earlier, one can see that similar phases are observed, but with different hydration boundaries. In the ternary mixture, the lamellar mesophase was formed at higher water concentrations, 10–20 wt.%, compared with the binary mixture (5–20 wt.%), while the cubic mesophase was detected at lower water content, 25–40 instead of 30–40 wt.%. However, the penetration of even small amounts of short- and medium-chain TAGs (C_4–C_8) into the binary mixture was very instrumental in destabilizing lamellar and cubic liquid crystals and in forming reverse hexagonal phase. Tricaprylin was found to be the most effective in altering the *CPP* value and in inducing the formation of reverse hexagonal liquid crystals. Figure 8.18 shows the ternary phase diagram of GMO, tricaprylin and water that contains the largest area of reverse hexagonal phase. The largest region of reverse hexagonal was detected even with a minimum tricaprylin concentration of 3 wt.% in line 96/4 at 20–25 wt.% water and up to a maximum concentration of 13.5 wt.% tricaprylin in line 85/15 at 10–20 wt.% water. Long-chain TAG (C_{12}, C_{14} and C_{18}) only at distinct elevated temperatures (at *ca.* 40 °C, yet lower than in the binary system) affected the *CPP* and R_0 values. Their disordering effect was detected only after a prolonged aging time in com-

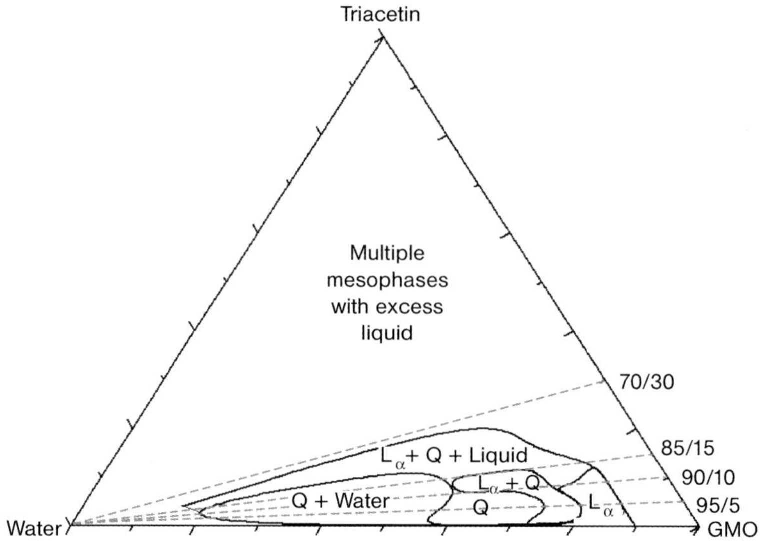

Fig. 8.17 Ternary phase diagram of GMO–triacetin–water at 25 °C. The dilution lines represent the surfactant-to-oil weight ratio.

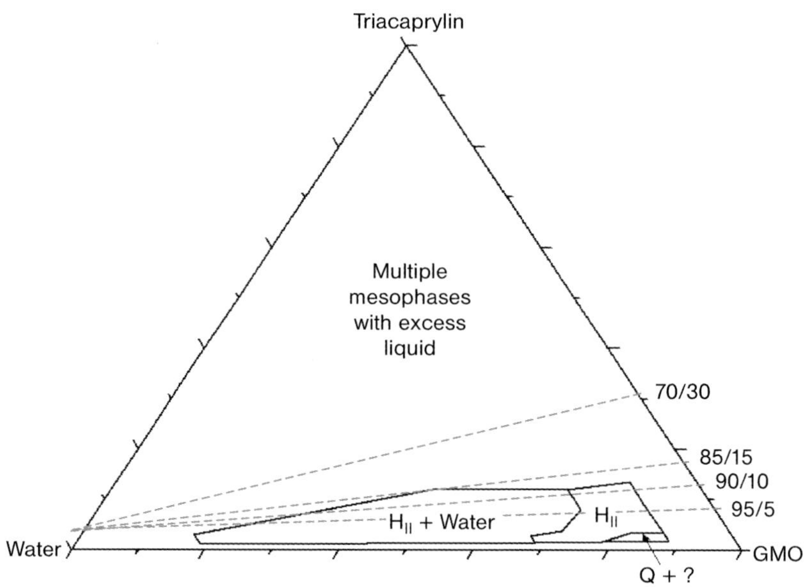

Fig. 8.18 The ternary phase diagram of GMO–tricaprylin–water at 25 °C. The dilution lines represent the surfactant-to-oil weight ratio.

parison with shorter chain length TAG. For example, the incorporation of trilaurin within the GMO chains required *ca.* 4 months, in comparison with just an hour or minutes when short- or medium-chain TAGs were used.

It was suggested by Amar-Yuli and Garti [32] that the very short-chain TAG triacetin only partially filled the void volumes between the cylindrical micelles and hydrated the GMO head group. As a result, it slightly increased the *CPP* and decreased R_0 values, which led to induction of the L_α–Q transformation and allowed the formation of cubic LLC at a lower water content than in a binary system. However, short- and medium-chain TAG were apparently incorporated between the GMO hydrophobic chains filling the void volumes and therefore they were able to affect dramatically the *CPP* and R_0 values without any dependence on temperature. As a result of their effective incorporation, they transformed the lamellar mesophase directly to hexagonal, not via a cubic mesophase (at very low temperatures, *ca.* 10 °C compared with *ca.* 85 °C in the binary system).

The incorporation of a hydrophobic guest molecule has been tested also in more hydrophilic surfactants such as sucrose esters. Aramaki et al. [33] were interested in finding the conditions for cubic phase formation on addition of hydrocarbon to water–mixed sucrose mono- and multi-dodecanoates (L-1695 and L-595, respectively). In the L-1695 (high HLB)–water system aqueous micelles and normal hexagonal phases were formed. On the other hand, L-595 (moderate HLB) with water formed only a lamellar LLC. In the mixed surfactants, with increasing L-595 content the molecular packing of surfactant lipophilic tails became denser and the average number of lipophilic tails per sucrose head group increased and resulted in an $H_I \rightarrow L_\alpha$ phase transformation. One can see that the addition of *n*-hexadecane (Fig. 8.19) to the ternary system (1 : 1 L-1695–L-595 and water) leads to the formation of additional phases [33]. On the water and mixed surfactants axis, an L_α phase appeared as expected. On addition of *n*-hexadecane, the L_α phase was changed to an oil-swollen microemulsion phase (W_m), whereas it changed to a cubic phase (I_I) through a hexagonal LLC (H_I) at *ca.* 60 wt.% of surfactant. In this system, the hydrophilic–lipophilic property of mixed surfactant is rather moderate, hence a lamellar phase with interfacial curvature zero was formed. However, upon addition of a hydrophobic molecule, the curvature changed to positive.

The dependence of cubic phase formation on the hydrophobic guest molecule geometry has been also studied by Aramaki et al. [33]. The added hydrophobic molecules were *n*-heptane, *n*-octane, *n*-decane, *n*-hexadecane and squalane. They found that the amount of oil required to induce the formation of cubic structure decreased in the order *n*-heptane > *n*-octane > *n*-decane > *n*-hexadecane > squalane. The concentrations of water needed to form these phases were similar for all systems. As expected, the bulkiest and longest molecule (squalane) was the most effective in increasing the curvature. However, no cubic phases were found in systems that contained guest molecules with larger molecular volume, such as *m*-xylene or cyclohexane. Upon addition of these two oils, the repulsion between neighboring hydrophilic moieties decreased and less positive curved aggregates became stable. Therefore, a hexagonal or lamellar phase was formed instead of an I_I phase.

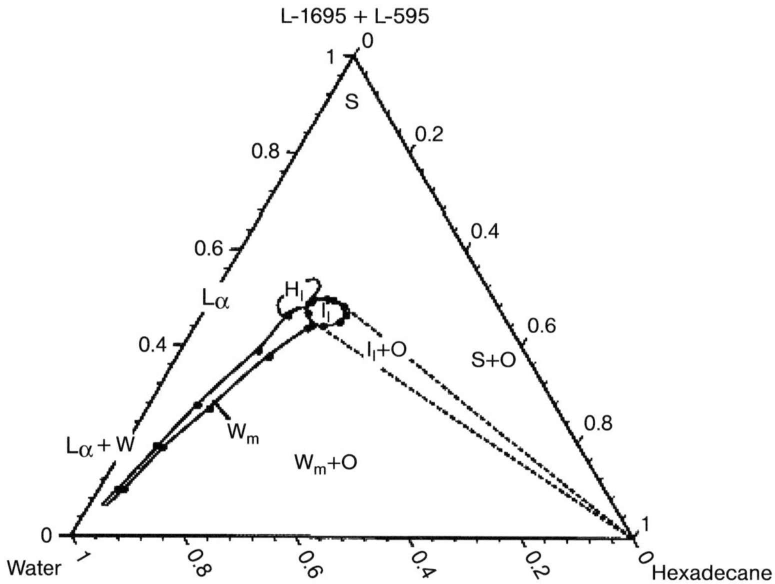

Fig. 8.19 Phase diagram in a pseudo-ternary water–L-1695–*n*-hexadecane system at L-1695 : L-595 = 1 : 1 and at 30 °C.

8.3.5
Co-surfactant

A co-surfactant, by definition, is expected to contribute to the structure formation more than the guest molecule owing to its miscibility in both of the liquid crystalline components, surfactant and water. Therefore, its influence on phase behavior should be easier and more pronounced.

Hossain et al. [22] studied the effect of dodecanoyl-*N*-methylethanolamide (NMEA-12) on the phase behavior of polyoxyethylene cholesteryl ether–water systems. The phase diagrams of the cholesterol ethoxylated with 10 and 15 ethylene oxide units are shown in Fig. 8.20 a and b, respectively. One should note that upon addition of NMEA-12, all phases transformed to L_2 via wide range of L_α. These results suggest that the incorporation of the co-surfactant in the palisade layer reduced the interfacial curvature from I_I, H_I and R_I to L_α phase. In both systems, the R_I phase was the most effective in solubilizing the co-surfactant up to the point where the interfacial curvature (of R_I) was flattened. In the cholesterol ethoxylated (10)–NMEA-12–water system the R_I phase dissolved NMEA-12 up to *ca.* 10 wt.%. Increasing the ethylene oxide units to 15 caused the formation of a looser LLC structure (oil-in-water) and allowed the incorporation of higher NMEA-12 concentrations, up to *ca.* 14 wt.%.

The addition of smaller co-surfactant molecules, such as linear and branched alcohol (1-butanol and 3,3-dimethyl-1-butanol) has been also studied and showed similar results. The linear and branched alcohols were added to the

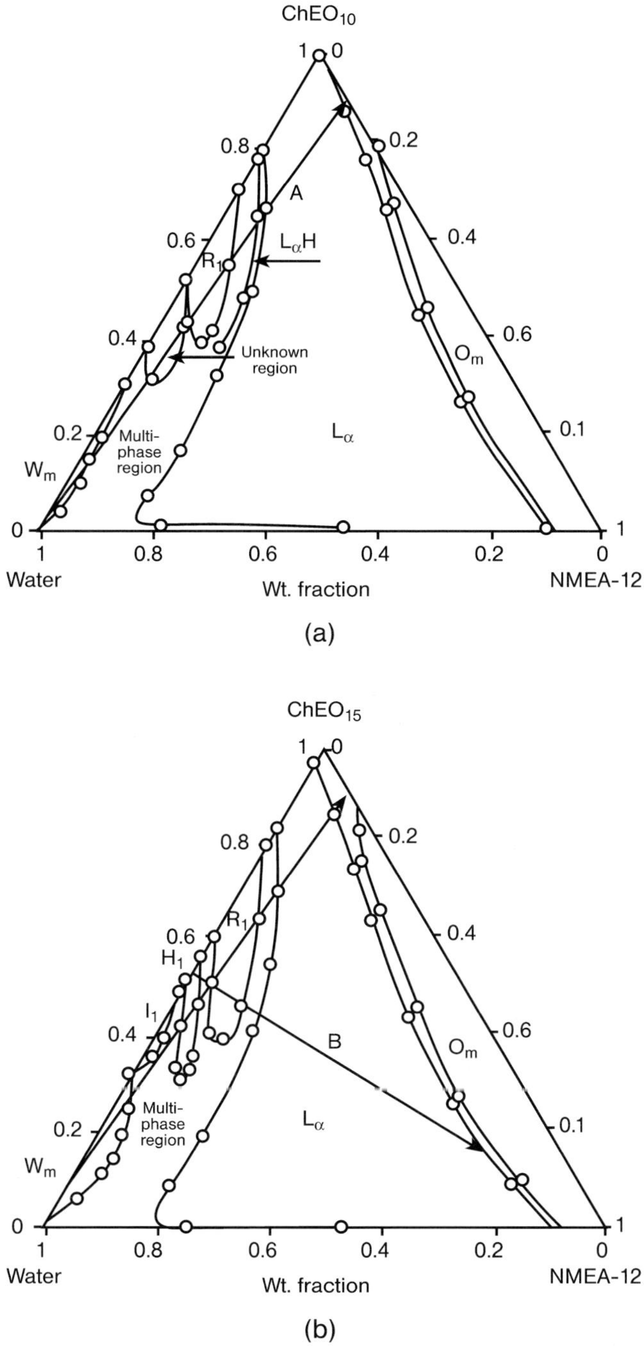

Fig. 8.20 Ternary phase diagrams of (a) cholesterol ethoxylated (10) ($ChEO_{10}$)–NMEA-12–water and b) cholesterol ethoxylated (15) (ChEO15)–NMEA-12–water, both at 25 °C.

laurylsulfobetaine–water system and were tested by Valiente and Álvarez [34]. The binary system formed L_1 (up to 50 wt.% of surfactant), I_I (52–65 wt.%) and H_I (66–95 wt.%) and upon addition of either branched or linear alcohol an $I_I \rightarrow H_I \rightarrow L_\alpha$ transformation occurred. I_I was very poor in solubilizing the alcohol, up to ~3 wt.%. Above this value, additional co-surfactant flattened the interfacial curvature. The extension and appearance of the H_I and L_α phases, respectively, were different depending of the alcohol structure. The incorporation of both alcohols into the palisade layer favored the formation of the hexagonal at lower surfactant concentrations; however, the hexagonal region was larger for the system that contained linear alcohol. For 3,3-dimethyl-1-butanol the extension was up to 46 wt.% and for 1-butanol up to 40 wt.%. Moreover, the solubilization of the linear alcohol was nearly double that of the branched alcohol in the hexagonal phase.

The formation of the lamellar phase was favored for the branched alcohol, hence its region in the phase diagram was larger.

Engström et al. [3] studied the phase behavior of more hydrophobic surfactants with the addition of polar co-surfactants. The GMO–water system was analyzed in the presence of co-surfactants such as dimethyl sulfoxide (DMSO), propylene glycol (PG), polyethylene glycol (PEG 400) and ethanol. At room temperature, at water concentrations of 30–60 wt.%, an L_3, isotropic LLC phase was observed in all four systems instead of a cubic phase that was found in the GMO–water system. The L_3 phase consists of a bicontinuous network of highly interconnected non-oriented surfactant bilayer. In general, the co-surfactants decreased the interfacial curvature which caused the Q–L_3 transformation. The water content of the L_3 sponge phase varies with each co-surfactant, i.e. 30 wt.% for DMSO and PEG 400, 40 wt.% for PG and 60 wt.% for ethanol. The extension of the narrow one-phase region or penetration also varied with the co-surfactant molecule, being much less for ethanol (*ca.* 22–26 wt.%) than for the other three molecules (*ca.* 22–62 wt.%). It is interesting that the amount of water needed to form the sponge phase for the co-surfactant used seemed to correlate well with the co-surfactant–water partitioning behavior. The more lipophilic the molecule was (lower co-surfactant–water partitioning), the more water was needed to transform the cubic phase into a sponge phase [3].

In summary, it was demonstrated that the addition of a third component, either hydrophilic or hydrophobic guest molecules or co-surfactant, influenced the phase behavior and transformations. For example, in monoglyceride-based systems one can obtain the same transformation sequence, $L_\alpha \rightarrow H_{II}$ or $Q \rightarrow H_{II}$, by increasing the *CPP* value upon addition of hydrophilic or hydrophobic guest molecules. However, on addition of hydrophobic guest molecule, its molecular volume or length determines the extent of its influence. Hydrophilic guest molecules, such as mono-, oligo- and polysaccharides, would decrease the parameter a_0 owing to a decrease in the number of water molecules hydrating the surfactant polar head, by introducing stronger hydrogen bond acceptors. Hydrophobic guest molecules, such as alkanes or TAGs, would increase the tail's volume. However, with a limited chain length of the hydrophobic guest molecule an opposite effect occurred, the parameter l increased. In order to control the extent of the effect, one can choose the guest molecule's molecular volume and length.

Co-surfactants, on the other hand, were found to decrease the interfacial curvature and were especially effective in destabilizing cubic phases.

8.4 Microstructure and Transformation Identification Techniques

Different instruments have been used to detect the phases and the phase transitions, with each technique providing unique information. The techniques can be divided into two categories: *direct techniques* – measurements which provide the phase identification (phase symmetry) such as small-angle X-ray scattering and polarized microscopy – and *indirect techniques* – where the data grants characterization of the phases such as nuclear magnetic resonance and rheology.

8.4.1 Optical Microscopy

The optical texture of liquid crystals under a polarized microscope in most cases provides effective and simple phase identification. The lamellar phase displays a distinct mosaic pattern whereas the hexagonal phase shows a fan-like and angular texture. On the other hand, the isotropic micellar solution phase and the stiff transparent isotropic cubic phases are non-birefringent and only a black background with no well-characterized texture is displayed.

8.4.2 X-ray Diffraction

The principle behind this technique is that X-rays are scattered by the electrons in the sample. The characterization of lipid mesophases by diffraction is based first on symmetry. Information on the long-range organization of the phase is contained in the low-angle region of the diffraction pattern: the long-range translational ordering of the lipid–water aggregates (bilayer, cylinder, micelles, etc.) on to one-, two- or three-dimensional lattices gives rise to Bragg reflections whose reciprocal spacings ($S_{hkl} = 1/d_{hkl}$) are in a characteristic ratio, for example:

- for lamellar LLC:

$$S_1 = \frac{1}{d}$$

- for hexagonal LLC:

$$S_{\text{hk}} = \frac{2(h^2 + k^2 - hk)^{\frac{1}{2}}}{\sqrt{3}a}$$

- for cubic LLC:

$$S_{\text{hkl}} = \frac{(h^2 + k^2 + l^2)^{\frac{1}{2}}}{a}$$

where a is the lattice parameter and d_{hkl} is the spacing of the set of lattice planes (h, k, l), characterized by the Miller indices h, k and l. The ratios of the different phases which observed from the small-angle X-ray scattering (SAXS) are for a lamellar structure $1, \sqrt{4}, \sqrt{9}, \sqrt{16}, \ldots,$ for a hexagonal structure $1, \sqrt{3}, \sqrt{4}, \sqrt{7}, \ldots$ and for a specific cubic structure $1, \sqrt{2}, \sqrt{3}, \sqrt{4}, \ldots$ [4].

The SAXS technique has been utilized the most since it provides both identification and characterization of the liquid crystalline structure. Kunieda and co-workers [22–25] used SAXS techniques for examination of phase transitions upon modifications of the surfactant's structure (hydrophilic moiety) and/or upon addition of a guest molecule.

In order to study the microstructure of $H_I \rightarrow R_I \rightarrow L_\alpha$ transformations in the cholesterol ethoxylated (15)–NMEA-12–water system, SAXS measurements were performed along line B in Fig. 8.20 by Hossain et al. [22]. They followed the planes symmetry of the several LLC phases in this system, while keeping the surfactant:water weight ratio constant and increasing the co-surfactant concentration.

As mentioned earlier, with addition of co-surfactant to this system, the H_I phase transformed into the R_I and further into the L_α phase owing to the increase in the packing constraint in the lipophilic core and, consequently, decrease in interfacial curvature of the aggregate. Figure 8.21 shows the SAXS patterns of each phase and their transformations upon addition of NMEA-12 while keeping the cholesterol ethoxylated (15):water ratio at 53:47. Up to *ca.* 2 wt.% of NMEA-12 a SAXS pattern of H_I was observed. Increasing the co-surfactant content, up to 4.5 wt.%, caused the appearance of an additional low-intensity peak at low angle corresponding to an additional symmetric plane, suggesting the beginning of a structure transformation. The additional peak developed as an intense peak of plane (02) of the R_I phase. A gradual shift of the (02) peak to low angle (increase in the unit cell width, b) upon further addition of the co-surfactant suggests a progressive deformation or elongation of the cross-section of the aggregates. Similarly, a gradual shift of the (11) peak to large angle (decrease in the unit cell height, a) corresponds to a decrease in the separation between the aggregates. With further increase in NMEA-12 concentration, the intensity of the low-angle (02) peak gradually decreased; however, the (11) peak remained intense. This tendency may be attributed to the gradual breaking of the rectangular symmetry, leading to the formation of a lamellar structure. The (11)

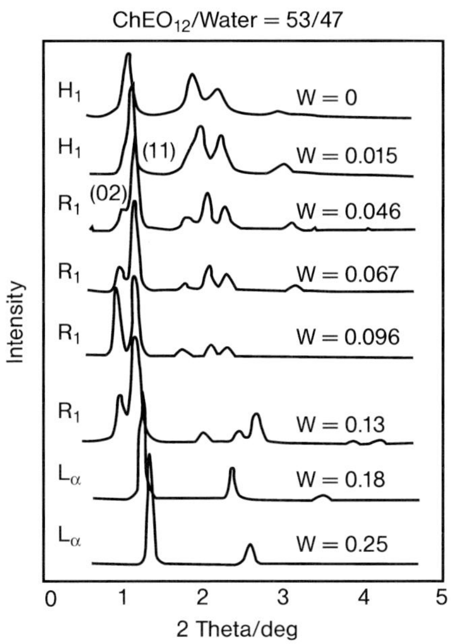

Fig. 8.21 SAXS patterns development for the $H_I \rightarrow R_I \rightarrow L_\alpha$ transformation on addition of NMEA-12 at a fixed surfactant-to-water ratio of 53:47 (cholesterol ethoxylated (15) denoted $ChEO_{15}$).

peak of the R_I phase transformed to the (10) plane of the L_α phase, suggesting that the phase transformation occurred by the lateral fusion of the rod-like aggregate of the (11) plane of the R_I phase [22].

The values of the structural parameters which were calculated along the same line are in agreement with these results. The interlayer spacing (d) of H_I, R_I and L_α gradually decreased with increasing co-surfactant weight fraction, W. In each liquid crystal, the average effective cross-sectional area of an amphiphile, a_s, decreased continuously with increasing co-surfactant content, indicating that the surfactant layer became more compact. Since NMEA-12 incorporates into the palisade layer of the surfactant aggregates, the smallest thickness of the lipophilic core of the ribbon phase (r_S) and half of the hydrophobic film thickness of L_α (d_L) decreased with increasing NMEA-12 content.

By additional SAXS measurements, Hossain et al. [22] studied the microstructure of the phase transformations with increasing water content (along line A in Fig. 8.20). It can be seen in Fig. 8.22 that the interlayer spacing (d) of H_I, R_I and L_α gradually increased with increasing water content. The water penetration increased the distance between the micelles of type I, instead of swelling in type 2, therefore the tendency is lower than in type II. In each liquid crystal, a_s increased slightly with increasing water content. This can be attributed to an increase in the hydration of EO chains, which increased the repulsion between head groups. As expected, the hydrophobic cylinder radius of H_I (r_H), the smallest thickness of the lipophilic core of the ribbon phase (r_S) and half of the hydrophobic film thickness of L_α (d_L) increased slightly with increasing total surfactant content, a reflection of the increase in the hydrophobic volume fraction.

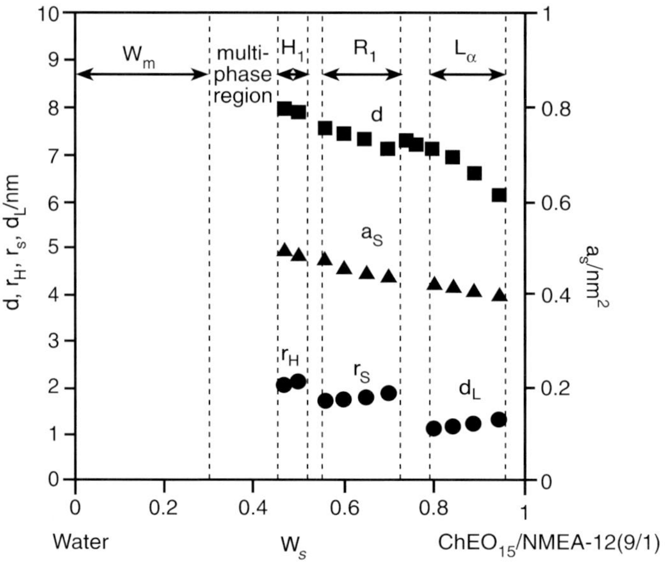

Fig. 8.22 Change in the interlayer spacing, d; average effective cross-sectional area of an amphiphile, a_s; apolar cylinder radius, r_H, for the hexagonal liquid crystal; smallest dimension of the hydrocarbon core of the aggregate in ribbon phase, r_s; and half length of the hydrophobic part in bilayer, d_L, in the cholesterol ethoxylated (15) ($ChEO_{15}$)–NMEA-12–water system as a function of the total surfactant weight fraction in the system, W_s. The $ChEO_{15}$-to-NMEA-12 mass ratio is kept fixed at 9 : 1. $T = 25\,^{\circ}C$.

It should be stressed that r_H in the H_I phase was close to the fully extended length of the hydrophobic chain of cholesterol, which is energetically unfavorable. The hydrophobic chains tend to shrink, hence the order of the parameters was $r_H > r_S > d_L$.

In the hydrophobic surfactant-based system of GMO, TAGs and water, the lattice parameter and composition dependences were studied by Amar-Yuli and Garti [32].

It was found that the lattice parameter of the H_{II} phase was dependent on the composition and on the TAG chain length. The TAG concentration effect on the lattice parameter was examined by focusing on the GMO–tricaprylin–water system (Fig. 8.18). Table 8.1 shows the lattice parameter of the H_{II} structure in several dilution lines at 10–35 wt.% water. At 10 wt.% water, the change in tricaprylin concentrations, in lines 85/15, 87/13 and 90/10 (13.5–9 wt.% tricaprylin), did not change the lattice parameter (44.74, 44.69 and 45.05 Å, respectively). A similar tendency was observed at 15 wt.% water. However, at 20 wt.%, when there was sufficient water to form well-defined hexagonal packing of the micelles, decreasing the tricaprylin concentration (and increasing GMO concentration) caused the formation of better defined structures with larger aggregation numbers and therefore the lattice parameter increased. At very high water

Table 8.1 Effect of tricaprylin and GMO content on the lattice parameter a in GMO–tricaprylin–water mixtures at different dilution lines.

Dilution line	Water (wt.%)	GMO (wt.%)	Tricaprylin (wt.%)	a (Å)
85/15	10	↓ 75.57	↑ 13.54	↓ 44.74
87/13	10	77.12	11.44	44.69
90/10	10	81.0	9.0	45.05
85/15	15	71.46	13.5	47.85
90/10	15	76.64	8.29	47.81
95/5	15	79.5	5.2	48.7
85/15	20	67.88	12.05	52.83
90/10	20	76	4.0	54.4
96/4	20	76.2	3.2	54.06
85/15	25	63.02	11.35	52.05
90/10	25	67.34	7.65	56.18
96/4	25	70.21	3.15	58.17
85/15	30	59.16	11.09	50.42
90/10	30	63.04	6.98	56.18
95/5	30	66.36	3.75	54.28
85/15	35	55.54	9.91	48.8
90/10	35	58.33	6.72	50.46
95/5	35	61.16	3.39	58.88
96/4	35	61.6	3.21	59.64

content (35 wt.%, full hydration), additional effects, such as swelling of the cylindrical structures, prevailed. As a result, the influence of decreasing TAG concentration or increasing GMO content on the lattice parameter was more dramatic at high water concentrations. For example, at lines 85/15, 90/10 and 95/5, the lattice parameter was 48.8, 50.46 and 58.88 Å respectively.

The lattice parameter–temperature dependence was further investigated using SAXS techniques and can be elucidated with the assistance of the theories of Israelachvili et al. [9] and Gruner et al. [7, 11]. Increasing the temperature increases the molecular mobility of the hydrocarbon chains, resulting in a demand for an increased molecular cross-section. This puts a strain on the stability of the phase. This strain may be appeased by a mesophase transition, so that close molecular packing at the polar/non-polar interfaces can persist while the relative volume of the non-polar hydrocarbon region is increased. These effects increase the hydrophobic chain volume, V_s, and decrease the cross polar head group area, a_0. Both parameters increase the *CPP* value. Gruner et al. [7, 11] proposed that at elevated temperatures the value of R_0 should decrease, resulting in the formation of an H_{II} mesophase. They also showed that in certain cases, the size of the H_{II} core in excess water (full hydration) may be taken as a good measure of R_0.

The observations of Amar-Yuli and Garti [32] show that with increase in temperature, the lattice parameter, a, decreased (Table 8.2). The GMO–tricaprylin–water system was examined at two dilution lines, 95/5 and 90/10, up to full hy-

Table 8.2 Effect of temperature on the lattice parameter (a) in GMO–tricaprylin–water mixtures at dilution lines 95/5 and 90/10.

Dilution line	Water (wt.%)	Temperature (°C)	a (Å)
95/5	20	10	54.5±0.24
		25	54.7±0.44
		41	L_2
	25	25	59.7±0.60
		41	53.3±0.34
90/10	10	10	45.7±0.70
		25	45.05±0.30
		41	L_2
	20	10	54.7±0.50
		25	54.4±0.55
		41	L_2
	25	10	59.4±0.62
		25	56.2±0.22
		41	53.1±0.66

dration, at 25 wt.% water and showed once more that at high water content the mesostructure was more sensitive to modifications. At line 95/5, at 20 wt.% water, raising the temperature from 10 to 25 °C caused a slight decrease in a. Similar behavior was detected in dilution line 90/10 at 10–20 wt.% water. However, at full hydration (25 wt.% water), at either dilution line 95/5 or 90/10, the decrease in R_0 with increase in temperature accounts for most of the decrease in the lattice parameter. At lower tricaprylin concentrations (line 95/5), a decreased from 59.7 Å at 25 °C to 53.3 Å at 41 °C. However, at higher tricaprylin concentrations (line 90/10), a decreased more moderately over a wider region of temperatures, from 59.4 Å at 10 °C, through 56.2 Å at 25 °C to 53.1 Å at 41 °C.

8.4.3 Differential Scanning Calorimetry (DSC)

In the L_α–H_{II} transformation, massive structural rearrangement occurs; however, it is achieved without major energetic changes. The low-enthalpy transition that can be observed by DSC may be due to opposing effects (contributions) which might be present. For example, the positive contribution to the enthalpy from activation of increased chain disorder might be partially abolished by a negative enthalpy change from the interfacial region, if the head-group packing becomes tighter in the H_{II} phase.

Inoue et al. [35] utilized DSC measurements to investigate the phase behavior of the heptaethylene glycol dodecyl ether ($C_{12}E_7$)–water system in the temperature range from –20 to 70 °C. According to the phase diagram, the mixture with a composition of 31 wt.% water exhibits the following phase sequence with increasing temperature: solid → H_I → V_I → L_α liquid. DSC thermograms of the ($C_{12}E_7$)–water system are shown in Fig. 8.23 a–c. The endothermic peak observed for pure $C_{12}E_7$ at about 25 °C is ascribed to its melting-point (first transition). The addition of water (6.12 wt.%) led to the appearance of another endothermic peak at *ca.* –5 °C. In addition, small endothermic peaks are superimposed between the two large endothermic peaks, as indicated by asterisks and arrows in Fig. 8.23 a and b. Inoue et al. [35] suggested that the small endothermic peaks observed at a fixed temperature (indicated by asterisks or arrows) are isothermal phase changes in the binary mixture which correspond to the coexistence of three phases at a specific temperature. Figure 8.23 c shows various peaks with small heat absorption at higher temperature for the mixture composition range 20–60 wt.%. The transition enthalpies for these small peaks are in the range 0.14–0.69 kJ mol^{-1}, which are much smaller than the heat of fusion of pure $C_{12}E_7$, 87.0 kJ mol^{-1}. In order to identify these small peaks, heating rates of 2 and even 1 °C min^{-1} were needed. These endothermic peaks should be ascribed to the phase transition between mesophases. For example, in Fig. 8.23 c at the thermograph representing 29.1 wt.% of water, there is one

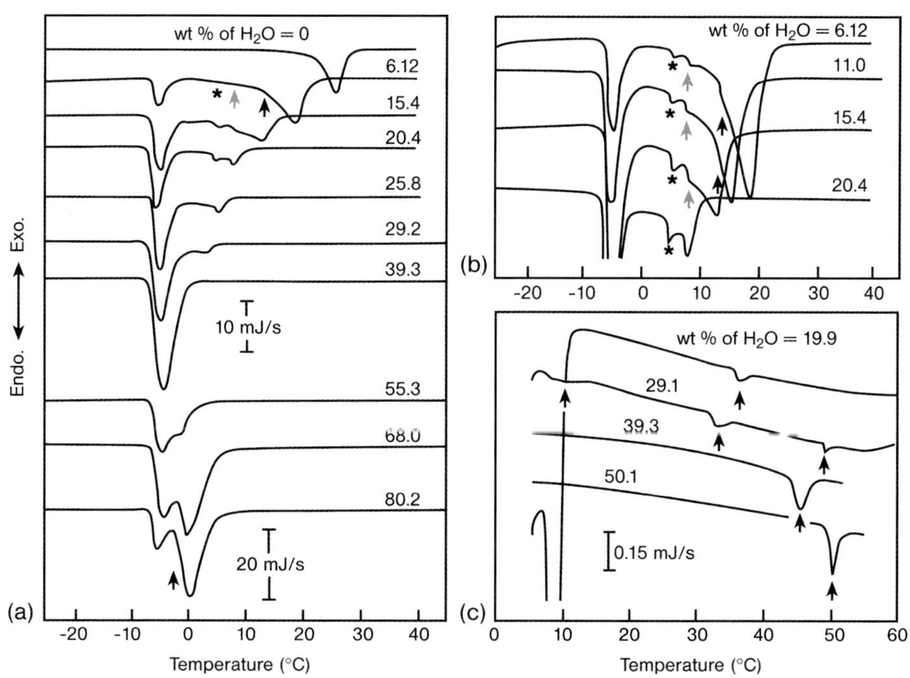

Fig. 8.23 DSC thermograms obtained for $C_{12}E_7$–water mixtures with various compositions (a and b) and those drawn in an expanded scale for a higher temperature range (c). The heating rate is 2 °C min^{-1}.

large endothermic peak at *ca.* –5 °C which is attributed to the melting of $C_{12}E_7$–water. On increasing the temperature, three small peaks observed at 7, 32 and 50 °C, which are ascribed to the $H_I \rightarrow V_I \rightarrow L$ liquid transformations.

Qiu and Caffrey [17] used the DSC technique to observe other phase transitions in a GMO–water binary system at several water concentrations. As seen and discussed earlier, in the binary phase diagram at 14.5, 38.5 and 87 wt.% water the $L_C \rightarrow L_\alpha$, $L_C \rightarrow Ia3d$ and $L_C \rightarrow Pn3m$ transformations, respectively, take place. On increasing the temperature to 40 and 50 °C, at 38.5 and 14.5 wt.% water, the transitions from *Ia3d* to *Pn3m* and from L_α to *Ia3d* occurred.

They found that the heat change associated with the liquid crystal phase transformation was highly endothermic regardless of the liquid crystal phase to which it converted. The corresponding enthalpy changes were as follows: 8 kcal mol^{-1} for the $L_C \rightarrow L_\alpha$, 9 kcal mol^{-1} for the $L_C \rightarrow Ia3d$ and 11 kcal mol^{-1} for the $L_C \rightarrow Pn3m$ transformations. Weaker endothermic peaks were revealed for liquid crystal transformations; however, a heating rate of 4 °C min^{-1} was sufficient for their identifications. The enthalpy change for the $L_\alpha \rightarrow Ia3d$ transformation was 0.04 kcal mol^{-1}. As expected, for transition between two kinds of cubic phases, $Ia3d \rightarrow Pn3m$, the enthalpy change was even lower, 0.005–0.01 kcal mol^{-1}. In contrast to the $L_\alpha \rightarrow Ia3d$ transformation, the phase transitions in cubic phases involved a change in lipid bilayer structure but no change in curvature of the bilayer itself.

8.4.4
Infrared (IR) Spectroscopy

Infrared spectroscopy has been used to clarify the polymorphism subject. Inoue et al. [35] investigated the phase behavior of the $C_{12}E_7$–water system by Fourier transform (FT) IR measurements, which were in agreement with the DSC measurements (discussed above). IR spectra obtained for the $C_{12}E_7$–water mixture with the composition of 31.1 wt.% D_2O (which exhibits the solid $\rightarrow H_I \rightarrow V_I \rightarrow L_\alpha \rightarrow$ liquid transformations) are shown in Fig. 8.24 a–e. They focused on five absorption bands in order to analyze the conformational structure of the surfactant molecule and its interaction with D_2O in the different phases. In Fig. 8.24 a, showing the spectrum obtained at 6.0 °C, the mixture is in the H_I phase (oil-in-water). The absorption band around 3300–cm^{-1}, which was attributed to the O–H stretching mode, ν_{OH}, was used to obtain information concerning the interaction of the terminal OH groups of POE chains. The frequency of ν_{OH} was expected to decrease when the hydrogen bond interactions become stronger. It can be seen that the absorption band appeared at around 3440 cm^{-1} instead of at 3310 cm^{-1}, suggesting that the hydrogen bonds between OH groups of the surfactant molecules are somewhat broken at this temperature. The conformational structure of the POE chain was examined in terms of the absorption bands around 1100 and 850 cm^{-1} which are ascribed to the coupled mode of C–O stretching, C–C stretching and methylene rocking ($\nu_{CO}+\nu_{CC}+\rho_{CH_2}$) and that of C–O stretching and methylene rocking ($\nu_{CO}+\rho_{CH_2}$)

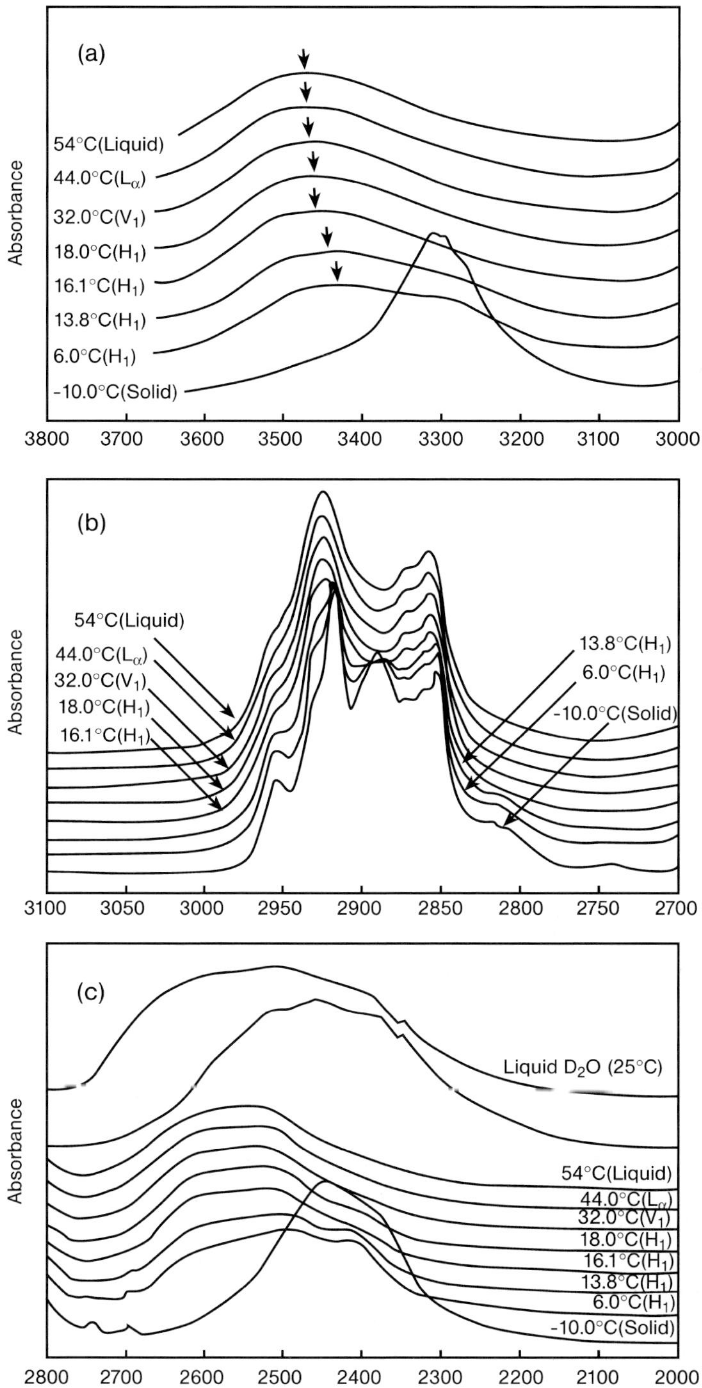

Fig. 8.24 (legend see page 238)

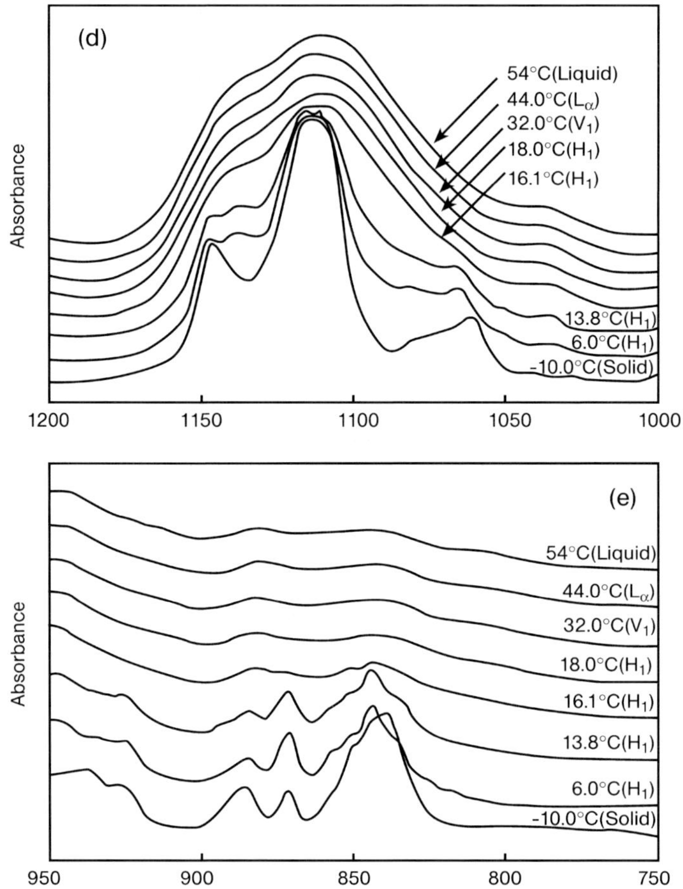

Fig. 8.24 IR spectra obtained for the $C_{12}E_7$–D_2O mixture at various temperatures in the frequency ranges (a) 3000–3800, (b) 2700–3100, (c) 2000–2800, (d) 1000–1200 and (e) 750–950 cm^{-1} and comparison of the IR spectra obtained for the $C_{12}E_7$–D_2O mixture and pure H_2O at 55 °C.

of the POE chain, respectively. These absorption bands are supposed to shift toward higher wavenumbers with increase in the gauche conformer fraction in the POE chain. From Fig. 8.24 d and e, one can conclude that the POE chain of the surfactant has a helix-like structure, due to dominant absorption bands (~1110 and ~850 cm^{-1}) at high wavenumbers. The O–D stretching mode (ν_{OD}), which is usually located at around 2450 cm^{-1}, implies that interactions between the POE chain and D_2O molecules take place. The stronger the hydrogen-bond interaction, the lower the frequency of ν_{OD} is. In Fig. 8.24 c, an absorption band around 2500 cm^{-1} emerged in addition to that around 2420 cm^{-1}. The high-frequency ν_{OD} may be an indication of a weakly hydrogen bond, suggesting partial dehydration of the POE chain at this temperature. The helical structure of the POE chain which was concluded from the $\nu_{CO}+\nu_{CC}+\rho_{CH_2}$ and $\nu_{CO}+\rho_{CH_2}$ modes may be facilitated owing to the release from a steric restriction forced on the POE chain by the bonded D_2O molecules. In contrast to the POE chain, the alkyl chain is considered to have a trans-zigzag structure, since the frequencies of the absorption ascribed to ν_{CH_2} modes were almost the same as those of the solid $C_{12}E_7$–D_2O mixture (Fig. 8.24 b).

Further increases in temperature, up to 16 °C, caused additional spectral changes even though the mixture is still in the H_I phase. The most pronounced change in the spectral pattern was seen in the frequency regions corresponding to the absorption due to the ν_{CH_2}, $\nu_{CO}+V_{CC}+\rho_{CH_2}$ and $\nu_{CO}+\rho_{CH_2}$ modes (Fig. 8.24 b, d and e, respectively). The absorption band due to the ν_{CH_2} mode shifted towards higher wavenumber together with broadening of the bandwidth. This indicates that the alkyl chain of the surfactant undergoes a transformation from an ordered trans-zigzag structure to a disordered structure containing the gauche conformer or, in other words, the melting of the alkyl chain occurs in this temperature range. Similarly, the spectral pattern in the $\nu_{CO}+\nu_{CC}+\rho_{CH_2}$ and $\nu_{CO}+\rho_{CH_2}$ mode regions became wider, suggesting that the POE chain also transformed from an ordered helix-like structure to a disordered structure at this temperature. The IR spectra obtained for the $C_{12}E_7$–D_2O mixture above 16 °C of the H_I, V_I, L_α and liquid phases are similar to each other, particularly in the frequency regions corresponding to the absorption bands ascribed to the ν_{CH_2}, $\nu_{CO}+\nu_{CC}+\rho_{CH_2}$ and $\nu_{CO}+\rho_{CH_2}$ modes. One can conclude that there is no essential difference in the conformational structure of the surfactant molecule among these phases. As for the absorption band due to the ν_{OH} and ν_{OD} modes, slight shifts towards higher wavenumbers are observed with increase in temperature. This suggests that the hydrogen-bond interaction between the POE chain of the surfactant and D_2O at an oil-in-water structure became somewhat weak on increasing the temperature from lower temperature phase, H_I, to the higher temperature phases, V_I, L_α and liquid [35].

8.4.5
Nuclear Magnetic Resonance (NMR) Spectroscopy

^2H NMR has been used in determination of the phase boundaries of isotropic and anisotropic LLC phases. ^2H NMR can also provide information concerning the hydration state of the amphiphiles owing to the deuterium nucleus possessing an electric quadrupole moment. To determine the monoolein–diolein–water phase diagram, which includes L_α and H_{II} phases, Borné et al. [36] examined the shape of ^2H NMR spectra as a function of composition and/or temperature. An isotropic micellar solution or an isotropic liquid crystalline phase was characterized by a sharp singlet for the quadrupolar interaction, since the rapid and isotropic molecular motion is averaged to zero. For an anisotropic phase, on the other hand, this interaction is averaged to a non-zero value and as a result generated a splitting of the signal into two peaks with equal intensity (seen in Fig. 8.25).

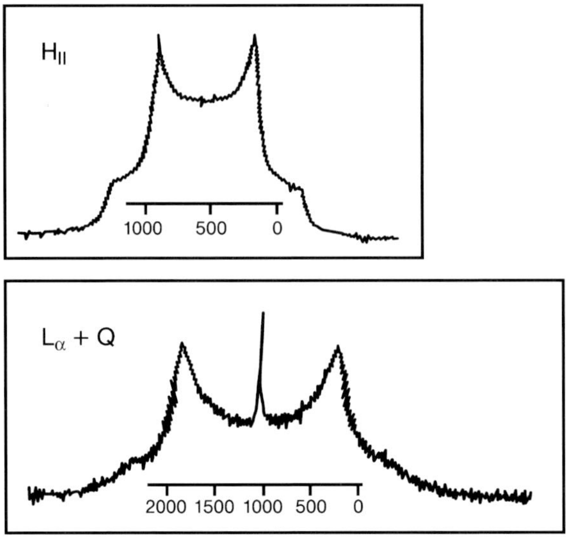

Fig. 8.25 ^{2}H NMR spectra for single hexagonal phase and two-phase region La + Q. The scale bars are in hertz.

The magnitude of the quadrupolar splitting was shown to follow Eq. (4) after a simplification by applying the two-site model:

$$\Delta = |P(\nu_Q)S| \tag{4}$$

where P is the fraction of bound water molecules, S is the order parameter describing the orientation of the fraction of ^{2}H and ν_Q is the quadrupolar coupling constant (220 kHz).

Further information from the measured ^{2}H NMR splitting values was achieved by changing P to the molar ratio of components:

$$\Delta_{obs} = (X_{lipid}/X_{water})n\nu_Q \tag{5}$$

where X_{lipid} and X_{water} are the mole fractions of lipid and water, respectively, and n is the (average) number of water molecules bound per lipid molecule. By plotting Δ versus $X_{(monoolein+diolein)}/X_{water}$ (Fig. 8.26), Borné et al. [36] obtained information on the hydration properties of the lipid molecules and on the swelling characteristics of the liquid crystal phases. In Fig. 8.26, Δ values for the H_{II} phase were found to increase linearly with decreased water content up to the molar fraction $X_{(monoolein+diolein)}/X_{water} \approx 0.36$, above which the increase of the splitting values was relatively small with decreasing water concentration. For the L_α phase, which exists within a very limited area at the water-poor part, the Δ values were almost independent of concentration. In a lamellar phase, the director is perpendicular to the lamellae and in a hexagonal phase it is perpendic-

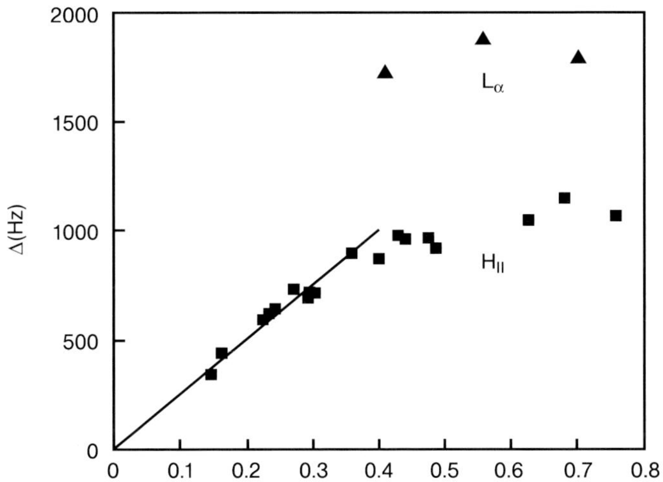

Fig. 8.26 Quadrupolar splitting (△) values of the liquid crystalline phases for the binary monoolein–water and monoolein–diolein–water systems, presented as a function of $X_{(monoolein+diolein)}/X_{water}$ at 25 °C; X = mole fraction; ■ represents H_{II} and ▲ represents L_{α}.

ular to the axes of the cylinders. In the H_{II} phase, the rapid diffusion of the deuterons in the aqueous cylinders caused a further averaging of the expression in Eq. (5). As a result of this effect, the absolute value of the order parameter for the H_{II} phase was reduced by half on transition from a lamellar to a hexagonal phase and the measured splitting for the lamellar phase was twice that for the hexagonal phase at an identical composition.

The linear relationship between Δ values and $X_{(monoolein+diolein)}/X_{water}$, up to $\Delta \approx 0.36$ for the H_{II} phase, indicates that the phase shows an ideal swelling behavior. Thus, an increase in the water concentration will not change the hydration of the lipid molecules but only the amount of free water.

8.4.6
Rheology

The rheological signatures of various liquid crystalline phases can be determined by rheological measurements. Mezzenga et al. [27] performed rheological measurements of the storage and loss moduli of LC phases during temperature scans. They used this technique in order to investigate the effect of mono-, oligo- and polysaccharide concentration on the $Q \rightarrow H_{II}$ transitions in monoglyceride (~62 wt.% monolinolein and ~25 wt.% monoolein)–water mixtures. In their method, the heating rate was slow enough (0.2 °C min^{-1}) and has been shown to be particularly suitable for detecting cubic-to-cubic and cubic-to-hexagonal transitions, allowing detailed insight into the region of coexistence of liquid crystalline phases, which cannot be investigated by optical cross-polariza-

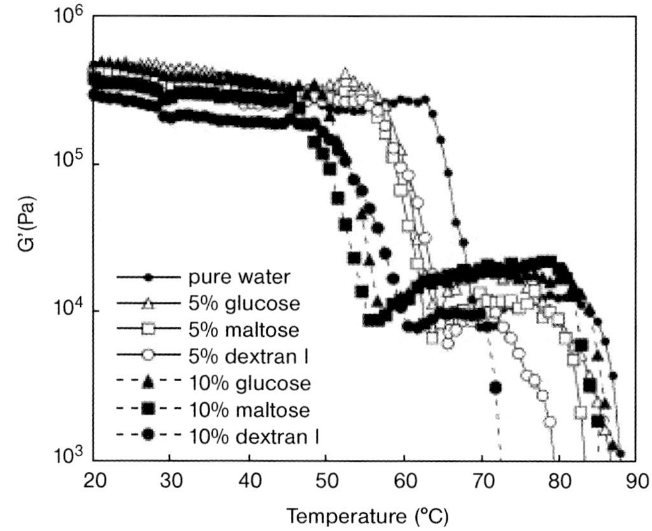

Fig. 8.27 Storage modulus, G', versus temperature, during temperature scans of 0.2 °C min, for monoglyceride with pure water and for monoglyceride with water–carbohydrate solutions. Different concentrations and types of carbohydrates are reported in the same graphics.

tion. Figure 8.27 shows the evolution of the storage modulus, G', as a function of temperature, for a system formed by 20 wt.% water solution and 80 wt.% monoglyceride. The water solutions, which contained 5 and 10 wt.% of glucose, maltose and dextran I (mono-, oligo- and polysaccharide, respectively), were compared with analogous cubic phases that were obtained in the presence of 20 wt.% pure water.

At this water content, the observed cubic phase above 45 °C was *Pn*3*m* for all types of sugars, whereas at lower temperatures it can be either *Ia*3*d* or *Pn*3*m*, depending on the type of carbohydrate. The sharp drop in G' (from a *Pn*3*m* phase) corresponds to the appearance of the H_{II} phase, which is known to be less rigid than the *Pn*3*m* phase, by a few orders of magnitude. As long as *Pn*3*m* and H_{II} phases coexist, G' continues to decrease with temperature, whereas a plateau was reached as soon as the *Pn*3*m* phase disappeared.

The curves expressing the storage modulus of the different solution-based cubic phases follow three master curves: the first corresponds to the pure water-based cubic phase, the second groups the solutions containing 5 wt.% sugar (open symbols) and the third groups the solutions containing 10 wt.% sugar (full symbols). Larger quantities of sugars in the solution result in a systematic decrease in the *Pn*3*m*–H_{II} transition temperature, as a consequence of the Hofmeister effect [28–30]. However, no dependence on the molecular weight of the sugar was observed.

All the saccharide molecules have the same density but differ slightly in their repeat unit molecular weights (180 g mol^{-1} for glucose and 162 g mol^{-1} for polymeric repeat units). Therefore, identical concentrations of sugar solutions result in identical volumes occupied by sugar in the water channels but with different amounts of hydrophilic groups. In oligomers and polysaccharides a hydroxyl group is converted into an ether bond via condensation with hydrogen from another glucose ring. In terms of hydrogen bond acceptor sites, a repeat unit of polymeric sugar has three hydroxyl groups and two ether bonds, compared with glucose, which has five hydroxyl groups and one ether bond. Therefore, equal weights of dissolved sugar result in a decrease in the number of hydrogen acceptor sites in the case of oligo- and polysaccharides and, hence, in a globally more hydrophobic compound; therefore, they are less efficient in altering the curvature of the lipid/water interface (*Pn*3*m* should be favored over H_{II}).

8.5 Conclusions

The relationship between surfactant geometry includes tail volume, tail length and area per head group and the corresponding mesophase formation has been demonstrated and analyzed.

In systems containing monoglycerides with one double bond, the stabilization or preferred formation of the H_{II} phase can be achieved within several modifications. Increasing the hydrophobic chain length and volume either by changing the surfactant molecule or by addition of hydrophobic guest molecule will increase the *CPP* value. However, the molecular volume or length of the hydrophobic guest molecule determines its influence tendency. Interestingly, addition of a hydrophilic guest molecule will have the same result, but due to a different effect. Hydrophilic guest molecules would decrease the parameter a_0 and as a result will increase the *CPP* value, owing to a decrease in the number of water molecules hydrating the surfactant polar head.

It has been shown that modifications in the unsaturation site or degree also had an effect on the phase behavior, but less pronounced. The hydration boundaries of the cubic phase changed owing to changes in the double bond site which affected the *CPP* value as a result of a formation with a more wedge-shaped surfactant. In addition, in systems containing PC molecules, the amount of oil required to induce the formation of non-lamellar phases was dependent on the degree of acyl chain unsaturation of the PC. It was found that the degree of unsaturation facilitated the formation of non-lamellar phases; however, there was no significant difference between mono- and diunsaturated molecules.

On the other hand, the formation of normal highly curved interfacial structures such as normal hexagonal or cubic phases could be preferred by altering the head chain length. Thus the *CPP* value will decrease. This phenomenon has been seen in systems that contained surfactant with different hydrophobic chains from hydrocarbon to the bulkier sterols.

If the formation of a flatter interface (L_α) is required, methylation in the PE surfactant's head, or addition of a co-surfactant such as a branched medium-chain alcohol can done. Short-chain alcohols, DMSO, PG or PEG 400 can be utilized to form an L_3 phase.

In addition, it was demonstrated that in order to identify a liquid crystal symmetry, one can use SAXS or an optical cross-polarized microscope. Furthermore, in order to confirm the identification or if further characterization of the phases or the transformation phases is needed, one can utilize FT-IR, NMR, DSC, SAXS and rheological techniques.

References

1 J.W. Jensen, J.S. Schutzbach, Biochemistry, *23* (**1984**) 1115.
2 L. Rilfors, Biochim. Biophys. Acta, *813* (**1985**) 151.
3 S. Engström, K. Alfons, M. Rasmusson, H. Ljusberg-Wahren, Prog. Colloid Polym. Sci., *108* (**1998**) 93.
4 S.T. Hyde, in K. Holmberg (ed.), Handbook of Applied Surface and Colloid Chemistry, Wiley, New York, **2001**, Chapter 16.
5 K. Larsson, J. Phys. Chem., *93* (**1989**) 7304.
6 S.Y. Lin, H.L. Lin, M.J. Li, Adsorption, *8* (**2002**) 197.
7 S.M. Gruner, M.W. Tate, G.L. Kirk, P.T.C. So, D.C. Turner, D.T. Keane, C.P.S. Tilcock, P.R. Cullis, Biochemistry, *27* (**1988**) 2853.
8 G. Rummel, A. Hardmeyer, C. Widmer, M.L. Chiu, P. Nollert, K.P. Locher, I. Pedruzzi, E.M. Landau, J.P. Rosenbusch, J. Struct. Biol., *121* (**1998**) 82.
9 J.N. Israelachvili, D.J. Mitchell, B.W. Ninham, J. Chem. Soc. (**1976**) 1525.
10 G. Kirk, S.M. Gruner, D.L. Stein, Biochemistry, *23* (**1984**) 1093.
11 S.M. Gruner, Proc. Natl. Acad. Sci., *82* (**1985**) 3665.
12 D.P. Siegel, Chem. Phys. Lipids, *42* (**1986**) 279.
13 G. Arvidson, I. Brentel, A. Khan, G. Lindblom, K. Fontell, Eur. J. Biochem., *152* (**1985**) 753.
14 J.M. Seddon, G. Cevc, R.D. Kaye, D. Marsh, Biochemistry, *23* (**1984**) 2634.
15 J.M. Seddon, Biochim. Biophys. Acta, *1031* (**1990**) 1.
16 K. Larsson, K. Fontell, N. Krog, Chem. Phys. Lipids, *27* (**1980**) 321.
17 H. Qui, M. Caffrey, Biomaterials, *21* (**2000**) 223.
18 Y. Misquitta, M. Caffrey, Biophys. J., *81* (**2001**) 1047.
19 M. Caffrey, Biochemistry, *26* (**1987**) 6349.
20 H. Qui, M. Caffrey, J. Phys. Chem. B, *102* (**1998**) 4819.
21 B.M. Folmer, M. Svensson, K. Holmberg, W. Brown, J. Colloid Interface Sci., *213* (**1999**) 112.
22 M.K. Hossain, D.P. Acharya, T. Sakai, H. Kunieda, J. Colloid Interface Sci., *277* (**2004**) 235.
23 H. Kunieda, K. Shigeta, K. Ozawa, M. Suzuki, J. Phys. Chem. B, *101* (**1997**) 7952.
24 H. Kunieda, Y. Yamaguchi, K. Ozawa, M. Suzuki, Colloids Surf. A: Physicochem. Eng. Aspects, *160* (**1999**) 15.
25 H. Kunieda, A. Akahane, J. Feng, M. Ishitobi, J. Colloid Interface Sci., *245* (**2002**) 365.
26 H. Kunieda, M. Horii, M. Koyama, K. Sakamoto, J. Colloid Interface Sci., *236* (**2001**) 78.
27 R. Mezzenga, M. Grigorov, Z. Zhang, C. Servais, L. Sagalowicz, A.I. Romoscanu, V. Khanna, C. Meyer, Langmuir, *21* (**2005**) 6165.
28 L.M. Crowe, J.H. Crowe, Biochim. Biophys. Acta, *946* (**1988**) 193.
29 P.W. Sanderson, L.J. Lis, P.J. Quinn, W.P. Williams, Biochim. Biophys. Acta, *1067* (**1991**) 43.
30 R. Koyonova, M. Caffrey, Chem. Phys. Lipids *62* (**1994**) 253.

31 M. Sjölund, L. Rilfors, G. Lindblom, Biochemistry, *28* (**1989**) 1323.

32 I. Amar-Yuli, N. Garti, Colloids Surf. B: Biointerfaces, *43* (**2005**) 72.

33 K. Aramaki, M. H. Kabir, N. Nakamura, H. Kunieda, M. Ishitobi, Colloids Surf. A: Physicochem. Eng. Aspects, *183* (**2001**) 371.

34 M. Valiente, M. Álvarez, Colloids Surf. A: Physicochem. Eng. Aspects, *183* (**2001**) 235.

35 T. Inoue, M. Matsuda, Y. Nibu, Y. Misono, M. Suzuki, Langmuir, *17* (**2001**) 1833.

36 J. Borné, T. Nylander, A. Khan, Langmuir, *16* (**2000**) 10044.

9
Particle Deposition as a Tool for Studying Hetero-interactions

Zbigniew Adamczyk, Katarzyna Jaszczółt, Aneta Michna, Maria Zembala, and Jakub Barbasz

Abstract

Colloid deposition kinetics as a possible tool for evaluating dynamic interactions was analyzed theoretically and experimentally. Methods of calculating the energy of interactions between dissimilar particles bearing electric double-layers are presented. They include the generalized Derjaguin summation approach applicable for any short-range interaction potential and the linear superposition approximation (LSA) describing the electrostatic interactions. The van der Waals interactions are discussed by exploiting the Hamaker theory based on pairwise additivity. Explicit expressions for the dispersion interaction energy of various macro-bodies, including rough objects, are compared with the expressions derived from the Derjaguin theory. The classical DLVO energy profiles originating from the superposition of the electrostatic and van der Waals interactions are discussed. These specific interactions are incorporated in an exact manner into the phenomenological equation describing transport to various interfaces discussed next. Limiting forms of this equation are considered, in particular the convective diffusion regime, the pure diffusion regime and the near-surface transport regime governed by the specific interactions. In the last case, the one-dimensional transport of particles throughout the surface layer adjacent to interfaces is solved. This allows one to formulate the foundations of the surface boundary layer (SFBL) theory and to establish the appropriate boundary conditions for bulk transport equations. Using these boundary conditions, limiting analytical solutions for the convective diffusion transport to interfaces are derived in particular for the bulk transport-controlled and the barrier-controlled regimes. Numerical solutions illustrating a significant role of the attractive electrostatic interaction in particle deposition processes are also presented. Illustrative experimental results are then discussed, obtained in well-defined systems, for example, in impinging-jet cells and packed-bed columns. The experiments proved that the initial adsorption flux of particles is considerably increased in dilute electrolytes and at higher flow rates owing to attractive electrostatic inter-

Colloids and Interface Science Series, Vol. 2
Colloid Stability: The Role of Surface Forces, Part II. Edited by Tharwat F. Tadros

ISBN: 978-3-527-31503-1

actions. This is found to be in quantitative agreement with the convective diffusion theory. It is suggested that this effect can be exploited for the quantitative evaluation of the interaction range and magnitude. A different behavior was observed in systems where repulsive double-layer interactions appeared. In this case the experimental deposition rates were found to be many orders of magnitude larger than the theoretical predictions stemming from the DLVO theory. It is suggested that this discrepancy could be explained in terms of the surface heterogeneity hypothesis. This was confirmed by experimental results obtained for model heterogeneous surfaces, produced by pre-adsorption of colloid particles or polyelectrolytes, where the apparent zeta potential of substrate and particle deposition rates were measured simultaneously. It is therefore concluded that the kinetic measurements of particle deposition furnish valuable information on the interactions between colloid particles and substrates, especially in hetero-systems, where direct force measurements are not feasible.

9.1
Introduction

Stability of colloid systems both of natural origin, for example clay suspensions, or synthetic, such as polystyrene lattices, is an essential factor in many practical applications such as diagnostic tests, controlled drug delivery, painting, coating, xerography and in the cosmetic, pharmaceutical and electronic industries. The quality of such basic products as drug carriers, pigments, toners, inks, binders for the casting industry, creams, emulsions and gels is critically dependent on the kinetics of particle aggregation processes.

A better understanding and control of these phenomena can only be achieved by acquiring information on specific interactions between particles, preferably under dynamic conditions, for example, in a flowing system. Yet, despite the essential significance of such data for colloid science, biophysics, medicine and soil chemistry, it is rather impractical to study the kinetics of colloid aggregation in its dynamic aspect. The main problem consist in the inevitable appearance of aggregates composed of more than two particles whose complex topology induces non-linear effects, for example, multiple light scattering. This reduces the precision of many experimental methods, preventing the collection of proper statistics for large aggregate populations having the same composition. This is especially critical in aggregation processes occurring in dissimilar particle systems, often called heterocoagulation, which is of great practical interest.

It seems that the dynamic interactions between particles in both homo- and hetero-systems can effectively be studied via the particle deposition route using substrates with well-defined and controlled surface properties. Another advantage of deposition studies over aggregation measurements is that the particle transport mechanism, for example, diffusion, convection and kinetics, can easily be varied and its intensity can be controlled. Also, statistics from large particle populations can be collected, which is advantageous for studying surface heterogeneity effects.

It should be mentioned that the significance of particle deposition studies reaches far beyond simply mimicking aggregation phenomena. Effective attachment of particles to surfaces, involving transport, adsorption and adhesion steps, is important for many practical processes such as water and waste water filtration, electroflotation, coating formation, paper making, xerography, production of magnetic tapes, catalysis, colloid lithography, protein and cell separation (affinity chromatography), foam stabilization, immobilization and separation of bio-particles such as DNA, proteins, viruses, bacteria, pathological cells and so forth. Controlled assembly of colloid particles into organized structures has potential applications for the production of nano- and micro-structured materials of desired functionality, for example, biocompatible coatings. Ordered arrays of colloid particles, for example silver particles, are exploited as narrow band optical filters and they can be used as optical switches, photonic bandgap materials, waveguides and other electro- and magneto-optical devices.

In other processes, for example, membrane filtration, biofouling of membranes and artificial organs, flotation (slime coating formation), production of microelectronic or optical devices, particle adsorption is highly undesirable.

Besides these practical applications, studies of colloid particle deposition, carried out using direct observation experimental methods, can furnish fundamental information on dynamic interactions between particles and interfaces and between adsorbed and moving particles.

Hence the aim of this chapter is to discuss the theoretical aspects of particle interactions governing both the aggregation and deposition phenomena. Then, by analyzing experimental results obtained for well-defined systems, analogies are pointed out between adsorption phenomena and particle aggregation in homo- and hetero-systems.

The chapter is organized as follows: in the first section, the electrostatic interactions governed by the classical Poisson-Boltzmann equation are discussed. Exact expressions describing interactions of planar interfaces are given. The approximate models for calculating interactions of non-spherical particles, e.g. spheroids, are presented, with emphasis on the extended Derjaguin summation method and the linear superposition method (LSA). The van der Waals interactions between macroscopic particles are discussed next. Exact expressions for the interaction energy of particles of various shape calculated from the microscopic Hamaker theory and the Derjaguin approximation are given. Analogies between the particle–particle and particle–interface interactions are pointed out. The DLVO energy profiles originating from the superposition of electrostatic and dispersion contributions are discussed, as well as the influence of surface roughness. In the next section, the role of these interactions in adsorption and deposition of colloid particles on solid surfaces is considered. The governing continuity equation is formulated, incorporating convective transport in the bulk and the specific force-dominated transport at the surface. Analytical results are discussed, describing transport in the near-surface layer governed by the specific interactions. The analytical results describing the convective diffusion of particles in the bulk are presented and the surface boundary layer approximation

(SFBLA) is discussed, which couples both transport regimes. Exact numerical results are also presented, illustrating the significance of attractive double-layer interactions, enhancing many-fold deposition rates for larger particles. In the last section, the experimental data are discussed, obtained in well-defined systems, for example, impinging jet cells and the packed-bed columns, with the emphasis focused on particle deposition on heterogeneous surfaces.

9.2 Specific Interactions

9.2.1 Electrostatic Interactions

Because of the non-linearity of the Poisson-Boltzmann (PB) equation governing interactions of charged particles in electrolytes, no analytical solutions can be derived that describe the energy for two dissimilar particles, including the limiting case of particle–interface [1, 2]. The situation becomes even more complicated for non-spherical particles because of the lack of an orthogonal coordinate system for expressing the PB equation. There exist, however, a few approximate approaches suitable for predicting the force and energy of particle interactions, which exploit simpler solutions obtained for a two-plate system or for a single particle. The most useful seem to be the Derjaguin summation [3, 4] and the linear superposition (LSA) [5] methods.

There are two major advantages of the Derjaguin method that have led to its widespread use, namely the separation of the geometric and electrostatic contributions and the possibility of its application for anisotropic particles. The disadvantage of the Derjaguin method is that it becomes less accurate for separation distances between particles larger than the double-layer thickness [6]. For larger distances and thicker double layers, the LSA method becomes more efficient. It seems, therefore, that the LSA and Derjaguin methods are complementary and can be used in combination to describe particle interactions for an arbitrary range of distances.

According to the original Derjaguin method, the interactions of spheres are calculated as a sum (integral) of corresponding interactions of infinitesimal surface elements (rings) with a planar geometry. The summation is carried out in the region close to the minimum separation distance h_m (see Fig. 9.1) by assuming a fast decay of interaction potential away from this region. The Derjaguin method is valid for arbitrary interaction potentials provided that the sphere radii a_1 and a_2 are both much larger than the characteristic range of interaction. By virtue of these assumptions, the interaction energy of two spheres can be calculated as the surface integral of the energy of two planar double layers discussed above

$$\phi = \int \Phi \mathrm{d}S = \int \Phi(y) 2\pi y \mathrm{d}y \tag{1}$$

where ϕ is the interaction energy of the two particles, Φ is the interaction energy of the two-plate system (per unit area), $\mathrm{d}S = 2\pi y\mathrm{d}y$ is the surface element and $y(h)$ is the radius of the ring, depending on the distance h (see Fig. 9.1).

From simple geometry, one can deduce that

$$h = h_\mathrm{m} + a_1 + a_2 - \sqrt{a_1^2 - y^2} - \sqrt{a_2^2 - y^2} \tag{2}$$

If $y << a_1$ and $y << a_2$, Eq. (2) reduces to the quadratic form

$$h = h_\mathrm{m} + \frac{1}{2}\frac{y^2}{a_1} + \frac{1}{2}\frac{y^2}{a_2} \tag{3}$$

Hence, $y\mathrm{d}y = \mathrm{d}h a_1 a_2/(a_1 + a_2) = G_\mathrm{D}\mathrm{d}h$, where $G_\mathrm{D} = a_1 a_2/(a_1 + a_2)$ is the geometric Derjaguin factor. It is interesting to note that $G_\mathrm{D} = 0.5$ for two equal spheres and $G_\mathrm{D} = a$ for the plane/sphere configuration.

By considering this, Eq. (1) can be expressed in the form

$$\phi(h_\mathrm{m}) = 2\pi Le G_\mathrm{D} \int_{h_\mathrm{m}}^{\infty} \Phi(h)\mathrm{d}h \tag{4}$$

where h_m is the minimum distance between the particle surfaces (see Fig. 9.1) and

$$\kappa^{-1} = Le = \left(\frac{\varepsilon kT}{2\mathrm{e}^2 I}\right)^{\frac{1}{2}} \tag{5}$$

is the Debye screening length, a parameter of primary importance for all particle interaction problems, ε is dielectric permittivity of the solvent, k is the Boltz-

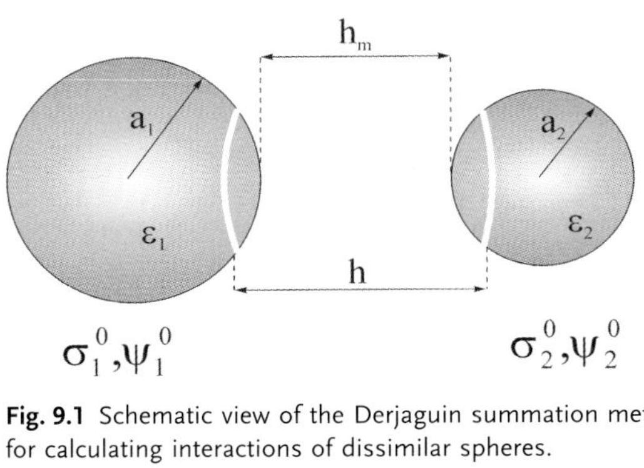

Fig. 9.1 Schematic view of the Derjaguin summation method used for calculating interactions of dissimilar spheres.

mann constant, T is the absolute temperature, e is the elementary charge, $I = \frac{1}{2}\sum_i z_i^2 n_i^{\mathrm{b}}$ is the ionic strength of the electrolyte solution and z_i, n_i^{b} are the valence and bulk concentration of the ith ion, respectively.

The interaction force of spheres, which is the derivative of the interaction energy upon the distance h, can be expressed as

$$\mathbf{F} = 2\pi Le G_{\mathrm{D}} \Phi(h_{\mathrm{m}}) \frac{\mathbf{r}_{12}}{r_{12}} \tag{6}$$

where $\mathbf{r}_{12}$ is the vector connecting particle centers and r_{12} is the length of this vector.

By using the linearized version of the PB equation, we derived the following equation for the interaction energy per unit area of plates [7]:

$$\Phi = kTLe\, I \left[\mp (1 - \coth \bar{h}) \left(\bar{\psi}_1^{0^2} + \bar{\psi}_2^{0^2} \right) + \frac{2\bar{\psi}_1^0 \bar{\psi}_2^0}{\sinh \bar{h}} \right] \tag{7}$$

where the upper sign denotes the constant charge (c.c.) boundary conditions and the lower sign the constant potential (c.p.) boundary conditions, $\bar{\psi}_1^0$, $\bar{\psi}_2^0$ are the electric potential values at the surface of the first and of the second particle, respectively, often identified with the electrokinetic potential [8].

Equation (7) was first derived for the c.p. case by Hogg et al. [7] and will be referred to as the HHF model. Wiese and Healy [9] and Usui [10] considered the c.c. model, whereas Kar et al. [11] derived analogous formula for the interaction energy in the case of the "mixed" case, that is, c.p. at one plate and c.c. at the other.

By substituting Eq. (7) into Eq. (4), one obtains upon integration the interaction energy of two dissimilar spheres:

$$\phi = \pi\varepsilon \left(\frac{kT}{\mathrm{e}} \right)^2 G_{\mathrm{D}} \left[\mp \left(\bar{\psi}_1^{0^2} + \bar{\psi}_2^{0^2} \right) \ln \left(1 - \mathrm{e}^{-2h_{\mathrm{m}}/Le}\right) + 2\bar{\psi}_1^0 \bar{\psi}_2^0 \ln \frac{1 + \mathrm{e}^{-h_{\mathrm{m}}/Le}}{1 - \mathrm{e}^{-h_{\mathrm{m}}/Le}} \right] \tag{8}$$

where the upper sign denotes the c.c. boundary condition. Note that in contrast to Eq. (7), the interaction energy for spheres does not depend explicitly on the ionic strength I.

By differentiation, the expression for the interaction force can be formulated as

$$F = 2\pi G_{\mathrm{D}} kTLe\, I \left[\mp (1 - \coth \bar{h}) \left(\bar{\psi}_1^{0^2} + \bar{\psi}_2^{0^2} \right) \frac{+2\bar{\psi}_1^{0^2} \bar{\psi}_2^{0^2}}{\sinh h} \right] \tag{9}$$

where $\bar{h} = h_{\mathrm{m}}/Le$.

Equations (8) and (9) were first derived by Hogg et al. [7] for the c.c. model and Wiese and Healy [9] and Usui [10] for the c.p. model.

For short separations $\bar{h} \ll 1$, Eq. (8) reduces to

$$\phi = \pm 4\pi\varepsilon \left(\frac{kT}{e}\right)^2 G_D \left(\bar{\psi}_1^0 \pm \bar{\psi}_2^0\right)^2 \ln \bar{h} \tag{10}$$

where the upper sign denotes the c.c. model.

It can be easily deduced that the interaction energy for the c.c. model tends to plus infinity (repulsion) for short separations, whereas the c.p. model predicts diametrically different behavior, that is, the interaction energy tends to minus infinity (attraction) for the same combination of surface potentials. The divergence between both models appearing at short separations is caused by the violation of the low potential assumption. Indeed, in order to observe the c.c. boundary conditions, the surface potential of the plates should tend to infinity when they approach closely each other, even if at large separations these potentials were very low [12]. As a consequence, $\bar{\psi} \gg 1$ for $\bar{h} \to 0$ and the linear P.B. equation is not valid.

On the other hand, for larger separations, the expression for the interaction energy, Eq. (8) for both models reduces to the same asymptotic form:

$$\phi = 4\pi\varepsilon \left(\frac{kT}{e}\right)^2 G_D \left(\bar{\psi}_1^0 \pm \bar{\psi}_2^0\right)^2 e^{-h_m/Le} \tag{11}$$

As one can see, the interaction energy between plates decreases exponentially at large separations, the rate of decay being proportional to $\kappa = 1/Le$.

For equal sphere potentials and the c.p. model, Eq. (8) simplifies to the form derived originally by Derjaguin [3]:

$$\phi = 2\pi\varepsilon a \left(\frac{kT}{e}\right)^2 \bar{\psi}^{0^2} \ln(1 + e^{-h_m/Le}) \tag{12}$$

Equations (11) and (12) are commonly used in the literature for determining stability criteria of colloid suspension and for describing the plane–particle interactions in particle deposition problems.

The Derjaguin method was generalized by White [13] to non-spherical bodies of arbitrary shape. The first step in these calculations was to determine the minimum separation distance h_m between the two particles involved. Then, the four principal radii of curvature R_1', R_1'', R_2'' and R_2', at the minimum separation region, are evaluated. This allowed one to formulate the *general* expression for the generalized Derjaguin factor in the form [14–16]

$$G_D = \sqrt{\frac{R_1' R_1'' R_2' R_2''}{(R_1' + R_2')(R_1'' + R_2'') + (R_1'' - R_1')(R_2'' - R_2')\sin^2\vartheta}} \tag{13}$$

where ϑ is the angle between the x_1 and x_2 axes of the local Cartesian coordinate system with the common z axis.

In the case of particle–plane or two coplanar particle configurations, one has $\vartheta = 0$ and Eq. (13) simplifies to [14–16]

$$G_{\mathrm{D}} = \sqrt{\frac{R_1' R_1'' R_2' R_2''}{(R_1' + R_2')(R_1'' + R_2'')}} \tag{14}$$

Knowing G_{D}, one can also calculate from Eq. (4) the interaction energy of arbitrary-shaped (convex) particles for other types of interactions (for example, van der Waals interaction as discussed later). The only condition is that the effective range of these interactions should remain smaller than the particle size involved, so they can effectively be treated as surface interactions.

However, despite the apparent simplicity, it is rather difficult to apply Eqs. (13) and (14) for three-dimensional situations and bodies of arbitrary shape. The main problem is finding the points of the minimum separation of the two bodies involved, h_{m}, as a function of their mutual orientation. Even for such simple particle shapes as spheroids, one has to solve high-order non-linear trigonometric equations, which can only be done in an efficient way by iterative methods [14].

However, useful analytical results can be derived for limiting orientations of prolate and oblate spheroids as shown in Table 9.1. These results are of practical significance because they shape many globular proteins such as lysozyme, bovine serum albumin (BSA) or fibrinogen resembles prolate spheroid [17]. It is interesting to observe that the ratio between the Derjaguin factors (and hence of the interaction energy) for the parallel and perpendicular orientations of prolate spheroids (against a planar boundary) equals $1/A^2$, where $A = b/a$ is the shorter to longer axis ratio. This means that the electrostatic attraction will be much larger for the parallel orientation (at the same separation distance h_{m}) so the particles will tend to adsorb parallel. In the case of electrostatic repulsion (adsorption against an electrostatic barrier), the particles will preferably adsorb under the orientation perpendicular to the surface. The same pertains to the oblate spheroid adsorption.

The results in Table 9.1 indicate that for spheres and spheroids the particle–particle interaction energy is just half of the particle–interface energy, which is the case for arbitrary surface potential of particles and interfaces. This has major practical significance because the interaction energy of particles with surfaces can be simply derived from experimental measurements of particle deposition as discussed later. It is interesting that in the case of spheroid–plane interactions, the Derjaguin factor can be evaluated analytically as a function of the orientation angle ϑ. For prolate spheroids one has [14]

$$G_{\mathrm{D}} = a \frac{A^2}{A^2 \cos^2 \vartheta + \sin^2 \vartheta} \tag{15}$$

Table 9.1 The Derjaguin $\bar{G}_D$ and the LSA $\bar{G}_e^0$ geometric factors for limiting spheroid orientations

Prolate spheroids					
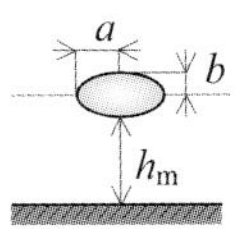	$\bar{G}_D = 1$ $\bar{G}_e^0 = \dfrac{2A}{A^2+1}$	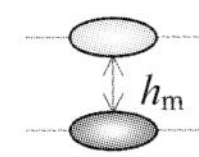	$\bar{G}_D = \dfrac{1}{2}$ $\bar{G}_e^0 = \dfrac{A}{A^2+1}$	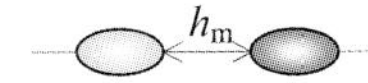	$\bar{G}_D = \bar{G}_e^0 = \dfrac{1}{2}A^2$
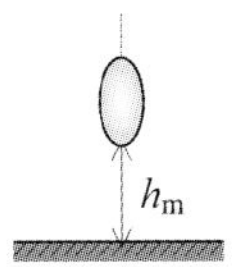	$\bar{G}_D = \bar{G}_e^0 = A^2$	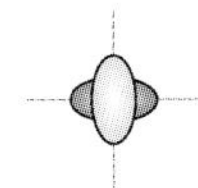	$\bar{G}_D = \bar{G}_e^0 = \dfrac{A}{A^2+1}$	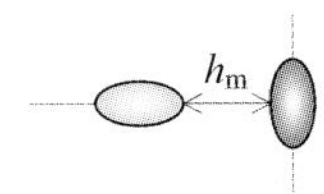	$\bar{G}_D = \dfrac{A^2}{\sqrt{(A^3+1)(A+1)}}$ $\bar{G}_e^0 = \dfrac{2A^2}{A^3+A+2}$
Oblate spheroids					
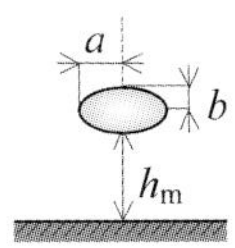	$\bar{G}_D = \bar{G}_e^0 = \dfrac{1}{A}$	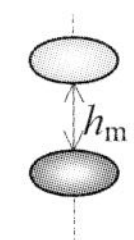	$\bar{G}_D = \bar{G}_e^0 = \dfrac{1}{2A}$	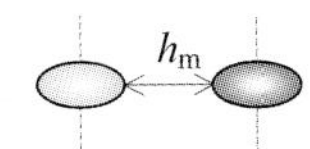	$\bar{G}_D = \dfrac{1}{2}A$ $\bar{G}_e^0 = \dfrac{A^2}{A^2+1}$
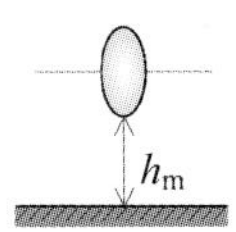	$\bar{G}_D = A$ $\bar{G}_e^0 = \dfrac{2A^2}{A^2+1}$	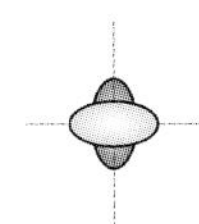	$\bar{G}_D = \bar{G}_e^0 = \dfrac{A^2}{A^2+1}$	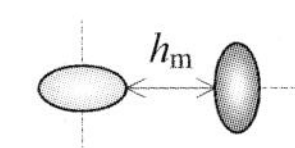	$\bar{G}_D = \dfrac{A}{\sqrt{(A^3+1)(A+1)}}$ $\bar{G}_e^0 = \dfrac{2A^2}{2A^3+A^2+1}$

whereas for the oblate spheroids the solution is

$$G_D = a\frac{A}{A^2\cos^2\vartheta + \sin^2\vartheta} \tag{16}$$

As mentioned, particle interactions at larger separations can be better described in terms of the LSA, introduced originally by Bell et al. [5]. The main postulate of this method is that the solution of the PB equation for a two-particle system can be constructed as a linear superposition of the solutions for isolated particles. This is justified because the electrostatic potential at separations larger than Le decreases to small values and can be described by the linear version of the PB equation. As a consequence, the solution of the PB equation in this region for a two-particle configuration can be obtained by postulating the additivity of potentials and fields stemming from isolated particles.

The LSA method can, in principle, be applied to arbitrary particle shape provided that a solution of the PB for isolated particles exists. At present, however, such solutions are known for spheres in a simple 1:1 electrolyte only.

Using the LSA method of Bell et al. [5] we derived the following analytical expression for the interaction force between two dissimilar particles:

$$\mathbf{F} = \phi_0 a_1 \frac{1 + r/Le}{r^2} \mathrm{e}^{-h_{\mathrm{m}/\mathrm{Le}}} \frac{\mathbf{r}_{12}}{r} \tag{17}$$

where $\phi_0 = 4\pi\varepsilon a_2(\mathrm{kT/e})\bar{Y}_1^0\bar{Y}_2^0$, $r = a_1 + a_2 + h_m$ is the distance between the particle centers and $\bar{Y}^0, \bar{Y}^0$ are the effective potentials of the two particles, which can be calculated for a 1:1 electrolyte from

$$\bar{Y}^0 = 4\tanh\left(\frac{\mathrm{e}\psi^0}{4kT}\right)\frac{2}{1 + \left[1 - \frac{[2a/Le+1}{(a/Le+1)^2}\tanh\frac{\mathrm{e}\psi^0}{4kT}\right]^{\frac{1}{2}}} \tag{18}$$

It can be deduced that in the case of $\psi^0\mathrm{e}/kT \gg 1$ Eq. (18) reduces to

$$\bar{Y}' = 8\frac{a + Le}{2a + Le} \tag{19}$$

On the other hand, for a/Le >>1 (planar interface), Eq. (18) reduces to

$$\bar{Y}^0 = 4\tanh\frac{\mathrm{e}\psi^0}{4kT} \tag{20}$$

The interaction energy can be easily calculated by integration of Eq. (17), which results in the expression [7]

$$\phi = \phi_0\frac{a_1}{r}\mathrm{e}^{-\mathrm{h}_{\mathrm{m}/\mathrm{Le}}} \tag{21}$$

In the case of sphere–plane interactions, Eqs. (17) and (21) reduce to the simpler form

$$F = \frac{\phi_0}{a_1} e^{-h_m/Le}$$
$$\phi = \phi_0 e^{-h_m/Le} \tag{22}$$

As can be seen, Eqs. (21) and (22) assume a simple two-parametric form, analogous to the Yukawa potential used widely in statistical mechanics. Since an exponential decay of the interaction force and energy with the distance is predicted, for $h_m/Le \gg 1$ they become negligible. An additional advantage of Eq. (22) is that it does not diverge to infinity in the limit $h_m \to 0$, but approaches the constant value $a_1/(a_1 + a_2)\phi_0$, which can be treated as the energy at contact. It can be calculated that for two equal spheres with the radius of 10^{-7} m (100 nm) and potential $\psi^0 = 100$ mV, the energy at contact (at room temperature) equals 65 kT units. This value increases proportionally to the particle size.

Because of their simple mathematical shape, Eqs. (17) and (22) are extensively used in numerical simulations of colloid particle adsorption problems. However, the disadvantage of the LSA method is that it can only be used in the original form exclusively for spherical particles. Owing to the increasing importance of anisotropic particle interactions, an approximate method has been proposed [14]. The essence of this approach, being in principle a mutation of the LSA, consists in replacing the interactions of convex bodies by analogous interactions of spheres with appropriately defined radii of curvature R_1', R_1'', R_2' and R_2''. As postulated in [14], these radii should be calculated as the geometric means of the principal radii of curvature evaluated at the point of minimum separation between the bodies, that is,

$$R_1 = \frac{2R_1' R_1''}{R_1' + R_1''} \tag{23}$$

$$R_2 = \frac{2R_2' R_2''}{R_2' + R_2''}$$

The advantage of this method, referred to as the equivalent sphere approach (ESA), consists in that the known numerical and analytical results concerning sphere interactions can be directly transferred to anisotropic particles. Thus, the LSA results, Eq. (21), can be expressed for spheroidal particles in the form

$$\phi = \phi_0 \frac{R_1 R_2}{a(R_1 + R_2 + h_m)} e^{-h_m/Le} = \phi_0 \frac{\bar{G}_e^0}{1 + \bar{G}_e \frac{h_m}{a}} e^{-h_m/Le} \tag{24}$$

where $\phi = 4\pi\varepsilon_0\varepsilon a\left(\frac{kT}{e}\right)^2 \bar{Y}_1^0 \bar{Y}_2^0$, a is the longer semiaxis of the spheroid and

$$\bar{G}_e^0 = \frac{R_1 R_2}{a(R_1 + R_2)} = \frac{2R_1' R_1'' R_2' R_2''}{a(R_1' R_1''(R_2' + R_2'') + R_2' R_2''(R_1' + R_1''))}$$
$$\bar{G}_e = \frac{a}{(R_1 + R_2)} = \frac{a(R_1' + R_1'')(R_2' + R_2'')}{2[R_1' R_1''(R_2' + R_2'') + R_2' R_2''(R_1' + R_1'')]} \tag{25}$$

are the two geometric factors.

Although Eq. (24) has the simple Yukawa-type form, its application in the general case of spheroid interaction in space is not straightforward owing to the necessity for a numerical evaluation of the geometric functions $\bar{G}_e^0$ and $\bar{G}_e$ [14]. However, analogously to the Derjaguin model, these functions can be evaluated analytically for some limiting orientations compiled in Table 9.1.

It is interesting that for the spheroid–plane interactions, because $\bar{G}_e = 0$, the energy is described by the equation analogous to the Derjaguin formula, Eq. (11) (at large separations), that is,

$$\phi = \phi_0 \bar{G}_e^0 e^{-h_m/Le} \tag{26}$$

where the geometric factor $\bar{G}_e^0$ can be evaluated analytically for prolate spheroids in terms of the inclination angle ϑ [14]. In the case of arbitrary orientation of spheroids, one has to use numerical methods for evaluating the minimum distance and calculating the local radii of curvature. The use of efficient iterative schemes makes this task fairly simple [14]. Even with this complication, the use of the ESA is more efficient than any attempt at solving the PB equation for anisotropic particles.

9.2.2 Van der Waals Interactions

The short-range, van der Waals forces are the net result of the Keesom orientational forces, the inductive Debye forces and the dispersion (London) forces arising because of the presence of spontaneous electron density fluctuations in atoms and molecules. The energy of interaction of molecules due to these forces can be in general expressed in terms of a power series expansion [1, 18]:

$$\phi_a = -\sum_l \frac{C_l}{r_0^l} \tag{27}$$

where C_l are appropriate constants and l are various exponents.

The exponent of the leading term in Eq. (27) is equal to six, but there also appear terms with $l=7$, describing the retarded interactions at larger distances, and $l=8$, describing the effect of multipole interactions [1, 19].

By knowing the intermolecular potential, one can calculate interaction of macro-bodies (particles) by using the classical microscopic approach developed by Hamaker [20]. The basic assumption of this approach is the additivity principle. A major advantage of the microscopic approach is that useful analytical expressions can be derived for complicated geometries of the interacting particles, including the case of rough surfaces.

According to the additivity principle, the interaction energy of two bodies 1 and 2 can be calculated by summing them in a pair for all atoms or molecules. For macro-bodies containing many atoms, the summation procedure can be replaced with integration introducing the number density of atoms. In the general case, these densities are position-dependent quantities (functions of the space variables in both bodies). Accordingly, the general expression for the energy of interaction of two particles of arbitrary shape can be formulated as a volume integral:

$$\phi = \int\limits_{v_1}\int\limits_{v_2} \varphi_{a}\rho_{a1}\rho_{a2}\mathrm{d}v_1\mathrm{d}v_2 = -\sum_{1} C_l \int\limits_{v_1}\int\limits_{v_2} \frac{\rho_{a1}(\mathbf{r}_1)\rho_{a2}(\mathbf{r}_2)}{r_0^l}\mathrm{d}v_1\mathrm{d}v_2 \tag{28}$$

where ρ_{a1}, ρ_{a2} are the number densities of atoms (molecules) in the two bodies involved and v_1, v_2 are the volumes of the bodies.

Various cases of practical interest can be derived from Eq. (28) by evaluating the volume integrals assuming a uniform atom density. For example, the energy of interaction of a circular disk of radius $R(h)$ and thickness dh is given by

$$\phi = -2\pi^2\rho_{a1}\rho_{a2}\sum_{1}\frac{C_l}{(l-2)(l-3)}\frac{R^2(h)\mathrm{d})h}{h^{l-3}} \tag{29}$$

On the other hand, the energy of interaction of two half-spaces (plates) per unit area is given by the expression

$$\Phi = \phi/\Delta S = -2\pi\rho_{a1}\rho_{a2}\sum_{1}\frac{C_n}{(l-2)(l-3)(l-4)}\frac{1}{h^{l-4}} \tag{30}$$

where ΔS is the surface area of the plates.

Equation (30) has major significance because by knowing interactions of half-spaces, one can calculate interactions of arbitrary shaped convex bodies by using Eq. (1) originating from the Derjaguin method. Thus, the van der Waals interaction energy of two convex bodies is given by the general expression

$$\phi = -2\pi G_\mathrm{D}\int\limits_{h_m}^{\infty}\Phi(h)\mathrm{d}h = -4G_\mathrm{D}\pi^2\rho_{a1}\rho_{a2}\sum_{1}\frac{C_l}{(l-2)(l-3)(l-4)(l-5)}\frac{1}{h_m^{l-5}} \tag{31}$$

where G_D is the geometric factor discussed above.

The interaction force (derivative of the energy upon the separation h) is given by the expression

$$F = -4G_D\pi^2\rho_{a1}\rho_{a2}\sum_l \frac{C_l}{(l-2)(l-3)(l-4)}\frac{1}{h_m^{l-4}} \tag{32}$$

Let us now consider some concrete cases that can be derived from the above equation for $l=6$. In the case of the two half spaces, the energy of interaction per unit area is given by the expression

$$\Phi = -\frac{\pi\rho_{a1}\rho_{a2}C_l}{12h_m^2} = -\frac{A_{12}}{12\pi h_m^2} \tag{33}$$

where $A_{12} = \pi^2 C_l\rho_{a1}\rho_{a2}$ is the Hamaker constant for two bodies 1 and 2 interacting across a vacuum.

Accordingly, by using Eq. (1) the Derjaguin expression for the energy of interaction of arbitrary-shaped bodies, for example, dissimilar spheres, can be formulated as

$$\phi = -G_D\frac{A_{12}}{6}\frac{1}{h_m} \tag{34}$$

Consequently, the interaction force is given by the expression

$$F = -G_D\frac{A_{12}}{6}\frac{1}{h_m^2} \tag{35}$$

As can be seen from Eqs. (34) and (35), the energy and force of interactions for the sphere–sphere case are just half of those of the sphere–half space interactions (because $G_D = 0.5a$ for two equal spheres and $G_D = a$ for the sphere–half space case).

On the other hand, the interactions of a disk of infinitesimal thickness dh and surface area $S(h)$ with a half space in the case of $n=6$ is described by

$$\phi = -\pi\rho_{a1}\rho_{a2}\frac{C_l}{6}\frac{S(h)\mathrm{d}h}{h^3} = -\frac{A_{12}}{6\pi}\frac{S(h)\mathrm{d}h}{h^3} \tag{36}$$

Equation (36) can be used for evaluating exact equations describing interactions of axis-symmetric bodies. For example, the sphere–half-space energy of interaction is given by the expression [1, 2]

$$\phi = -\frac{A_{12}}{6}\left[\frac{a}{h_m} + \frac{a}{2a+h_m} + \ln\left(\frac{h_m}{2a+h_m}\right)\right] = \phi_s(h_m, a) \tag{37}$$

If $h \ll a$, Eq. (37) reduces to the simple form

$$\phi = -\frac{A_{12}}{6}\frac{a}{h_m} \tag{38}$$

The interaction force is given by

$$F = -\frac{A_{12}}{6}\frac{a}{h_{\mathrm{m}}^2} \tag{39}$$

As can be seen, Eqs. (38) and (39) become identical with Eqs. (34) and (35) derived from the Derjaguin approach by substituting $G_{\mathrm{D}} = a$.

Using Eq. (36), one can also evaluate interactions of spheroids, both prolate and oblate, by simple integration. For a prolate spheroid oriented with its longer axis perpendicular to an interface (see Table 9.2), the cross-section area is given by the equation $S(h) = b^2[1 - (h - h_0)^2/a^2]$, where $h_0 = h_{\mathrm{m}} + a$ is the position of the spheroid center, a is the longer axis and b the shorter axis.

Integration of Eq. (36) with this expression for $S(h)$ gives

$$\phi = -\frac{b^2}{a^2}\phi_{\mathrm{s}}(h_{\mathrm{m}}, a) \tag{40}$$

Equation (40), representing an exact result, indicates that the interaction of a prolate spheroid with a flat interface is described by the same equation as that of a sphere, except for the geometric pre-factor b^2/a^2. Obviously, the same concerns the interaction force.

For short separations, Eq. (40) reduces to the form

$$\phi = -\frac{A_{12}}{6h_{\mathrm{m}}}\frac{b^2}{a} \tag{41}$$

In the case where a prolate spheroid is oriented parallel to the interface, the expression for the interaction energy becomes

$$\phi = \frac{a}{b}\phi_{\mathrm{s}}(h_{\mathrm{m}}, a) \tag{42}$$

For oblate spheroids with the shorter axis b perpendicular to the interface, the energy is given by

$$\phi = \frac{a^2}{b^2}\phi_{\mathrm{s}}(h_{\mathrm{m}}, b) \tag{43}$$

For short separations, Eq. (43) becomes

$$\phi = -\frac{A_{12}}{6h_{\mathrm{m}}}\frac{a^2}{b} \tag{44}$$

It is interesting that in all of the above equations, the asymptotic form of the interaction energy in the limit of $h_{\mathrm{m}} \to 0$ tends to infinity proportionally to h_{m}^{-1}. Accordingly, the interaction force changed as h_{m}^{-2}. Moreover, both quantities were proportional to the interacting particle dimension rather than to its surface area.

Table 9.2 Double-layer interaction energy expression for various configurations.

System	Expression	Remarks, References
plate/plate	$\Phi_0 \bar{Y}_1^0 \bar{Y}_2^0 e^{-\bar{h}}$	LSA [16], valid for arbitrary potential , $\bar{h} > 1$
plate/plate	$\Phi_0 \left[\mp(\bar{\psi}_1^{02} + \bar{\psi}_2^{02})(1 - \coth\bar{h}) + \dfrac{2\bar{\psi}_1^0 \bar{\psi}_2^0}{\sinh\bar{h}} \right]$	linear model [7] $\bar{\psi}_1^0 < 1$, $\bar{\psi}_2^0 < 1$ upper sign c.c., lower sign c.p.
sphere/plate	$\phi_0' \bar{Y}_1^0 \bar{Y}_2^0 e^{-\bar{h}}$	LSA [5], valid for arbitrary potential, $\bar{h}_m > 1$
sphere/sphere	$\phi_0' \bar{Y}_1^0 \bar{Y}_2^0 \dfrac{a_1}{a_1 + a_2 + h_m} e^{-\bar{h}}$	LSA [5], valid for arbitrary potential, $\bar{h}_m > 1$
sphere/plate	$\dfrac{1}{4}\phi_0' \left[m(\bar{\psi}_1^{02} + \bar{\psi}_2^{02}) \ln(1 - e^{-2\bar{h}_m}) + 2\bar{\psi}_1^0 \bar{\psi}_2^0 \ln \dfrac{1 + e^{-\bar{h}_m}}{1 - e^{-\bar{h}_m}} \right] \dfrac{1}{4}\phi_0' f(\bar{h}_m)$	linear [7], $\bar{\psi}_1^0 < 1$, $\bar{\psi}_2^0 < 1$ upper sign, c.c., lower sign, c.p.
sphere/sphere	$\dfrac{1}{4}\phi_0' \dfrac{a_1}{a_1 + a_2} f(\bar{h}_m)$	linear c.c., c.p. [7] valid for $\bar{\psi}_1^0 < 1$, $\bar{\psi}_2^0 < 1$
identical spheres	$\dfrac{1}{2}\phi_0' \bar{\psi}_0^2 \ln(1 + e^{-\bar{h}_m})$	Derjaguin formula
convex particle/plate	$2\pi G_D \int \Phi(h) dh$	generalized Derjaguin model valid for arbitrary interaction law $\Phi(h)$ [13], $a/Le > 1$
prolate spheroid/ plate	$\dfrac{1}{4}\phi_0' \dfrac{(b/a)^2}{\left(\dfrac{b}{a}\right)^2 \cos^2\vartheta + \sin^2\vartheta} f(\bar{h}_m)$	linear [14], valid for $\bar{\psi}_1^0 < 1$, $\bar{\psi}_2^0 < 1$

$\bar{\psi}_1^0 = \psi_1^0 \dfrac{e}{kT} \qquad \bar{\psi}_2^0 = \psi_2^0 \dfrac{e}{kT},$

$\bar{h} = h/Le \qquad \bar{h}_m = h_m/Le \qquad Le = \left(\dfrac{\varepsilon kT}{2e^2 I}\right)^{1/2}$

$\Phi_0 = kTLe\, I = \left[\dfrac{\varepsilon (kT)^3 I}{2e^2}\right]^{1/2} \qquad \phi_0' = 4\pi\varepsilon \left(\dfrac{kT}{e}\right)^2 a_2$

$\bar{Y}_1^0 = 4\tanh\left(\dfrac{e\psi_1^0}{4kT}\right) \qquad \bar{Y}_2^0 = 4\tanh\left(\dfrac{e\psi_2^0}{4kT}\right)$ for spheres see Eq. (18)

$f(\bar{h}_m) = \left[\mp(\bar{\psi}_1^{02} + \bar{\psi}_2^{02}) \ln(1 - e^{-2\bar{h}_m}) + 2\bar{\psi}_1^0 \bar{\psi}_2^0 \dfrac{1 + e^{-\bar{h}_m}}{1 - e^{-\bar{h}_m}} \right]$ for arbitrary $\bar{h}_m$

$f(\bar{h}_m) = 4\bar{\psi}_1^{02} \bar{\psi}_2^{02} e^{-\bar{h}_m}$ for $\bar{h}_m \gg 1$

The same integration procedure can be applied for deriving exact equations describing retarded interactions of particles with interfaces.

The integration procedure can also be applied for deriving exact equations describing interactions of two particles. For two dissimilar spheres, the integration method (performed in the spherical coordinate system) was first applied by Hamaker [20] who derived the equation

$$\phi = -\frac{A_{12}}{6}\left[\frac{2a_1a_2}{h_m(h_m+2a_1+2a_2)} + \frac{2a_1a_2}{(h_m+2a_1)(h_m+2a_2)} + \ln\frac{h_m(h_m+2a_1+2a_2)}{(h_m+2a_1)(h_m+2a_2)}\right] \tag{45}$$

At short separations, when $h/a << 1$, Eq. (45) reduces to

$$\phi = -\frac{A_{12}}{6h_m}\left(\frac{a_1a_2}{a_1+a_2}\right) \tag{46}$$

This is again identical with the Derjaguin result because for two dissimilar spheres, G_D equals $a_1a_2/(a_1+a_2)$.

For more complicated geometries of particles such as spheroids under arbitrary orientations, evaluation of exact results by integration becomes too complicated. In this case, the Derjaguin method becomes especially useful because the geometric factors are known for various orientations (see Table 9.3).

Another area of application of the Derjaguin method is the calculation of interactions of rough particles or interfaces. This problem has vital practical significance because all colloid particles, whether of natural origin or synthesized, exhibit an appreciable degree of surface heterogeneity of geometric nature. In order to illustrate the significance of this effect, let us consider a planar interface with a small particle of size δ_r attached to it. The interaction of a larger particle of arbitrary shape with the interface bearing this model roughness element can be expressed according to the Derjaguin equation as

$$\phi = -G_D\frac{A_{12}}{6(h_m+\delta_r)} - G'_D\frac{A_{12}}{6h_m} \tag{47}$$

where G'_D is the Derjaguin factor for the interactions of a large particle with roughness and h_m is the minimum separation distance between the particle and the roughness.

Using Eq. (47), one can calculate that the ratio of the interaction energy of the particle with smooth to rough surfaces at the same minimum separation distance is

$$\frac{\phi}{\phi_s} = \frac{h_m}{h_m+\delta_r} + \frac{G'_D}{G_D} \cong \frac{h_m}{h_m+\delta_r} \tag{48}$$

because $G'_D << G_D$.

One can deduce from this simple equation that for $h=1$ nm, a micro-roughness of the size of 1 nm will reduce the interaction energy of a larger sphere two-fold. For a 10 nm roughness, this reduction is 11-fold. This means that the van der Waals interactions of rough particles are considerably reduced over smooth particles, which can prevent their efficient adhesion [21].

This problem has been studied in much detail [22–31]. Model rough surfaces were generated either by depositing spherical particles of polydisperse size dis-

Table 9.3 Van der Waals energy expressions for various macro bodies $A_{12} = \pi^2 \rho_{a1} \rho_{a2} C_1$ (Hamaker constant for interactions in vacuum)

Configuration	Energy expression (leading term)	Energy expression short separations	Energy expression Derjaguin
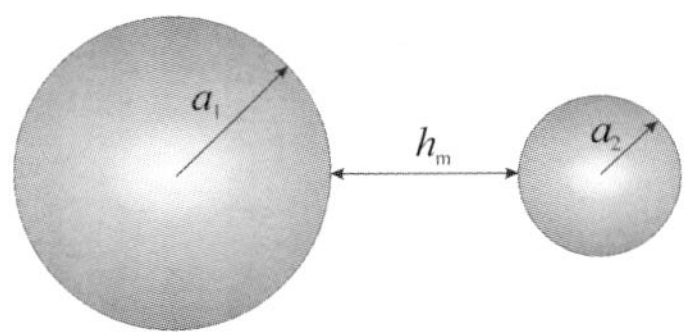 sphere sphere	$-\frac{A_{12}}{6}\left[\frac{2a_1 a_2}{h_m(h_m + 2a_1 + 2a_2)} + \frac{2a_1 a_2}{(h_m + 2a_1)(h_m + 2a_2)} + \ln\frac{h_m(h_m + 2a_1 + 2a_2)}{(h_m + 2a_1)(h_m + 2a_2)}\right] = \phi_{ss}(h_m, a_1, a_2)$	$-\frac{A_{12}}{6h_m}\left(\frac{a_1 a_2}{a_1 + a_2}\right)$	$-\frac{A_{12}}{6h_m}\left(\frac{a_1 a_2}{a_1 + a_2}\right)$
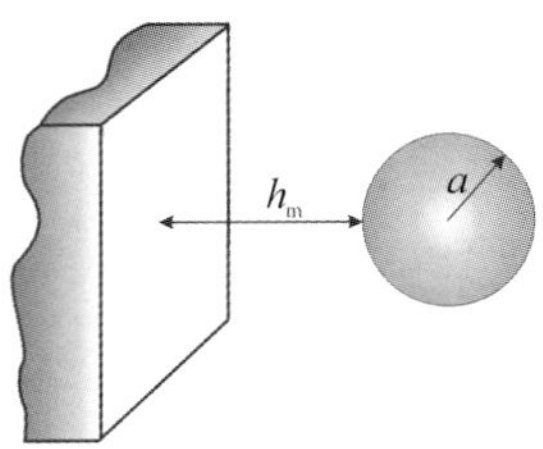 half space sphere	$-A_{12}\left[\frac{2a(a + h_m)}{h_m(2a + h_m)} + \ln\frac{h_m}{h_m + a}\right] = \phi_s(h_m, a)$	$-\frac{A_{12}a}{6h_m}$	$-\frac{A_{12}a}{6h_m^2}$
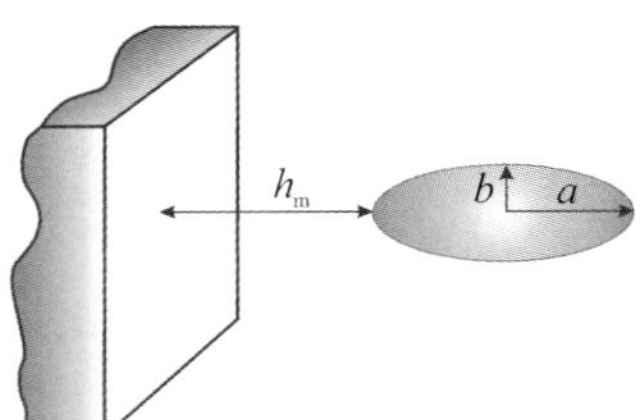 half space prolate spheroid	$-\frac{b^2}{a^2}\phi_s(h_m, a)$	$-\frac{A_{12}b^2}{6h_m a}$	$-\frac{A_{12}b^2}{6h_m a}$

Table 9.3 (continued)

Configuration	Energy expression (leading term)	Energy expression short separations	Energy expression Derjaguin
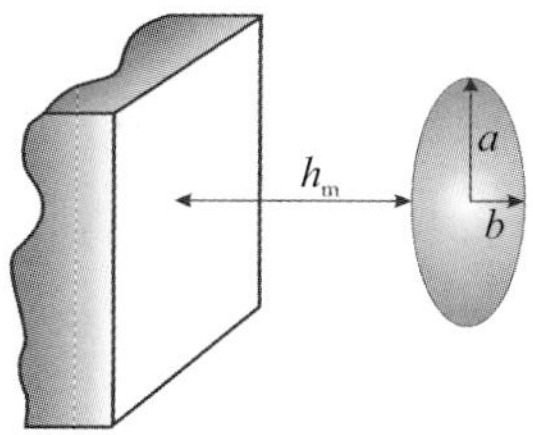 half space / prolate spheroid	$\frac{a}{b}\phi_s(h_m, b)$	$-\frac{A_{12}a}{6h_m}$	$-\frac{A_{12}a}{6h_m}$
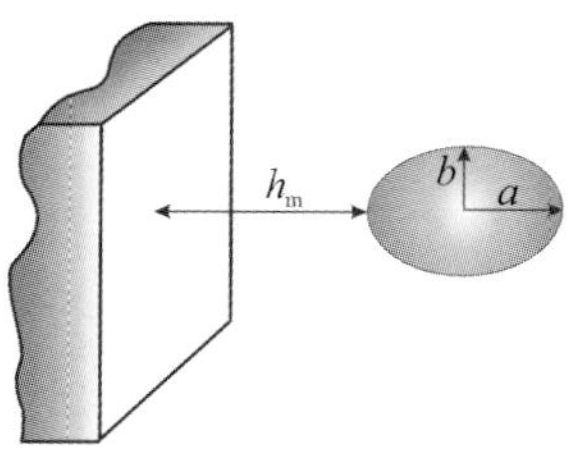 half space / oblate spheroid	$\frac{b}{a}\phi_s(h_m, a)$	$-\frac{A_{12}b}{6h_m}$	$-\frac{A_{12}b}{6h_m}$

Table 9.3 (continued)

Configuration	Energy expression (leading term)	Energy expression short separations	Energy expression Derjaguin
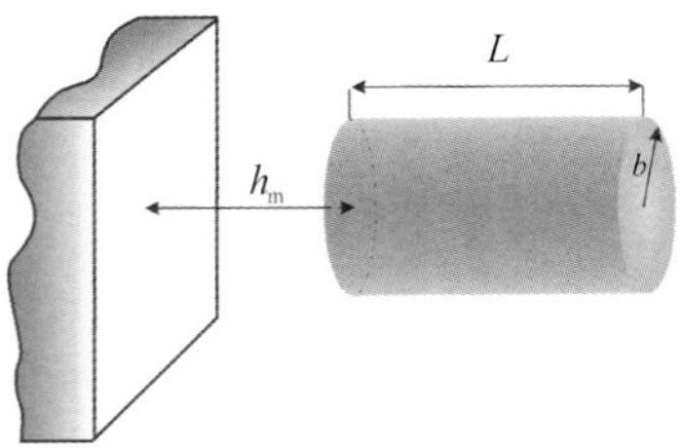 half space cylinder	$-\frac{A_{12}b^2}{12h_m^2}\left[1-\left(\frac{h_m}{L}\right)^2\right]$	$-\frac{A_{12}b^2}{12h_m^2}$	–
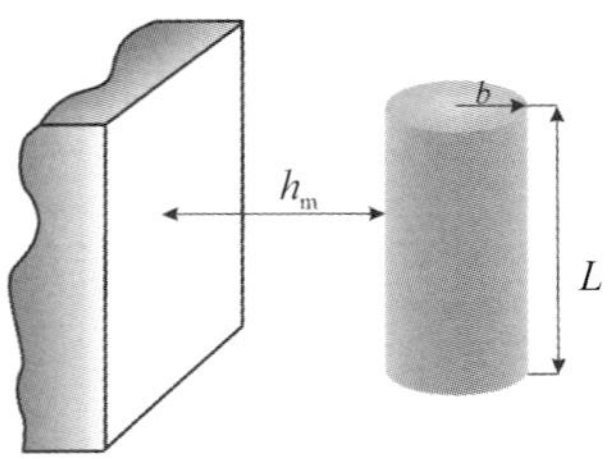 half space cylinder	$\frac{-A_{12}b^2L}{3\pi[h_m(h_m+2b)]^{3/2}}x$ $\left[\mathrm{arctg}\frac{h_m+2b}{\sqrt{h_m(h_m+2b)}}+\mathrm{arctg}\frac{h_m}{\sqrt{h_m(h_m+2b)}}\right]$ $=\phi_s(h_m b)\frac{L}{2\sqrt{2bh_m}}$	$-\frac{A_{12}b}{6h_m}\frac{L}{(h_m^{1/2}b^{1/2}2^{3/2})}$	–

tribution or by covering the smooth core with a rough shell of well-defined density [28, 29]. Although exact results for arbitrary statistical distribution of microroughness were found to be rather complicated, Czarnecki [28] and Czarnecki and Dąbroś [29] derived a simple interpolating function for ϕ_r valid for both sphere–sphere and sphere–plane interactions:

$$\phi_r = \phi_s \left(\frac{h_m}{S}\right)^c \tag{49}$$

where ϕ_s is the energy for the smooth particle interface and the distance h_m is measured between the two outermost points at the particle surfaces, $S = h_m + (b_1 + b_2)/2$, b_1 and b_2 are the thicknesses of the rough layer at particle 1 and 2, respectively, and c is the exponent, close to 1 for the unretarded case and 1.5 for the retarded case [28, 29]. Thus, in the limit $h_m \to 0$, Eq. (49) reduces to the following simple form (for unretarded interactions):

$$\phi_r = \phi_s H \tag{50}$$

where $H = 2h_m/(b_1 + b_2)$ is the scaled distance between particle surfaces.

9.2.3
Interactions in Dispersing Media, Hamaker Constant Calculations

The expressions presented above are strictly valid only for interactions of particles across a vacuum (air). In this case, the Hamaker constant can be easily calculated because the data on atomic densities and the London constant C_1 are known for many substances [1, 18].

However, for most of the experimentally relevant situations, interacting particles are immersed in a dispersing medium (most frequently water), which significantly modifies their van der Waals interactions. As shown by Hamaker [20], the role of the intervening medium can be evaluated by exploiting the energy additivity principle, which results in the following expression:

$$A_{132} = A_{12} + A_{33} - A_{13} - A_{23} \tag{51}$$

where A_{132} is the composite Hamaker constant describing interactions of particle 1 with particle 2 across the medium denoted by the subscript 3 and A_{ij} are the Hamaker constants characterizing various interactions in a vacuum.

In the case of two bodies of the same material, Eq. (51) simplifies to the form

$$A_{131} = A_{11} + A_{33} - 2A_{13} \tag{52}$$

It should be mentioned, however, that the additivity assumption pertinent to the Hamaker theory in general and Eqs. (51) and (52) in particular has certain limitations, especially when dealing with condensed phases. It is expected that

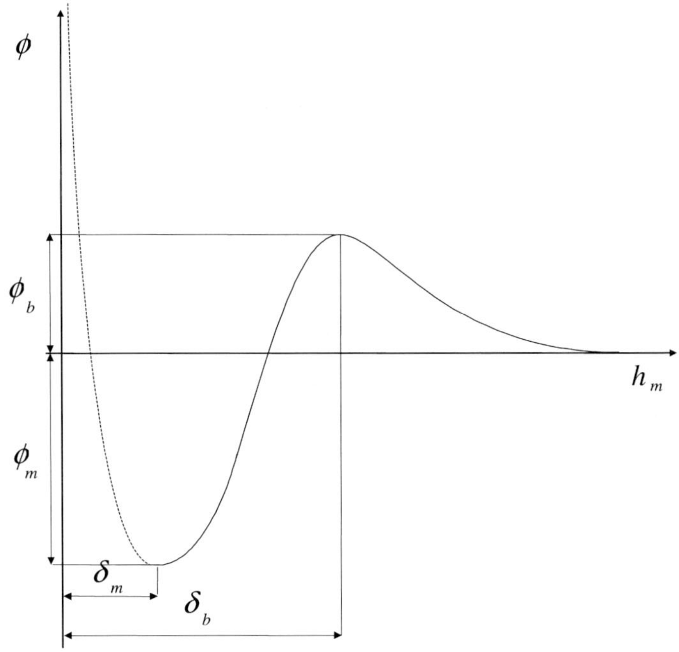

Fig. 9.2 Energy profile of Type II, characterized by an energy minimum of depth ϕ_m and an energy barrier of height ϕ_b.

the van der Waals interactions will be underestimated. This is so because the fluctuating electric field generated within the interacting pair of molecules (atoms) is reflected from other molecules that enhances the pair interactions. These many-body effects have been considered in a more universal way in the so-called macroscopic theory of van der Waals interactions developed by Lifshitz [32] and Dzialoshinsky and Lifshitz [33] for the case of two planar interfaces (half spaces) separated by a third medium. It was shown that the interaction energy between two half spaces is given in this limit by an expression analogous to that previously derived using the Hamaker theory:

$$\phi_{132} = -\frac{A_{132}}{12\pi}\frac{1}{h_m^2} \tag{53}$$

where the composite Hamaker constant A_{132} can be explicitly evaluated from the equation [18]

$$A_{132} = \frac{3}{2}kT\sum_n \left[\frac{\varepsilon_1(i\tilde{\nu}_n) - \varepsilon_3(i\tilde{\nu}_n)}{\varepsilon_1(i\tilde{\nu}_n) + \varepsilon_3(i\tilde{\nu}_n)}\right]\left[\frac{\varepsilon_2(i\tilde{\nu}_n) - \varepsilon_3(i\tilde{\nu}_n)}{\varepsilon_2(i\tilde{\nu}_n) + \varepsilon_3(i\tilde{\nu}_n)}\right] = \frac{3}{2}kT\sum_n \bar{\varepsilon}_{13}(i\tilde{\nu}_n)\bar{\varepsilon}_{23}(i\tilde{\nu}_n) \tag{54}$$

where $\varepsilon_1(i\tilde{\nu}_n), \varepsilon_2(i\tilde{\nu}_n)$ and $\varepsilon_3(i\tilde{\nu}_n)$ are the real components of dielectric permittivities of the three phases involved, which are functions of the imaginary frequency $i\tilde{\nu}_n$. Equation (54) can be used for calculating the Hamaker constant for many systems because dielectric permittivities of many substances of practical interest are fairly well known from experiments [18]. For example, for the polystyrene–water–polystyrene system the Hamaker constant equals 1.4×10^{-20} J. It is interesting to observe that values of the Hamaker constant for composite systems can be calculated by exploiting the so-called combining relation [18] being, in principle, semi-empirical interpolating functions:

$$A_{132} = \pm\sqrt{A_{131}A_{232}} \tag{55}$$

On the other hand, the Hamaker constant for interactions of two different particles across an arbitrary solvent when the corresponding interactions across vacuum (air) are known can be calculated from the dependence

$$A_{132} = \left(\sqrt{A_{11}} - \sqrt{A_{33}}\right)\left(\sqrt{A_{22}} - \sqrt{A_{33}}\right) \tag{56}$$

9.2.4 Superposition of Interactions and the Energy Profiles

Once the electrostatic and van der Waals interactions for various particle–particle and particle–interface configurations are known, one can attempt to construct the overall interaction energy profile. This is a prerequisite for estimating colloid particle stability, adsorption, deposition and adhesion phenomena, which are of vital practical interest. An approach of appealing simplicity would be to treat these interactions as independent of each other and to construct the energy profile as a sum of electrostatic and van der Waals contributions. This was precisely the idea behind the DLVO theory (abbreviation from Derjaguin and Landau [34] and Vervey and Overbeek [35], who are credited with its foundation) extensively used over the decades in the field of colloid sciences [8, 12, 36]. The sum of the electrostatic and van der Waals interactions is, therefore, often called the DLVO potential, used as a reference potential. In consequence, all interactions except these two are referred to as non-DLVO interactions, examples being the Born repulsion, steric interactions due to adsorbed polymer layers, depletion interactions, hydrogen and chemical bonding, external forces such as gravity or magnetic forces, hydrodynamic forces and so forth.

Because, for most known systems, the Hamaker constant is positive and confined within the range 10^{-21}–10^{-20} J (with the exception of metals, where it is slightly larger), one can deduce from the above equations that the van der Waals contribution to the interaction energy becomes negative and significantly larger (in absolute value) than a kT unit. This is so for particle–particle or particle–interface separations much smaller than particle dimensions, approximately 10–100 nm. For larger distances, the retardation effect makes these interactions negligible.

In contrast, the electrostatic interactions can be either positive or negative depending on surface potentials of particles, boundary conditions at their surfaces and on the separation distance. Additionally, their range can be varied between broad limits (1–1000 nm) by simply changing the ionic strength of electrolyte solutions. As a result, at separations comparable with particle dimensions, the electrostatic interactions dominate over the van der Waals contribution except for high ionic strength solutions. Because of the diversified range and magnitude of electrostatic interactions, the resulting energy profiles may become quite complicated, as discussed in Ref. [37].

However, for the sake of convenience, these profiles can be classified into some basic categories. The Type I profiles reflects interactions of similarly charged particles in not too concentrated electrolyte solutions when the overall interaction energy is dominated by the electrostatic interactions. In this case, the energy increases monotonically when the particles approach each other attaining values much higher than the kT unit for $h_m \to 0$. The appearance of this profile, typical for particle sizes >100 nm, excludes the possibility of particle aggregation, so the colloid suspension remains indefinitely stable in time. This is advantageous for performing particle adsorption experiments since the bulk particle concentration remains constant and no aggregates appear in the colloid suspension. Obviously, a profile of this type may appear for the particle–interface interactions when both are equally charged. In this case, particle adsorption remains negligible unless charge heterogeneity (either natural or artificially introduced) appears at the surface.

It is often convenient when analyzing particle adsorption under non-linear conditions (surface blocking effects) to replace the energy profile of Type I (soft interaction profile) by the idealized profile called the hard-particle interaction. According to this model, introduced originally by Barker and Henderson [38], the interaction energy remains zero except for the critical distance h_c, where it tends to infinity. Physically this means that the interacting particles can be treated as hard particles having the equivalent dimensions increased over the true geometric dimensions by the small value, which can be treated as the effective interaction range. Therefore, this concept is often referred to as the effective hard particle (EHP) model. It was demonstrated [1, 12, 14, 37] that $h^* = 1/2h_c$, is strictly related to the double layer thickness Le. In practice, for $a/Le << 1$, h^* remains proportional to Le with the proportionality constant being about 2 for colloid suspensions [14]. The numerical calculations discussed later demonstrate that the EHP concept is a powerful method of analyzing adsorption of interacting particles of spherical and non-spherical shape.

On the other hand, particle adsorption phenomena are reflected by the energy profile of Type Ib, where the interaction energy decreases monotonically from zero attaining large negative values (attraction) when $h_m \to 0$. This profile appears in systems when particle and interface bear opposite surface charges. In order to simplify the mathematical analysis of particle transport phenomena, this energy profile is often idealized by introducing the perfect sink (PS) model, as done originally by Smoluchowski [39] in his fast coagulation theory. According to this approach, the interaction energy remains zero up to a characteristic distance δ_m where it becomes minus infinity. Smoluchowski originally assumed $\delta_m=0$. This model can be improved upon by assuming that $\delta_m=2h^*$, analogously to that of the particle–particle interactions. One can interpret in this way the enhanced particle transport due to attractive double-layer interactions as discussed in Refs. [1, 14, 15].

Obviously, both the energy profile of Type I and the PS model should be treated as an idealization of any real situation because, at very small separations, the interaction energy must become positive due to the Born repulsion preventing particle–wall penetration.

In any case, the appearance of the repulsive interactions fixes the minimum value of the interaction energy, which remains finite in accordance with intuition. This minimum energy value is often referred to as the primary minimum denoted by ϕ_m and the distance where it appears is called the primary minimum distance δ_m (see Fig. 9.2). One may expect that δ_m is of the order of the range of the Born repulsive forces, that is, 0.5–1 nm. However, the extension of the region where the interaction energy assumes a negative value can be much larger, comparable to the Debye screening length, that is, about 96 nm for a 10^{-5} M electrolyte solution at a temperature of 293 K.

In the case where the interface and particle bear a surface charge of similar sign, except for the above monotonic energy profiles, another, more complicated, energy profile of Type II is likely to appear for higher ionic strengths. The characteristic feature of the profile, shown schematically in Fig. 9.2, is the presence of a maximum energy (barrier) of height ϕ_b at the distance δ_b. This energy profile corresponds to the activated transport conditions in chemical kinetics. Obviously, the height of the barrier is very sensitive to the electrolyte concentration and composition (presence of polyvalent ions), the Hamaker constant, particle size, shape and orientation. The presence of the energy barrier considerably reduces the particle adsorption rate, which may become inaccessible for accurate measurements. Therefore, in this case, the experimental measurements are often difficult to interpret in terms of the classical DLVO theory. It should also be mentioned that owing to a large transport resistance induced by the barrier, the bulk transport conditions are less important in this case.

9.3 Phenomenological Transport Equations

Knowing the specific interactions discussed above, one can formulate the phenomenological equations describing particle transport to various interfaces. Solutions of these equations allow one to interpret properly experimental data on particle deposition. This opens up the possibility of learning about the true energy profile for a concrete system, which is a prerequisite for predicting its stability under dynamic conditions.

Within the framework of the phenomenological approach based on irreversible thermodynamics, the mass balance (continuity) equation for the dispersed particle phase is given by the expression [1, 15, 40]

$$\frac{\partial n}{\partial t} + \nabla \cdot \tilde{\mathbf{j}} = 0 \tag{57}$$

where n is the local particle number concentration, t is the time and $\tilde{\mathbf{j}}$ is the generalized flux vector incorporating the translational and the rotary fluxes given by

$$\tilde{\mathbf{j}} = -\tilde{\mathbf{D}} \cdot \nabla n + (\tilde{\mathbf{M}} \cdot \mathbf{F})n = -\tilde{\mathbf{D}} \cdot \nabla n + \frac{1}{kT}(\tilde{\mathbf{D}} \cdot \tilde{\mathbf{F}})n \tag{58}$$

where $\tilde{\mathbf{D}}$ is the generalized diffusion coefficient and $\tilde{\mathbf{F}}$ is the generalized force.

Because the translational and rotary fluxes are usually coupled for non-spherical particles, in the general case the mass balance equation, Eq. (57) becomes a system of partial differential equations of the parabolic type expressed in three spatial and three orientation coordinates.

In the case of spherical particles having homogeneous properties, the rotary flux becomes irrelevant and the translation flux becomes

$$j = -\mathbf{D} \cdot \nabla n + \frac{1}{kT}(\mathbf{D} \cdot \mathbf{F}) = -\mathbf{D} \cdot \nabla n - \frac{1}{kT}(\mathbf{D} \cdot \nabla \phi)n + \mathbf{U}_{\mathrm{h}} n \tag{59}$$

where ϕ is the interaction potential of the net conservative force $\mathbf{F}=-\nabla\phi$ incorporating the specific interaction potential ϕ_{s} discussed above and the external force potential and $\mathbf{U}_{\mathrm{h}}=\mathbf{M}\cdot\mathbf{F}_{\mathrm{h}}+\mathbf{M}_{\mathrm{r}}\cdot\mathbf{T}_{0\mathrm{h}}$ is the particle velocity resulting from hydrodynamic force $\mathbf{F}_{\mathrm{h}}$ and torque $\mathbf{T}_{0\mathrm{h}}$.

Using the expression for the flux, Eq. (59), one can formulate the continuity equation for spherical particles in the form

$$\frac{\partial n}{\partial t} = \nabla \cdot \left[\mathbf{D} \cdot \nabla n + \frac{1}{kT}(\mathbf{D} \cdot \nabla \phi)n - \mathbf{U}_{\mathrm{h}} n\right] \tag{60}$$

It should be remembered, however, that in formulating Eq. (60) all hydrodynamic and specific interactions among particles were neglected and no coupling was assumed between hydrodynamic and specific interactions. Moreover, the diffusion coefficients occurring in Eq. (60) were assumed to be independent of the particle concentration n and its gradient. Therefore, this diffusion coefficient is often referred to as the self-diffusion coefficient [40, 41]. As a consequence of these simplifying assumptions, Eq. (59) remains strictly valid for diluted suspensions of non-interacting particles only. However, Eq. (60) remains a useful approximation for calculating deposition and aggregation phenomena because the volume fraction of particles in these processes $\Phi_{\mathrm{v}}=nv_1$ (where v_1 is the volume of a single particle) is usually of the order of 0.01 and lower, so the corrections to the diffusion coefficient remain of this order of magnitude.

The initial conditions for the above mass conservation equations are usually specified in the form

$$n = n_0(\mathbf{r}) \text{ for } t = 0 \tag{61}$$

where n_0 is the initial distribution of particles in the suspension. Most often one assumes that the initial concentration in the suspension is uniform, that is, $n = n_b$ where n_b is the bulk concentration of particles.

As far as the boundary conditions at the interface are concerned, it is usually postulated, in accordance with the Smoluchowski approach [39], that they are in the form of the perfect sink:

$$\delta_m = 0, h = \delta_m \tag{62}$$

where $h = \delta_m$ is the primary minimum distance.

This boundary condition simplifies any numerical handling of conservation equations. However, the postulate $n = 0$ is physically improper since particles concentration at the interface remains always finite and increases with time in deposition processes. An appropriate formulation of the boundary condition at the collector surface can only be attained when considering the specific energy profile at the surface. This leads to the kinetic boundary conditions, expressed in terms of the surface concentration of particles rather than bulk concentration. This aspect of particle deposition will be discussed in detail later.

It is useful to derive limiting forms of Eq. (60) because its analytical solution is impractical in the general case as a result of a complicated dependence of the specific interaction potential on the distance from the interface. One such limiting forms can be derived when the flow vanishes and particle transport can be assumed one-dimensional. This situation occurs in the case of particle aggregation and colloid particles transport in the thin region adjacent to the interface of the thickness δ_a, often called the surface boundary layer [42–45]. For a spherical geometry Eq. (60) becomes

$$\frac{\partial n}{\partial t} = \frac{1}{r^2}\frac{\partial}{\partial r}\left[r^2 D(r)\left(\frac{\partial n}{\partial_r} + \frac{\nabla \Phi}{kT} n\right)\right] \tag{63}$$

where r is the radial coordinate, $D(r)$ is the position-dependent diffusion coefficient and Φ is the net interaction potential stemming from the particle surface interactions, incorporating also the part originating from pre-adsorbed particles [14].

For a planar geometry, one obtains

$$\frac{\partial n}{\partial t} = \frac{\partial n}{\partial z}\left[D(z)\left(\frac{\partial n}{\partial z} + \frac{\nabla \Phi}{kT} n\right)\right] \tag{64}$$

where z is the coordinate perpendicular to the interface.

Equations (63) and (64) are important because their solutions can be used to formulate proper boundary conditions for the bulk transport equation.

If all specific interactions are neglected together with the hydrodynamic boundary effects, Eq. (44) simplifies to the form called the Smoluchowski-Levich equation [1, 40]:

$$\frac{\partial n}{\partial t} = D\nabla^2 n - \frac{D}{kT}\nabla \cdot (\mathbf{F}n) - \mathbf{V} \cdot \nabla n \tag{65}$$

where $\mathbf{V}$ is the unperturbed (macroscopic) fluid velocity vector.

On the other hand, under the no-convection conditions, when $\mathbf{V}=0$, Eq. (65) reduces to the form called the Smoluchowski equation:

$$\frac{\partial n}{\partial t} = D\nabla^2 n - \frac{D}{kT}\nabla \cdot (\mathbf{F}n) = D\nabla^2 n - \nabla \cdot (\mathbf{U}n) \tag{66}$$

where $\mathbf{U}=(D/kT)\mathbf{F}$ is the migration velocity of particles due to external forces.

If the external force $\mathbf{F}$ vanishes, the Smoluchowski equation is reduced to a form that is, in principle, a mathematical formulation of Fick's second law:

$$\frac{\partial n}{\partial t} = D\nabla^2 n \tag{67}$$

The non-stationary Eq. (67), being of a linear, parabolic type, can be solved analytically using, for example, the Laplace transformation method for many situations of practical interest, for example, adsorption on a spherical interface from a finite or infinite volume [46, 47].

The boundary conditions for Eqs. (65) and (67) are usually specified in the form of the perfect sink model, expressed by Eq. (62).

Under steady-state conditions, when $\partial n/\partial t = 0$ and $\mathbf{F} = 0$, Eq. (65) simplifies to the form

$$\bar{\nabla}^2 n - Pe\bar{V} \cdot \bar{\nabla} n = 0 \tag{68}$$

where $Pe = V_{\rm ch}L_{\rm ch}/D$ is the dimensionless Peclet number, $V_{\rm ch}$ is the characteristic velocity and $L_{\rm ch}$ is the characteristic length scale of the flow.

Equation (68), often called the convective diffusion equation [48], has been exploited widely for describing transfer of particles to collectors of a simple geometry [1, 15, 40, 49, 50]. However, this equation becomes inappropriate for larger colloid particles when the specific interactions and hydrodynamic wall effects play an important role. In this case, one has to use either numerical techniques or approximate approaches combining the bulk transport with the surface boundary layer transport, as discussed next.

9.3.1
Near-surface Transport

Let us consider particle transport through a thin layer adjacent to the interface where the convection effects can be neglected. The thickness of this layer δ_m is comparable to the range of specific interactions, usually being much smaller than particle and the interface dimensions. Therefore, the local curvature effects can be neglected and particle transport can be considered one-dimensional, governed by Eq. (63). If, moreover, steady-state conditions are assumed and the specific interactions between particles are neglected, Eq. (66) can be expressed in the convenient form

$$\frac{\partial}{\partial z}\left[D(z)e^{-\frac{\phi(z)}{kT}}\frac{\partial}{\partial z}\left(ne^{\frac{\phi(z)}{kT}}\right)\right] = -\frac{\partial}{\partial z}j = 0 \tag{69}$$

The boundary conditions for this equation are

$$\begin{aligned} n &= 0 \qquad \text{at} \quad h = z - a \to 0 \\ n &= n(\delta_a) \quad \text{at} \quad h = z - a = \delta_a \end{aligned} \tag{70}$$

where δ_a is the adsorption layer thickness.

Assume that the interaction potential is of Type 2, analyzed earlier (see Fig. 9.2), that is, there appears an energy barrier of arbitrary shape and a primary minimum of depth ϕ_m at the distance $h=\delta_m$. The particle concentration at $h=\delta_m$ equals $n(\delta_m)$ and at $h=\delta_a$ equals $n(\delta_a)$. Under these assumptions, Eq. (69) can be integrated, which results in the following expression for the constant particle flux j_a through the boundary layer [1, 15, 51]:

$$-j_a = \frac{n(\delta_a)e^{\phi(\delta_a)/kT} - n(\delta_m)e^{\phi(\delta_m)/kT}}{\int\limits_{\delta_m}^{\delta_a} \frac{e^{\phi(z')/kT}dz'}{D(z')}} \tag{71}$$

where

$$\begin{aligned} k'_a &= e^{\phi(\delta_a)/kT} \Big/ \int\limits_{\delta_m}^{\delta_a} \frac{e^{\phi/kT}}{D(z)}dz = \frac{e^{\phi(\delta_a)kT}}{R_b} \\ k'_d &= k'_a e^{[\phi_m - \phi(\delta_a)]/kT} \end{aligned} \tag{72}$$

are the adsorption and desorption rate constants and

$$R_b = \int_{\delta_m}^{\delta_a} \frac{e^{\phi/kT} dz}{D(z)} \tag{73}$$

can be treated as the barrier "resistance".

It is interesting to observe that the particle concentration distribution around the region of the primary minimum is well approximated by a quasi-Boltzmann distribution [1, 52]:

$$n(z) \cong n_m e^{-\phi(z)/kT} \tag{74}$$

where $n_m = n(\delta_m) e^{\phi(\delta_m)/kT}$.

Equation (74) has a major theoretical significance indicating that under non-equilibrium conditions, when a stationary value of particle flux is maintained through a layer, particle concentration remains locally equilibrated governed by the Boltzmann distribution with the appropriate normalization constant.

One can convert Eq. (74) to a more convenient form by introducing the surface concentration, defined as

$$N = \int_{\delta_m}^{\delta_a} n dz = n'_m = n_m I_m \tag{75}$$

By substituting this relationship into Eq. (71), one obtains

$$-j_a = k_a n(\delta_a) - k_d N \tag{76}$$

where

$$k_a = \frac{1}{\int_{\delta_m}^{\delta_a} \frac{e^{\phi(z)}/kT}{D(z)} dz} \tag{77}$$

$$k_d = \frac{k_a}{I_m} = \frac{k_a}{\int_{\delta_m}^{\delta_a} e^{-\phi(z)/kT} dz} \tag{78}$$

Instead of the surface concentration N one often introduces the dimensionless coverage Θ, defined as

$$\Theta = S_g N \tag{79}$$

where S_g is the characteristic cross-section of the particle, equal to πa^2 for spherical particles.

By using this definition, Eq. (77) becomes

$$-j_a = k_a n(\delta_a) - \frac{k_d}{S_g}\Theta \tag{80}$$

Equation (80) represents a general expression describing the reversible adsorption of particles under linear conditions (for negligible surface blocking effects). It is often used as the "kinetic" boundary condition for bulk transport problems [46, 47, 51, 52]. Under equilibrium conditions when $j_a=0$, Eq. (80) becomes

$$\Theta = S_g K_a n(\delta_a) \tag{81}$$

where $K_a = k_a/k_d = \int_{\delta_m}^{\delta_a} e^{-\phi/kT} dh$. Equation (81) express Henry's adsorption law that the amount adsorbed is proportional to the bulk concentration.

On the other hand, in the case when $|\phi_m| >> kT$ the constant k_d vanishes and particle adsorption becomes practically irreversible. The expression for particle flux Eq. (80) becomes

$$-j_a = k_a n(\delta_a) \tag{82}$$

When $k_a \to \infty$, which is the case when the depth of the primary energy minimum tends to infinity, Eq. (82) simplifies to

$$n(\delta_a) \to 0 \quad \text{at} \quad h = \delta_a \tag{83}$$

In this way, one obtains the perfect-sink boundary condition introduced by Smoluchowski.

One can evaluate analytically k_a, k_d and K_a for simple shape of the specific interaction energy profile. For example, in the case of the triangular barrier, where $\phi_b >> kT$ one obtains the expression [51]

$$\begin{aligned} k_a &= \frac{D}{a}\left(\frac{\phi_b}{kT}\right) e^{-\phi_b/kT} \\ k_d &= \frac{k_a}{\delta_m} e^{\phi_m/kT} \\ K_a &= \delta_m e^{-\phi_m/kT} \end{aligned} \tag{84}$$

In a more general case, the energy distributions around the primary minimum and the barrier region can be approximated by a parabolic distribution. Then, these constants are given by [51]

$$k_a = D(\delta_b)\left(\frac{\gamma_b}{2\pi kT}\right)^{\frac{1}{2}} e^{-\phi_b/kT} \cong \frac{D}{a}\left(\frac{\phi_b}{2\pi kT}\right)^{\frac{1}{2}} e^{-\phi_b/kT}$$

$$k_d = k_a\left(\frac{\gamma_m}{2\pi kT}\right)^{\frac{1}{2}} e^{-\phi_b/kT} \cong \frac{k_a}{\delta_m}\left(\frac{\phi_m}{2\pi kT}\right)^{\frac{1}{2}} e^{-\phi_m/kT} \tag{85}$$

$$K_a = \left(\frac{2\pi kT}{\gamma_m}\right)^{\frac{1}{2}} e^{-\phi_m/kT} \cong \left(\frac{\pi kT}{\phi_m}\right)^{\frac{1}{2}} \delta_m e^{-\phi_m/kT}$$

$$\text{where } \gamma_b = -\left(\frac{d^2\phi}{dh^2}\right)_{\delta_b} \cong \frac{2\phi_b}{\delta_b^2} \qquad \gamma_m = \left(\frac{d^2\phi}{dh^2}\right)_{\delta_m} \cong \frac{2\phi_m}{\delta_m^2}$$

and $D(\delta_b)$ is the value of the diffusion coefficient in the barrier region [51].

Equation (80) and its irreversible counterparts Eqs. (82) and (83) have high significance because they can be used as boundary conditions for bulk transport problems governed by the Smoluchowski-Levich (SL) equation. This is justified by the fact that the thickness of the surface boundary layer δ_a is generally much smaller than the particle dimension and diffusion boundary layer defined as the region where particle concentration changes occur. Hence, except for particle sizes above micrometers, the surface and bulk transport steps can be decoupled.

9.3.2
Limiting Solutions for the Perfect Sink Model

In the general case, the kinetics of particle transfer to interfaces can be predicted by solving the governing mass balance equation (Eq. 60), which incorporates the effects of specific, external and hydrodynamic forces in an exact manner. However, this is only possible by numerical techniques, for example, the finite difference method, which become complicated for non-stationary transport and multi-dimensional problems. This is so because of the disparity of the specific force range scale and the diffusion boundary layer thickness, differing by orders of magnitude, which makes it necessary to use special functions for transforming the grid. Numerical solutions become fairly efficient for the one-dimensional case only, for example, for collectors of a simple geometry and stagnation point flows as discussed later.

When dealing with transport problems of particles of submicrometer size range, it is considerably more efficient to use approximate approaches providing analytical solutions with a broad range of application. These methods consist in splitting the entire transport path of particles from the bulk to the interface into two separate regions:

1. The bulk diffusion boundary layer where the specific forces and hydrodynamic wall effects are neglected.
2. The surface boundary layer where the fluid convection effects are neglected but the specific interactions are taken into account as well as hydrodynamic wall corrections.

The dividing plane of these two regions is located at the distance $h=\delta$ from the collector where the specific surface forces vanish. Usually, the value of δ, referred to as the surface force boundary layer, is much smaller than particle dimensions. At this plane, the boundary conditions for the bulk transport equation are specified by exploiting the near-surface transport results discussed above. In the general case, these are the kinetic boundary conditions, given by Eq. (76), which are most appropriate for describing surfactant and molecule adsorption. For protein and colloid particles one usually assumes irreversible adsorption, so the boundary condition is given by Eq. (82). The approach exploiting this type of boundary condition is referred to in the literature as the surface force boundary layer (SFBL) method [43–45]. By using these approaches, we will analyze in this section the non-stationary transport of particles to a spherical interface under diffusion-controlled transport conditions, then transport under a uniform external force and finally convective transport under quasi-stationary conditions for various interfaces of various shapes.

Let us consider particle transfer to a spherically shaped homogeneous interface (often called a collector or an adsorber) of radius R which may represent either a liquid drop, a gas bubble or a solid particle, in contact with a dispersion of particles of arbitrary size.

Since the specific force, convection and external force are neglected the governing equation, Eq. (63), for the spherical geometry assumes the form

$$\frac{\partial n}{\partial t} = D \frac{1}{r^2} \frac{\partial}{\partial r}\left(r^2 \frac{\partial n}{\partial r}\right) \tag{86}$$

In accordance with Eq. (76), the boundary conditions at the edge of the surface boundary layer are expressed as [51]

$$-j_a = \left(D \frac{\partial n}{\partial r}\right)_{\delta_a} = k_a n_a - k_d N = k_a n_a - k_d \frac{\theta}{S_g} \tag{87}$$

The external boundary condition for the particle adsorption problem is formulated as

$$n = n_b \text{ at } r \to \infty \tag{88}$$

where n_b is the bulk concentration of particles.

The initial conditions for this problem are usually postulated to have the form of uniform particle distribution, that is

$$n = n_b \text{ for } R + \delta_a < r < \infty \text{ at } t = 0 \tag{89}$$

The boundary value problem, expressed by Eqs. (86) to (89), describes non-stationary adsorption kinetics for the Henry isotherm. It can be solved analytically in the limiting case of major practical significance, when $k_a/k_d = K_a >> 1/S_g n_a$

and so adsorption can be treated as irreversible and proceeding from an infinite volume. The solution describing the flux of particles, in this case, was first derived by Smoluchowski in relation to particle aggregation kinetics [39]. The expression for particle flux assumes the form

$$-j_a = \left[\frac{D_{12}}{a+R} + \left(\frac{D_{12}}{\pi t}\right)^{1/2}\right] n_b \tag{90}$$

where

$$D_{12} = D + D_c = kT/6\pi\eta\left(\frac{1}{a} + \frac{1}{R}\right) = D(1 + a/R)$$

is the diffusion coefficient of the particle relative to the collector (larger particle), which, in the general case, may also undergo diffusion, $D = kT/6\pi\eta a$ is the particle diffusion coefficient, D_c is the collector diffusion coefficient and η is the dynamic viscosity of the fluid.

As one can deduce from Eq. (90), for longer times, when $(a+R)^2/D_{12}t < 1$, the flux attains the steady-state value:

$$-j_a = \frac{D_{12}}{a+R} n_b = k_c n_b \tag{91}$$

where $k = D_{12}/(a+R)$ is the mass transfer rate constant.

It can be estimated by assuming $a = R = 100$ nm that the critical relaxation time equals about 10^{-2} s. For $a = 1000$ nm the relaxation time becomes about 10 s.

Equation (91) has a major significance since it can be used to calculate the coagulation rate constants, as done originally by Smoluchowski [39]. The rate of coagulation r_{12} of two particles assumed to have a spherical shape and the radii a_1 and a_2 is given by

$$r_{12} = -k_{12} n_1 n_2 \tag{92}$$

where n_1 and n_2 are the number concentrations of the particles in the bulk.

The coagulation rate constant k_{12} is given by

$$k_{12} = -4\pi(a_1 + a_2)^2 \frac{j_a}{n_b} = 4\pi(a_1 + a_2)(D_1 + D_2) = \frac{2}{3}\frac{kT}{\eta}\left(2 + \frac{a_1}{a_2} + \frac{a_2}{a_1}\right) = \frac{2}{3}\frac{kT}{\eta} f_{12}^0 \tag{93}$$

where D_1 and D_2 are the diffusion coefficients of particles 1 and 2, respectively, and

$$f_{12}^0 = 2 + \frac{a_1}{a_2} + \frac{a_2}{a_1}$$

The function f_{12}^0 in Eq. (93) characterizes the geometric effect, which is weakly dependent on particle dimension. For example, a two-fold increase in particle size ratio causes only a 12.5% increase in the value of f_{12}^0. For $a_1/a_2 = 4$, the increase equals 56.2%.

These solutions have been derived by neglecting the specific interactions between particles and by assuming that the diffusion coefficient remains a position-independent quantity. This can reflect practically occurring situations where the attractive interactions compensate for the decrease in the diffusion coefficient at small separations [1]. In systems when an energy barrier appears, the above equations fail and should be generalized by considering the real energy profile. This can be achieved by matching the near-surface flux with the bulk diffusion flux expressed by Eq. (91). In this way one obtains the equation [1, 15, 51]

$$-j_a = \frac{k_c}{1 + \frac{k_c}{k_a}} n_a \tag{94}$$

where the adsorption constant k_a is given by

$$k_a = \frac{1}{\displaystyle\int_{\delta_m+a+R}^{\delta_a+a+R} \frac{e^{\phi(r)/kT}}{D_{12}(r)} \mathrm{d}r} \tag{95}$$

and $D_{12}(r) = D_{12}F_s(H)$ is the position-dependent diffusion coefficient, $H = (r - a - R)/a$ is the gap width and F_s is the hydrodynamic correction function evaluated in [51]. For small distances, where $H << 1$, this correction function can be approximated by

$$F_s(H) \cong \left(1 + \frac{a}{R}\right) H \tag{96}$$

By using this expression, one can calculate the adsorption constant for the triangular and parabolic barrier as previously done (cf. Eq. 85). In this way, one can formulate the expression for the particle flux (Eq. 94), for the triangular barrier in the explicit form

$$-j_a = \frac{k_c n_b}{1 + \frac{k_a}{k_c}} = \frac{D n_b}{R} \frac{1}{1 + \frac{a}{R\left(1 + \frac{a}{R}\right)^2} \frac{kT}{\phi_b} e^{\phi_b/kT}} \tag{97}$$

For a parabolic barrier one has

$$-j_{\mathrm{a}} = \frac{Dn_{\mathrm{b}}}{R} \frac{1}{1 + \frac{a}{R(1+\frac{a}{R})^2}\left(\frac{2kT}{\phi_{\mathrm{b}}}\right)^{\frac{1}{2}} \mathrm{e}^{\phi_{\mathrm{b}}/kT}} \tag{98}$$

It can be deduced from Eqs. (97) and (98) that for high energy barriers when the inequality $\phi_{\mathrm{b}} >> 1$ is fulfilled, the role of the bulk transport becomes negligible and the net flux is given by

$$-j_{\mathrm{a}} = k_{\mathrm{a}} n_{\mathrm{b}} \tag{99}$$

For a triangular and parabolic barrier Eq. (99) becomes

$$-j_{\mathrm{a}} = \frac{Dn_{\mathrm{b}}}{a}\left(1+\frac{a}{R}\right)^2 \left(\frac{\phi_{\mathrm{b}}}{kT}\right) \mathrm{e}^{-\phi_{\mathrm{b}}/kT}$$

$$-j_{\mathrm{a}} = \frac{Dn_{\mathrm{b}}}{a}\left(1+\frac{a}{R}\right)^2 \left(\frac{\phi_{\mathrm{b}}}{2\pi kT}\right)^{1/2} \mathrm{e}^{-\phi_{\mathrm{b}}/kT} \tag{100}$$

Using Eq. (100), one can express the aggregation rate constant k_{12} for the triangular and parabolic barrier in the following form

$$k_{12} = 4\pi(a+R)^2\left(\frac{-j_{\mathrm{a}}}{n_{\mathrm{b}}}\right) = k_{12}^0\left(2+\frac{R}{a}+\frac{a}{R}\right)\left(\frac{\phi_{\mathrm{b}}}{kT}\right) e^{-\phi_{\mathrm{b}}/kT} \quad \text{(triangular barrier)}$$

$$k_{12} = k_{12}^0\left(2+\frac{R}{a}+\frac{a}{R}\right)\left(\frac{\phi_{\mathrm{b}}}{2\pi kT}\right)^{1/2} e^{-\phi_{\mathrm{b}}/kT} \quad \text{(parabolic barrier)}$$

$$\text{where} \quad k_{12}^0 = \frac{2}{3}\frac{kT}{\eta}\left(2+\frac{R}{a}+\frac{a}{R}\right) \tag{101}$$

Equations (100) and (101), indicating that the flux decreases exponentially with the barrier height and is independent of the flow intensity, describe the so-called activated or barrier-controlled aggregation of particles. One can also deduce from these equations that by measuring the aggregation rate of particle suspensions one can determine the barrier height. However, these measurements are rather impractical because one cannot prevent forming aggregates composed of more than two particles. It is considerably more efficient, as suggested in the Introduction, to perform particle deposition experiments, which can also be used for evaluating the interaction energy profile.

9.3.3 Convective-diffusion Transport to Various Interfaces

In the case of convection-driven transport, particle deposition kinetics can be calculated using the stationary Levich equation (Eq. 68), which is valid after a short transition time of the order of seconds [1, 54]. Useful analytical solutions can be derived for collector geometries of practical interest where the normal component of the flow remains independent of the tangential coordinate and the tangential flow component vanishes. In this case there is no gradient of particle concentration parallel to the interface (if it is homogeneous). Such a property is possessed by the flow in the vicinity of the stagnation point, near the rotating disk, the impinging jet cells or sphere and cylinder placed in uniform flow in the region close to the stagnation point and so forth [1, 15, 51]. The surface exposed to this type of flow is called uniformly accessible. In this case, the solution of Eq. (68) with irreversible adsorption boundary conditions (Eq. 82), leads to the following expression for particle flux [1, 51]:

$$j = -\frac{k_c}{1+\dfrac{k_c}{k'_a}} n_b = \frac{j_0}{1+\dfrac{j_0}{n_b k'_a}} \tag{102}$$

where k'_a is the modified adsorption constant:

$$k'_a = \frac{1}{\displaystyle\int_{\delta_m}^{\delta_a} \frac{e^{\phi/kT}-1}{D(z)}\,dz} \tag{103}$$

$$-j_0 = k_c n_b = \frac{D n_b}{L_{ch}} Sh_0 \tag{104}$$

is the stationary flux from the bulk to the primary minimum, called the limiting flux, and $Sh_0 = j_0 L_{ch}/D n_b$ is the dimensionless Sherwood number characterizing the mass transfer rate from the bulk to the primary minimum in the case of no barrier.

The mass transfer rate for uniformly accessible surfaces is given by the expression

$$k'_c = k_c = \frac{D}{L_{ch}\Gamma\left(\dfrac{4}{3}\right)} \left(\frac{C_\perp Pe}{3}\right)^{\frac{1}{3}} \tag{105}$$

where $\Gamma(4/3) = 0.893$ is the Euler's gamma function value for 4/3 and $C_\perp$ is a dimensionless constant depending on the collector geometry [1].

Analytical expressions for k_c can be found in Refs. [1, 15, 49, 51] for both uniformly accessible surfaces and other collectors of practical interest. For the sake of convenience, selected results are compiled in Table 9.4. The definitions of the Peclet number are also given, calculated by assuming $L_{ch} = a$.

The definite integral in Eq. (103) can be evaluated analytically for a triangular and parabolic barrier whose height exceeds a few kT units as done previously (cf. Eq. 85). In this case, one can express Eq. (102) for a triangular barrier by assuming $L_{ch} = a$ in the form

Table 9.4 *Pe* definitions and bulk transfer rate constants (reduced flux) k_c for uniformly accessible surfaces in the case $a = R$

Collector and flow configuration	***Pe* definition microscopic $L_{ch} = 2a$**	***Pe* definition macroscopic**	$k_c = -j/n_b$
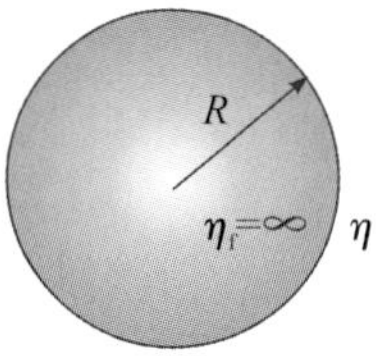 Sphere in quiescent fluid	0	0	$\frac{D}{R}$
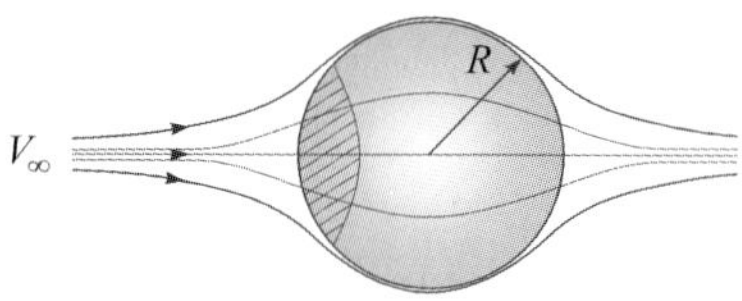 Sphere in uniform flow (near stagnation point) [a]	$\frac{3A_f(Re)V_\infty}{R^2}\frac{a^3}{D}$	$2A_f(Re)V_\infty\frac{R}{D}$	$0.889\frac{A_f^{1/3}V_\infty^{1/3}D^{2/3}}{R^{2/3}}$
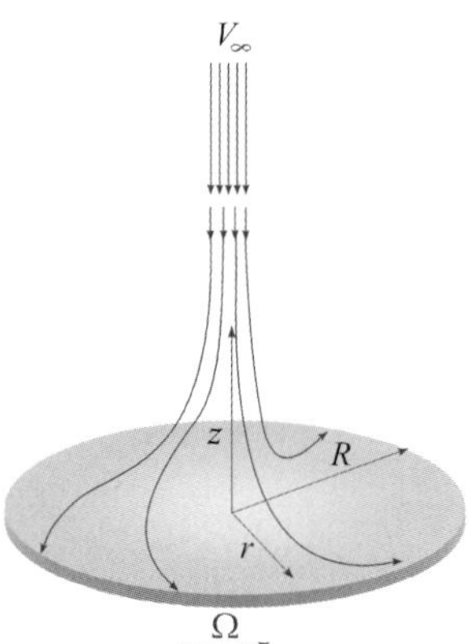 The rotating disk	$\frac{1.021\Omega^{3/2}}{\nu^{1/2}}\frac{a^3}{D}$	$\frac{2V_\infty\delta_h}{D} = 6.38\frac{\nu}{D}$	$0.620\frac{\Omega^{1/2}D^{2/3}}{\nu^{1/6}}$

a) $A_f(Re) = \frac{3}{2}\left(1 + \frac{0.19Re}{1 + 0.25Re^{0.56}}\right)$ valid for $Re = \frac{2RV_\infty}{\nu} < 300$

b) α_r for various (h/R) and Re are given in Refs. [1, 61, 77]
$V_\infty = Q/\pi R^2$ Q – volumetric flow rate $Re = 2V_\infty R/\nu$

c) α_s for various d and Re are given in Refs. [1, 60, 69]
$V_\infty = Q/2dl$ Q – volumetric flow rate $Re = 2V_\infty d/\nu$

d) $A_f = (\beta - 0.48\beta^3)$, $\beta = (2 - \ln Re)^{-1}$, for $Re < 1$
$A_f = 0.44Re^{0.52}$, $Re = 2V_\infty R/\nu$, for $1 < Re < 200$

Table 9.4 (continued)

Collector and flow configuration	***Pe* definition microscopic $L_{ch} = 2a$**	***Pe* definition macroscopic**	$k_c = -j/n_b$
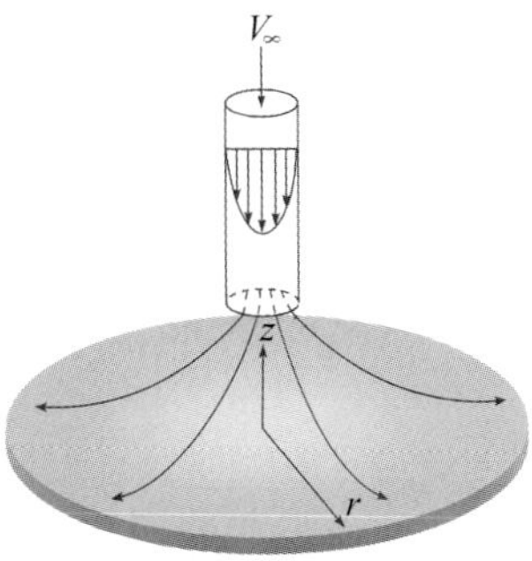 Radial impinging-jet RIJ (near stagnation point) b)	$\frac{2\alpha_r V_\infty}{R^2}\frac{a^3}{D}$	$\frac{2V_\infty R}{D}$	$0.776\frac{\alpha_r^{1/3}V_\infty^{1/3}D^{2/3}}{R^{2/3}}$ $0.530\frac{\alpha_r^{1/3}Q^{1/3}D^{2/3}}{R^{4/3}}$
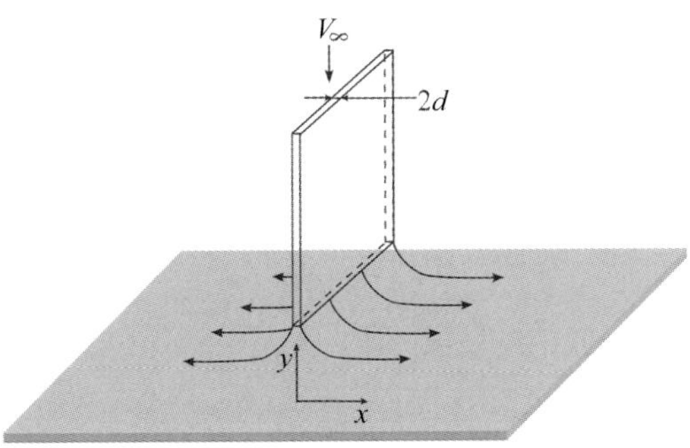 Slot impinging-jet cell (near stagnation point) c)	$\frac{2\alpha_s V_\infty a^3}{d^2 D}$	$\frac{2V_\infty d}{D}$	$0.776\frac{\alpha_s^{1/3}V_\infty^{1/3}D^{2/3}}{d^{2/3}}$ $0.616\frac{\alpha_s^{1/3}Q^{1/3}D^{2/3}}{dl^{1/3}}$
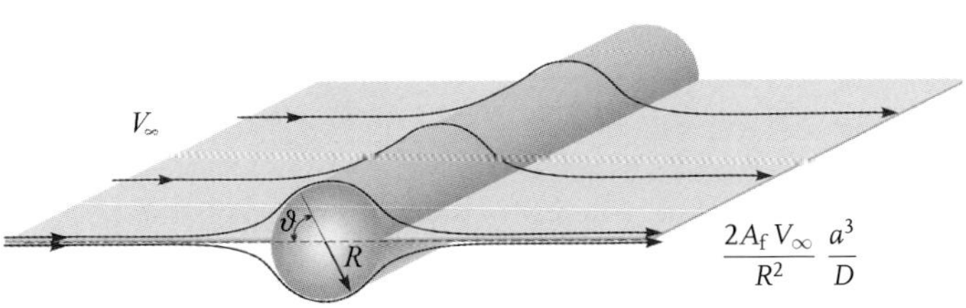 Cylinder in uniform flow (near stagnation point) d)	$\frac{2A_f V_\infty}{R^2}\frac{a^3}{D}$	$\frac{A_f V_\infty R}{D}$	$0.776\frac{A_f^{1/3}V_\infty^{1/3}D^{2/3}}{R^{2/3}}$

$$j = j_0 \frac{1}{1 + Sh_0 \dfrac{kT}{\phi_b} e^{\phi_b/kT}} \tag{106}$$

For a parabolic barrier one has

$$j = j_0 \frac{1}{1 + Sh_0 \left(\dfrac{2\pi kT}{\phi_b}\right)^{\frac{1}{2}} e^{\phi_b/kT}} \tag{107}$$

It can be deduced from Eqs. (106) and (107) that for a high energy barrier, when the inequality $\phi_b/kT > -\ln Sh + \ln(\ln Sh)$ is fulfilled, the role of the bulk transport becomes negligible and the net flux is given by

$$\begin{aligned} &\phi_b/kT > -\ln Sh \\ &-j = -k_a n_b \\ &k/k'_a << 1 \end{aligned} \tag{108}$$

Equation (108), analogously to diffusion-controlled transport to a spherical interface, for a triangular and parabolic barrier becomes

$$\begin{aligned} -j &= \frac{Dn_b}{a}\frac{\phi_b}{kT} e^{-\phi_b/kT} \\ -j &= \frac{Dn_b}{a}\left(\frac{\phi_b}{2\pi kT}\right)^{\frac{1}{2}} e^{-\phi_b/kT} \end{aligned} \tag{109}$$

Equations (109) indicate that the flux decreases exponentially with the barrier height and is independent of the flow intensity.

On the other hand, in the case when $k_c/k'_a << 1$ (fast transfer rate through the adsorption layer), Eq. (106) simplifies to

$$-j = k_c n_b \tag{110}$$

In this case, particle adsorption is governed by the bulk transport alone, which is referred to as the barrierless adsorption regime.

From Eqs. (106) and (107), one can deduce that by measuring particle deposition rate as a function of flow intensity (influencing the value of the Peclet number), one can extract information on the specific interaction energy profile (height and shape of the energy barrier).

It should be mentioned that the above results are well suited for barrier-controlled deposition regimes and for particles having a size below 500 nm. In this case, the range of hydrodynamic wall effects and specific surface interactions remains negligible in comparison with bulk transfer distances. For larger particles, both the hydrodynamic and the specific forces, especially if they are attractive, are coupled in a complicated way. This makes it necessary to use the exact transport equation (Eq. 60), for calculating mass transfer rates for particles of arbitrary size under the simultaneous action of specific and external forces.

A major difficulty when dealing with this exact transport equation is the lack of appropriate analytical methods for their solution, because of the complicated dependence of the specific interaction potential ϕ on the distance from the collector. Therefore, exact solutions can only be derived by numerical techniques, which become inefficient for multi-dimensional problems.

However, for most cases involving nano-sized particles, the governing transport equation can be simplified considerably by applying a dimensional analysis as shown in Ref. [1]. This is so because the concentration gradients in the direction perpendicular to the collector are much larger than those in the direction parallel to the surface, and therefore the tangential diffusion terms can be neglected. Moreover, the collector surface can be treated as locally planar, which allows one to derive expressions for the mobility matrix in terms of the hydrodynamic correction functions. By virtue of these assumptions, one can formulate the one-dimensional transport equation for the uniformly accessible surfaces in the dimensionless form [1]:

$$\frac{\mathrm{d}}{\mathrm{d}H}\left[F_1(H)\left(\frac{\mathrm{d}\bar{n}}{\mathrm{d}H}-\bar{F}_{\mathrm{H}}\bar{n}\right)+\frac{1}{2}PeF_1(H)F_2(H)(H+1)^2\bar{n}\right]=PeF_3(H)(H+1)\bar{n} \tag{111}$$

where $H = h/a$ is the dimensionless gap width, $F_1(H)$ and $F_3(H)$ are the hydrodynamic correction functions accounting for the wall effects [1] and $\bar{F}_{\mathrm{H}}$ is the net force acting on particles (dimensionless).

$$-\frac{\mathrm{d}\bar{\phi}}{\mathrm{d}H}=\frac{a}{kT}\bar{F}_{\mathrm{H}}=\bar{F}_{\mathrm{ext}}+\bar{F}_{\mathrm{el}}+\bar{F}_{\mathrm{s}} \tag{112}$$

where $\bar{F}_{\mathrm{ext}}$ is the external force, most often the gravitational force, leading to migration effects, $\bar{F}_{\mathrm{el}}$ is the electrical double-layer force and $\bar{F}_{\mathrm{s}}$ is the van der Waals force discussed above.

For most situations of practical interest, the electrical double-layer forces can be approximated by the LSA model, yielding the two-parametric expression

$$\bar{F}_{\mathrm{el}}=\frac{\kappa a}{kT}\phi_0 \mathrm{e}^{-\kappa a\mathrm{H}}=Dl\,\mathrm{e}^{-\kappa a\mathrm{H}} \tag{113}$$

where $Dl = \kappa a\phi_0/kT$ is the electrostatic double-layer number and κa is a parameter describing the ratio of particle radius to the double-layer thickness.

The van der Waals interactions can be described by

$$\bar{F}_{\mathrm{s}}=-\frac{A_{132}a}{kT}f_{\mathrm{a}}(z,\lambda_{\mathrm{r}})=-Ad\bar{f}_{\mathrm{a}}(H,\bar{\lambda}_{\mathrm{r}}) \tag{114}$$

where $Ad = A_{132}/6kT$ is the dimensionless adhesion number, $\bar{\lambda}_{\mathrm{r}} = \lambda_{\mathrm{r}}/a$ is a dimensionless parameter describing the retardation effect and $\bar{f}_{\mathrm{a}}(H,\bar{\lambda}_{\mathrm{r}})$ is the dimensionless function to be found in Table 9.1 (in the case $\bar{\lambda}_r = 0$), which can well be approximated for various particles by the Derjaguin model.

The boundary condition at the surface for Eq. (111) is usually expressed in the form of the perfect sink, that is, $\bar{n} = 0$ at $H = \delta_m/a$ (where δ_m is the primary minimum distance).

Most of the above transport equations can be effectively solved by using the finite-difference methods discussed in some detail elsewhere [50]. For the one-dimensional, non-stationary transport equations pertinent to uniformly accessible surfaces, one can effectively use the implicit Crank–Nicholson scheme with appropriate net transformation functions [55–59]. The set of linear algebraic equations originating from the discretization procedure is solved directly by Gauss elimination. This allows one to treat problems having large mesh point numbers.

Calculations for both transient and stationary conditions have been performed for the rotating disk [54, 55], cylindrical [57] and impinging jet collectors [60, 61]. The same method can be effectively used to solve the stationary transport equations for spherical [62] and cylindrical [57] collectors, parallel-plate and cylindrical channels [56], and plates in uniform flow [58, 59]. This is so because in all these cases the governing transport equation under stationary conditions can be reduced to the quasi-one-dimensional form as shown above.

It was demonstrated in these calculations that the most decisive role played the double-layer interactions, whose range and magnitude is governed by two major parameters $\kappa a = a/Le = (2e^2 Ia^2/\varepsilon kT)^{\frac{1}{2}}$ and Dl. For the LSA model, Dl can be expressed as

$$Dl = 4\pi \left(\frac{kT}{e}\right)^2 a \times 4\tanh\left(\frac{\zeta_c e}{4kT}\right) \times 4\tanh\left(\frac{\zeta e}{4kT}\right) \tag{115}$$

where ζ_c is the zeta potential of the collector and ζ is the zeta potential of the particle.

Because the ionic strength of the electrolyte can be varied between wide limits (for aqueous suspensions practically between 10^{-6} and 5 M), the parameter κa can be varied between 0.1 and 10^3. Also, the sign and magnitude of Dl can be regulated within wide limits by changing the ionic strength, pH and adsorption of surfactants. For $Dl < 0$, the zeta potentials of the particle and the interface are opposite, which produces attractive interactions, and for $Dl > 0$, one has to deal with repulsive interactions leading to an energy barrier analyzed above.

Examples of theoretical results, derived by numerical solution of Eq. (111) illustrating the effect of the EDL forces are shown in Figs. 9.3 and 9.4. In Fig. 9.3, the dependence of Sh on Pe, calculated numerically for uniformly accessible surfaces in the case when attractive electrostatic interactions appeared, is shown [63]. One can observe that the electrostatic interactions play a significant role already for $Pe > 10^{-2}$. For $Pe > 1$, Sh increases almost proportionally to Pe, which is caused by the interception effect enhanced by the attractive electrostatic interactions. This can be well reflected by the modified interception equation [1, 63]:

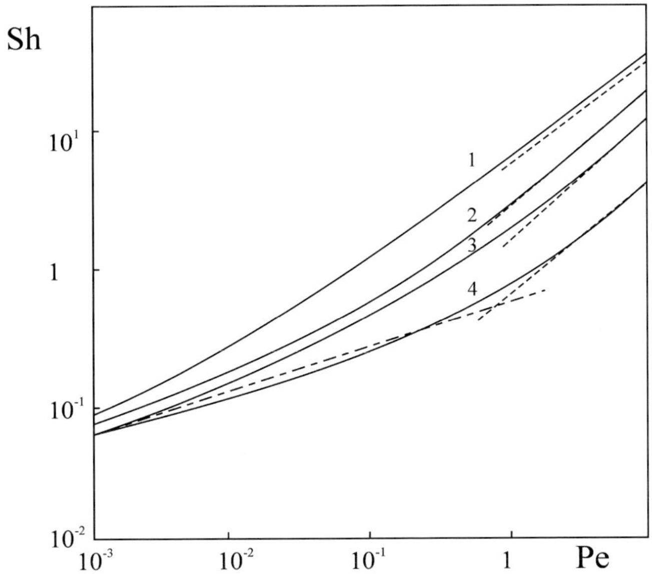

Fig. 9.3 Dependence of the Sherwood number *Sh* (reduced bulk transfer rate $k_c a/D$) on the Peclet number *Pe* calculated numerically for uniformly accessible surfaces in the case of no external force, $Ad=0.2$, $\phi_0=-10^3\ kT$. (1) $\kappa a=2$; (2) $\kappa a=5$; (3) $\kappa a=10$; (4) $\kappa a=100$. The limiting analytical results calculated from $Sh=0.616\ Pe^{\frac{1}{3}}$ from (diffusion-controlled transport) are denoted by the dashed-dotted line and those calculated from Eq. (116) (interception-controlled flux) by the dashed line (from [63]).

$$k_c = \frac{D}{a}\left(1+\frac{h^*}{a}\right)^2 \times \frac{1}{2}Pe \tag{116}$$

where $h^*/a = H^*$ is the effective range of the attractive double-layer interactions.

By equating the electrostatic forces and the hydrodynamic forces due to fluid convection, it was shown [1, 63] that H^* can be well approximated by the expression

$$H^* = h^*/a = \frac{1}{\kappa a}\left[\ln\frac{2|Dl|}{\kappa a Pe} - 2\ln\left(1+\frac{1}{\kappa a}\ln\frac{2|Dl|}{\kappa a Pe}\right)\right] \tag{117}$$

As can be seen in Fig. 9.3, Eq. (117) reflects fairly well the exact numerical data for $Pe > 1$ and all values of κa studied. On the other hand, for $Pe < 10^{-2}$, the role of electrostatic interaction becomes negligible and the dimensionless flux is given by the limiting Levich equation, Eq. (105). These theoretical predictions suggest that it would be possible to evaluate the magnitude of the attractive interaction potential by measuring the deposition rate of particles as a function of flow rate (governing the value of *Pe*). Then, by using Eq. (116), one can evaluate h^*/a and consequently, by inverting Eq. (117), it is possible to calculate *Dl*.

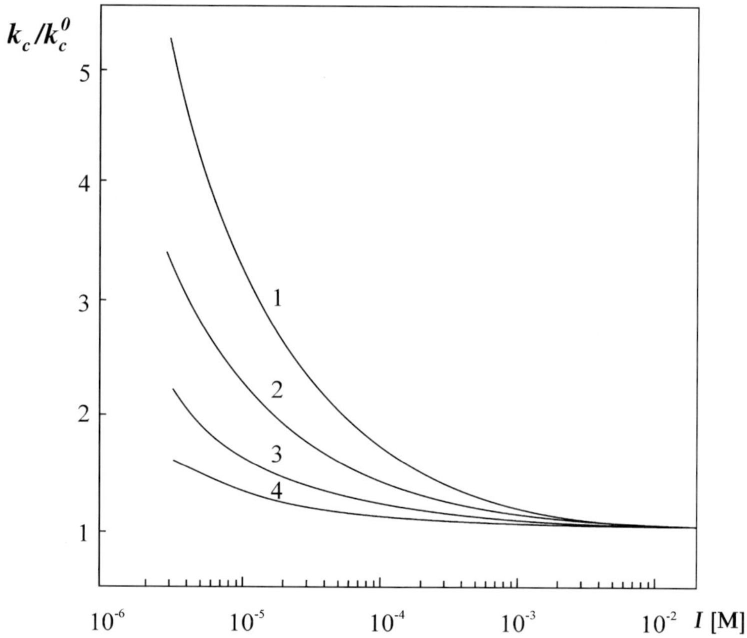

Fig. 9.4 Dependence of normalized mass transfer rate $k_c/k_c^0 = Sh/Sh^0$ (where k_c^0 is the mass transfer rate for negligible electrostatic interactions) vs. the ionic strength I of a 1:1 electrolyte, calculated numerically (solid lines) for a packed bed of spherical particles of radius $R=1.5\times10^{-4}$ m, $V_\infty=10^{-3}$ m s^{-1}, bed porosity 0.36, $T=298$ K, $Ad=0.4$ ($A_{123}=10^{-20}$ J), zeta potentials of particles and collectors +40 and −30 mV, respectively. (1) $a=0.8$ μm; (2) $a=0.4$ μm; (3) $a=0.2$ μm; (4) $a=0.1$ μm (from [64]).

Analogous deposition rate enhancement caused by ionic strength variations (governing κa) was predicted for spherical collectors forming a packed bed [64]. In order to make this effect more visible, the deposition rate was expressed in the reduced form $k_c/k_c^0 = Sh/Sh^0$ (where k_c is the mass transfer rate for negligible electrostatic interactions) as a function of the ionic strength I of a 1 : 1 electrolyte. The calculations were done for $R=1.5\times10^{-4}$ m, $V_{ch}=10^{-3}$ m s^{-1}, $T=298$ K $Ad=0.4$ ($A_{123}=10^{-20}$ J), and zeta potentials of particles and collector of +40 and −30 mV, respectively. As can be observed in Fig. 9.4 for particles of diameter 0.8 μm, the deposition rate is predicted to increase more than five-fold by reducing the ionic strength from 10^{-3} to 10^{-6} M. For a particle size of 0.1 μm, the deposition rate enhancement is limited to 1.8-fold.

A deposition rate increase due to the electrostatic interactions is also predicted for other collector geometries such as the cylindrical collector [57] and the parallel-plate channel [56], widely used in practice and in experiments aimed at particle and protein deposition [51].

These theoretical results suggest that in contrast to aggregation kinetic studies, particle deposition experiments can furnish information on the specific energy profiles also in the case when the energy becomes attractive.

9.4
Illustrative Experimental Results

9.4.1
Initial Deposition Rates

As the above theoretical considerations suggest, particle deposition studies can be exploited as an efficient tool for deriving the interaction energy profile under dynamic conditions, which is essential for predicting stability of colloid systems. In contrast to bulk system behavior, particle deposition kinetics can be studied *in situ* using direct experimental methods. Hence, the results obtained for colloid deposition can also be used for calibrating indirect methods such as reflectometry, ellipsometry, streaming potential, radiotracer techniques and so forth, In this way, precise measurements of molecular adsorption kinetics, for example of DNA, proteins and polyelectrolytes, become feasible.

In this section, we discuss representative experimental results obtained under well-defined transport conditions, which allowed us to evaluate in a precise way the essential role of the electrostatic interactions and to study the surface heterogeneity effects.

Most of the results discussed were obtained by direct observations of adsorbed particles using optical microscopy coupled with the image analysis technique [65–72]. Surface coverage of particles below 0.01 can be effectively measured by this technique, which enables one to determine accurately initial deposition rates of particles. The well-defined transport conditions are realized using impinging jet cells [15, 66, 67, 69, 70, 73–79], parallel-plate channel [43, 72, 80, 81], the rotating disk [82–84] arrangements. Impinging jet cells, most often the radial impinging jet (RIJ) [15, 61, 66, 67, 73–79] or the slot impinging jet (SIJ) [69], allow one to produce uniform transport conditions over macroscopic surface areas. Recently the oblique impinging jet cell (OBIJ) [70] has been applied for direct studies of colloid particle deposition on non-transparent substrates, such as metallic gold [85]. The use of impinging jet cells is advantageous because the drying step is eliminated and the experimental data can be effectively interpreted theoretically since the governing transport equation becomes one-dimensional.

On the other hand, particle deposition under negligible convection was studied using the diffusion cell described in Refs. [15, 71].

The electrokinetic properties of substrate surfaces either bare or covered by particles and polyelectrolytes were determined by the streaming potential method using the parallel-plate channel set-up [86, 87]. The high sensitivity of the method is due to the fact that adsorbed particles change significantly the electrokinetic properties of a solid/electrolyte solution interface. As discussed in Refs. [86–89], there are two main reasons why this effect occurs universally for arbitrary particle size and surface charge, including neutral particles. The first is the damping of macroscopic flow near the interface by adsorbed particles, which reduces the convective current of ions. The second is the outflow of the counter-ions from the double-layer surrounding adsorbed particles. For oppositely charged particle and substrate surfaces (which is usually the case in deposition experiments), these two effects act in parallel. This ensures a high sensitivity of the electrokinetic method, especially within a low coverage range, which exceeds considerably any other indirect method. Particle coverage as low as 0.01 can be detected by this technique [86–89]. However, the disadvantage of the electrokinetic method is that its accuracy decreases for high (low signal) and for low electrolyte concentrations owing to the appearance of surface conductivity. In the latter case, when the surface conductance dominates, for example for metallic substrates [90] and in the case of heterogeneous surfaces [91], streaming current measurements are more appropriate.

The quantity determined directly in the deposition experiments is the number of particles N_p on various surface areas of equal size ΔS [14, 61, 69, 71, 79]. Because N_p is a statistical variable that obeys the Poisson fluctuation law for low coverage [67, 92, 93], consequently the accuracy of determining the average value of $\langle N_p \rangle$ is inversely proportional to the square root of N_t, where N_t is the total number of particles counted. In the experiments discussed hereafter, N_t was usually above 1000, which gives a standard deviation of $\langle N_p \rangle$ of about 3%. For higher coverage, the fluctuations in N_p are considerably reduced owing to surface exclusion effects [92, 93]. These effects increase significantly the precision of measurements. On the other hand, for barrier-controlled deposition regimes, the number of particles adsorbed is generally very low, so N_p is subject to considerable fluctuations, increased by surface heterogeneity. In these cases the standard deviation of $\langle N_p \rangle$ may well exceed 10%.

By knowing $\langle N_p \rangle$ as a function of time, one can determine experimentally the kinetic curves, that is, $\Theta = \pi a^2 \langle N_p \rangle$ versus the adsorption time t. According to the theoretical predictions discussed above, under convective transport conditions and low coverage, particle adsorption kinetics are linear with respect to the adsorption time t and the bulk suspension concentration n_b. This has been widely confirmed experimentally for various particle size and transport conditions [15, 61, 67, 69–71, 79]. Once such linear kinetic runs have been performed, the experimental mass transfer rate constant k_c can be calculated from the constitutive equation

$$k_c = |j_0/n_b| = \frac{\Delta\langle N_p \rangle}{n_b \Delta S \Delta t} = \frac{1}{n_b \pi a^2} \frac{\Delta\Theta}{\Delta t} \tag{118}$$

where $\Delta\langle N_p \rangle$ is the change in the average number of particles adsorbed over ΔS within the time interval Δt.

The accuracy of the mass transfer rate (particle flux) determination can be increased by averaging over many experiments carried out at different bulk suspension concentration n_b. Equation (118) can also be used for determining the local flux in the case when it depends on the position over the interface. This procedure can be applied, in principle, for arbitrary coverage. It is, however, most accurate for the linear adsorption regime where the particle deposition rate remains independent of time and proportional to the bulk suspension concentration n_b.

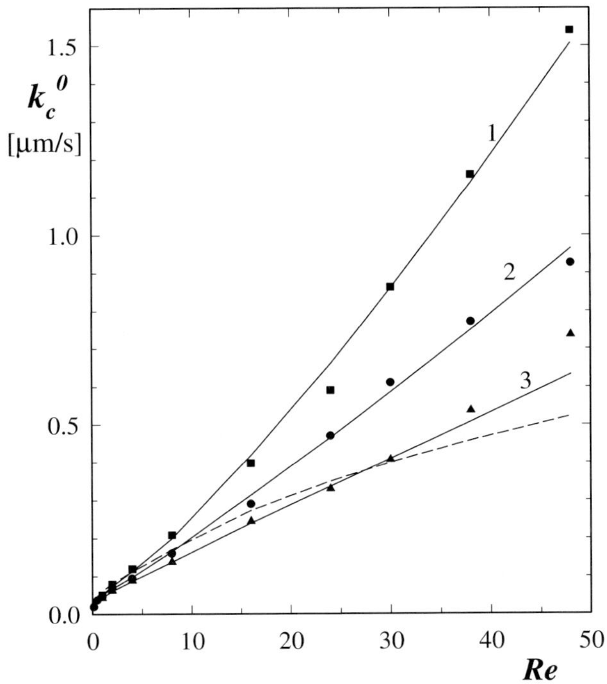

Fig. 9.5 Dependence of k_c^0 on *Re*; the points denote the experimental results obtained in the RIJ cell for latex particles (diameter 0.87 μm) adsorbing at mica. (1) $I=2\times10^{-5}$ M; (2) $I=10^{-4}$ M; (3) $I=10^{-3}$ M; the continuous lines denote the theoretical results obtained by solving numerically Eq. (111) and the dashed line represents the analytical results calculated from Eq. (105) using the Smoluchowski-Levich approximation (from [61]).

Using direct microscope observation methods, extensive measurements of the initial flux as a function of particle size, flow intensity and ionic strength have been performed, described in some detail in review papers [12, 14, 15, 37]. Here we report some representative data illustrating a major role of double-layer interactions for both homogeneous and heterogeneous systems.

The results obtained for polystyrene latex suspension using the RIJ cell (average particle diameter 0.87 μm, modified mica as the substrate surface) are shown in Fig. 9.5. The solid lines denote the exact theoretical results derived by a numerical solution of Eq. (111). The interaction force was assumed to consist of the electrostatic double-layer force described in terms of the LSA model and the retarded van der Waals forces. The zeta potential of the latex particles ζ_p as a function of the ionic strength was determined by the micro-electrophoretic method. It varied between −44 and −77 mV for the ionic strength range 10^{-5}–10^{-3} M. The zeta potential of the mica modified by adsorption of silanes ζ_c was varying between 60 and 30 mV for the ionic strength range 10^{-5}–10^{-3} M.

As can be seen in Fig. 9.5, for the entire range of the *Re* studied (0.15–48) the experimental data are in quantitative agreement with the theoretical calculations. An important feature of the results is that the adsorption rate is considerably enhanced by a decrease in the ionic strength of the suspension. This is so because an increase in the double-layer interaction range causes an increase in the effective interaction range h^*/a as described by Eq. (117). This leads in turn to an increased interception effect responsible for an enhanced particle deposition rate at lower ionic strength. These results suggest that the range of electrostatic interactions in hetero-systems can be effectively evaluated by measuring particle deposition rate for a higher Reynolds number range.

On the other hand, for $I=10^{-3}$ M, the electrostatic interactions are effectively eliminated and the adsorption rate can well be approximated by the Smoluchowski-Levich model (shown by dashed lines in Fig. 9.5), which is described by the equation

$$k_\mathrm{c} = 0.530\frac{a_\mathrm{r}^{\frac{1}{3}} Q^{\frac{1}{3}} D^{\frac{2}{3}}}{R^{\frac{4}{3}}} \tag{119}$$

where a_1 is the flow intensity parameter, Q is the volumetric flow rate of the suspension and R is the radius of the capillary.

The dependence of the flow intensity parameter on *Re* was calculated numerically for the cell geometry ($R=0.1$ cm, $h_\mathrm{c}/R=1.6$, where h_c is the distance of the outlet of the capillary from the substrate surface). The numerical results can be well fitted by the interpolating polynomial [61], $a_\mathrm{r}=c_0+c_1 Re+c_2 Re^2$, where $c_0=1.78$, $c_1=0.186$ and $c_2=0.034$ valid for $Re<20$.

For $Re>20$, a better fit was attained using the formula: $a_\mathrm{r}=4.96\cdot Re^{1/2}–8.41$.

The agreement of the experimental and theoretical data observed in Fig. 9.5 suggests that by measuring particle deposition rates, preferably for higher flow rates, one can fairly accurately estimate the parameters describing the attractive double-layer interactions or to be more precise the function

$$\phi_0 = 4\pi\varepsilon\left(\frac{kT}{\mathrm{e}}\right)^2 a \times 4\tanh\left(\frac{\zeta_\mathrm{c}\mathrm{e}}{4kT}\right) \times 4\tanh\left(\frac{\zeta_\mathrm{p}\mathrm{e}}{4kT}\right)$$

The initial flux enhancement due to attractive double-layer interactions was also observed for other types of cells, for example the oblique impinging jet cell, especially useful for studying particle deposition on non-transparent substrates. The results shown in Fig. 9.6, obtained for polystyrene latex particles of size 1.48 μm adsorbing on modified mica, are in good agreement with theoretical predictions for the whole *Re* range and various ionic strengths [70]. Analogously to the RIJ cell, the deposition rate of particles was significantly enhanced by the decrease in the ionic strength. This increases the range of the attractive electrostatic interactions, which leads to the enhancement of the interception effect predicted by Eq. (116).

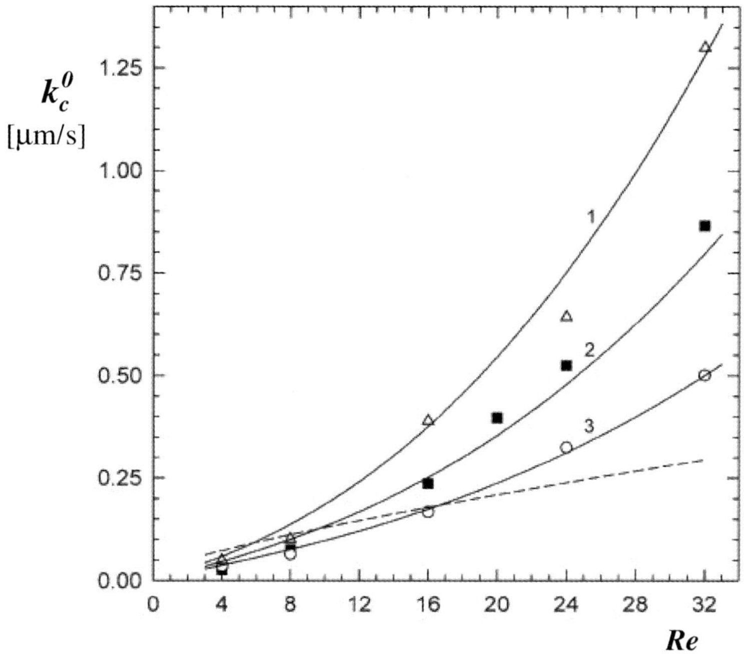

Fig. 9.6 Dependence of k_c^0 on *Re* measured near the stagnation point. The points denote the experimental results obtained in the oblique impinging jet cell (OBIJ) for polystyrene latex particles (diameter 1.48 μm) adsorbing at mica. (1) $I=2\times10^{-5}$ M; (2) $I=10^{-4}$ M; (3) $I=10^{-3}$ M. The continuous lines denote the theoretical results obtained by solving numerically Eq. (111) and the dashed line represents the analytical results calculated from Eq. (105) using the Smoluchowski–Levich approximation (from [70]).

The increase in the initial deposition rate (limiting flux) in dilute electrolyte solutions caused by the interception effect is a universal phenomenon occurring for other flow configurations more related to practice. For example, Elimelech [64] measured the particle adsorption (filtration) rate in columns packed with glass beads (having an average diameter of 0.046 cm). He used suspensions of positively charged latex particles of various sizes ranging from 0.08 to 2.51 μm and the ionic strength was varied between 5×10^{-6} M (deionized water) and 0.1 M. The number of particles adsorbed was determined indirectly by monitoring the particle concentration change (by turbidimetry) at the inlet and outlet of the column. The results shown in Fig. 9.7 confirm, analogously to those obtained in the RIJ cell, that the initial flux increased substantially (almost fourfold) when performing experiments in deionized water. This effect was interpreted quantitatively in terms of the numerical solutions of the two-dimensional transport equation (Eq. 68), formulated for spherical geometry [37, 64]. Similar results were obtained for larger particle sizes although the measured particle deposition rates were generally smaller than predicted theoretically.

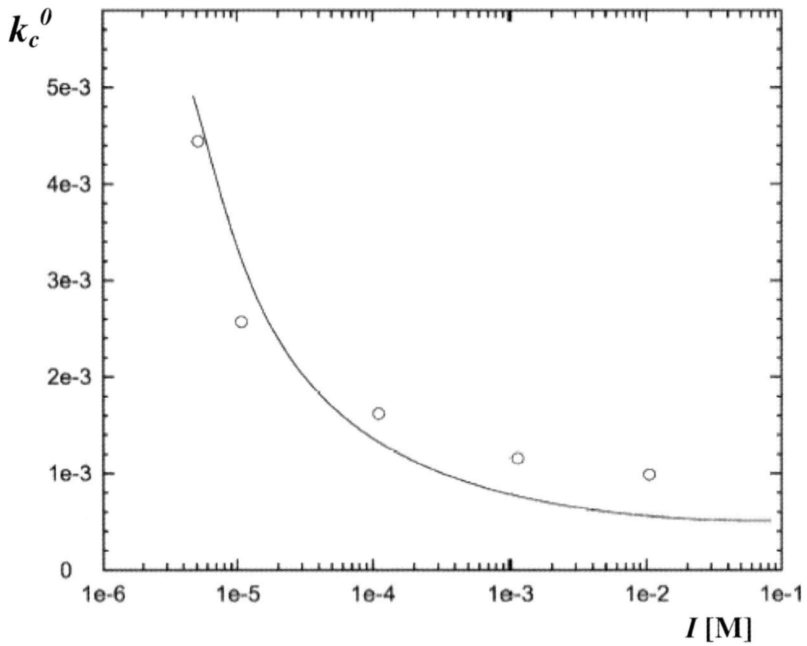

Fig. 9.7 Dependence of k_c^0 on the ionic strength I determined experimentally for positive polystyrene latex particle (average diameter 1.15 μm) adsorbing on negatively charged glass beads of diameter 0.046 cm forming a packed bed column. The solid line denotes the theoretical results calculated by solving numerically Eq. (68) for spherical geometry (from [64]).

The above results and others discussed elsewhere [14, 15, 49, 94] confirmed quantitatively the validity of the convective diffusion theory in which the coupling between the specific (electrostatic) and hydrodynamic force fields was considered in an exact manner.

A different behavior is observed in systems where repulsive electrostatic interactions appear, which leads to the appearance of an energy barrier. The energy profile in this case is shown schematically in Fig. 9.2. Because the height of the energy barrier is very sensitive to the ionic strength of the suspension and the substrate zeta potential, one can expect that obtaining reliable data for barrier-controlled deposition is significantly more difficult. One of few systematic studies of this type was performed by Hull and Kitchener using the rotating disk method [84]. Polystyrene latex particles of diameter 0.31 μm were used and the substrates were microscope slides covered with a polyvinylpyridine copolymer. The zeta potential of the latex was varied between –78 and –38 mV by increasing the ionic strength of the suspension. The zeta potential of the substrate varied between –51 and –32 mV, respectively. As can be seen in Fig. 9.8, the experimentally observed deposition rate of particles was considerably larger than theoretically predicted by using Eq. (102).

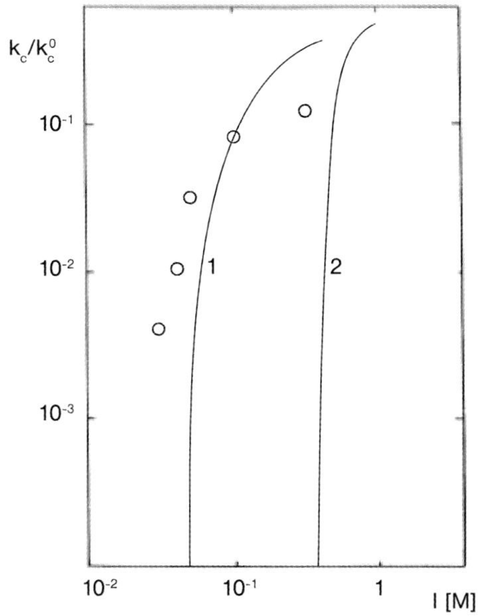

Fig. 9.8 Influence of ionic strength I on the relative deposition rate $k_c/k_c^0 = j/j_0$ (where k_c^0 is the rate value in the absence of electrostatic interactions). The points represent the experimental results obtained using the rotating disk method [84]. The solid lines are the theoretical predictions for a homogeneous charge distribution (curve 2) and a Gaussian charge distribution characterized by a relative standard deviation of 0.35 (curve 1).

Similar behavior was observed by Varrenes and van de Ven [95], who measured deposition rates of polystyrene latex particles (average diameter 3 µm) in the impinging jet cell. They detected measurable particle deposition for ionic strengths as low as 10^{-5} M, where the theory predicts practically no deposition. Analogous results were reported by Litton and Olson [96], who measured the deposition rate of positively charged latex particles (0.25 µm in diameter) in a packed bed column formed by glass beads 0.0275 cm in diameter. Again, the measured deposition rate of particles was found to be orders of magnitude higher than theoretically predicted.

These positive deviations from theoretical flux values were interpreted in terms of the surface heterogeneity hypothesis proposed in Refs. [21, 97]. The heterogeneity may appear because of natural fluctuation in the charge density on particles and interfaces. For example, local micro-patches can be formed, characterized by more favorable deposition conditions than the average surface. Even if these favorable areas occupy a small fraction of the entire surface, the overall deposition rate will be much larger than theoretically predicted for uniform surfaces because of a high sensitivity of particle flux to local surface charge. This hypothesis is supported by kinetic curves exhibiting saturation at low particle coverage of the order of percent and an unevenness of the particle coverage distribution [84, 95]. Also the micro-electrophoretic measurements indicate that a natural distribution of zeta potential appears within particle populations, in particular polystyrene latex suspensions [98, 99]. This will lead to enhanced particle deposition because particles bearing smaller charges will be selectively deposited from suspensions. This hypothesis was exploited to interpret the data shown in Fig. 9.8. When it was assumed that the relative standard deviation of particle zeta potential distribution equaled 0.35, the agreement of the theoretical and experimental data was found to be significantly better.

These results suggest that in the case of barrier-controlled deposition, the classical DLVO energy profiles calculated for homogeneous surfaces are not sufficient for a theoretical interpretation of particle adsorption data. In order to test this hypothesis in more detail, experimental work has been recently carried out on particle deposition on heterogeneous surfaces [31, 71, 79, 100–103].

9.4.2
Particle Deposition on Heterogeneous Surfaces

Results obtained for two types of heterogeneous surfaces will be discussed:

1. substrates produced by pre-adsorption of colloid particles having opposite surface charge sign;
2. substrates obtained by irreversible adsorption of cationic polyelectrolytes.

In the former case, well-defined heterogeneous surfaces were produced by controlled deposition of latex particles on uniform substrates, such as mica [71, 79, 100, 102, 103]. The degree of heterogeneity was expressed in terms of the coverage $\Theta_s = \pi a_s^2 N_s$ (where a_s is the particle radius and N_s is their surface concentration).

Typical configurations of heterogeneities produced in this way on a mica substrate (using the RIJ cell) are shown in Fig. 9.9 [104]. In Fig. 9.9a the results obtained for high ionic strength, $I = 10^{-3}$ M, and Θ_s are shown, and in Fig. 9.9b the results obtained for low ionic strength, $I = 10^{-5}$ M, and $\Theta_s = 0.5$. Positively charged latex particles of average diameter 0.47 μm were used in these experiments. The distribution of heterogeneities was quantitatively characterized in terms of the pair correlation function defined as [104]

$$g(r) = \frac{\pi a_s^2}{\Theta_s} \left\langle \frac{\Delta N_p}{2\pi r \Delta r} \right\rangle \tag{120}$$

where $\langle\rangle$ means the ensemble average and N_p is the number of particles adsorbed within the ring $2\pi r \Delta r$ drawn around a central particle. Usually the number of particles used for evaluating the pair correlation function was 1000–2000 [12, 37, 49, 104].

As can be seen in Fig. 9.9, the distribution of particles on the surface can be well reflected in terms of the Boltzmann distribution:

$$g(r) = e^{-\phi(r)/kT} \tag{121}$$

where $\phi(r)$ is the effective potential comprising the electrostatic interactions between particle pairs and the attraction between the particle and the interface.

By inverting Eq. (121), one can calculate in principle the effective interaction profile from the dependence

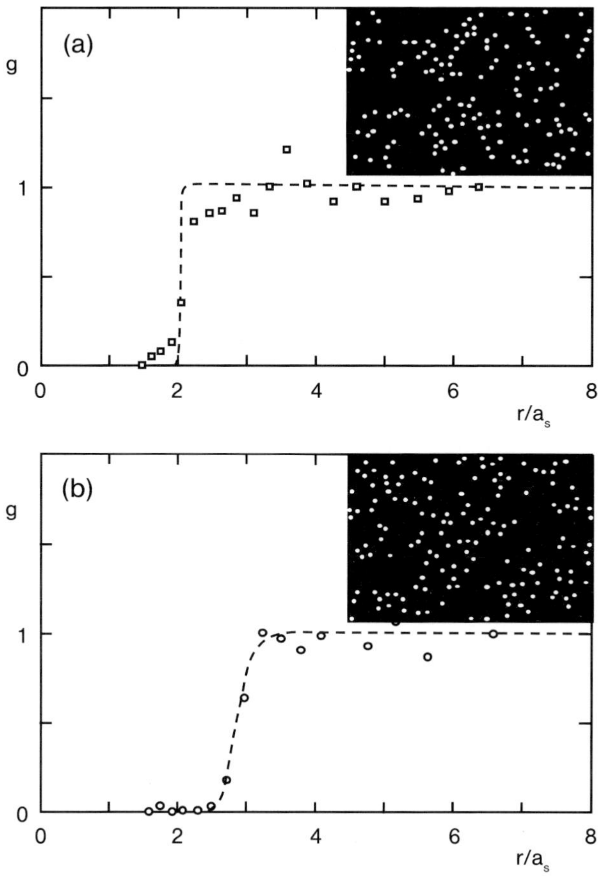

Fig. 9.9 Structure of monolayers of polystyrene latex particles adsorbed on mica expressed in terms of the pair-correlation function g: (a) $I=10^{-3}$ M, $\Theta=0.05$; (b) $I=10^{-5}$ M, $\Theta=0.05$. The dashed lines represent the pair-correlation function calculated from the Boltzmann distribution $g=e^{-\phi/kT}$; the insets show the adsorbed particles forming a two-dimensional gas phase (from [104]).

$$\phi(r) = -kT \ln g(r) \tag{122}$$

As discussed in Ref. [12], this is a fairly effective procedure, although rather tedious because of the necessity for measuring the positions of thousands of particles.

The electrokinetic characteristics of heterogeneous substrates covered with particles were determined by using the streaming potential method in the parallel-plate arrangement described in Refs. [86–89]. The apparent zeta potential of the substrate (mean value of the zeta potential averaged for the entire heterogeneous surface) was calculated from the modified Smoluchowski dependence accounting for the finite depth of the channel:

$$\zeta_c = \zeta_1 \left(1 - \frac{16\alpha_c}{\pi^3} + \frac{16\alpha_c}{\pi^3}\frac{\zeta_2}{\zeta_1}\right) = \frac{4\pi\eta}{\varepsilon}\left(\frac{E_s}{\Delta P}\right)\lambda_{eff} \tag{123}$$

where $a_c = b/c$ is the ration of the thickness of the channel to its width, ζ_1 is the zeta potential of the substrate, ζ_2 is the zeta potential of the side walls of the channel, E_s is the streaming potential measured for a given hydrostatic pressure drop ΔP and λ_{eff} is the effective conductivity of the electrolyte. The dependence of the apparent zeta potential determined in this way ζ_c on the coverage Θ_s of the positive latex (particle diameter 0.47 μm, the same as above) is shown in Fig. 9.10. It is interesting that the particle coverage was determined directly by using the microscope counting procedure. As can be observed, the apparent zeta potential, initially being –96 mV (for bare mica), changed abruptly with the particle coverage, reaching zero for a particle coverage as low as 0.13. For coverage reaching 0.3, the apparent zeta potential reached 40 mV, which is close to the value of the zeta potential of latex particles in bulk ($\zeta_p = 42$ mV). It is interesting that the experimental dependence of ζ_c on Θ was quantitatively described by the theoretical model discussed in Refs. [86–89]:

$$\zeta_c = \zeta_1 e^{-|C_i^0|\Theta} + \zeta_p \left(1 - e^{-C_p^0 \Theta}\right) \tag{124}$$

where ζ_p is the zeta potential of the particles in a bulk and C_i^0 and C_p^0 are the dimensionless functions characterizing the role of the flow damping by adsorbed

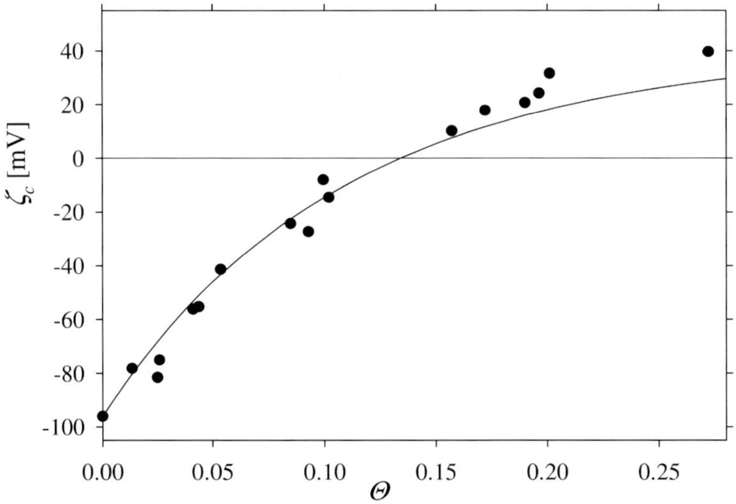

Fig. 9.10 Dependence of zeta potential of mica covered by positively charged latex particles on Θ. The points represent experimental data obtained by the streaming potential method in the plate-parallel channel. The solid line was calculated according to Eq. (124). $I = 10^{-4}$ M, $2a = 0.47$ μm, $\kappa a = 7.7$.

particles and counter-ion outflow from the double layer, respectively. For $\kappa a > 5$ these functions become practically independent of the ionic strength, reaching the limiting values $C_i^0 = -10.21$ an d $C_p^0 = 6.51$.

Further experiments performed for various ionic strengths and other particle–substrate systems, including neutral particles on charged substrates and vice versa, confirmed the validity of Eq. (124) for predicting the apparent zeta potential of heterogeneous surfaces [86–89].

Using the substrate surfaces prepared and characterized as described above, kinetic experiments were performed in Refs. [71, 79, 100, 101] with the aim of determining the initial particle deposition rate k_c^0 as a function of Θ_s. Measurements were made both under convection deposition conditions in the RIJ cell [79, 100, 101] and under diffusion transport conditions [71, 102]. The dependence of the coverage $\Theta_p = \pi a^2 N_p$ of larger polystyrene latex particles (of average size $2a = 0.9$ nm) on surfaces of degree of heterogeneity characterized by various Θ_s was studied systematically. Selected results obtained for $I = 10^{-3}$ M are shown in Fig. 9.11. As can be seen, in the case of the convective transport (Fig. 9.11 a), these dependencies are linear with the slope increasing abruptly with the degree of heterogeneity. Practically, for $\Theta_s > 0.05$ the kinetics of particle adsorption became identical with those pertinent to homogeneous surfaces (shown by the dashed line in Fig. 9.11 a). The latter kinetics was calculated exactly by solving numerically the governing mass transfer equation, Eq. (111).

A very similar trend was observed in the case of the diffusion-controlled deposition of particles (Fig. 9.11 b). In this case the kinetic runs were expressed in terms of the dependence of Θ_p on the square root of the deposition time $t^{\frac{1}{2}}$. As can be seen, for $\Theta_s > 0.042$, the kinetic runs became linear in this coordinate system with a slope very similar to the homogeneous surfaces case when the deposition rate in the limit of low particle coverage is given by [71, 102]

$$k_c^0(t) = \left(\frac{D}{\pi t}\right)^{\frac{1}{2}} \tag{125}$$

and the coverage

$$\Theta_p = \int_0^t k_c^0(t')\mathrm{d}t' = 2\left(\frac{Dt}{\pi}\right)^{\frac{1}{2}} n_b \tag{126}$$

The results shown in Fig. 9.11 suggest unequivocally that for surface heterogeneity of the order of a few percent only, the initial deposition kinetics of particles become very similar to those for homogeneous surfaces. This experimental evidence is significant, because they indicate that the well-known results for homogeneous surfaces (see Table 9.2, for example) can be exploited for predicting deposition on heterogeneous surfaces, characterized by $\Theta_s > 0.05$. In order to emphasize this point, the dependence of the reduced deposition rate $k_c(\Theta)/k_c^0$ (where k_c^0 is the deposition rate for a homogeneous surface in the limit of low

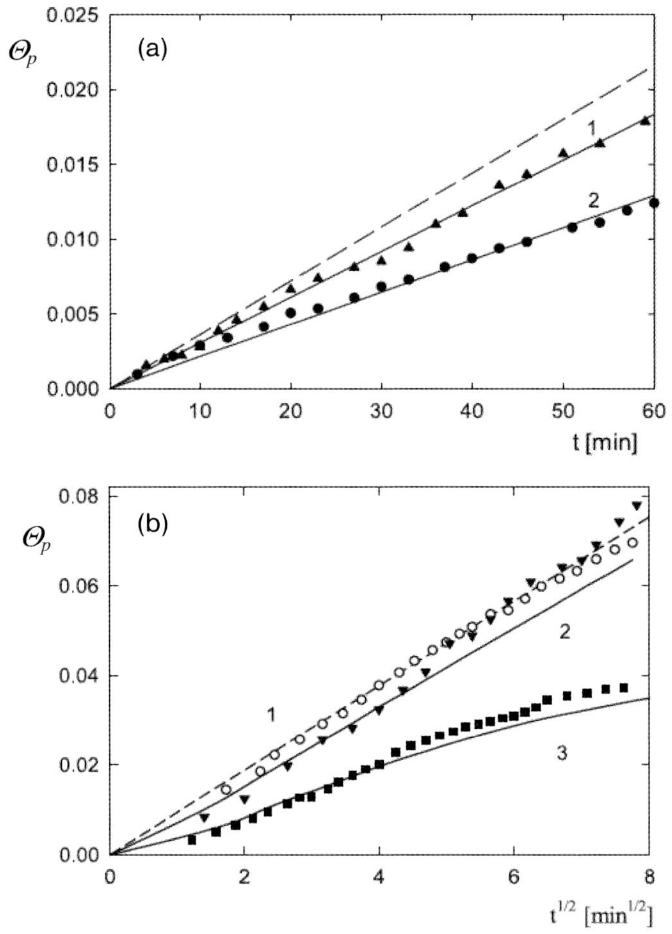

Fig. 9.11 Kinetics of deposition of polytyrene latex particles, (average diameter $2a = 0.9$ μm, $I = 10^{-3}$ M) on surfaces pre-covered by smaller sized latex particles ($2a = 0.47$ μm). (a) The impinging jet cell (convection-controlled deposition), $Re = 4$, (1) $\Theta_s = 0.032$, (2) $\Theta_s = 0.018$. (b) The diffusion cell (diffusion-controlled deposition): (1) $\Theta_s = 0.10$, (2) $\Theta_s = 0.042$, (3) $\Theta_s = 0.0068$. The dashed lines represent the theoretical results predicted for homogeneous surfaces (from [71]).

coverage) on the degree of heterogeneity Θ_s is shown in Fig. 9.12. A quantitative interpretation of these results can be performed by exploiting the surface boundary layer concept expressed by Eq. (102), which can be rearranged to the form [15, 79]

$$\frac{k_c(\Theta_s)}{k_c^0} = \frac{Kp_0(\Theta_s)}{1 + (K-1)p_0(\Theta_s)} \tag{127}$$

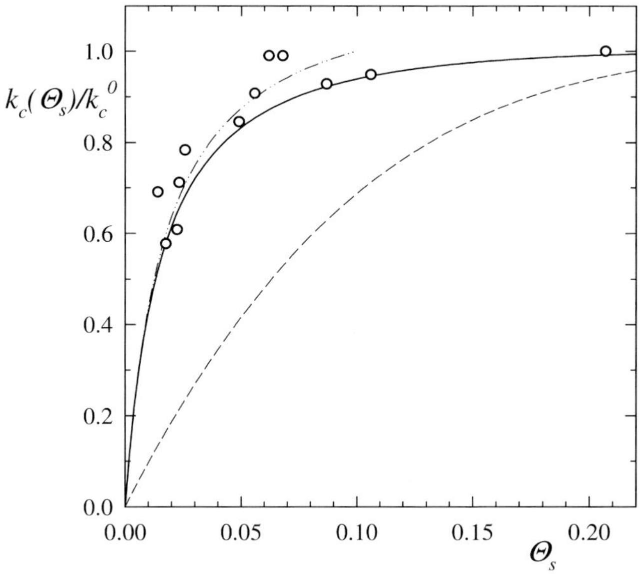

Fig. 9.12 Dependence of the normalized deposition rate of larger particles at heterogeneous surfaces $k_c(\Theta_s)/k_c^0$ (where k_c^0 is coefficient for an uncovered surface) on Θ_s (the degree of heterogeneity). The points denote the experimental results obtained for the latex particles on mica in the RIJ cell and the continuous line shows the analytical results calculated from Eq. (219) according to the generalized RSA method. The dashed line shows the results calculated by neglecting the coupling ($K=1$) and the dashed-dotted line presents the data calculated from the Eq. (231) (from [79]).

where $K = k_a/k_c^0$ is the coupling constant characterizing the ratio of particle transport rate through the surface layer to the rate of the transport through the bulk of the suspension and k_a in the adsorption rate constant given by Eq. (77). If the transport of particles from the bulk to the interface remains much slower than the transport through the surface adsorption layer, $K{>>}1$ and particle transport is solely governed by the bulk

The function $p_0(\Theta_s)$ in Eq. (127) describes the probability of particle adsorption on the heterogeneous surface provided that it has appeared at the edge of the surface boundary layer. It can be well described by the analytical equation derived in Refs. [15, 105]:

$$p_0(\Theta_s) = 1 - (1 - \Theta_s)e^{\frac{(4\lambda-1)\Theta_s}{1-\Theta_s} - \left[\frac{(2\sqrt{\lambda}-1)\Theta_s}{1-\Theta_s}\right]^2} \tag{128}$$

where $\lambda = a_p/a_s$ is the size ratio of the particle to the heterogeneity.

For $K \gg 1$ and $\lambda\Theta_s \ll 1$, by combining Eqs. (127) and (128) one can derive a simple relationship for the reduced flux of larger particles (adsorption rate):

$$\frac{k_c(\Theta)}{k_c^0} = \frac{4\lambda K \Theta_s}{1 + 4\lambda(K-1)\Theta_s} \tag{129}$$

As can be seen in Fig. 9.12, the theoretical predictions stemming from Eq. (127) reflect well the experimental data for the entire range of the degree of heterogeneity (with $K=7$, which corresponds to the experimental conditions). As expected, Eq. (129) also gives satisfactory agreement with the experimental data for $\Theta_s < 0.10$. This suggests that the basic features of particle adsorption at heterogeneous surfaces are reflected by this equation, which has practical significance in view of its simplicity. In particular, one can deduce from Eq. (129) that the adsorption rate for heterogeneous surfaces attains the limiting value if the following criterion is fulfilled:

$$\Theta_s > \frac{1}{4\lambda K} = \frac{a_s k_c^0}{4 a_p k_a} \tag{130}$$

As can be deduced, this limiting value is proportional to the heterogeneity size and the rate constant of the bulk transport k_c^0. Hence the increase in the flux is most dramatic for smaller heterogeneity size (at fixed coverage) and low flow rate (Reynolds number) when the thickness of the diffusion boundary layer becomes considerably larger than the particle dimensions. This observation suggests that the coupling between the bulk and surface transport plays a decisive role for particle and protein adsorption at heterogeneous surfaces. Additionally, the experimental results presented in Fig. 9.12 strongly support the heterogeneity hypothesis by indicating that the presence of local charge heterogeneities of a patchy form, occupying a few percent of the entire interface area, significantly increase the particle deposition rate.

It is interesting to correlate the deposition rate on heterogeneous surfaces with the apparent zeta potential of the substrate. Experimental results obtained for the negative latex of size 1.38 μm are plotted in this way in Fig. 9.13. As can be seen, the particle deposition rate (more precisely its reduced value k_c/k_c^0) becomes comparable to unity for a highly negative apparent zeta potential. This disagrees with the DLVO theory for homogeneous surfaces, which predicts that particle deposition rate, calculated from Eq. (102) by considering the electrostatic interactions via Eq. (113) and the van der Waals interactions from Eq. (38), should be reduced to zero for an apparent zeta potential as low as –1 mV (see the dashed line in Fig. 9.13).

In contrast, the theoretical model described by Eqs. (128) and (129), depicted by the solid line in Fig. 9.13, reflects well the experimental results for the entire range of Θ_s. This is so because in this model the local topology of the charge distribution associated with the presence of heterogeneities is considered.

One can expect that by decreasing the size of the heterogeneities, when the quasi-homogeneous charge distribution is approached, the correlation between the apparent zeta potential and the rate of particle deposition should be better described by the DLVO model. This hypothesis was tested by studying particle

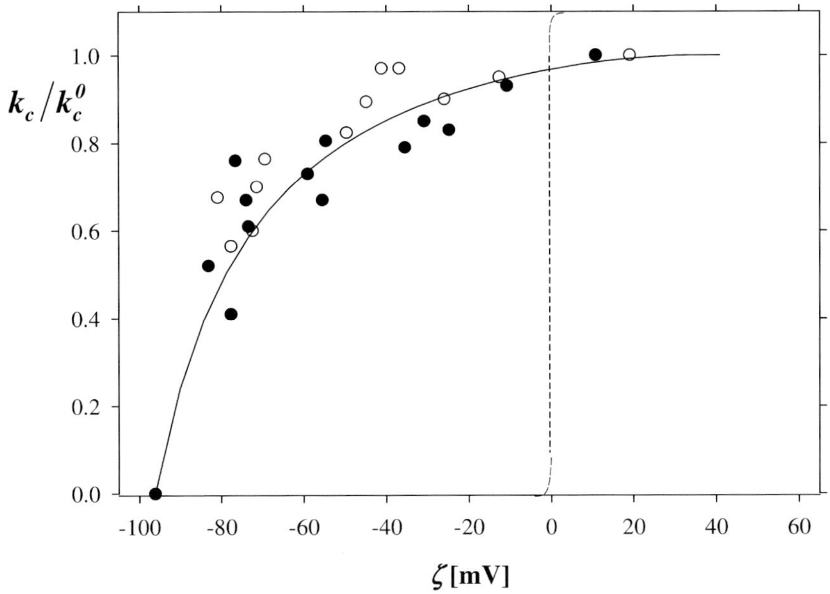

Fig. 9.13 Dependence of the reduced initial deposition rate on the apparent zeta potential of heterogeneous surface. The points represent the experimental results obtained for negative latex particles ($2a = 1.38$ μm) deposited on positively charged centers [positive latex particles ($2a = 0.47$ μm)]. The empty points were measured for $Re = 2$, full points for $Re = 8$, $I = 10^{-4}$ M. The solid line represents the theoretical results calculated from Eq. (128), the dashed line represents the dependence predicted from DLVO theory for homogeneous surfaces.

deposition on heterogeneous surfaces of the second kind, produced by the controlled deposition of polyelectrolytes. A cationic (PAH) polyelectrolyte was used having a molecular weight of 70000, a hydrodynamic radius (measured by dynamic light scattering) of 17.5 and 18.7 nm for ionic strength of 5×10^{-3} and 0.15 M, respectively, and a bulk zeta potential value of 50 mV at an ionic strength of 10^{-2} M [106].

The electrokinetic characteristics of the mica surface covered by the polyelectrolyte is shown in Fig. 9.14 [107]. It is interesting that in the case of the polyelectrolyte, its coverage Θ_s, characterizing the degree of heterogeneity, was calculated assuming convection-controlled transport. As can be observed in Fig. 9.14, the apparent zeta potential of mica changed abruptly with the polyelectrolyte coverage, analogously as for the particle covered surfaces, attaining zero for Θ_s as low as 0.13. For coverage reaching 0.4, the apparent zeta potential reached 62 mV, which is close to the value of the zeta potential for latex particles in bulk. It is interesting to observe that the experimental dependence of ζ_c on Θ was fairly well described by Eq. (124). However, a definite tendency to overestimate the experimental data is visible. This can be attributed to the fact

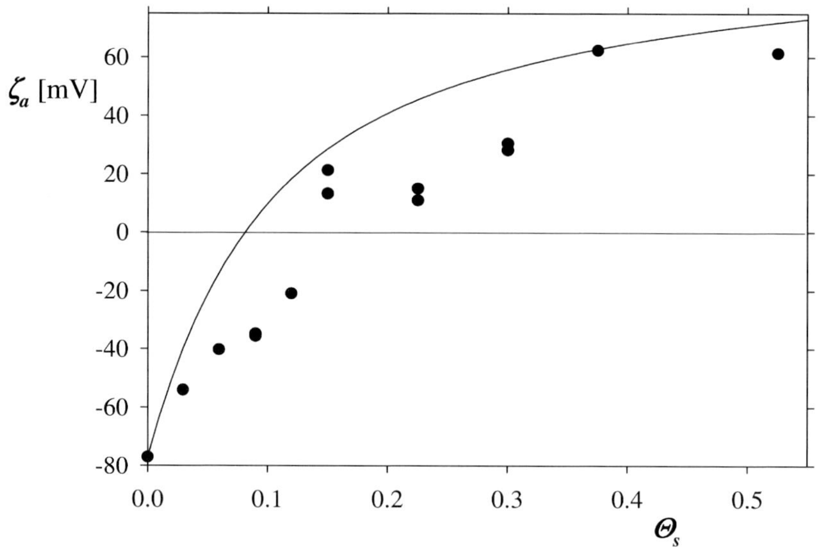

Fig. 9.14 Zeta potential of mica covered by PAH (MW ≈ 70 000, $I=10^{-2}$ M). The points represent the experimental data obtained by streaming potential technique. The solid line denotes theoretical results calculated from Eq. (124) for $\kappa a=0.2$: $C_i^0=-3.2$, $C_p^0=14.25$, $\zeta_i=-77$ mV, $\zeta_p=85$ mV (from [107]).

that κa for the polyelectrolyte (where a is the bare chain radius, estimated to be about 0.4 nm [106]) was below unity. This causes a significant overlap of the double layers surrounding the chains. Another explanation is that part of the polymer is adsorbed in the form of loops and trains rather than lying flat on the surface [108]. These effects makes Eq. (124) less accurate for $\kappa a<1$.

Using the heterogeneous surfaces prepared by the adsorption of the polymer, kinetic runs were performed with the aim of determining the initial deposition rate of the latex as a function of Θ_s [109]. Measurements were made under convection deposition conditions in the RIJ cell by using the negatively charged latex of average size $2a=0.66$ μm. The dependence of particle coverage Θ_p on the deposition time t determined for $Re=2$ and $I=10^{-3}$ M is shown in Fig. 9.15. As can be seen, these dependences are linear with the slope (deposition rate of particles) increasing systematically with the initial polyelectrolyte coverage Θ_s. However, the increase in the deposition rate in this case is definitely less abrupt that in the previous case of colloid particle heterogeneities. For polymer-covered surfaces the maximum deposition rate was attained for a degree of heterogeneity $\Theta_s>0.2$ rather than 0.05 as was the case previously (see Fig. 9.12). The difference between these two types of heterogeneous surfaces can be seen in Fig. 9.16, where the dependences of the reduced deposition rate k_c/k_c^0 on the apparent zeta potential are collected. For both types of surfaces a finite deposition rate appears for negative apparent zeta potential, that is, of the same sign as

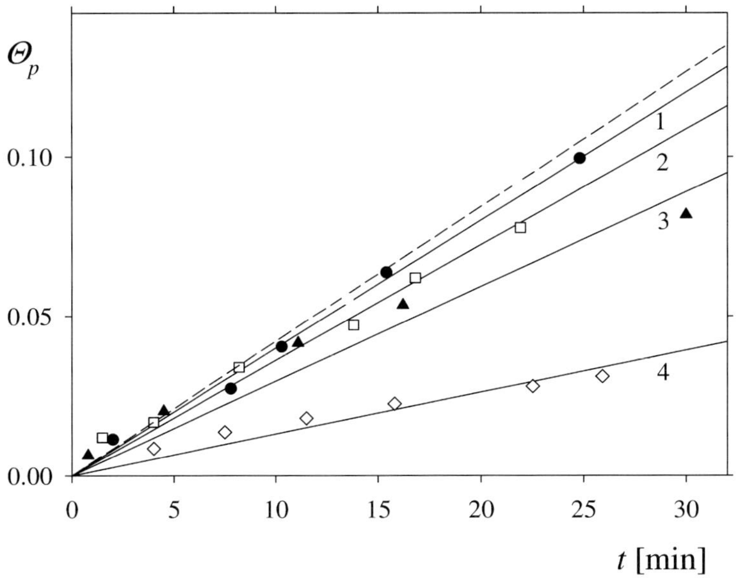

Fig. 9.15 Initial rate of particle deposition (the linear regime) at surfaces with various amounts of adsorbed PAH. Particle deposition performed at $Re=2$, $I=10^{-3}$ M, PAH adsorbed under diffusion controlled regime from solutions of $I=10^{-2}$ M: (1) (●) $\theta_s=0.299$; (2) (□) $\theta_s=0.254$; (3) (▲) $\theta_s=0.156$; (4) (◇) $\theta_s=0.129$. The dashed line denotes the theoretical results calculated for homogeneous surfaces (from [107]).

the zeta potential of the particles. For example for $\zeta_c=-20$ mV, k_c/k_c^0 was 0.2 for polyelectrolyte pre-covered surfaces and 0.9 for the particle-covered surfaces. Interestingly, in the case of a polyelectrolyte the ratio k_c/k_c^0 deviates from unity for positive potential range of the substrate. For example, $k_c/k_c^0=0.6$ for $\zeta_c=20$ mV and $k_c/k_c^0=0.9$ for $\zeta_c=40$ mV. This differs from the results predicted by the DLVO theory, postulating a continuous charge distribution (depicted by the dashed line in Fig. 9.16), although the deviation is significantly smaller than for particle-covered surfaces. The explanation of the effect shown in Fig. 9.16 can be sought in the non-uniform charge distribution over the surface. Evidently, spots appear on the substrate surface of a size comparable to the deposition particle dimensions, where the local surface charge density differs significantly (in both the positive and negative directions) from the average value.

Therefore, the results in Fig. 9.16, indicating unequivocally that the DLVO theory breaks down on the microscopic scale, provide a natural explanation of the discrepancies observed in kinetic studies of colloid particle deposition on heterogeneous surfaces. However, further systematic studies involving various polyelectrolytes and a wider range of particle sizes are necessary to elaborate this hypothesis quantitatively.

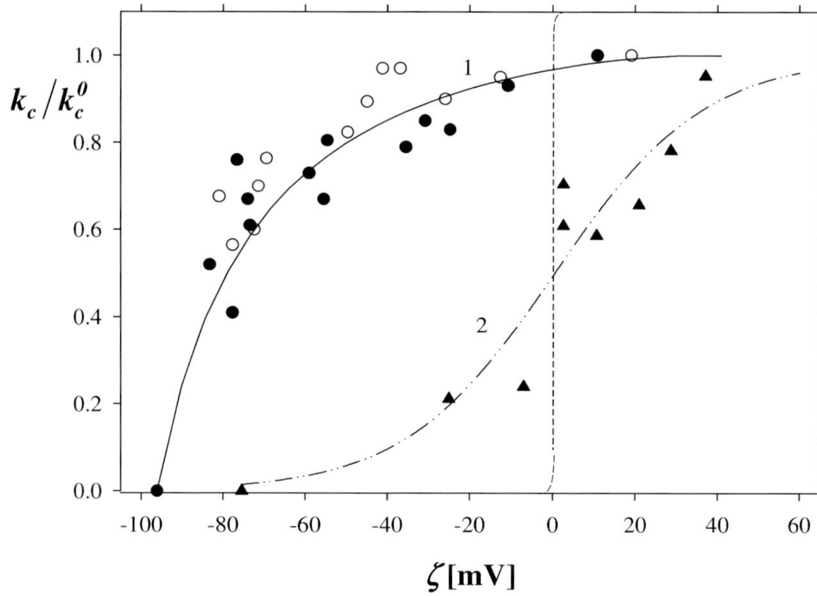

Fig. 9.16 Dependence of the reduced initial deposition rate on the apparent zeta potential of heterogeneous surfaces. 1, Surfaces produced by pre-adsorption of latex particles. The solid line denotes the theoretical results. 2, Surfaces produced by polyelectrolyte adsorption. The dashed-dotted line represents the fitting curve. The dashed line denotes predictions of DLVO theory for a homogeneous surface (from [107]).

9.5 Conclusions

The analysis of experimental data collected under well-defined transport conditions confirmed that electrostatic interactions play an essential role in adsorption and deposition phenomena of colloid particles. In the case of homogeneous surfaces, the limiting deposition rates of particles, for a low coverage range, are considerably enhanced by the attractive electrostatic interactions, especially for low ionic strength and larger *Pe*. This behavior was quantitatively interpreted in terms of the convective diffusion theory incorporating the DLVO theory for homogeneous surfaces. It was suggested, therefore, that by measuring particle deposition rates under well-defined transport conditions the magnitude and the range of the double-layer interactions can be derived for hetero-systems (dissimilar surfaces).

In the case of repulsive double-layer interactions when an energy barrier appears, significant deviations from theoretical predictions based on the classical DLVO theory were found in experimental studies. The discrepancy was accounted for by formulating a surface heterogeneity hypothesis, which was confirmed in model experiments involving heterogeneous surfaces produced by pre-adsorption of particles and polyelectrolytes. According to this theory, for barrier-controlled and heterogeneous systems, the DLVO theory is applicable in a local sense only. Therefore, in order to characterize real systems unequivocally one should know not only the apparent average zeta potential of the surface but also its distribution over the substrate surface or within particle population.

Acknowledgment

This work was financially supported by the Ministry of Science and Information, Grants 4 T09A 07625 and 4 T08B 03425.

References

1 Z. Adamczyk, *Particles at Interfaces*, Academic Press, New York, **2006**.
2 J. Lyklema, *Fundamentals of Interface and Colloid Science*, Academic Press, London, **1993**.
3 B.V. Derjaguin, *Kolloid Z.*, **1934**, *69*, 155–164.
4 B.V. Derjaguin, *Trans. Faraday Soc.*, **1940**, *36*, 203–215.
5 G.M. Bell, S. Levine, L.N. McCartney, *J. Colloid Interface Sci.*, **1970**, *33*, 335–359.
6 S.L. Carnie, D.Y.C. Chan, J. Stankovich, *J. Colloid Interface Sci.*, **1994**, *165*, 116–128.
7 R. Hogg, T.W. Healy, D.W. Fuerstenau, *Trans. Faraday Soc.*, **1973**, *62*, 1638–1651.
8 R.J. Hunter, *Foundation of Colloid Science*, 2nd edn., Oxford University Press, New York, **2001**.
9 G.R. Wiese, T.W. Healy, *Trans. Faraday Soc.*, **1970**, *66*, 490–499.
10 S. Usui, *J.Colloid Interface Sci.*, **1973**, *44*, 107–113.
11 G. Kar, S. Chander, T.S. Mika, *J. Colloid Interface Sci.*, **1973**, *44*, 347–355.
12 Z. Adamczyk, P. Warszyński, *Adv. Colloid Interface Sci.*, **1996**, *63*, 41–149.
13 L.R. White, *J. Colloid Interface Sci.*, **1983**, *95*, 286–288.
14 Z. Adamczyk, P. Weroński, *Adv. Colloid Interface Sci.*, **1999**, *83*, 137–226.
15 Z. Adamczyk, Irreversible adsorption of particles, in *Adsorption: Theory, Modeling and Analysis*, J. Toth (ed.), Marcel Dekker, New York, **2002**, p. 251.
16 Z. Adamczyk, P. Belouschek, D. Lorenz, *Ber. Bunsenges. Phys. Chem.*, **1990**, *94*, 1483–1492.
17 B. Lassen, M. Malsten, *J. Colloid Interface Sci.*, **1996**, *180*, 339–349.
18 J.N. Israelachvili, *Intermolecular and Surfaces Forces*, 2nd edn., Academic Press, New York, **1992**.
19 H. Morgenau and N.R. Kestner, *Theory of Intermolecular Forces*, Pergamon Press, New York, **1969**.
20 H.C. Hamaker, *Physica*, **1937**, *4*, 1058–1072.
21 Z. Adamczyk, *Colloids Surf.*, **1989**, *39*, 1–37.
22 H. Krupp, *Adv. Colloid Interface Sci.*, **1967**, *1*, 111–239.
23 M.C. Herman, K.D. Papadopoulos, *J. Colloid Interface Sci.*, **1990**, *136*, 385–392.
24 M.C. Herman, K.D. Papadopoulos, *J. Colloid Interface Sci.*, **1991**, *142*, 331–342.
25 L. Suresh, J.Y. Walz, *J. Colloid Interface Sci.*, **1996**, *183*, 199–213.
26 L. Suresh, J.Y. Walz, *J. Colloid Interface Sci.*, **1997**, *196*, 177–190.
27 M. Kostaglou, A.J. Karabelas, *J. Colloid Interface Sci.*, **1992**, *151*, 534–545.
28 J. Czarnecki, *J. Colloid Interface Sci.*, **1979**, *72*, 361–362.
29 J. Czarnecki, T. Dąbroś, *J. Colloid Interface Sci.*, **1980**, *78*, 25–30.
30 J. Czarnecki, V. Itschenskij, *J. Colloid Interface Sci.*, **1984**, *98*, 590–591.
31 J.Y. Waltz, *Adv. Colloid Interface Sci.*, **1998**, *74*, 119–168.

32 E. M. Lifshitz, *Sov. Phys. JETP*, **1956**, *2*, 73–83.
33 I. E. Dzyaloshinski, E. M. Lifshitz, L. P. Pitaevski, *Adv. Phys.*, **1961**, *10*, 165–209.
34 B. V. Derjaguin, L. D. Landau, *Acta Physicochem. URSS*, **1941**, *14*, 633–662.
35 E. J. W. Vervey, J. Th. G. Overbeek, *Theory of the Stability of Lyophobic Colloids*, Elsevier, Amsterdam, **1948**.
36 W. B. Russel, D. A. Saville, W. R. Schowalter, *Colloidal Suspensions*, Cambridge University Press, Cambridge, **1993**.
37 Z. Adamczyk, *Adv. Colloid Interface Sci.*, **2003**, *100–102*, 267–347.
38 J. A. Barker, D. Henderson, *J. Chem. Phys.*, **1967**, *47*, 4714–4721.
39 M. Smoluchowski, *Phys. Z.*, **1916**, *17*, 557–571, 587–599.
40 T. G. M. van de Ven, *Colloid Hydrodynamics*, Academic Press, New York, **1989**.
41 W. B. Russel, D. A. Saville, W. R. Schowalter, *Colloidal Suspensions*, Cambridge University Press, Cambridge, **1993**.
42 L. A. Spielman, S. K. Friedlander, *J. Colloid Interface Sci.*, **1974**, *46*, 22–31.
43 B. D. Bowen, S. Levine, N. Epstein, *J. Colloid Interface Sci.*, **1976**, *54*, 375–390.
44 E. Ruckenstein, *J. Colloid Interface Sci.*, **1978**, *66*, 531–543.
45 D. C. Prieve, M. M. J. Lin, *J. Colloid Interface Sci.*, **1980**, *76*, 32–47.
46 Z. Adamczyk, J. Petlicki, *J. Colloid Interface Sci.*, **1987**, *118*, 20–49.
47 Z. Adamczyk, *J. Colloid Interface Sci.*, **1989**, *133*, 23–56.
48 V. G. Levich, *Physicochemical Hydrodynamics*, Prentice-Hall, Englewood Cliffs, NJ, **1962**.
49 Z. Adamczyk, B. Siwek, M. Zembala, P. Belouschek, *Adv. Colloid Interface Sci.*, **1994**, *48*, 151 –280.
50 M. Elimelech, J. Gregory, X. Jia, R. A. Williams, *Particle Deposition and Aggregation*, Butterworth-Heinemann, Oxford, **1996**.
51 Z. Adamczyk, Kinetics of particle and protein adsorption, in *Surface and Colloid Science*, Vol. 17, E. Matijevic, M. Borkovec (eds.), Kluwer, New York, **2004**, Chap. 5.
52 Z. Adamczyk, *J. Colloid Interface Sci.*, **2000**, *229*, 477–489.
53 L. A. Spielman, *J. Colloid Interface Sci.*, **1977**, *62*, 529–541.
54 Z. Adamczyk, *J. Colloid Interface Sci.*, **1980**, *78*, 559–562.
55 T. Dąbroś, Z. Adamczyk, J. Czarnecki, *J. Colloid Interface Sci.*, **1977**, *33*, 562–571.
56 Z. Adamczyk, T. G. M. van de Ven, *J. Colloid Interface Sci.*, **1981**, *80*, 340–356.
57 Z. Adamczyk, T. G. M. van de Ven, *J. Colloid Interface Sci.*, **1981**, *84*, 497–518.
58 Z. Adamczyk, T. G. M. van de Ven, *Chem. Eng. Sci.*, **1982**, *37*, 869–880.
59 Z. Adamczyk, T. Dąbroś, T. G. M. van de Ven, *Chem. Eng. Sci.*, **1982**, *37*, 1513–1522.
60 Z. Adamczyk, L. Szyk, P. Warszyński, *Colloids Surf.*, **1993**, *75*, 185–193.
61 Z. Adamczyk, B. Siwek, P. Warszyński, E. Musiał, *J. Colloid Interface Sci.*, **2001**, *242*, 14–24.
62 D. C. Prieve, E. Ruckenstein, *AIChE J.*, **1974**, *20*, 1178.
63 Z. Adamczyk, B. Siwek, M. Zembala, P. Warszyńki, *J. Colloid Interface Sci.*, **1989**, *130*, 578–587.
64 M. Elimelech, *J. Colloid Interface Sci.*, **1994**, *164*, 190–199.
65 T. Dąbros, T. G. M. van de Ven, *J. Colloid Interface Sci.*, **1982**, *89*, 232–244.
66 T. Dąbroś, T. G. M. van de Ven, *Colloid Polym. Sci.*, **1983**, *261*, 694–707.
67 Z. Adamczyk, M. Zembala, B. Siwek, J. Czarnecki, *J. Colloid Interface Sci.*, **1986**, *110*, 188–200.
68 Z. Adamczyk, B. Siwek, M. Zembala, *Colloids Surf. A*, **1993**, *76*, 115–124.
69 Z. Adamczyk, L. Szyk, P. Warszyński, *J. Colloid Interface Sci.*, **1999**, *209*, 350–361.
70 Z. Adamczyk, E. Musiał, B. Siwek, *J. Colloid Interface Sci.*, **2004**, *269*, 53–61.
71 Z. Adamczyk, B. Siwek, P. Weroński, K. Jaszczół, *Colloids Surf. A*, **2003**, *222*, 15–25.
72 M. Meinders, J. Noordmans, H. J. Busscher, *J. Colloid Interface Sci.*, **1992**, *152*, 265–280.
73 T. Dąbroś, T. G. M. van de Ven, *Physicochem. Hydrodyn.*, **1981**, *8*, 161–172.
74 J. C. Dijt, M. A. Cohen Stuart, J. E. Hofman, G. J. Fleer, *Colloids Surf.*, **1990**, *51*, 141–158.
75 Z. Adamczyk, B. Siwek, M. Zembala, *J. Colloid Interface Sci.*, **1992**, *151*, 351–369.

76 Z. Xia, T. G. M. van de Ven, *Langmuir*, **1992**, *8*, 2938–2946.

77 Z. Adamczyk, L. Szyk, P. Warszyński, *Bull. Pol. Acad. Chem.*, **1998**, *46*, 425–439.

78 A. Göransson, Ch. Trägårdh, *J. Colloid Interface Sci.*, **2000**, *231*, 228–237.

79 Z. Adamczyk, B. Siwek, E. Musiał, *Langmuir*, **2001**, *17*, 4529–4533.

80 W. Norde, E. Rouwendal, *J. Colloid Interface Sci.*, **1990**, *139*, 169–176.

81 R. A. Hayes, M. R. Boehmer, L. G. J. Fokkink, *Langmuir*, **1999**, *15*, 2865–2870.

82 Z. Adamczyk, A. Pomianowski, *Powder Technol.*, **1980**, *27*, 125–136.

83 J. K. Marshall, J. A. Kitchener, *J. Colloid Interface Sci.*, **1966**, *22*, 342–351.

84 M. Hull, J. A. Kitchener, *Trans. Faraday Soc.*, **1969**, *65*, 3093–3104.

85 Z. Adamczyk, E. Musiał, B. Jachimska, L. Szyk-Warszyńska, A. Kowal, *J. Colloid Interface Sci.*, **2002**, *254*, 283–286.

86 M. Zembala, Z. Adamczyk, *Langmuir*, **2000**, *16*, 1593–1601.

87 M. Zembala, *Adv. Colloid Interface Sci.*, **2004**, *112*, 59–92.

88 M. Zembala, Z. Adamczyk, P. Warszyński, *Colloids Surf. A*, **2001**, *195*, 3–15.

89 M. Zembala, Z. Adamczyk, P. Warszyński, *Ann. Univ. M. Curie-Skłodowska, Sect. AA*, **2001**, *56*, 242–260.

90 R. Schweiss, P. B. Weltzel, C. Werner, W. Knoll, *Langmuir*, **2001**, *17*, 4304–4311.

91 D. Erickson, D. Li, *J. Colloid Interface Sci.*, **2001**, *237*, 283–289.

92 Z. Adamczyk, B. Siwek, L. Szyk, M. Zembala, *J. Chem. Phys.*, **1996**, *105*, 5552–5561.

93 Z. Adamczyk, L. Szyk, B. Siwek, P. Weroński, *J. Chem. Phys.*, **2000**, *113*, 11336–11342.

94 Z. Adamczyk, P. Warszyński, L. Szyk-Warszyńska, P. Weroński, *Colloids Surf. A*, **2000**, *165*, 157–187.

95 S. Varennes, T. G. M. van de Ven, *Physicochem. Hydrodyn.*, **1987**, *9*, 537–559.

96 G. M. Litton, T. M. Olson, *Colloids Surf. A*, **1994**, *87*, 39–48.

97 L. Song, P. R. Johnson, M. Elimelech, *Environ. Sci. Technol.*, **1994**, *28*, 1164–1171.

98 D. Velegol, J. D. Feick, L. R. Collins, *J. Colloid Interface Sci.*, **2000**, *230*, 114–121.

99 J. D. Feick, D. Velegol, *Langmuir*, **2002**, *18*, 3454–3458.

100 Z. Adamczyk, B. Siwek, P. Weroński, E. Musiał, *Appl. Surf. Sci.*, **2002**, *196*, 250–263.

101 Z. Adamczyk, B. Siwek, E. Musiał, *Colloids Surf. A*, **2003**, *214*, 219–229.

102 Z. Adamczyk, K. Jaszczółt, B. Siwek, P. Weroński, *J. Chem. Phys.*, **2004**, *120*, 11155–11162.

103 Z. Adamczyk, K. Jaszczółt, B. Siwek, P. Weroński, *Colloids Surf. A*, **2004**, *249*, 95–98.

104 Z. Adamczyk, P. Weroński, B. Siwek, M. Zembala, *Top. Catal.*, **2000**, *11/12*, 435–449.

105 Z. Adamczyk, P. Weroński, E. Musiał, *J. Colloid Interface Sci.*, **2002**, *248*, 67–75.

106 Z. Adamczyk, M. Zembala, P. Warszyński, B. Jachimska, *Langmuir*, **2004**, *20*, 10517–10525.

107 Z. Adamczyk, A. Michna, M. Zembala, to *Langmuir*, **2006**.

108 S. Stoll, DNA-like polyelectrolyte adsorption onto polymer colloids, in *Colloidal Biomolecules, Biomaterials and Biomedical Applications*, A. Elaissari (ed.), *Surfactant Science Series*, Vol. 116, Marcel Dekker, New York, **2004**, pp. 211–252.

109 A. Michna, Z. Adamczyk, M. Zembala, *Colloid Surf. A* (in press).

10
Recent Developments in Dilational Viscoelasticity of Surfactant Layers

Libero Liggieri, Michele Ferrari, and Francesca Ravera

10.1
Introduction

The mechanical behavior of liquid interfaces has high relevance for many technological and natural processes involving multiphase systems subjected to dynamic conditions. This behavior is primarily linked to the instantaneous value of the interfacial tension and to its instantaneous variation ("interfacial tension" will be used as a generic term throughout this chapter referring to both liquid–liquid and liquid–vapor interfaces). The dynamic aspects related to the adsorption of surface-active molecules are then important to understanding the behavior of these systems. Moreover, when systems characterized by a high specific area, such as emulsions of foams, are concerned, these aspects becomes the main driving force for the processes involved [1–4]. Low interfacial tensions favor the formation of emulsions and foams, while the variations of interfacial tension are concerned with their stability, since they are strongly involved in the properties of liquid films [5–10].

In order to control these properties, surface-active molecules are utilized, such as surfactants, polymers and proteins, which show an amphiphilic character and are variably soluble in water and in the oil phases [11, 12]. For all these systems, then, it is important not only to investigate equilibrium properties, but also to obtain information on the dynamic interfacial tension, in particular on the response of interfacial tension to variations of the interfacial area. This latter is generally referred to as the interfacial rheology of adsorption layers.

A close link exists between dynamic interfacial tension and adsorption mechanisms, which, in turn, reflect some molecular characteristics of surfactants. Hence the development of suitable theoretical models to interpret experimental data is an effective tool for investigating these features.

Currently, awareness of the relevance of surface rheology for significant scientific and practical purposes, such as those related to emulsion and foam stabilization, liquid film deposition and multiphasic flows, is growing in the scientific and technical community. It is therefore important to focus on the development

Colloids and Interface Science Series, Vol. 2
Colloid Stability: The Role of Surface Forces, Part II. Edited by Tharwat F. Tadros

ISBN: 978-3-527-31503-1

of suitable models to describe the response to dilational stimuli on different time-scales.

10.2 Surface Rheology of Surfactant Layers

10.2.1 Adsorption Kinetics and Interfacial Rheology

In order to understand interfacial rheology, one has first to look into the processes involved in restoring the equilibrium state of interfacial systems.

For soluble surfactants, the interface exchanges surfactant molecules with the adjacent volume. As a consequence, a dynamic equilibrium is established between the bulk concentration, c_0, and the surfactant excess concentration (or adsorption), Γ. The quantities defining the thermodynamic equilibrium state – Γ, c_0 and the interfacial tension, γ – are linked by the Gibbs isotherm [13]:

$$\Gamma = -RT\frac{\mathrm{d}\gamma}{\mathrm{d}\ln c_0} \tag{1}$$

Interfacial tension and adsorption are instead linked by the surface equation of state, which can be equivalently expressed by using the so-called Gibbs elasticity, ε_0:

$$\varepsilon_0(\gamma) = -\frac{\mathrm{d}\gamma}{\mathrm{d}\ln \Gamma} \tag{2}$$

The above two equations set at equilibrium a univocal relationship between γ and c_0, often referred as the γ–c isotherm, which is the most common subject of experimental investigation for soluble surfactants.

Whereas the Gibbs isotherm has general thermodynamic validity, the equation of state and thus $\varepsilon_0(\gamma)$ and the γ–c isotherm, are material properties, that is, specific to the system under consideration. Several adsorption models have been developed in order to predict the form of the surface equation of state. Among them, the most utilized are the Langmuir and Frumkin models, which have been widely utilized to interpret γ–c isotherms.

For soluble surfactants, both Eqs. (1) and (2) hold at equilibrium, hence an univocal equilibrium status is defined after providing the value for a single variable among c_0, γ and Γ.

The equilibrium relations between c_0, γ and Γ are exemplified in the plots in Fig. 10.1. There a Langmuir model is assumed, but other models do not change the main features of these dependences: surface tension decreases with adsorption and adsorption increases with bulk concentration.

For insoluble surfactants there is no exchange between the interface and the bulk, hence the only equilibrium relationship is the surface equation of state,

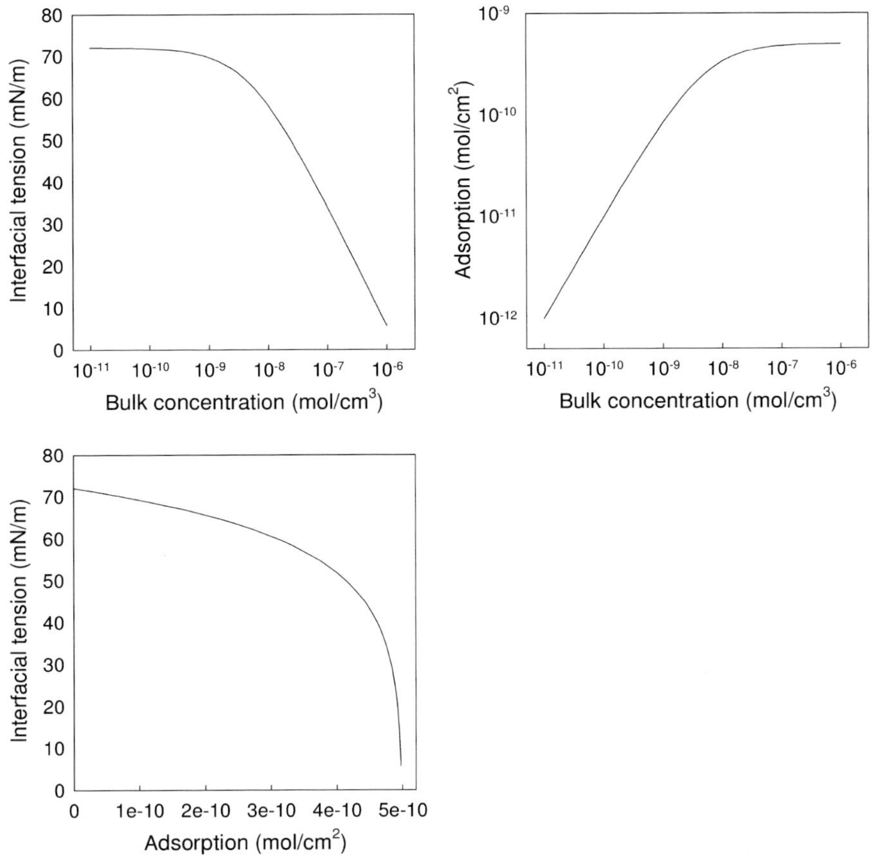

Fig. 10.1 Plot of the equilibrium relations between the quantities defining the status of an adsorption layer in surfactant solutions: γ (interfacial tension), Γ (adsorption) and c (bulk concentration).

which is normally utilized in the form of a relation between γ and the interfacial area, A, considering the number of adsorbed molecules to be fixed.

A perturbation of the adsorption equilibrium arises from a change in the interfacial area, which, in fact, imposes a dilution or enrichment of the adsorption layer. As a consequence, different processes start to contribute to the re-equilibration. The resulting whole relaxation process bringing the system back to equilibrium is called adsorption dynamics or adsorption kinetics.

Different relaxation phenomena are involved in adsorption dynamics [14–16], connected with surfactant transport and rearrangements of the adsorption layer, which have different characteristic times. A sketch of the structure of the liquid air system in the presence of soluble surfactants is shown in Fig. 10.2. The adsorption layer at the interface exchanges molecules with the volume just adjacent to the interface, called the sublayer. When a perturbation of the layer–sub-

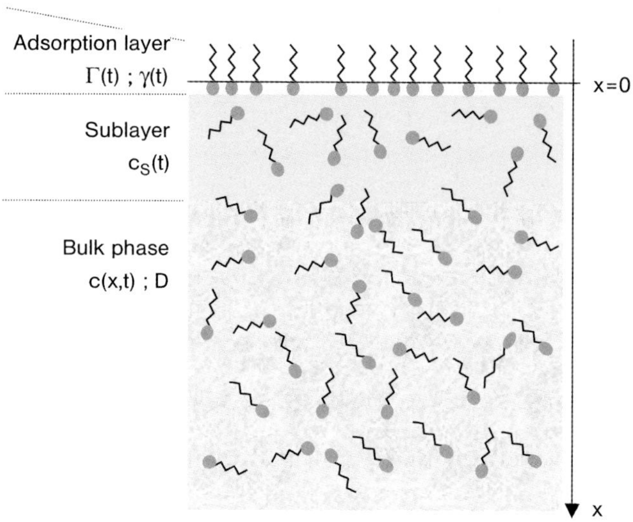

Fig. 10.2 Sketch of the interfacial layer structure, illustrating the transport of surfactant from the bulk to interface.

layer equilibrium occurs, for example owing to dilution of the interfacial layer, the concentration in the sublayer changes because surfactant molecules are transported to/from the interface. This then triggers the diffusion process in the bulk, to re-establish the bulk concentration equilibrium. Diffusion is in most cases the controlling step, since surfactant diffusion proceeds slowly compared with the layer–sublayer exchange.

As will be discussed in more detail later, the surface equation of state holds in some form also out of equilibrium. Hence, during adsorption equilibration, the interfacial tension relaxation follows a path on the equation of state. If, for example, an instantaneous dilation of the surface area is imposed, adsorption is displaced by the same relative amount as the area change. Interfacial tension reaches instantaneously a corresponding value established by the equation of state (see Fig. 10.1), along which progresses the subsequent relaxation of Γ and γ.

Other kinetic processes can occur inside the adsorbed layer depending on the nature of the surface-active species and of the liquid bulk phases. Thus for some non-ionic surfactants molecular surface reorientation or surface aggregation has been observed [17–20], whereas in other cases surface reactions, molecular conformational changes [21] or compressibility of the adsorption layer [22, 23] have been argued in order to explain experimental observations. For liquid–liquid interfaces, the surfactant transfer across the interface, with consequent depletion of one solution, may become the controlling process for adsorption kinetics [24–26].

Even if the phenomena involved in adsorption dynamics occur at the molecular scale, they always correspond to a transient of the interfacial tension that can be monitored at the macroscopic scale. Hence there are different ways to investigate surfactant adsorption, based on the measurement of the equilibrium and dynamic interfacial tension, γ, and the surface dilational viscoelasticity, ε.

First, measurements of the equilibrium interfacial tension versus the surfactant bulk concentration provide the thermodynamic properties of the surfactant system, i.e. the adsorption isotherm. Already using these equilibrium data with appropriate thermodynamic modeling it is possible to make a hypothesis about a possible relaxation process in the monolayer and to obtain values of the molecular characteristics of adsorbed molecules and their interaction, such as surface molecular coverage and interaction molecular force. This approach has been utilized, for example, to describe the equilibrium features in systems showing surfactant reorientation [27] and 2D phase transitions [28, 29].

Measuring the dynamic interfacial tension, $\gamma(t)$, during the aging of a fresh surface obtained, for example, with a large and fast expansion of the surface, is a classical way to investigate adsorption kinetics.

In most cases, these data are well interpreted by the diffusion-controlled model first established by Ward and Tordai [30]. In this approach, the increase in the adsorption is assumed to equal the diffusion flux at the surface:

$$\frac{d\Gamma}{dt} = \Phi_{\text{diff}} \tag{3}$$

which, for the a planar interface located at $x=0$ in a reference with the x-axis directed towards the solution, reads

$$\frac{d\Gamma}{dt} = D\frac{\partial c}{\partial x}\bigg|_{x=0^+} \tag{4}$$

where $c \equiv c(x, t)$ is the concentration profile at the time t and D is the surfactant diffusion coefficient.

This approach provides [30–32] an expression for adsorption, Γ, versus time, t, during the aging of a fresh interface, which, for the particular case of a single adsorbing component from a hemi-infinite phase, reads

$$\Gamma(t) = \sqrt{\frac{D}{\pi}}\left[2c_0\sqrt{t} - \int_0^t \frac{c_s(\tau)}{\sqrt{t-\tau}}\,dt\right] \tag{5}$$

where c_0 is the initial bulk concentration and c_s is the concentration in the subsurface, that is, in the layer just adjacent the interface. Coherently with the assumption of diffusion-controlled adsorption, c_s can be assumed at local equilibrium with Γ. Hence the adsorption isotherm Γ–c can be utilized to express c_s vs. Γ in Eq. (5), which can then be solved. Once $\Gamma(t)$ is known, $\gamma(t)$ can be obtained using the equation of state.

The concepts of diffusion-controlled adsorption have been widely utilized to interpret $\gamma(t)$ data over a wide range of experimental conditions [33–36] and surfactant properties. Some developments account for the solubility in both liquids

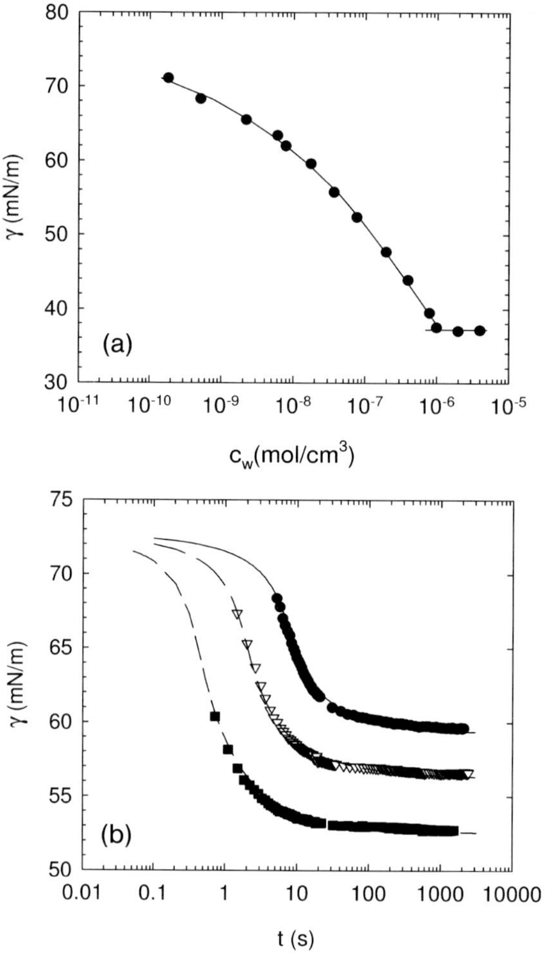

Fig. 10.3 $C_{10}E_8$ aqueous solution: surface isotherm (a) and dynamic surface tension (b), at $T=20\,^{\circ}C$. In (b) the data correspond to different surfactant concentrations: from the top, $2 \times 4 \times 8 \times 10^{-8}$ mol/cm^3.

[37] and for multiple adsorbing species [38]. Exact and numerical solutions of the problem of diffusion-controlled adsorption in spherical geometry have been reported to describe more realistic conditions such as adsorption on a drop surface [39–41] and adsorption from finite volumes [25].

A critical step for the interpretation of adsorption data is the choice of the appropriate equation of state. Thus, apparent deviations from the diffusion-controlled behavior [42] have re-entered this scheme after introducing new isotherms [43] instead of the classical Langmuir or Frumkin isotherms.

Therefore, for an accurate investigation of a surfactant system, it is very important to couple equilibrium and dynamic interfacial tension measurements.

The former versus the surfactant concentration allows the correct thermodynamic model to be established and provides the adsorption isotherm parameter; the latter can be treated with the diffusion-controlled adsorption model in order to establish if the adsorption is diffusion controlled and to evaluate the diffusion coefficients. In Fig. 10.3, data concerning adsorption of $C_{10}E_8$ at a water/air interface [20] are reported as an example. The measured equilibrium surface tension versus concentration (Fig. 10.3 a) is in good agreement with the prediction of a model accounting for two possible adsorption states [27]. The best fit values provide physically sound values of the isotherm parameters. In particular, the values of the two surface molar areas correspond to two orientations of the adsorbed molecules. For the same surfactant, the dynamic interfacial data consequent to adsorption kinetics at a fresh interface are shown in Fig. 10.3 b. The theoretical curve corresponds to diffusion-controlled adsorption where the same isotherm obtained from the equilibrium measurement is utilized. The only free parameter was in this case the diffusion coefficient, the best fit value of which is completely reasonable. From this analysis, one can conclude that whatever dynamic transformation exists in the monolayer, with an exchange of adsorbed molecules between the two states, its characteristic time is short compared with the diffusion characteristic time. Hence such processes can be considered instantaneously at equilibrium during the aging of the surface.

It is then evident that adsorption kinetic investigations do not allow the dynamic aspects of the transformation inside the adsorbed monolayer to be accessed. Investigation of the surface rheology allows this limitation to be overcome by experimental studies of dilational viscoelasticity versus frequency.

10.2.2 Main Surface Dilational Rheology Concepts

Dilational rheology concerns the response of surface tension to isotropic dilations of the interfacial area. This is in general a viscoelastic response since, as described above, time-dependent mechanisms are involved in restoring the equilibrium state.

The dilational stress of an adsorbed layer is the variation of the interfacial tension, γ, from its initial value γ^0 to a generic value at the time t, i.e. $\Delta\gamma=\gamma(t)-\gamma^0$. The surface deformation related to this stress is the expansion (or contraction) of the surface area A. For purely elastic dilational behavior, the surface stress is proportional to the relative variation of the area (or surface deformation), $\alpha=\Delta A/A_0=[A(t)-A_0]/A_0$. The presence of relaxation phenomena gives a viscoelastic character to the rheological response. Hence the dilational surface stress is also proportional to the rate of surface deformation [44]:

$$\Delta\gamma = E_0\alpha + \eta\frac{\mathrm{d}\alpha}{\mathrm{d}t} \tag{6}$$

and the coefficients E_0 and η are by definition dilational surface elasticity and viscosity, respectively.

If only small amplitude perturbations are considered, the system can be assumed to behave linearly and, by using the Fourier formalism, all time-dependent quantities can be expressed as a superposition of harmonic functions. Hence it is sufficient to describe the system behavior in the frequency domain, where Eq. (6) can be written as

$$\varepsilon = \frac{\Delta\gamma}{\Delta A / A_0} \tag{7}$$

where ε is the surface dilational viscoelasticity, which is a frequency-dependent complex quantity, whose real part is the dilational elasticity and the imaginary part is related to the dilational viscosity, i.e.

$$\varepsilon = \varepsilon_r + i\varepsilon_i = E_0 + 2\pi i \nu \eta \tag{8}$$

where $i = \sqrt{-1}$ is the imaginary unit.

Some synonyms are often utilized in the literature for the surface dilational viscoelasticity, such as surface elasticity modulus and dynamic surface elasticity.

By using Fourier analysis concepts, the response of such a system to an arbitrary dilational deformation of the interface can be expressed [45] by the following integral equation:

$$\Delta\gamma(t) = \int_0^t \hat{\varepsilon}(\tau) \ln\left[\frac{A(t-\tau)}{A^0}\right] \mathrm{d}\tau \tag{9}$$

where $\hat{\varepsilon}$ is the inverse Fourier transform of ε.

Hence ε completely characterizes the dilational behavior of a given interface. In fact, Eq. (9) is widely used in the interpretation of experiments involving non-harmonic perturbations of the interfacial area, such as trapezoidal or rectangular variations in stress-relaxation experiments.

On the other hand, the dependence of ε on the frequency of the input perturbation can be effectively utilized to investigate the relaxation processes such as diffusion–adsorption and surface reorganization.

Considering small-amplitude harmonic perturbations, Eq. (7) reads

$$\varepsilon = \frac{\mathrm{d}\gamma}{\mathrm{d}\ln A} \tag{10}$$

Starting from this equation and assuming an equilibrium surface model and a model for the relaxation/transport processes involved, the dependence of ε on the frequency can be expressed in terms of the principal parameters related to the adsorption dynamics and to the equilibrium properties of the system.

Hence for diffusion-controlled adsorption one finds the classical Lucassen-van den Tempel expression [46] for ε:

$$\varepsilon = \frac{\varepsilon_0}{1+(1-i)\xi} = \varepsilon_0 \frac{1+\xi+i\xi}{1+2\xi+2\xi^2} \tag{11}$$

where $\xi = \sqrt{\nu_D/2\nu}$, ν_D is the characteristic frequency of the diffusional exchange,

$$\nu_D = \frac{D}{2\pi}\left(\frac{dc}{d\Gamma}\right)^2_{eq}$$

and ε_0 is the Gibbs elasticity, as defined by Eq. (2). These latter parameters are then completely defined by the surface model.

Equation (11) has been widely utilized by many authors, often, coupled with the Langmuir adsorption model, also for liquid–liquid systems [47] and for surfactant mixtures [48].

More recently, other thermodynamic models, which account for different possible states of the molecules in the adsorption layer, have been introduced [49] in Eq. (11). Moreover, an approach considering the effect of two-dimensional compressibility of the adsorbed layer has been developed, always in the framework of the diffusion-controlled adsorption [22].

In this approach, the molar surface area is assumed to decrease linearly with the surface pressure. This provides a plateau in the Gibbs elasticity at large surface coverage, i.e. large surfactant concentration, which agrees with most experimental observations [50–52].

However, it is worth noting that in the case of insoluble monolayer, which corresponds to a vanishing diffusion flux (i.e. $\xi=0$), Eq. (11) provides ε equal to the Gibbs elasticity. Hence the Lucassen-van den Tempel approach is not able to account for viscoelastic effects arising from internal transformation of the adsorbed layer. Instead, it can be utilized only if any other process is at equilibrium with respect both to the diffusion exchange with the bulk and to the perturbation of the surface.

10.3 Dilational Rheology with Multiple Relaxation Processes

10.3.1 General Approach

To find expressions for ε suitable for the description of viscoelastic effects arising from processes internal to the interface, the dynamic aspects of these processes must be explicitly described and incorporated into the model. Here a general method set up by the authors for extending the Lucassen approach for that purpose is sketched.

First the surface equation of state must be considered, which give an expression for Π as a function of the independent variables X_i, defining the state of tension of the adsorption layer according to the adopted adsorption model:

$$\Pi = \Pi(T_1, T_2, \ldots, T_M; X_1, X_2 \ldots X_N) \tag{12}$$

where T_k is the set of thermodynamic parameters involved in the model. Assuming the X variables to be independent, the above equation of state holds also outside the adsorption equilibrium.

The number, N, of independent variables is equal to the number of relaxation processes. In fact, for each jth relaxation process at equilibrium with respect to the others, an additional relation applies between the Xs:

$$X_j^{eq} = X_j^{eq}(T_1, T_2, \ldots, T_M; X_1, X_2 \ldots X_N) \tag{13}$$

The interface is at adsorption equilibrium when all relaxation processes are at equilibrium. In this case, these relations unequivocally provide a set of equilibrium values, X_j^{eq}, for any given value of Π.

Hence, if one considers a 2D phase transition of the adsorbed molecules in addition to the diffusive exchange with the bulk, two independent variables are needed to describe the state of the system. These may be, for example, the adsorptions corresponding to the two phases. If the phase transition is at equilibrium, a relation between these two variables exists and only one variable can describe the system, e.g. the total adsorption. This latter case corresponds to diffusion-controlled adsorption, where, in fact, always an unequivocal relation between the sublayer concentration and the total adsorption exists.

All the quantities describing the adsorption dynamics and in particular, the dilational viscoelasticity, are then functions of X_i. Hence it is possible to recast Eq. (10) as

$$\varepsilon = -\sum_j \varepsilon_{0X_j} \frac{d \ln X_j}{d \ln A} \tag{14}$$

The second factor in the summation is the frequency-dependent contribution:

$$\varepsilon_{0X_j} = \left(\frac{\partial \Pi}{\partial \ln X_j} \right)_{eq} \tag{15}$$

In fact, since small perturbations are considered, ε_{0X_j} are equilibrium thermodynamic quantities which can be calculated from the equilibrium relationships describing the adsorption layer (surface isotherms, etc.).

To obtain the expression for the frequency-dependent term, it is necessary to introduce explicitly the mass balance at the interface and the kinetic expression of the relaxation process.

Assuming only one adsorbing species, the mass balance at the interface is

$$\frac{d\Gamma}{dt} + \Gamma \frac{d \ln A}{dt} = \Phi_{\mathrm{diff}} \tag{16}$$

where Φ_{diff} is the diffusion flux from the solution. For a planar interface

$$\Phi_{\mathrm{diff}} = D \frac{\partial c}{\partial x}\bigg|_{x=0^+} \tag{17}$$

which is zero when insoluble surfactants are involved.

The kinetic equation of the relaxation process completes the mass balance at the interface. Since small deviations from equilibrium are considered, these relaxation processes can always be suitably described as first-order reaction-like processes, i.e. where the flux depends linearly on the deviations $\Delta X_j = (X_j - X_{j,\mathrm{eq}})$. Hence for the jth process one has

$$\frac{dX_j}{dt} = \Phi_j(K_j, \Delta X_j) \tag{18}$$

where K_j is the rate of the process. The equilibrium values, $X_{j,\mathrm{eq}}$, are those corresponding to the actual (dynamic) value of Π according to the equilibrium relationship (13).

Because the system shows linear behavior and small-amplitude perturbations are considered, a first-order perturbative approach can be utilized to write Eq. (14) in terms of the frequency ν of the area perturbation.

The interfacial area oscillates harmonically around its reference value, i.e.

$$A = A^0 + \tilde{A} e^{2\pi i \nu t} \tag{19}$$

the total adsorption Γ and all the independent variables oscillate with the same frequency:

$$\Gamma = \Gamma^0 + \tilde{\Gamma} e^{2\pi i \nu t} \tag{20}$$

$$X_j = X_j^0 + \tilde{X}_j e^{2\pi i \nu} \tag{21}$$

Assuming these forms for the variables in Eq. (14), the dilational viscoelasticity reads

$$\varepsilon = -\sum_j \varepsilon_{0X_j} \frac{A^0 \tilde{X}_j}{X_j^0 \tilde{A}} \tag{22}$$

A harmonic response is assumed also for all the other dependent variables, $Y_i = Y_i(X_1, \ldots, X_N)$, involved in the process. In particular, it is very important to note that for a given value of the dynamic Π, the driving force for the equilibration of the internal processes in Eq. (18) is the deviation between the current

X_js and the set of equilibrium values $X_{j,\mathrm{eq}}$ corresponding to Π. Since this latter is oscillating owing to oscillation of the X_js, corresponding harmonic variation of the $X_{j,\mathrm{eq}}$ are set:

$$X_{j,\mathrm{eq}} = X^0_{j,\mathrm{eq}} + \tilde{X}_{j,\mathrm{eq}} e^{2\pi i\nu t} \tag{23}$$

Also for the sublayer concentration one has

$$c_\mathrm{s} = c^0_\mathrm{s} + \tilde{c}_\mathrm{s} e^{2\pi i\nu t} \tag{24}$$

By solving the diffusion problem with this harmonic boundary condition [53], the diffusion flux reads

$$\Phi_\mathrm{diff} = -(i+1)\sqrt{\pi D\nu}\,\tilde{c}_\mathrm{s} e^{2\pi i\nu t} \tag{25}$$

All dependent variables, Ys, can be developed at the first order in terms of X_i, which allows one to obtain their amplitudes as

$$\tilde{Y}_i = \sum_j^N \left(\frac{\partial Y_i}{\partial X_j}\right)^0 \tilde{X}_j \tag{26}$$

which allows the whole problem to be expressed in terms of the $\tilde{X}_j$s.

By using Eqs. (19) to (20) and (23) to (26) in the mass balances (16) and (18) and solving the resulting set of equations, $\tilde{X}_j$ is obtained in terms of the frequency, of the thermodynamic parameters T_k and of the characteristics of the reference state. Then Eq. (22) provides the expression of the viscoelasticity versus the frequency ν:

$$\varepsilon = \varepsilon_\mathrm{r}(\Pi^0, T_\mathrm{k}, \nu_j, \nu_\mathrm{D}; \nu) + i\,\varepsilon_i\,(\Pi^0, T_\mathrm{k}, \nu_j, \nu_\mathrm{D}; \nu) \tag{27}$$

with $1 \le j \le N$, $1 \le k \le M$, where N is the number of relaxation processes and M the number of adsorption isotherm parameters. Π^0 is the equilibrium surface pressure of the reference state which, together the T_k parameters, completely defines the reference state; ν_j is the characteristic frequency of the jth relaxation process, which is defined from the jth rate K_j.

The effect of each relaxation process on the surface dilational viscoelasticity is the appearance of a maximum in the imaginary part of the dilational modulus versus frequency and an inflection point in the real part. This aspect has general validity because it comes directly from the assumption of the relaxation phenomena as a superposition of linear kinetic processes.

The characteristic frequency of the process is, by definition, the frequency corresponding to the peak, which depends on K_j.

The procedure sketched above can be applied, sometimes with small adjustments, to calculate the dilational viscoelasticity in different cases, as summarized below.

10.3.2
Adsorbed Layers with Variable Average Molar Area

Adsorbed layers with a variable average molar area, in particular depending on the surface pressure, arise from different physical situations. Variable molar area can be the consequence either of an intrinsic compressibility of the adsorbed monolayer [22] or of the possibility of the adsorbed molecules changing their orientation in the monolayer [17, 18]. In more complex amphiphiles, such as proteins or polymers, folding/unfolding processes may be at the origin of molar area changes.

Many of these situations have been suitably described by assuming the equation of state [54]

$$\Pi = \Pi(\Gamma, \Omega) = -\frac{RT}{\Omega}\ln(1 - \Omega\Gamma) \tag{28}$$

The different thermodynamic models specify then the dependence of Ω on Γ at adsorption equilibrium [55]. In particular, if Ω is a constant, this reduces to the classical Langmuir relationship.

When dilational stresses are imposed on the layer, the adsorbed molecules can have a delay in the achievement of adsorption equilibrium. During this transient stage, the state of the surface is defined by the dynamic values of Γ and Ω.

Correspondingly, the expression for the dilational viscoelasticity is

$$\varepsilon = -\varepsilon_{0\Gamma}\frac{\mathrm{d}\ln\Gamma}{\mathrm{d}\ln A} - \varepsilon_{0\Omega}\frac{\mathrm{d}\ln\Omega}{\mathrm{d}\ln A} \tag{29}$$

with $\varepsilon_{0\Gamma}$ and $\varepsilon_{0\Omega}$ being defined by Eq. (15). The mass balances at the interface is

$$\frac{\mathrm{d}\Gamma}{\mathrm{d}t} + \Gamma\frac{\mathrm{d}\ln A}{\mathrm{d}t} = \Phi_{\mathrm{diff}} \tag{30}$$

which is complemented by an equation describing the re-equilibration of the average molar area:

$$\frac{\mathrm{d}\Omega}{\mathrm{d}t} = -K_c(\Omega - \Omega_{\mathrm{eq}}) \tag{31}$$

Ω_{eq} is the equilibrium average molar area corresponding to the current value of the surface pressure, i.e. related to it through the equilibrium relation, given by the thermodynamic model. Again, when the system is out of equilibrium, a variable Π provides a variable value of Ω_{eq}. In particular for harmonic perturbation, one has

$$\Omega_{\mathrm{eq}} = \Omega_{\mathrm{eq}}^0 + \tilde{\Omega}_{\mathrm{eq}}e^{2\pi i\nu t} \tag{32}$$

where, with a first-order development, one obtains

$$\tilde{\Omega}_{eq} = \left(\frac{\partial \Omega_{eq}}{\partial \Gamma}\right)^0 \tilde{\Gamma} + \left(\frac{\partial \Omega_{eq}}{\partial \Omega}\right)^0 \tilde{\Omega} \tag{33}$$

In the same way, the amplitude of the oscillating sublayer concentration can be developed as

$$\tilde{c}_s = \left(\frac{\partial c_s}{\partial \Gamma}\right)^0 \tilde{\Gamma} + \left(\frac{\partial c_s}{\partial \Omega}\right)^0 \tilde{\Omega} \tag{34}$$

Using the expressions of the oscillating variables and introducing all these relations into Eqs. (30) and (31), one obtains a linear system in θ and Ω. The solution of this system introduced in Eq. (29) provides, after some rearrangements,

$$\varepsilon = \frac{\varepsilon_{0\Gamma}(1 - i\lambda) + \varepsilon_{0\Omega} G i\lambda}{\dfrac{1 + \xi - i\xi - G'}{1 - G'} - (1 + \xi - i\xi) i\lambda} \tag{35}$$

where ξ is the same as defined by Eq. (11), $\lambda = \nu_c/\nu$ and ν_c is the characteristic frequency associated with the average area relaxation:

$$\nu_c = \frac{K_c}{2\pi}\left(1 - \frac{\partial \Omega_{eq}}{\partial \Omega}\right) \tag{36}$$

G and G' are quantities depending on the thermodynamic characteristics of the system and on the reference state:

$$G = -\frac{\Gamma^0}{\Omega^0} \frac{\partial \Omega_{eq}}{\partial \Gamma}\left(1 - \frac{\partial \Omega_{eq}}{\partial \Omega}\right)^{-1} \tag{37}$$

$$G' = \frac{\Omega^0}{\Gamma^0} G \frac{\partial c_s}{\partial \Omega}\left(\frac{\partial c_s}{\partial \Gamma}\right)^{-1} \tag{38}$$

In the limiting case of fast equilibration of the relaxation process with respect to diffusion and surface perturbation, the terms in λ are dominant and Eq. (35) reduces to

$$\varepsilon_D = \frac{\varepsilon_{0\Gamma} - G\varepsilon_{0\Omega}}{1 + (1 - i)\xi} \tag{39}$$

which is equal to the Lucassen-van der Tempel equation, Eq. (11). In fact,

$$\varepsilon_{0\Gamma} - G\varepsilon_{0\Omega} = \frac{\partial \Pi}{\partial \ln \Gamma} + \frac{d \ln \Omega}{d \ln \Gamma} \frac{\partial \Pi}{\partial \ln \Omega} = \frac{d\Pi}{d \ln \Gamma} \tag{40}$$

is the Gibbs elasticity.

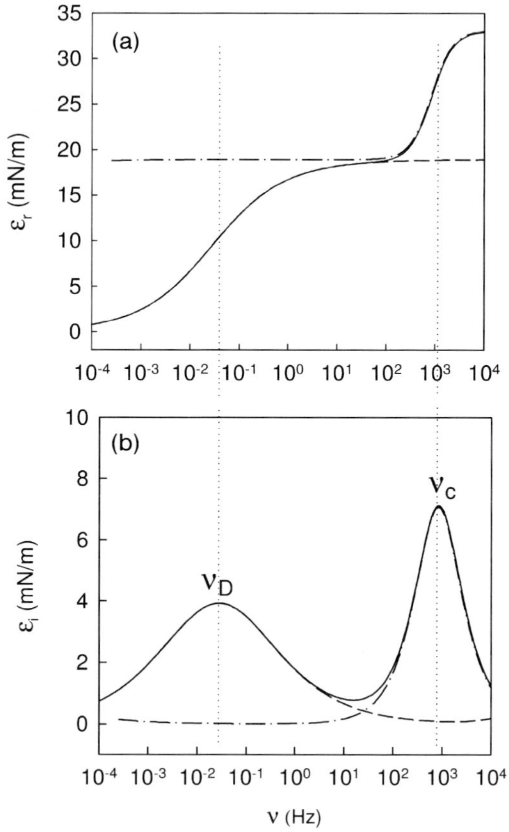

Fig. 10.4 Real (a) and imaginary (b) parts of ε: reorientation + diffusion relaxation (solid line, Eq. 35), diffusion relaxation (dashed line, Lucassen-van der Tempel, Eq. 11) and reorientation relaxation (dot-dashed line, limit of insoluble layer, Eq. 63). Calculated with the two-state isotherm parameters typical for $C_{10}E_4$ aqueous solutions $\Omega_1 = 6.7\times10^9$ mol cm^{-2}, $\Omega_2 = 2.6\times10^9$ mol cm^{-2}, $b_2 = 1.35\times10^8$ cm^3 mol^{-1}, $\beta = 7.73$ and $D = 5\times10^{-6}$ cm^2 s^{-1}, $k_c = 200$ Hz.

A second limiting case (ε_c), corresponding to insoluble layers, is obtained for fast equilibration of diffusion in respect of the internal process and it will be treated in more detail in the following sections.

In Fig. 10.4, an example of the viscoelasticity versus the frequency as derived by Eq. (35) is reported together with the limiting cases concerning diffusional exchange (ε_D) alone and in the insoluble layer (ε_c). Each process originates a maximum in the imaginary part of the viscoelasticity and a corresponding inflection point in the real part.

The characteristic frequencies ν_D and ν_c are those of the maxima of the curves corresponding to ε_D and ε_c respectively.

Under the conditions of Fig. 10.4, where the characteristic frequencies of the processes differ by some orders of magnitude, the plot of the sum $\varepsilon_D + \varepsilon_c$ is practically coincident with that of ε. Hence ε is in practice the simple superposition of the two processes, which are nearly not interacting. This circumstance can be useful for practical purposes, since the expressions of ε_D and ε_c does not depend on G'. Figure 10.5 shows a situation where the characteristic frequencies are instead much closer. From the plot, it is apparent that in this case ε

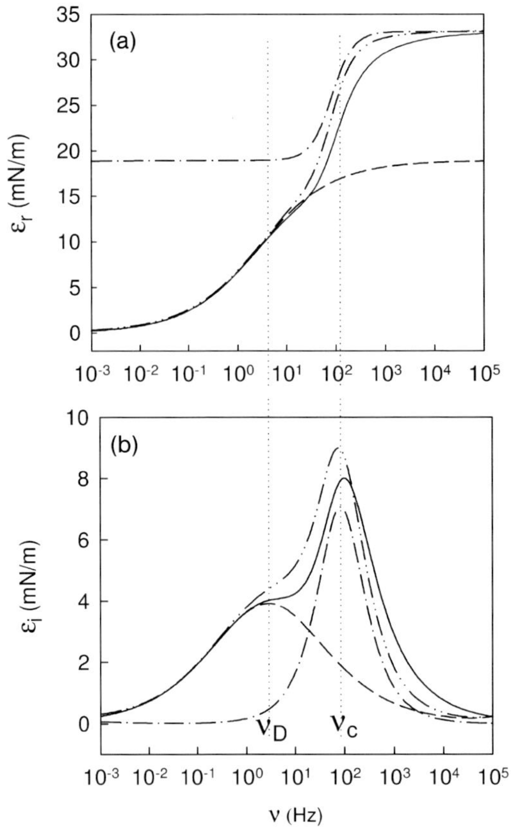

Fig. 10.5 The same as Fig. 10.4, with D and k_c values arranged in order to obtain closer values of ν_D and ν_c. The dash-dotted line represents the superposition of the diffusion relaxation curve plus the reorientation relaxation curve (see text).

cannot be straightforwardly considered as the superposition of the two processes. It is interesting that the position of the maximum corresponding to the internal process is slightly different from ν_c, which indicates non-linear coupling of the two processes. The coupling is, however, rather weak, since the deviation between ε and $\varepsilon_D + \varepsilon_c$ is always limited.

Equation (35) was obtained only assuming that the variation of Ω is in phase with the variation of Π, without assuming a particular mechanism for the dynamic aspect of the average molar area. The results obtained are then applicable to all cases where the system presents a relaxation process driving back the average molar area to an equilibrium value determined by the surface pressure. For this reason, this expression of ε is quite general and a more specific modeling of the dependence of the molar area on the state of the surface enters the expression of ν_c, ν_D, $\varepsilon_{0\Omega}$, $\varepsilon_{0\Gamma}$, G and G'.

For example, some authors have proposed an adsorption model accounting for an intrinsic compressibility of the adsorbed layer by stating a linear dependence of the molar area on the surface pressure according to a two-dimensional compressibility κ:

$$\Omega = \Omega_{\text{void}}(1 - \kappa\Pi) \tag{41}$$

where Ω_{void} is the molar area at vanishing surface pressure. The model has been applied so far to describe the surface layers of simple surfactants without considering any time dependence for the compression process. However when complex amphiphiles are considered, such as polymers or proteins, a relaxation of the occupation area could occur, which can be accounted for by the above simple linear relation with the dynamic Π.

In the framework of that model, the Langmuir equation of state, Eq. (28), and the Γ–c isotherm:

$$c_s = a\frac{\Omega\Gamma}{1 - \Omega\Gamma} \tag{42}$$

are utilized together with Eq. (41). Thus one trivially calculates

$$\varepsilon_{0\Omega} = \Pi^0\left[\frac{RT}{\Omega^0\Pi^0}\frac{1 - \exp\left(-\frac{\Omega^0\Pi^0}{RT}\right)}{\exp\left(-\frac{\Omega^0\Pi^0}{RT}\right)} - 1\right] \tag{43}$$

$$\varepsilon_{0\Gamma} = \frac{RT}{\Omega^0}\frac{1 - \exp\left(-\frac{\Pi^0\Omega^0}{RT}\right)}{\exp\left(-\frac{\Pi^0\Omega^0}{RT}\right)} \tag{44}$$

$$\nu_c = \frac{K_c}{2\pi}\left(1 + \frac{\kappa\Pi^0}{1 - \kappa\Pi^0}\right) \tag{45}$$

$$G = \frac{1 - \exp\left(-\frac{\Pi^0\Omega^0}{RT}\right)}{1 - \left(1 + \frac{\Pi^0\Omega^0}{RT}\right)\exp\left(-\frac{\Pi^0\Omega^0}{RT}\right)} \tag{46}$$

$$G' = \frac{\Omega^{02}}{1 - \exp\left(-\frac{\Pi^0\Omega^0}{RT}\right)}G \tag{47}$$

where Ω^0 can be expressed in terms of Π^0 by using Eq. (41).

Another interesting class of adsorption models that can be described by means of Eq. (35) are those where adsorbed molecules are distributed in a certain number of states [27], so that $\Gamma = \sum_i \Gamma_i$. The molecules in each state are characterized by a constant molar area Ω_i and a given surface activity. These states can then refer, for example, to different molecular orientations of the adsorbed surfactants or to different steric arrangements of the molecule, as may occur, for example, for adsorbing proteins or polymers [21, 56].

In the framework of these models, it is possible to define a variable molar area as the average on all possible surface states:

$$\Omega = \sum_i \Omega_i \Gamma_i \Big/ \sum_i \Gamma_i \tag{48}$$

Thus in dynamic conditions, Π is unequivocally defined by the set of Γ_i values according to the equation of state (Eq. 28), to that definition of Ω. It is worth examining in more detail the model where only two adsorption states are considered. The latter has been widely utilized to describe the adsorption kinetics and the rheological properties of different polyoxyethylated non-ionic surfactants at both liquid–air and liquid–liquid interfaces [14, 19, 20, 57, 58].

Considering only two states, one has by definition

$$\Omega = \frac{g\Omega_1 + \Omega_2}{1+g} \tag{49}$$

where g is the ratio between the adsorptions in the two states, Γ_1/Γ_2.

At equilibrium, g and Π are linked by [27]

$$g = \beta \exp\left[-\frac{\Pi(\Omega_1 - \Omega_2)}{RT}\right] \tag{50}$$

which allows Ω_{eq} to be expressed versus Π.

The Γ–c isotherm is

$$c_s = \frac{\Omega\Gamma}{b[\beta(1-\Omega\Gamma)^{\Omega_1/\Omega} + (1-\Omega\Gamma)^{\Omega_2/\Omega}]} \tag{51}$$

where b and β are isotherm parameters linked to surface activity of molecules adsorbed in the two states.

With these relationships, one obtains

$$\nu_c = \frac{K_c}{2\pi}\left\{1 - \frac{g^0(\Omega_1 - \Omega^0)(\Omega_1 - \Omega_2)}{(1+g^0)\Omega^{02}}\left[1 + \frac{\Pi^0\Omega^0}{RT} - \exp\left(\frac{\Pi^0\Omega^0}{RT}\right)\right]\right\} \tag{52}$$

and

$$G = \frac{1 - \exp\left(-\frac{\Pi^0\Omega^0}{RT}\right)}{1 - \left[1 + \frac{\Pi^0\Omega^0}{RT} - \frac{(1+g^0)\Omega^{02}}{g^0(\Omega_1 - \Omega^0)(\Omega_1 - \Omega_2)}\right]\exp\left(-\frac{\Pi^0\Omega^0}{RT}\right)} \tag{53}$$

where Ω^0 and g^0 are linked to Π^0 by the equilibrium relationships (28), (49) and (50). $\varepsilon_{0\Gamma}$ and $\varepsilon_{0\Omega}$ are the same as in Eqs. (43) and (44), with Ω^0 given by Eq. (49).

The expression for G' for the two states is too long to be reported; however, it can be trivially calculated according to its definition.

The curves in Figs. 10.4 and 10.5 were derived in the framework of this model assuming values of the isotherm parameters typical of surfactants such as C_iE_j at the liquid–air interface.

An approach similar to that described here has been applied to obtain kinetic information from ε measurements on that class of surfactants adsorbed at liquid–air and liquid–liquid interfaces. The results have been reported and widely discussed elsewhere [57, 58] and lead to values of the characteristic frequency for the layer re-orientation process in the millisecond range.

10.3.3 Interfacial Phase Transition with Aggregation

Phase transition with surfactant aggregation is a common phenomenon in the adsorption of insoluble surfactants. More recently, this phenomenon has been observed for sparingly soluble surfactants, such as long-chain alcohols [28, 29].

Direct observations of the interface of *n*-dodecanol solutions by BAM microscopy show that above a critical surface pressure, Π_c, 2D dendritic aggregates form in the monolayer. With increasing surface pressure, the aggregates grow and eventually merge and form a continuous monolayer. Similar behavior is also observed for mixed SDS–*n*-dodecanol layers [59]. Π_c is a function of temperature and for *n*-dodecanol the phase transition disappears above 20 °C. To model this system, one can consider that above Π_c, corresponding to a critical adsorption Γ_c, two phases coexist in the adsorption layer. One of the two phases is constituted by monomer surfactant or small aggregates. The other phase is instead that of the large 2D surfactant aggregates. The monomeric and aggregate surface phases are characterized by the adsorptions Γ_m and Γ_m^* respectively.

To describe the equilibrium properties of the system [28], a Langmuir-like model has been suggested. Below the critical pressure the equilibrium properties are suitably described according to the ordinary Langmuir model, whereas generalized equations were instead derived [28] to describe the zone above Π_c:

$$\Pi = -\frac{RT}{\omega_m}\frac{\Gamma_m}{\Gamma}\ln(1-\omega_m\Gamma) \tag{54}$$

$$bc_s = \frac{\omega_m\Gamma}{[1-\omega_m\Gamma]^{\theta}} \tag{55}$$

where $\theta = \Gamma_m/\Gamma$ and $\Gamma = \Gamma_m + \Gamma_m^*$.

To close this latter set of equations, a relation for Γ_m above Π_c is needed. One could simply state $\Gamma_m = \Gamma_c$, but the comparison of this model with the isotherm of *n*-dodecanol at the water interface [29] suggests packing of the molecules in the condensed phase. Thus a packing parameter $\delta < 1$ was introduced, such that the molar area of the molecules in the aggregates is $\omega_m^* = (1 - \delta)\omega_m$, where ω_m is the molar area of the surfactant in the monomeric phase. The packing makes the distribution of surfactant between the two phases shift towards the aggregated phase on increasing the surface pressure, which is modeled according to

$$\Gamma_m = \Gamma_c \exp\left(-\frac{\Pi - \Pi_c}{RT}\omega_m\delta\right) \tag{56}$$

To find an expression for the surface viscoelasticity, a scheme similar to that utilized in the previous section can be adopted.

After assuming Γ_m and Γ_m^* to be independent variables to describe the system out of equilibrium, Eq. (10) can be written as

$$\varepsilon = -\varepsilon_{0m}\frac{d\ln\Gamma_m}{d\ln A} - \varepsilon_{0m}^*\frac{d\ln\Gamma_m^*}{d\ln A} \tag{57}$$

where ε_{0m} and ε_{0m}^* are defined according to Eq. (15).

Assuming aggregate molecules to be insoluble, the mass balance at the surface is given by the set

$$\frac{1}{A}\frac{d(\Gamma_m A)}{dt} = \Phi_D - \Phi_{agg} \tag{58}$$

$$\frac{1}{A}\frac{d(\Gamma_m^* A)}{dt} = \Phi_{agg} \tag{59}$$

where Φ_D is the diffusion flux of monomer. According to its definition, the diffusion flux is assumed to be incident perpendicular to the interface, which may conflict with the presence of insoluble aggregates. The assumption holds provided that the size of the aggregates is negligible with respect to the diffusion length, $(D\tau_D)^{1/2}$ [60].

The aggregation flux, Φ_{agg}, can again be described assuming the aggregation process to follow first-order reaction-like kinetics [61]:

$$\Phi_{agg} = k_{agg}(\Gamma_m - \Gamma_{m,eq}) \tag{60}$$

where $\Gamma_{m,eq}$ is the equilibrium adsorption as provided by the surface equation of state; k_{agg} is the aggregation rate constant, from which the characteristic aggregation frequency can be defined as

$$\nu_{agg} = \frac{k_{agg}}{2\pi}(s * +1 - s) \tag{61}$$

Adopting the usual perturbation scheme and solving the mass balance equations to obtain the adsorption amplitudes, from Eq. (57) one obtains

$$\varepsilon = \frac{(\varepsilon_{0\mathrm{m}}/q)\left[(1-i)w*\xi + i(1+q)\dfrac{s^*}{s*+1-s}\lambda - q\right] - \varepsilon_{0\mathrm{m}}^*\left[(1-i)w\xi - i(1+q)\dfrac{1-s}{s*+1-s}\lambda + 1\right]}{(i+1)\left[\dfrac{s*w+(1-s)w*}{s*+1-s}\right]\xi\lambda + (i-1)w\xi + i\lambda - 1} \qquad (62)$$

where q is the ratio between the adsorptions in the reference state, $q = (\Gamma_{\mathrm{m}}^0/\Gamma_{\mathrm{m}}^{*0})$, and $s = (\partial\Gamma_{\mathrm{m,eq}}/\partial\Gamma_{\mathrm{m}})^0$, $s^* = (\partial\Gamma_{\mathrm{m,eq}}/\partial\Gamma_{\mathrm{m}}^*)^0$, $w = (\partial c_{\mathrm{s}}/\partial\Gamma_{\mathrm{m}})^0/(\mathrm{d}c_{\mathrm{s}}/\mathrm{d}\Gamma)^0$, $w^* = (\partial c_{\mathrm{s}}/\partial\Gamma_{\mathrm{m}}^*)^0/(\mathrm{d}c_{\mathrm{s}}/\mathrm{d}\Gamma)^0$. $\lambda = \dfrac{\nu_{\mathrm{agg}}}{\nu}$ and ξ are the dimensionless

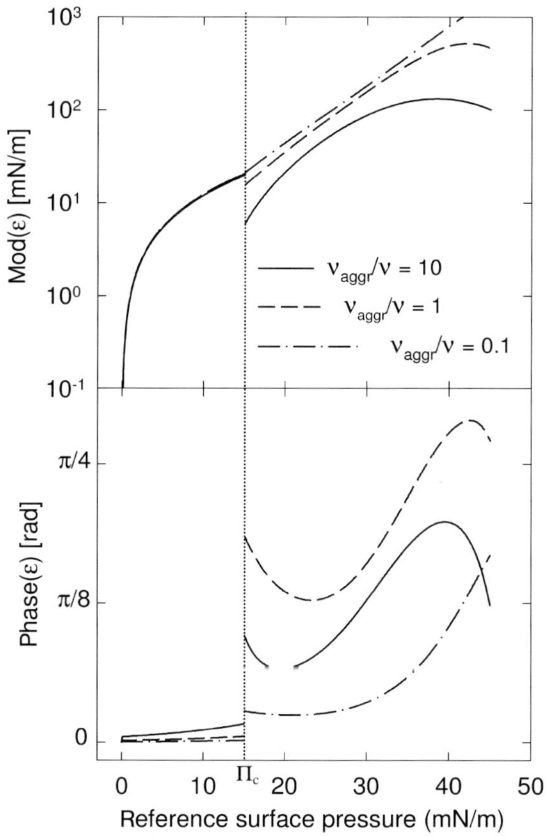

Fig. 10.6 Dilational viscoelasticity in the presence of diffusion plus surface aggregation, as a function of the surface pressure of the reference state, for $\nu_{\mathrm{agg}} = 0.1$. Calculated from Eq. (62) above Π_{c} and by the Gibbs elasticity of the Langmuir model below Π_{c}. Isotherm parameter values are those typical for *n*-dodecanol (see the text for references). The discontinuity is located at the critical surface pressure, Π_{c}.

frequencies of the aggregation and diffusional exchange processes, the latter being defined in the usual way.

The specificity of the model is given by Eqs. (58) to (60). Therefore, it is worth noting that the results apply to any soluble adsorption layer where a first-order rate process at the surface produces an insoluble species. The application is then specified by the expressions of q, s, s^*, w and w^*.

The overall features of $\varepsilon(\nu)$, as predicted by Eq. (62), are similar to those reported in Figures 10.4 and 10.5 for layers with variable adsorption molar area, with two peaks in the imaginary part located at ν_D and ν_{agg}.

Figure 10.6 shows the corresponding dependence of ε on the surface pressure at a given frequency; ε is described by Eq. (62) above Π_c, whereas below Π_c the Lucassen elasticity is utilized with the Langmuir isotherm. The main characteristic is the discontinuity at Π_c, with a significant fall of the ε modulus and phase, which is confirmed by experimental observations [62].

10.3.4
Insoluble Surfactant Layers

Surface rheology of insoluble monolayers is an important topic because of the large number of practical applications, especially concerning multicomponent adsorbed layers [63, 64] or surface phase transitions [65].

In the absence of relaxation processes, an insoluble layer has a purely elastic behavior and the dilational modulus coincides with the Gibbs elasticity which is provided by the thermodynamics. Many experimental observations show, however, that an insoluble monolayer can present slow relaxation processes which introduces a surface viscosity and which are originated by phase transitions, layer reorganization or steric rearrangements of the adsorbed amphiphiles, etc.

Theoretical approaches to describe the surface rheology of these systems have been developed [64] and applied specifically to systems undergoing phase transitions or where surface chemical reactions occur.

The expressions of ε for soluble surfactants derived in the previous sections can be applied to insoluble monolayers in the limiting case of vanishing diffusion flux. This can be obtained by assuming the characteristic frequency for diffusional exchange to be vanishing, that is, in the limit $\xi \to 0$.

For example, in the case of insoluble surfactants with a variable molar area, Eq. (35) reduces to

$$\varepsilon = \varepsilon_{0\Gamma} + \varepsilon_{0\Omega} G \frac{i\lambda}{(1 - i\lambda)} \tag{63}$$

This equation coincides with the general expression found by other authors [66] for insoluble monolayers with a given number of surface relaxation processes.

Figure 10.5 shows the corresponding plot of the real and imaginary parts of ε versus the frequency. Since there is only one relaxation process, a single maximum exists in the imaginary part, located at $\nu = \nu_c$. Indeed, owing to the definition of ν_c, the imaginary part of ε in Eq. (63) has a maximum at $\lambda=1$. At low frequency ($\lambda \to +\infty$), a finite purely elastic response is obtained; ε is in that case given by the Gibbs elasticity, $\varepsilon_0 = \varepsilon_{0\Gamma} - \varepsilon_{0\Omega} G$, whereas it was vanishing for the process of diffusional exchange alone (see Eq. 11). At high frequency ($\lambda \to 0$), the response is again purely elastic, with elasticity equal to $\varepsilon_{0\Gamma}$. Note that G is always positive, hence the real part of ε is always increasing with ν.

For an insoluble layer undergoing an aggregation process, an expression for ε can be derived for $\xi \to 0$ from Eq. (62):

$$\varepsilon = \frac{\varepsilon_{0\mathrm{m}} + \varepsilon_{0\mathrm{m}}^{*} - (1+q)\left[\dfrac{(\varepsilon_{0\mathrm{m}}/q)s^{*} + \varepsilon_{0\mathrm{m}}^{*}(1-s)}{s^{*}+1-s}\right] i\lambda}{1 - i\lambda} \tag{64}$$

Also this equation agrees with the theoretical approach of Ref. [66].

Accordingly, ε shows a purely elastic behavior at high ($\lambda \to +\infty$) and low ($\lambda \to 0$) frequency, with values

$$\varepsilon = \varepsilon_{0\mathrm{m}} + \varepsilon_{0\mathrm{m}}^{*} \tag{65}$$

$$\varepsilon = (1+q)\left[\frac{(\varepsilon_{0\mathrm{m}}/q)s^{*} + \varepsilon_{0\mathrm{m}}^{*}(1-s)}{s^{*}+1-s}\right] \tag{66}$$

respectively. The latter plays the role of the Gibbs elasticity for the system.

Data for the dilational elasticity of *n*-dodecanol at the water–air interface at frequencies below 0.2 Hz have been interpreted in this context [67], as shown in Fig. 10.7. In fact, at these frequencies and under the adopted experimental conditions, *n*-dodecanol layers behave as insoluble, since $\nu_D \ll \nu$, so that above Π_c, ε is described by Eq. (64).

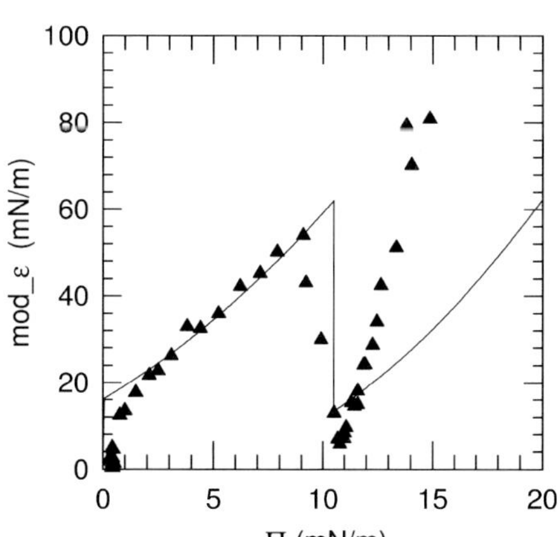

Fig. 10.7 Measured viscoelasticity module of *n*-dodecanol monolayer (c=12 μM, T=10 °C) at ν=0.0625 Hz and theoretical prediction from the model (Eqs. 62 and 69) assuming a surface aggregation process.

With this aim, a generalized Volmer equation of state was utilized to describe adsorption above the critical pressure:

$$\Pi + \Pi_{\mathrm{coh}} = \frac{RT}{n} \frac{\beta^* \Gamma_{\mathrm{m}}}{1 - \Omega_{\mathrm{m}}[(1-\delta)\Gamma + \delta\beta^* \Gamma_{\mathrm{m}}]} \tag{67}$$

where

$$\beta^* = 1 - \omega_{\mathrm{m}}(1-\delta)(\Gamma - \Gamma_{\mathrm{m}})$$

and $\Gamma = \Gamma_{\mathrm{m}} + \Gamma_{\mathrm{m}}^*$ is the total adsorption, Γ_{m} being given by Eq. (56). Π_{coh} is the so-called cohesion pressure (or Volmer constant) arising from surfactant interaction and n is an aggregation number, which accounts for the possibility of the surfactant forming small aggregates (dimer–trimer) also below the critical pressure. Below Π_{c}, in fact, a correspondingly modified version of the usual Volmer surface equation of state has been adopted:

$$\Pi + \Pi_{\mathrm{coh}} = \frac{RT\Gamma}{n(1 - \omega_{\mathrm{m}}\Gamma)} \tag{68}$$

Since the layer is insoluble, below Π_{c}, ε coincides with the corresponding Gibbs elasticity:

$$\varepsilon_0 = \frac{RT(\Pi + \Pi_{\mathrm{coh}}) + n\omega_{\mathrm{m}}(\Pi + \Pi_{\mathrm{coh}})^2}{RT} \tag{69}$$

These models have been proven to be useful for the description of the equilibrium properties of *n*-dodecanol layers at the water–air interface.

The curves in Fig. 10.7 were calculated according to Eqs. (64) and (69), with parameters in agreement with those derived from equilibrium investigations [28] and molecular quantum mechanics calculations [68]. The match in the subcritical pressure range is nearly perfect confirming the suitability of the equilibrium model adopted. Above the critical pressure, the agreement is just qualitative since the module of ε increases much rapidly than predicted by Eq. (64). As discussed in Ref. [67], this is most likely due to a strong interaction between the aggregate branches, which saturate the interface as the pressure increases.

Moreover, under these conditions the surfactant interfacial flow/diffusion is hindered, which causes a larger elastic response. Neither effect was considered in the model. Nevertheless, the analysis of the region just above the critical pressure, where aggregates are still relatively sparse, allows the estimation of kinetic parameters, since the initial value of ε after the transition depends on ν_{c}. The comparison with the experimental data provides ν_{c} of the order of 15 Hz.

10.3.5 Interfacial Reactions in Insoluble Monolayers

The composition of the interfacial layer under the dynamic conditions set by a chemical reaction can play an important role in many natural and artificial processes related to multiphasic systems. The variation of molecular structure of the adsorbed species can in fact deeply influence the response to external forces of the adsorbed layer.

Some experimental studies and theoretical approaches concerning dynamic interfacial tension and adsorption kinetics in the presence of surface reactions have been reported [69–72], which provide some examples about the role played by chemical reactions in interfacial phenomena.

As an example (see Fig. 10.8), the reactions between a hexane–soluble acid, palmitic acid, and a strong (NaOH) or a weak (NH_3) base have been investigated by interfacial tensiometry. This is an acid–base interfacial reaction with the formation of a new chemical product with different interfacial properties. With this aim, a drop of a hexane solution of palmitic acid was brought into contact with pure water until surface equilibrium was reached. After the equilibration, corresponding to the time $t=0$ in Fig. 10.8, a small amount of the base was introduced into the water phase. The dynamic interfacial tension was measured during the whole process until the re-equilibration.

For the first 200 s, the interfacial tension was constant and equal to the value of the hexane solution of palmitic acid in contact with pure water. After this time, during which the diffusion of the base reactant occurred, a decrease in γ was observed, due to the formation of palmitate. Desorption of palmitate and re-equilibration processes with the formation of a diffuse double layer in water produced the following interfacial tension response. When the added reactant was a weak base, a lower value of the equilibrium interfacial tension was achieved since the palmitate was only partially dissociated.

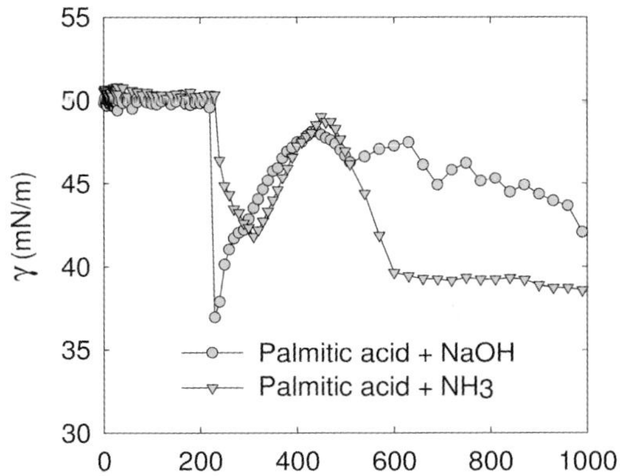

Fig. 10.8 Dynamic interfacial tension in the presence of the chemical reaction of palmitic acid with a strong (NaOH) or a weak (NH_3) base at the water–air interface (see text for details).

Despite its qualitative character, this experiment shows that the chemical modifications of an adsorbed layer caused by reactions can be investigated by tensiometric methods and, as for other relaxation processes, rheological studies could be an effective tool to access the kinetic aspect of surface reactions.

Nevertheless, despite the high potential of this kind of studies, there is a lack of experimental results on this topic and only a few theoretical models concern the relationship between surface viscoelasticity and surface reactions [66, 73]. In particular in this latter work it is rigorously shown as the chemical reactions occurring in the adsorbed layer introduce a dilational viscous behavior.

Here a simplified case is considered as a further example to show the applicability of the proposed method for modeling the $\varepsilon(\nu)$ dependence for insoluble surfactant systems.

With this aim, an interfacial reversible interfacial chemical reaction is considered involving N species:

$$a_1 A_1 + \ldots + a_M A_M \rightleftarrows a_{M+1} A_{M+1} + \ldots + a_N A_N \tag{70}$$

where a_j are the stoichiometric coefficients of reactants. By convention, $a_j < 0$ for $j = 1,\ M$ and $a_j > 0$ for $j = M+1,\ N$.

At equilibrium, the thermodynamic relation between Π and the distribution of the adsorption of all components is

$$\prod_{i=1}^{N} (\Omega_i \Gamma_i)^{a_i} = K^s_{\text{eq}} \exp(-\Pi \Omega_\Sigma / RT) \tag{71}$$

where Ω_i is the molar area of the ith reactant,

$$\Omega_\Sigma = \sum_{i=1}^{M} a_i \Omega_i \tag{72}$$

and K^s_{eq} is the reaction equilibrium constant [74].

After a dilational perturbation, the distribution of the adsorptions of the reactants is shifted from the equilibrium distribution and a relaxation process occurs described by the rate of the chemical reaction:

$$\Phi_{\text{R}} = k^+ \prod_{i=1}^{M} \Gamma_i^{|a_i|} - k^- \prod_{i=M+1}^{N} \Gamma_i^{a_i} \tag{73}$$

Also in this case, Eq. (28) is assumed to be an equation of state, valid also outside equilibrium. Γ and Ω are the independent variables describing the process.

According to Eq. (22), the dilational viscoelasticity is then

$$\varepsilon = -\varepsilon_{0\Gamma} \frac{\text{d} \ln \Gamma}{\text{d} \ln A} - \varepsilon_{0\Omega} \frac{\text{d} \ln \Omega}{\text{d} \ln A} \tag{74}$$

The mass balance at the surface is defined by the following three equations:

$$\frac{d\Gamma_i}{dt} + \Gamma_i \frac{d \ln A}{dt} = a_i \Phi_R \tag{75}$$

$$\frac{d\Omega}{dt} = -\frac{\Omega}{\Gamma}\frac{d\Gamma}{dt} + \frac{1}{\Gamma}\sum_i \Omega_i \frac{d\Gamma_i}{dt} \tag{76}$$

$$\frac{d\Gamma}{dt} + \Gamma \frac{d \ln A}{dt} = a_\Sigma \Phi_R \tag{77}$$

where $a_\Sigma = \sum_i a_i$. Combining Eqs. (75) and (76), one obtains

$$\Gamma \frac{d\Omega}{dt} = -\Phi_R (a_\Sigma \Omega - \Omega_\Sigma) \tag{78}$$

For small deviations from equilibrium, Φ_R can be developed at the first order on the adsorption amplitudes:

$$\Phi_R \cong -k_R \sum_{k=1}^{N} a_k \frac{\Delta\Gamma_k}{\Gamma_k^0} \tag{79}$$

where

$$k_R \equiv k^+ \prod_{i=1}^{M} \Gamma_{ieq}^{|a_i|} = k^- \prod_{i=M+1}^{N} \Gamma_{ieq}^{a_i} \tag{80}$$

By developing the Γ_k at first order in the variables Γ and Ω, Eq. (79) reads

$$\Phi_R = \Phi'_{R\Gamma} \Delta\Gamma + \Phi'_{R\Omega} \Delta\Omega \tag{81}$$

where

$$\Phi'_{R\Gamma} = \frac{d\Phi_R}{d\partial\Gamma} = -k_R \sum_{k=1}^{N} \frac{a_k}{\Gamma_k^0} \tag{82}$$

$$\Phi'_{R\Omega} = \frac{d\Phi_R}{d\Omega} = k_R \sum_{k=1}^{N} \frac{a_k \Gamma^0}{\Gamma_k^0 (\Omega^0 - \Omega_k)} \tag{83}$$

With this expression of Φ_R, introducing oscillating dependences for Γ, Ω and A in the mass balances (77) and (78) and in the dilational viscoelasticity (Eq. 74), and following the perturbative scheme introduced above, one eventually find the following expression for ε:

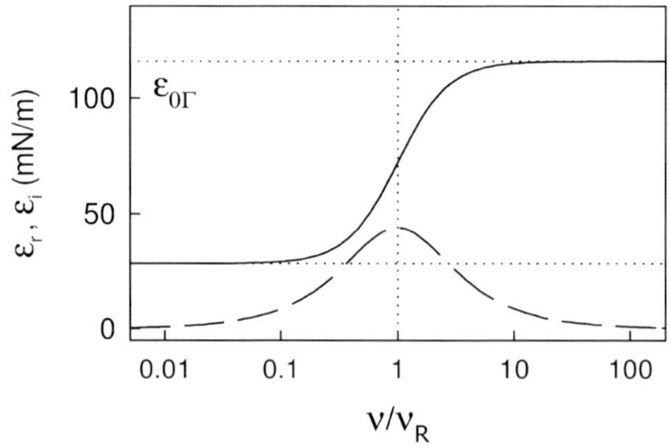

Fig. 10.9 Example of real (solid line) and imaginary (dashed line) parts of ε in the presence of an interfacial reaction, as calculated from Eq. (84).

$$\varepsilon = \frac{\varepsilon_{0\Gamma}}{1+\lambda_R^2} + \left(\frac{\varepsilon_{0\Gamma} - \varepsilon_{0\Omega} G_{R1}}{1 - G_{R2}}\right) \frac{\lambda_R^2}{1+\lambda_R^2} - \left(\frac{\varepsilon_{0\Gamma} G_{R2} - \varepsilon_{0\Omega} G_{R1}}{1 - G_{R2}}\right) \frac{i\lambda_R}{1+\lambda_R^2} \tag{84}$$

where $\lambda = \nu_R/\nu$ and ν_R is the characteristic relaxation frequency of the surface reaction, defined as

$$\nu_R = \frac{(a_\Sigma \Omega^0 - \Omega_\Sigma)\Phi'_{R\Omega} - a_\Sigma \Gamma^0 \Phi'_{R\Gamma}}{2\pi\Gamma^0} \tag{85}$$

This parameter is proportional to the rate of the chemical reaction and contains the kinetic aspect of the relaxation process. G_{R1} and G_{R2} are instead quantities depending on the thermodynamic characteristics of the system and on the parameters defining the reference state for the perturbation:

$$G_{R1} = \frac{\Gamma^0 \Phi'_{R\Gamma}}{\Omega^0 \Phi'_{R\Omega}} \tag{86}$$

$$G_{R2} = \frac{\Phi'_{R\Gamma}}{\Phi'_{R\Omega}} \frac{a_\Sigma \Gamma^0}{(a_\Sigma \Omega^0 - \Omega_\Sigma)} \tag{87}$$

Also in this case the relaxation process introduces a viscous aspect in the adsorbed layer behavior. For $\nu = \nu_R$, the imaginary part of the dilational modulus shows a maximum and the real part an inflection point, as shown in Fig. 10.9. The high limiting frequency ($\lambda \to 0$) provides $\varepsilon = \varepsilon_{0\Gamma}$, whereas for vanishing ν (i.e. $\lambda \to +\infty$) it is $\varepsilon = (\varepsilon_{0\Gamma} - \varepsilon_{0\Omega} G_{R1})/(1 - G_{R2})$.

10.4 Conclusions and Perspectives

As shown above, suitable models provide a link between the structural and kinetic properties of the adsorption layers and its viscoelastic response to dilational stresses. Hence a number of experimental studies [56, 57, 75–78] are available dealing with the rheological properties of surfactant systems in relation to the adsorption mechanisms.

The different experimental techniques available for the dynamic surface tension can be adapted, in principle, to evaluate also the surface viscoelasticity of adsorbed layers. A review of these methods is given in [58]. The major constraint for their utilization is accessible frequency. Hence the direct measurement of interfacial tension while imposing harmonic perturbations of the interfacial area in a Langmuir trough apparatus or in a drop shape tensiometer [79–82] is used for measurements in the range 10^{-3}–10^{-1} Hz. The method utilizing pendant drops is today the most suitable for this frequency range, since ε can be derived straightforwardly from its definition, without the need for a fluidodynamic model. A more recent method is based instead on the utilization of a capillary pressure tensiometer [20, 41, 77, 83–85] and is used to measure ε in the range from 10^{-1} to about 500 Hz [57, 58, 77, 86, 87]. Finally, for higher frequencies, from 10^2 to 10^4 Hz, techniques based on the damping of capillary waves [89–91] are utilized instead. These last two latter methods rely on a fluid dynamic model.

An alternative way of investigating the surface rheology is offered by stress-relaxation methods, which are based on the application of impulsive area perturbations. This technique has been used, especially for liquid–air systems, implemented either on a Langmuir-Blodgett balance [91, 92] or in drop tensiometers [93, 94], again providing information on the low-frequency range.

Most of the experimental studies reported so far interpret the results on the basis of the diffusion-controlled approach of Lucassen (Eq. 11). Indeed, most of these studies deal with simple surfactants and measurements refer to low frequencies (below 1 Hz), comparable to the characteristic frequency for diffusion in dilute and semi-dilute solutions. Another set of studies report data for high frequencies (above 1000 Hz), where these systems often behave as mostly elastic. Since the method based on capillary pressure measurement is relatively new, at present there is a lack of information on the viscoelastic properties of adsorption layers in the intermediate frequency range (1–1000 Hz). This range seems important for investigating and understanding relaxation processes internal to the adsorption layer, if any.

When investigating surface dilational rheology, it is very important to obtain ε data over a wide frequency window, ideally several decades. This requires investigating these systems with the different available techniques. Lacking this approach, one can derive only a partial picture, often based on diffusional relaxation.

Today, the relevance of surface rheology for purposes related to different natural and artificial phenomena (foams, emulsions, liquid films, multiphasic flows, etc.) is growing. Unfortunately, in spite of the fact that theoretical developments have been progressing very fast in recent years, experimental investigations are still relatively scarce. Indeed, apart from those for low frequency, the available methods are difficult to implement and require complex interpretative tools to obtain ε from the rough data, which are still a topic of some debate. Hence they are in practice available in only a restricted number of laboratories.

Furthermore, these methods need to be improved and developed in order to provide the most accurate measurements on a wider time-scale. They then have to be applied to measurements on a large number of systems, in order to create a solid base for deriving a complete picture of the interfacial rheological response and for supporting the needs of technology.

References

1 P.M. Kruglyakov, D.R. Exerowa, Foams and foam films, in *Studies in Interface Science*, Vol. 5, D. Möbius and R. Miller (eds.), Elsevier, Amsterdam, **1997**, 88–344.

2 I.B. Ivanov (ed.), *Thin Liquid Films: Fundamentals and Applications*, Surfactant Science Series, Vol. 29, Marcel Dekker, New York, **1988**.

3 A.J. Sjoblom (ed.), *Emulsions and Emulsion Stability*, Surfactant Science Series, Vol. 61, Marcel Dekker, **1996**.

4 I.B. Ivanov, P.A. Kralchevsky, *Colloids Surf. A, 128* (**1997**) 155.

5 J. Lucassen, in *Anionic Surfactants. Physical Chemistry of Surfactant Action*, Surface Science Series, Vol. 11, E.H. Lucassen-Reynders (ed.), Marcel Dekker, New York, **1981**, pp. 217–265.

6 D. Langevin, *Adv. Colloid Interface Sci., 88* (**2000**) 209.

7 D.S. Valkovska, K.D. Danov, I.B. Ivanov, *Adv. Colloid Interface Sci., 96* (**2002**) 101.

8 D.S. Valkovska, K.D. Danov, I.B. Ivanov, *Colloids Surf. A, 175* (**2000**) 179.

9 C. Stubenrauch, R. Miller, *J. Phys. Chem B, 108* (**2004**) 6412.

10 N.A. Mishchuk, A. Sanfeld, A. Steinchen, *Adv. Colloid Interface Sci., 112* (**2004**) 129.

11 R. Aveyard, B. Binks, J.H. Clint, *Adv. Colloid Interface Sci., 100* (**2003**) 503.

12 B. Binks, C.P. Whitby, *Colloids Surf. A, 253* (**2005**) 105.

13 J. Lyklema, *Fundamentals of Interfaces and Colloid Science, Vol. 1: Fundamentals*, Academic Press, New York, **1992**, p. 269.

14 F. Ravera, M. Ferrari, L. Liggieri, *Adv. Colloid Interface Sci., 88* (**2000**) 129.

15 J. Eastoe, J.S. Dalton, *Adv. Colloid Interface Sci., 85* (**2000**) 103.

16 J.K. Ferri, K.J. Stebe, *Adv. Colloid Interface Sci., 85* (**2000**) 61.

17 J.R. Lu, M. Hromadova, R.K. Thomas, J. Penfold, *Langmuir, 9* (**1993**) 2417.

18 J.R. Lu, Z.X. Lee, R.K. Thomas, E.J. Staples, L. Thompson, I. Tucker, J. Penfold, *J. Phys. Chem., 98* (**1994**) 6559.

19 L. Liggieri, F. Ravera, R. Miller, *Colloids Surf. A, 175* (**2000**) 51.

20 L. Liggieri, M. Ferrari, A. Massa, F. Ravera, *Colloids Surf. A, 156* (**1999**) 455.

21 J. Maldonado-Valderrama, A. Martin-Molina, M.J. Galvez-Ruiz, A. Martin-Rodriguez, M.A. Cabrerizo-Vilchez, *J. Phys. Chem. B, 108* (**2004**) 12940.

22 V.B. Fainerman, R. Miller, V.I. Kovalchuk, *Langmuir, 18* (**2002**) 7748.

23 V.I. Kovalchuk, G. Loglio, V.B. Fainerman, R. Miller, *J. Colloid Interface Sci., 270* (**2004**) 475.

24 M. Ferrari, L. Liggieri, F. Ravera, C. Amodio, R. Miller, *J. Colloid Interface Sci., 186* (**1997**) 40.

25 L. Liggieri, F. Ravera, M. Ferrari, A. Passerone, R. Miller, *J. Colloid Interface Sci., 186* (**1997**) 46.

26 M. Ferrari, L. Liggieri, F. Ravera, *J. Phys. Chem.*, *102* (**1998**) 10521.

27 V. B. Fainerman, R. Miller, R. Wüstneck, A. V. Makievski, *J. Phys. Chem.*, *100* (**1996**) 7669.

28 D. Vollhardt, V. B. Fainerman, G. Emrich, *J. Phys. Chem. B*, *104* (**2000**) 8536.

29 V. B. Fainerman, D. Vollhardt, *J. Phys. Chem. B*, *103* (**1999**) 145.

30 A. F. H. Ward and L. Tordai, *J. Chem. Phys.*, *14* (**1946**) 453.

31 R. S. Hansen, *J. Phys. Chem.*, *64* (**1960**) 637.

32 K. L. Sutherland, *Aust. J. Sci. Res.*, *A5* (**1952**) 685.

33 E. V. Aksenenko, A. V. Makievski, R. Miller, V. B. Fainerman, *Colloids Surf. A*, *143* (**1998**) 311.

34 C. T. Hsu, C. H. Chang, S. Y. Lin, *Langmuir*, *15* (**1999**) 1952.

35 S. Y. Lin, Y. C. Lee, M. J. Shao, C. T. Hsu, *J. Colloid Interface Sci.*, *244* (**2001**) 372.

36 J. P. Fang, P. Joos, *Colloids Surf.*, *65* (**1992**) 113.

37 F. Ravera, L. Liggieri, A. Passerone, A. Steinchen, *J. Colloid Interface Sci.*, *163* (**1994**) 309.

38 V. B. Fainerman, R. Miller, *Langmuir*, *13* (**1997**) 409.

39 K. J. Mysels, *J. Phys. Chem.*, *86* (**1982**) 4648.

40 J. K. Ferri, S. Y. Lin, K. J. Stebe, *J. Colloid Interface Sci.*, *241* (**2001**) 154.

41 C. A. MacLeod, C. J. Radke, *J. Colloid Interface Sci.*, *160* (**1993**) 435.

42 J. Eastoe, J. Dalton, P. G. A. Rogueda, E. R. Crooks, A. R. Pitt, E. A. Simister, *J. Colloid Interface Sci.*, *188* (**1997**) 423.

43 H. C. Chang, C. T. Hsu, S. Y. Lin, *Langmuir* 14 (**1998**) 2476.

44 D. A. Edwards, H. Brenner, D. T. Wasan, *Interfacial Transport Process and Rheology*, Butterworth-Heinemann, Boston, **1991**.

45 G. Loglio, R. Miller, A. Stortini, U. Tesei, N. Degli Innocenti, R. Cini, *Colloids Surf. A*, *95* (**1995**) 63.

46 J. Lucassen, M. van den Tempel, *Chem. Eng. Sci.*, *27* (**1972**) 1283.

47 R. Miller, G. Loglio, U. Tesei, *Colloid Polym. Sci.*, *270* (**1992**) 598.

48 P. R. Garret, P. J. Joos, *J. Chem. Soc., Faraday Trans.*, *69* (**1976**) 2161.

49 V. B. Fainerman, V. I. Kovalchuk, E. V. Aksenenko, M. Michel, E. Leser, R. Miller, *J. Phys. Chem. B*, *108* (**2004**) 13700.

50 K. D. Wantke, H. Fruhner, J. Fang, K. Lunkenheimer, *J. Colloid Interface Sci.*, *208* (**1998**) 34.

51 V. I. Kovalchuk, J. Krägel, A. V. Makievski, G. Loglio, F. Ravera, L. Liggieri, R. Miller, *J. Colloid Interface Sci.*, *252* (**2002**) 433.

52 K. D. Wantke, H. Fruhner, *J. Colloid Interface Sci.*, *237* (**2001**) 185.

53 P. R. Garrett, P. Joos, *J. Chem Soc., Faraday Trans. 1*, *72* (**1976**) 2161.

54 V. B. Fainerman, R. Miller, R. Wüstneck, A. V. Makievski, *J. Phys. Chem.*, 100 (**1996**) 7669.

55 V. B. Fainerman, R. Miller, E.V. Aksenenko, A. V. Makievski, J. Krägel, G. Loglio, L. Liggieri, *Adv. Colloid Interface Sci.*, *86* (**2000**) 83.

56 J. Maldonado-Valderrama, V. B. Fainerman, E. Aksenenko, M. J. Galvez-Ruiz, M. A. Cabrerizo-Vilchez, R. Miller, *Colloids Surf. A*, *261* (**2005**) 85.

57 L. Liggieri, M. Ferrari, D. Mondelli, F. Ravera, *Faraday Discuss.*, *129* (**2005**) 125.

58 F. Ravera, M. Ferrari, E. Santini, L. Liggieri, *Adv. Colloid Interface Sci.*, *117* (**2005**) 75.

59 D. Vollhardt, G. Brezesinski, S. Siegel, G. Emrich, *J. Phys. Chem. B*, *105* (**2001**) 12061.

60 W. Jost, *Diffusion in Solids, Liquids, Gases*, Academic Press, New York, **1952**, Chapter 1.

61 M. Van Uffelen, P. J. Joos, *J. Colloid Interface Sci.*, *158* (**1993**) 452.

62 M. Ferrari, F. Ravera, L. Liggieri, H. Motschmann, Z. Yi, J. Krägel, R. Miller, *J. Colloid Interface Sci.*, *272* (**2004**) 277.

63 D. Vollhardt, G. Emrich, S. Siegel, R. Rudert, *Langmuir*, *18* (**2002**) 6571.

64 B. A. Noskov, T. U. Zubkova, *J. Colloid Interface Sci.*, *170* (**1995**) 1.

65 D. Vollhardt, *Mater. Sci. Eng. C*, *22* (**2002**) 121.

66 B. A. Noskov, G. Loglio, *Colloids Surf. A*, *143* (**1998**) 167.

67 L. Liggieri, F. Ravera, M. Ferrari, *Langmuir*, *19* (**2003**) 10233.

68 Y. B. Vysotsky, V. S. Bryantsev, V. B. Fainerman, D. Vollhardt, R. Miller, *Colloids Surf. A, 209* (**2002**) 1.

69 R. P. Borwankar, D. T. Wasan, *AIChE J., 32* (**1986**) 455.

70 J. Rudin, D. T. Wasan, *Colloids Surf., 68* (**1992**) 67.

71 J. Rudin, D. T. Wasan, *Colloids Surf., 68* (**1992**) 81.

72 J. P. Fang, P. J. Joos, *Colloids Surf., 83* (**1994**) 63.

73 I. B. Ivanov, K. D. Danov, K. P. Ananthapadmanabhan, A. Lips, *Adv. Colloid Interface Sci., 114* (**2005**) 61.

74 A. Sanfeld, A. Passerone, E. Ricci, J.-C. Joud, *Nuovo Cimento, 12* (**1990**) 353.

75 L. Liggieri, F. Ravera, M. Ferrari, *Langmuir, 19* (**2003**) 10233.

76 R. G. Muñoz, L. Luna, F. Monroy, R. G. Rubio, F. Ortega, *Langmuir, 16* (**2000**) 6657.

77 L. Liggieri, V. Attolini, M. Ferrari, F. Ravera, *J. Colloid Interface Sci., 252* (**2002**) 225.

78 C. Stubenrauch, V. B. Fainerman, E. V. Aksenenko, R. Miller, *J. Phys. Chem. B, 109* (**2005**) 1505.

79 G. Loglio, P. Pandolfini, R. Miller, A. V. Makieski, F. Ravera, M. Ferrari, L. Liggieri, in *Novel Methods to Study Interfacial Layers*, D. Moebius, R. Miller (eds.), Studies in Interface Science Series, Vol. 11, Elsevier, Amsterdam, **2001**, pp. 439–483.

80 E. H. Lucassen-Reynders, A. Cagna, J. Lucassen, *Colloids Surf. A, 186* (**2001**) 63.

81 L. Liggieri, A. Passerone, *High Temp. Technol., 7* (**1989**) 80.

82 M. E. Leser, S. Acquistapace, A. Cagna, A. V. Makievski, R. Miller, *Colloids Surf. A, 261* (**2005**) 25.

83 R. Nagarajan, D. T. Wasan, *J. Colloid Interface Sci., 159* (**1993**) 164.

84 L. Liggieri, F. Ravera, A. Passerone, *J. Colloid Interface Sci., 169* (**1995**) 226.

85 L. Liggieri, F. Ravera, in *Drops and Bubbles in Interfacial Research*, D. Möbius, R. Miller (eds.), Studies in Interface Science Series, Vol. 6, Elsevier, Amsterdam, **1998**, pp 239–278.

86 H. Fruhner, K.-D. Wantke, *Colloids Surf. A, 114* (**1996**) 53.

87 J. Ortegren, K.D. Wantke, H. Motschmann, *Rev. Sci. Instrum., 74* (**2003**) 5167.

88 F. Monroy, F. Ortega, R. G. Rubio, *Phys. Rev. E, 58* (**1998**) 7629.

89 P. Cicuta, I. Hopkinsons, *Colloids Surf., 233* (**2004**) 97.

90 J. C. Earnshaw, E. McCoo, *Langmuir, 11* (**1995**) 1087.

91 P. Cicuta, E. M. Terentjev, *Euro Phys. J., 16* (**2005**) 147.

92 R. Wüstneck, N. Wüstneck, D. O. Grigoriev, U. Pison, R. Miller, *Colloids Surf. B, 15* (**1999**) 275.

93 P. Saunier, F. Boury, A. Malzert, D. Heurtault, Tz. Ivanova, A. Cagna, I. Panaiotov, J. E. Proust, *Langmuir, 17* (**2001**) 8104.

94 E. M. Freer, K. S. Yim, G. G. Fuller, C. J. Radke, *Langmuir, 20* (**2004**) 10159.

11
Rapid Brownian and Gravitational Coagulation*

Andrei S. Dukhin and Stanislav S. Dukhin

11.1
Introduction

Gravity is an important factor that affects the kinetics of particle ensemble evolution in time. For instance, it contributes to the rate of particle collision due to the differential settling velocity. It competes with the surface forces with regard to the collision efficiency of building up an aggregate. Importantly, there is no need for particles to be large from the very beginning. Increasing particle size in the process of Brownian coagulation eventually will bring the system to the time point when gravity becomes important.

This was all known even to Smoluchowski [1], who made the first attempt to incorporate this factor into his kinetic theory. There have been many other efforts towards refining the theory and collecting more experimental evidence on the role of gravity in coagulation kinetics. Many different systems have been involved in these studies. Consequently, all relevant papers are scattered in wide variety of different publications. The scope of the field is very broad: it involves general colloid science, aerosols, flotation, waste water treatment, general physics, emulsion science, nano-technology and many others. Unfortunately, this leads to the situation where scientists working in different fields have only limited knowledge about similar developments on other fields. We give a few examples here.

One of the most sophisticated analyses of the role of gravity in coagulation was been performed by Davis and co-workers in the early 1990s [2–7]. They worked with emulsions. As a result, their publications are not mentioned in modern studies that deal mostly with the fractal approach.

Another example is the theory of flotation. Differential settling causes the bubble–particle hydrodynamic interaction. This hydrodynamic interaction can be an important factor in the flotation process, as was shown by Derjaguin and Dukhin 40 years ago [8–13]. They developed a detailed theory of this interaction.

* Please find a list of symbols at the end of the text.

Colloids and Interface Science Series, Vol. 2
Colloid Stability: The Role of Surface Forces, Part II. Edited by Tharwat F. Tadros

ISBN: 978-3-527-31503-1

A similar effect occurs when solid particles or emulsion droplets interact. However, the original Derjaguin-Dukhin work remains practically unknown to scientists who are dealing with regular dispersions or emulsions.

In addition, some authors attribute certain basic derivations to the wrong sources. For instance, Fuchs calculated the critical particle size for the transition from Brownian to gravitational coagulation 40 years ago [14, 15], but there have been similar independent calculations in several subsequent papers with no reference to Fuchs' or each other's work. There is also controversy regarding the expression for rectilinear collision frequency. Many authors attribute it to Friedlander [16], whereas it was actually first given many years earlier in the 1940s by Camp and Stein [17] based on Smoluchowski's theory of ortho-kinetic coagulation [18].

The situation becomes even more complicated because some of the relevant papers had been published in obscure journals prior to the Internet era. For instance, the first analytical solution regarding the transition from Brownian to gravitational coagulation [19, 20] remains practically unknown.

There is also ambiguity regarding the terms *aggregation, flocculation* and *coagulation*. For instance, Meakin applies term "aggregate" to weak fractal structures [21, 22]. At the same time, there is a general understanding in colloid science that aggregates are compact particulates, in contrast to the weakly organized flocs that often coagulate in secondary minima.

Resolution of this ambiguity is now important because of new experimental methods that allow us to distinguish between these processes. For instance, we can define aggregation as the formation of a new, dense, rigid, compact particle, whereas flocculation leads to a weak, highly porous structure. Using these definitions as a basis, we can use ultrasound attenuation measurements for testing a particular system and verifying what type of coagulation phenomena occur in it, either aggregation or flocculation [23]. There was also a mention of the usefulness of ultrasound attenuation measurements for studying fractals by Meakin [22].

There is also theoretical justification for using these terms separately. Historically, aggregation is the first subject of the kinetic theory. It is much easier to build a theoretical model of this effect than flocculation starting with first physical principles, such as hydrodynamics and electrodynamics. From this viewpoint, coalescing emulsions or aerosol droplets are the best objects for theoretical modeling.

Kinetic description of flocculation began to progress rapidly after the introduction of fractal models [30]. At the present time it dominates this field. In principle, it should include a description of aggregation as an asymptotic case when the fractal dimension equals 3. This might be one of the reasons why old studies on aggregation kinetics fell into obscurity.

However, we believe that the introduction of a compact aggregate as a new particle offers considerable theoretical simplification, and that this notion deserves attention. It is very doubtful whether old theoretical results derived for aggregates could be reconstructed as asymptotic solutions of fractal modeling of flocculation phenomena. That is why we think that a modern review of the old work on the kinetics of aggregation should be combined with more recent papers on the kinetics of flocculation based on fractal modeling.

Summarizing this subject, we suggest the following terminology that is used in this chapter.

- *Aggregation* – build-up of compact new aggregates that move as a single entity and can be considered as new particles with almost same density. Coalescing emulsion droplets are the perfect example. Objects with large fractal dimension close to 3 would also belong to this group.
- *Flocculation* – build-up of weak structures that could deform under stress. The density is much lower than for primary particles owing to the high liquid content. Objects with fractal dimension below 2 belong to this group.
- *Coagulation* – general term for any form of particles association.
- *Aggregate* – this term is widely used in fractal science for weak structures. In principle these structures must be described as *flocs*. However, historical precedence suggests that the term *aggregate* could be used for any type of composite particle. We will use term *flocs* only when internal particle rheology is relevant.

We want to bring together studies from different fields and establish what is unique for the particular field and what is common for all of them.

Clearly we cannot present all of the publications on this subject. We limit the scope of this review to only the kinetic aspects of the coagulation process. We do not consider the influence of gravity on pair interactions when the effect of the electric surface charges becomes important. We also do not consider aspects of particle deformation due to gravity or interactions.

11.2 Population Balance Equations

The best known population balance equation (PBE) was derived by Smoluchowski [1] as a conservation law of the dispersed phase in every element of the system volume:

$$\frac{d\varphi(v,t)}{dt} = \frac{1}{2}\int_0^v \beta(v', v-v')\varphi(v-v',t)\varphi(v')dv' - \int_0^v \beta(v,v')\varphi(v,t)\varphi(v',t)dv' \qquad (1)$$

where φ is unary distribution function for particles and β is collision frequency. The first integral in this equation reflects increasing number of aggregates with volume v due to aggregation of particles with volumes v' and $v - v'$. The second integral describes a decreasing number of particles with volume v by aggregation with particles of volume v'.

There are several major limitations of this equation: first, only binary collisions are taken into account; second, fluctuations of the dispersed phase content are neglected; third, it neglects possibility of the aggregates breaking up; and fourth, it does not take into account directed motion of the particles. It assumes that particles undergo only random Brownian motion.

The third restriction was removed by Melzak [24], who suggested a more general PBE that takes into account aggregate break-up:

$$\frac{d\varphi(v,t)}{dt} = \frac{1}{2}\int_0^v \beta(v', v-v')\varphi(v-v',t)\varphi(v')dv' - \int_0^v \beta(v,v')\varphi(v,t)\varphi(v',t)dv + \int_0^v \delta(v+v')\varphi(v+v')dv' - \delta(v)\phi(v) \tag{2}$$

The two additional terms on the right-hand side of Eq. (2) reflect an increasing number of particles with volume v due to break-up of the aggregates with volume $(v+v')$ and a decreasing number of the particles with volume v due to their break-up.

Incorporation of the directed particle motion in addition to the Brownian motion is more complicated. The particle size distribution becomes a function of the position of the particular volume element in the system. Historically, it was considered such a big change that a special terminology was suggested for distinguishing between aggregation induced by only Brownian motion – "perikinetic" – and directed particles motion – "orthokinetic".

From a mathematical point of view, an additional term appears on the left-hand side of the PBE equation. This term would reflect variation of the number of particles in the element of volume due to particles passing through it with a certain velocity $\vec{V}(\vec{r},t)$. The vector $\vec{r}$ denotes the position of the element in the system. As a result, the PBE equation for the case of negligible aggregate break-up becomes

$$\frac{d\varphi(\vec{r},v,t)}{dt} + \text{div}J(\varphi,t) = \frac{1}{2}\int_0^v \beta(v', v-v')\varphi(\vec{r}, v-v'',t)\varphi(\vec{r},v')dv' - \int_0^v \beta(v,v')\varphi(\vec{r},v,t)\varphi(\vec{r},v',t)dv \tag{3}$$

where J is the flux density of the particles with volume v. It consists of two terms, diffusion and convection:

$$J(\varphi) = -D\frac{\partial\varphi}{\partial x} + V\varphi \tag{4}$$

In addition, gravity or any other directed motion of particles would also affect the "collision frequency" β and "break-up frequency" δ. There are several ways to take this into account, as will be described below.

Neither of the population balance equations mentioned above can be solved exactly. Any possible solution is associated with a certain "theoretical error". Clearly, we would like to minimize this error. However, it is important to recognize that there is also "modeling error", which we make when substituting a model heterogeneous system for the real one. When the "theoretical error" becomes less than the "modeling error", further improvement of theoretical solution becomes meaningless.

The modeling error of the complex heterogeneous systems is large in many cases. This means that even very approximate theoretical solutions with rather large theoretical errors could be useful for describing real aggregation phenomena. This is especially important for a few existing analytical solutions that are less accurate than numerical solutions, but much more general and simple. That is why we describe these analytical solutions in more detail.

The Smoluchowski solution for Brownian perikinetic aggregation serves as a fundamental basis for an analytical solution that takes gravity into account. That is why we present here a short overview of this historically first PBE solution.

11.3 Smoluchowski Solution for Brownian Coagulation

The classical understanding of coagulation kinetics was given by von Smoluchowski [1]. Computer simulations [41–43] serve as a means to test the validity of the mean field approach.

In order to solve the PBE, one must specify the "collision efficiency". Smoluchowski suggested an approximate way to achieve this for Brownian perikinetic aggregation. He calculated the number of particles colliding with an imaginary stationary central particle within unit time. He assumed that each collision would lead to aggregation. This is so-called "fast aggregation process".

Then, allowance can be made for the fact that the central particle itself is one of many similar particles undergoing Brownian motion and so the appropriate collision frequency can be derived. We present his results below for discrete presentation of the particle size distribution. The collision frequency between particles of the fraction i and j equals:

$$\beta_{ij}^{\mathrm{d}} = 4\pi R_{ij}(D_i + D_j) \tag{5}$$

where D_i and D_j and n_i and n_j are the diffusion coefficients and number concentrations, respectively. The term R_{ij} is the collision radius for the pair of particles and represents the center-to-center distance at which the particles may be assumed to be in contact. For spherical particles

$$R_{ij} = \frac{d_i + d_j}{2} \tag{6}$$

This assumption plus the Einstein expression for the diffusion coefficient (Eq. 1) yields the following expression for the "collision frequency":

$$\beta_{ij}^{d} = \frac{(2kT/3\eta_m)(d_i + d_j)^2}{d_i d_j} \tag{7}$$

If the dispersed phase is monodisperse, the initial collision frequency can be easily calculated because only identical primary particles contribute to it. This leads to the following equation for the total particle concentration $N(t)$:

$$-\frac{dN(t)}{dt} = \left(\frac{4kT}{3\eta_n}\right) N_T^2 = K_F N(t)^2 \tag{8}$$

where K_F is known as the coagulation rate constant and has a value of approximately 6.13×10^{-18} m^3 s^{-1} for aqueous dispersions at 25 °C.

The most noteworthy feature of Eq. (8) is that it does not include the particle size because the size terms cancel from Eq. (7) when $d_i = d_j$. As the particle size increases, the diffusion coefficient decreases, but the collision radius increases. These opposite effects balance each other exactly for equal particles.

The collision frequency of the Brownian coagulation is not very sensitive to the particles size in general, even for particles of different size. The term $(d_i + d_j)^2/d_i d_j$ in Eq. (7) is very close to 4, provided that the sizes do not differ too greatly. For instance, if the sizes are different by a factor of 3, the size term in Eq. (7) has a value of about 5.

The simple Eq. (8) yields the following expression for the total number of particles as a function of time:

$$N(t) = \frac{N_0}{1 + K_F N_0 t} \tag{9}$$

where N_0 is the initial concentration of the primary particles.

There is a characteristic coagulation time τ_{sm}, during which the number of particles reduces by half:

$$\tau_{sm} = \frac{1}{K_F N_0} \tag{10}$$

It is also possible to calculate the concentration of single particles and aggregates assuming the collision frequency to apply to all collisions. The derivation is reproduced in many text books, e.g. [44, 65].

The role of Brownian motion in the coagulation process decreases with increasing particle size. The influence of the differential settling begins to domi-

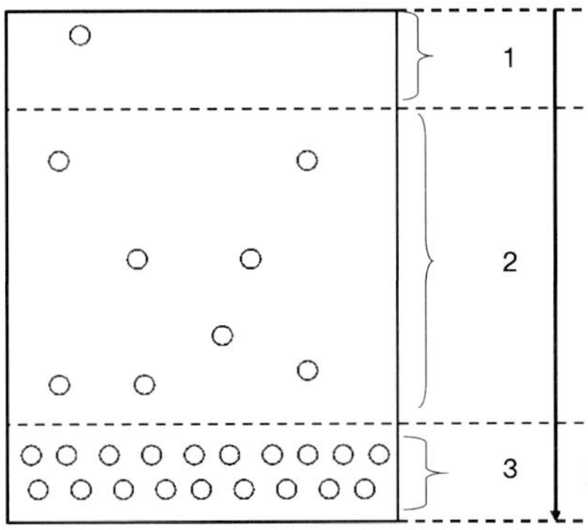

Fig. 11.1 Illustration of the various zones appearing during sedimentation: (1) depleted zone; (2) uniform zone; (3) deposit.

nate and a more general population balance equation must be used (Eq. 3). Unfortunately, there is no known analytical solution of this equation.

However, there is one particular simplification, which makes possible some analytical analysis of the orthokinetic gravitational coagulation, namely the introduction of the "zone with the uniform dispersed phase". The next section describes this important notion.

As mentioned above, the population balance equations of Smoluchowski and Melzak assume that the amount of the dispersed phase remains constant in every element of the system volume during aggregation. This assumption becomes invalid due to gravity. If the dispersed phase is denser than the dispersion medium, the top layers of the system become gradually depleted with particles owing to their sedimentation. At the same time, the bottom layers of the system become enriched with particles that build up deposit. If the dispersed phase is lighter than the dispersion medium (emulsions), the situation would be reversed. Figure 11.1 illustrates the existence of these three zones, which are important for the proper description of aggregation under the influence of gravity.

Consequently, the particle size distribution becomes a function of the space coordinate or a depth in the solution. The PBE gains an additional term (see Eq. 3).

There is, however, an intermediate zone where the flows of particles in and out of the each volume element compensate each other. The condition of constant and uniform dispersed phase content holds in this zone. This simplifies the PBE tremendously. The particle size distribution again becomes independent of the special coordinate and the "div" term on the left-hand side of the Eq. (3) becomes zero. This means that in this intermediate zone Eq. (3) simply reduces to the Smoluchowski equation (Eq. 1), if we can neglect aggregate break-up.

This concept was introduced for the first time by Dukhin in 1987 [19, 20]. It was re-introduced independently about 10 years later by Kumar et al. [45].

According to this concept, we can apply the Smoluchowski equation for describing the coagulation kinetics in the uniform zone even under the influence of gravity, which would only affect the value of the "collision frequency" β^{g}:

$$\frac{d\varphi(v,t)}{dt} = \frac{1}{2}\int_0^v \beta^g(v', v - v')\varphi(v - v', t)\varphi(v')dv' - \int_0^v \beta^g(v, v')\varphi(v, t)\varphi(v', t)dv' \quad (11)$$

This concept does allow an analytical description of at least some aspects of gravity-influenced coagulation phenomena, as we will show below. This analytical solution requires an appropriate expression for the "collision frequency" β. The next section describes several different approaches for determining this parameter with a contribution from gravity.

Introduction of the "uniform zone" is also very important for experimental verification of the theoretical developments. It was stressed by Kumar et al. [45], especially for emulsions. They pointed out that, "In fact, most investigators have not specified from where they drew the samples or which type of coalescence they were studying" [45]. Consequently, they made reasonable critical remarks about the state of experiments. They pointed out that the interpretation of many experimental studies of droplet coalescence is uncertain, because the coalescence role was not isolated from the trivial droplet loss due to sedimentation. After considering different means for eliminating this loss, they concluded that the utilization of the uniform coagulation zone provides the simplest and most reliable method for achieving this goal.

We can conclude that the introduction of the uniform zone yields substantial benefits for both theory and experiment. We illustrate this conclusion in the following sections.

11.4 Collision Frequency for Gravitational Aggregation

There are two completely different approaches for deriving the collision frequency expression. Historically, the first one was based on the first physical principles of hydrodynamics, electrodynamics and statistical mechanics. It usually leads to rather complicated mathematical expressions that do not allow a solution of the PBE. That is why the second approach was adopted. It is based on purely mathematical considerations. Workers pursuing this approach simply tried various simple mathematical expressions for the collision frequency that allow analytical or numerical solution of the PBE. We present here a short overview of both of these approaches.

11.4.1 Collision Frequency Derived from First Principles

The first attempt at estimating the collision frequency induced by directed motion of particles was made by Smoluchowski [18]. He derived the following equation relating β to the velocity gradient of the fluid du/dy:

$$\beta_{ij} = \frac{4}{3} R_{ij}^3 \frac{\mathrm{d}u}{\mathrm{d}y} \tag{12}$$

This work by Smoluchowski was the basis for incorporating gravity into the collision frequency. Apparently it was done first by Camp and Stein around 1943 [17], who presented the following widely used expression for the rectilinear collision frequency:

$$\beta_{ij}^{\mathrm{g}} = R_{ij}^2 |u_i - u_j| E_{ij} = \frac{2g\pi\Delta\rho}{9\eta_{\mathrm{m}}} R_{ij}^3 |d_i - d_j| E_{ij} \tag{13}$$

where u is the Stokes sedimentation velocity for particles of diameter d_i. This expression is often attributed mistakenly to other authors.

The parameter E_{ij} is "collision efficiency". It reflects the influence of the hydrodynamic interaction and surface forces on the coagulation kinetics. Coagulation follows a "fast kinetics" route when electrostatic repulsion does not hamper the particle attachment and each collision results in aggregation.

The rectilinear model ignores completely hydrodynamic particle interactions and the contribution of surface forces. There have been a number attempts to incorporate these factors into the "collision frequency". In this review, we ignore all attempts dedicated to electrostatic repulsion and particle deformation. We present here only results regarding the influence of gravity induced hydrodynamic interactions on the "collision frequency".

Historically, the first efforts had been made for aerosols and especially flotation. Gravitational effects are most pronounced in the bubble–particle interaction (flotation) owing to the large density contrast. Many of these results are easily extended to liquid-based dispersions and emulsions. For instance, a bubble with the surface retarded by surfactant is identical with a solid particle from a hydrodynamic viewpoint.

Initially, theoretical development applied simplification by considering only particles with very different sizes. This assumption yields a small parameter (particle size ratio), which can be used for simplifying the mathematics. Later developments eliminated this assumption. We present here these developments in chronological order.

Sedimentation of a large particle creates a liquid flow around it. This liquid flow does not affect much other large particles, which would move almost linearly owing to the inertia (Fig. 11.2, line 1). At the same time, fairly small particles move essentially along the corresponding liquid flow lines (Fig. 11.2, line 2). The long-range hydrodynamic interaction causes this deviation in the trajectory of small particles. It affects the trajectory when the distance to the surface of larger particles is comparable to its size.

This deviation of the particle trajectories from a linear path causes a change in the collision frequency. Corresponding models are generalized with one term – the curvilinear collision frequency. The dimensionless multiplier E in the expression for the collision frequency is reserved for characterizing this effect.

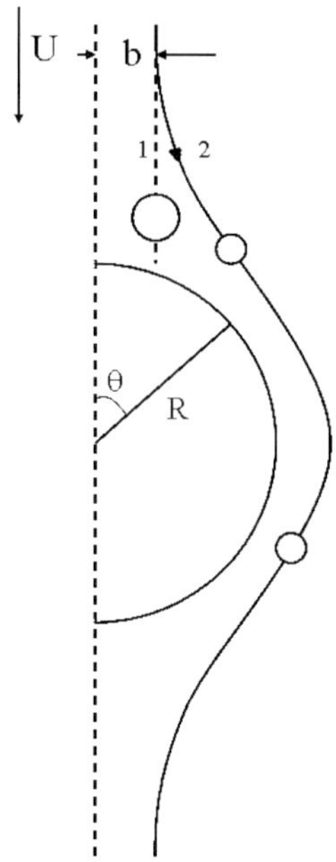

Fig. 11.2 Influence of the inertia of particles on their trajectory in the vicinity of a floating bubble. Trajectories of large (inertial) (line 1) and small (inertial-free) (line 2) particles at the same target distance *b*.

This parameter of the collision efficiency in the case the long-range hydrodynamic interaction equals:

$$E = \frac{b^2}{a^2} \tag{14}$$

where b is the maximum radius of the cylinder around the larger particle, which encompass all particles that would deposit on the larger one (Fig. 11.3). The particles moving along the streamline at the target distance b accumulate on the surface of a bigger particle (Fig. 11.3, indicated by a dashed line). Otherwise, the particles are carried off by the flow. It is evident that the calculation of collision frequency is essentially reduced to the so-called "limiting (grazing) trajectory" (continuous curve) and, correspondingly, to the target distance.

Such calculations were presented for the first time by Sutherland [25] and later by Derjaguin and Dukhin [26]. They were based on the consideration of liquid streamlines and of the finite size of spherical particle collisions due to interception:

$$E_0 = \frac{3a_1^2}{2a_2^2} \tag{15}$$

where a_1 and a_2 are the radii of smaller and larger particles, respectively.

Later, Dukhin and Derjaguin proved the theorem that the particle concentration would remain constant if the velocity field is solenoidal [11]. They quanti-

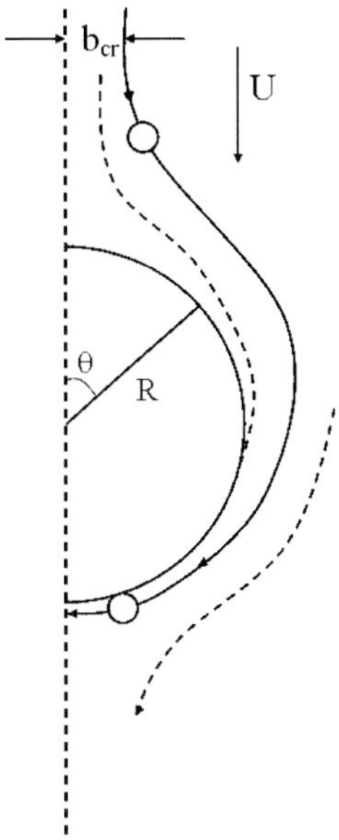

Fig. 11.3 The continuous line illustrates the concept of limiting trajectory of particles and the dashed lines indicate trajectories at $b<b_{cr}$ and $b>b_{cr}$.

fied the influence of the gravitational deposition of smaller particles on the surface of larger ones with the following expression:

$$E_0 = \frac{3a_1^2}{2a_2^2} - \frac{\Delta\rho a_1^2}{\rho a_2^2} \tag{16}$$

where $\Delta\rho$ is the density difference.

Further theoretical analysis showed that long-range hydrodynamic interaction becomes inadequate when the distance between two approaching particles diminishes. The hydrodynamic interaction at distances comparable to the size of the smaller particle is called the short-range hydrodynamic interaction [10].

It is in the equatorial plane that the closest approach of a streamline to the surface of a larger particle is attained. In Fig. 11.4, the broken line (curve 1) represents the liquid streamline whose distance from the surface of a larger particle in the equatorial plane is equal to the radius of the smaller particle. Under the influence of the short-range hydrodynamic interaction, the smaller particle is displaced from liquid streamline 1 so that its trajectory (curve 2) in the equatorial plane is shifted from the surface by a separation larger than its radius. Therefore, no contact with the surface occurs and, correspondingly, $b(a_1)$ is not a critical target distance.

Owing to the short-range hydrodynamic interaction, the distance from the smaller particle to the surface in the equatorial plane is larger than the distance from the surface to the liquid streamline with which the trajectory of the smaller particle coincides at large distances from the bigger particle. It may therefore

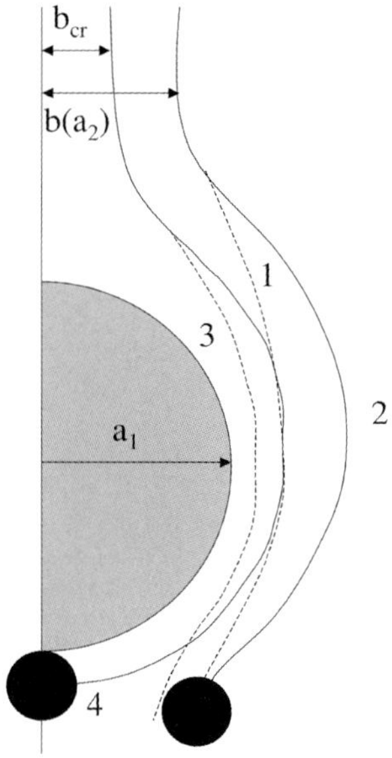

Fig. 11.4 Influence of finite dimension of particles in inertial-free flotation on their trajectory in the vicinity of a floating bubble. The liquid flow lines corresponding to target distances $b(a)$ and b_{cr} are indicated by dashed lines. The continuous lines are characteristic of the deviation of the trajectory of particles from the liquid flow lines under the influence of short-range hydrodynamic interaction.

be concluded that $b_{cr} < b(a_1)$. The limiting liquid streamline (curve 3) is characterized by the particle trajectory (curve 4), which, branching off under the influence of the short-range hydrodynamic interaction, runs in the equatorial plane at a distance a_1 from the surface of the bigger particle.

The value of b_{cr} decreases, first, owing to the deflection of the liquid streamline under the influence of the long-range hydrodynamic interaction and, second, owing to the deflection of the small drop trajectory from the liquid streamline under the influence of the short-range hydrodynamic interaction. Therefore, the collision efficiency is expressed as the product of two factors, E_0 and E_1, both smaller than unity. The former represents the influence of the long-range hydrodynamic interaction and the latter the influence of the short-range hydrodynamic interaction.

Later, Spielman [46] and Goren [47, 48] incorporated van der Waals attraction into the theory of small particle deposition from laminar flow on the surface of a larger particle. Their numerical results served as an important component of the microflotation theory derived by Derjaguin and co-workers [8, 9].

As a next step, Rulyov [12] obtained following approximate analytical equation for the collision efficiency using the same basic physics:

$$E = E_0 E_1(W); \; W = \frac{A a_2^2}{27 u_s \pi \eta_m a_1^4} \tag{17}$$

where A is the Hamaker constant. The function $E_1(W)$ is plotted in [12] and can be approximated according to [13]. It leads [8] to the equation:

$$E = 0.11 \frac{a_1^{1.4}}{a_2^2} A^{1/6} \tag{18}$$

Examination of Rulyov's approximation by Dukhin and Sjoblom [49] indicates that it agrees well with the results of Spielman and Fitzpatrick [27, 28]. The impression arises that coagulation is possible at any small (but finite) value of the Hamaker constant, because the dependence of E on A is very weak. However, further fragmentation is possible at sufficient small A.

Further development of the theory pursued a goal of removing the restriction of small sizes ratio. If the difference between particle sizes is not large, the description of their interaction requires different parameterization. The motion of two interacting particles is a superposition of the mass center motion and their relative motion in the frame of reference that is associated with the mass center. When the spheres are very close, the mobility functions have the asymptotic forms given by Jeffrey and Onishi [50].

Davis [2] incorporated these asymptotic forms into the equation for particle trajectories. His results agree with Rulyov's results for small particle size ratios $a_1/a_2 << 1$.

Davis [2] also suggested an equation for the collision frequency of the emulsion droplets that takes into account the difference in viscosity of the medium η_m and droplet η_p:

$$\beta_{ij}^{g} = R_{ij}^2 |u_i - u_j| E_{ij} = \frac{2\left(\frac{\eta_p}{\eta_m} + 1\right) g\pi\Delta\rho}{3\left(3\frac{\eta_p}{\eta_m} + 2\right)\eta_m} R_{ij}^3 |d_i - d_j| E_{ij} \tag{19}$$

In addition, Davis and co-workers [4, 6] developed a procedure for calculating the collision efficiency E_{ij} that takes into account simultaneously gravity and Brownian effects and hydrodynamic droplet interaction.

There is a version presented by Reed et al. [51], which is very similar to Sutherland's solution [25] but with no (at least not mentioned) restriction of the size ratio:

$$E_0 = \frac{3y_c^2}{2(a_1 + a_2)^2} \tag{20}$$

where y_c is the size of the smaller particle. It seems to us that it is actually identical with Sutherland's solution, which was published about 30 years earlier with all applied limitations discussed above.

Later, Zhang and Davis [6] used Batchelor's expression [52] for the speed of the particles' relative motion. They modified it for spherical emulsion drops. It turns out that the collision efficiency approaches a finite value as the drops become equi-sized ($a_1/a_2 \rightarrow 1$). However, the collision rate goes to zero in this limit because the relative velocity of two drops approaches zero. The unique peculiarity of the results is that the non-zero collision efficiency occurs in the absence of attractive molecular forces due to accounting for liquid mobility inside the droplets.

There are several more recent studies on interactions in fractal systems. Interaction between primary particles and as fractal aggregate is the primary goal. We mention here just two models that found their way to other studies.

This first is by Han and Lawler [53], who suggested the following numerical solution for the curvilinear collision efficiency between particle and aggregate:

$$E_{\mathrm{pa}} = \exp\left[-3.4 + 0.62 \log \gamma + \frac{d_{\mathrm{p}}}{d_{\mathrm{a}}}(3.5 - 1.2 \log \gamma)\right] \tag{21}$$

where $\gamma = 8A/3\pi u_{\mathrm{a}} d_{\mathrm{a}}^2$, A is the Hamaker constant and u_{a} is the settling velocity of the aggregate with size d_{a}. This model has been used in several modern studies related to fractals. It does not take into account flow through the fractal aggregate.

The second model which takes into account internal flow through the aggregate is advection, developed Li and Logan [54]. They used Brinkman's model for describing aggregate permeability. Their expression is valid for small particle size ratios:

$$\beta_{\mathrm{pa}} = \frac{9\pi d_{\mathrm{a}}^2 u\left(1 - \dfrac{\tanh \xi}{\xi}\right)\left[1 - \exp\left(-\dfrac{3}{2s}(1-\varepsilon)\alpha_{\mathrm{filt}}\eta d_{\mathrm{a}}^{1-\mathrm{b}}\right)\right]}{4\alpha\left[2\xi^2 + 3\left(1 - \dfrac{\tanh \xi}{\xi}\right)\right]} \tag{22}$$

where $\xi = d_{\mathrm{a}}/2\sqrt{\kappa}$. The permeability κ according to Brinkman is given by

$$\kappa = \frac{d_{\mathrm{p}}^2}{72}\left(3 + \frac{4}{1-\varepsilon} - 3\sqrt{\frac{8}{1-\varepsilon} - 3}\right) \tag{23}$$

where d_{p} is the diameter of primary particles and ε is porosity of the aggregate. The meaning of the other parameters is given in the original paper.

The difference in particle densities is accounted for in the theory of gravitational coagulation in [55].

11.4.2 Collision Frequency Assumed from Mathematical Reasoning

As mentioned above, there is also a completely different approach to specifying collision frequency. Instead of using physical principles one could just suggest some mathematical expression for this parameter that would allow further solution of the population balance equation.

This approach was very popular in aerosol science, especially in the former USSR. These studies were summarized by Voloshchuk and Sedunov about 30 years ago [56, 57], but unfortunately not in English. One of the expressions that they suggested was used [58] for describing the aggregation of erythrocytes.

It seems to us that this approach has reached its culmination in the work by Van Dongen and Ernst [31]. They suggested the following expression for the collision frequency:

$$\beta_{ij} = a^{\lambda} i^{\mu} j^{\lambda-\mu} \tag{24}$$

where $\lambda \leq 2$; $j \gg i$; $\lambda - \mu \leq 1$.

The two parameters λ and μ provide fundamental information about the reactivity between clusters of different sizes.

Some authors, e.g. [59], consider that the van Dongen and Ernst model is sufficient to specify most of the important aspects of coagulation kinetics; specifically the shape of the cluster size distribution and the time evolution of size classes and moments. They agree that there is a substantial cost in ignoring the analytical details of the collision frequency. They then express hope that some missing constants could be found from experiments. Several papers have reported results on the influence of gravity on coagulation kinetics using van Dongen-Ernst collision frequencies, e.g. [59, 60].

11.4.3 Collision Frequency for Simultaneous Brownian and Gravitational Coagulation

In comparison with the large amount of research devoted to calculation of the collision frequency in either Brownian or gravitational aggregation, very little work has been done when both Brownian motion and gravity are important simultaneously.

The first work on this subject, as far as we know, was done by Fuchs about 40 years ago [14, 15]. He formulated a question regarding the critical particle size that separates Brownian and gravitational coagulation, $d_{\mathrm{b\_g}}$. The coagulation kinetics would be only Brownian when the particle size is much smaller than $d_{\mathrm{b\_g}}$, whereas for larger sizes it would be pure gravitational. In order to estimate the value of this number, Fuchs compared diffusion and hydrodynamic fluxes of particles on to selected centers. He came up with 1 μm as an approximate particle size that separates Brownian and gravitational kinetic modes.

This simple calculation has been repeated in several more recent studies, apparently without knowing that it had been done 40 years ago. His conclusion regarding this value of the critical particle size has been confirmed experimentally recently by Morbidelli's group [38, 39].

This success of Fuchs' approach suggests that we can use a similar procedure for calculating the collision frequency. There is a dimensionless number that characterizes the ratio between diffusion and hydrodynamic fluxes: the Peclet number, *Pe* [40]. Diffusion dominates over convection when this number is small. We can determine the Peclet number of coagulation phenomena as a parameter that relates Brownian and sedimentation fluxes on the particular center particle. This yields the following expression:

$$Pe_0 = \frac{\pi \Delta \rho g d_1^4}{24kT} \tag{25}$$

The subscript 0 on the Peclet number and the particle diameter reflects the initial time moment. We determine this parameter at the initial stage of the aggregation process.

If particles are small, Brownian diffusion dominates aggregation kinetics and the Peclet number is small. Increasing the size of the particles leads to an increase in Peclet number owing to the larger contribution of sedimentation to the particles' collision frequency. We would obtain Fuchs' estimate for d_{b_g} assuming that $Pe=1$.

This parameter depends very strongly on the size. This means that the transition from Brownian to gravitational coagulation is very sharp on the size scale.

This also means that Fuchs' estimate is valid for practically any particles independent of their density. For instance, in the case of emulsions, the density contrast can be 100 times less than for rigid oxide particles. This would lead to an increase in d_{b_g} for emulsions, but only three-fold. It appears that 1 μm is a universal good estimate for the transition range.

The Peclet number remains small at the initial stages of the transition from Brownian to gravitational aggregation. We can use it as a small parameter and present the "collision frequency" as a two-term expansion [19]:

$$\beta^{g}(v, v') = \beta^{d}(v, v')[1 + Pe_0|\tilde{v} - \tilde{v}'|] \tag{26}$$

where the volume of particles is normalized by the volume of initial particles.

Expression (26) assumes additivity between Brownian and gravitational coagulation. The same assumption was used by Wang and Davis [7] 9 years after Dukhin [19] and by Reddy et al. [61]. Expression (26) also ignores hydrodynamic interactions. In the earlier work by Reddy et al. [61], hydrodynamic particle interaction was also disregarded.

There is theoretical justification for both assumptions. It comes from the most detailed numerical calculation of the collision frequency, which was performed by Zinchenko and Davis [4]. They took into account both gravity and Brownian effects, with complete hydrodynamic interactions. Unlike previous studies on the collision frequency, a numerical solution was valid for arbitrary Peclet number, thus covering the whole range of sizes in typical dispersed systems (Fig. 11.5).

A key finding is that the synergistic effects of Brownian motion and gravity sedimentation give much higher coagulation rates than either mechanism acting alone or predicted by the additivity approximation in the Peclet range $10 < Pe < 10^4$ with its maximum around 10^2.

At the same time, this numerical calculation confirmed that the error of the "additivity approximation" is negligible for small Peclet numbers <1. This is justification for describing transition from Brownian to gravitational coagulation when $Pe < 1$ using the additivity approximation, as it was done in [19].

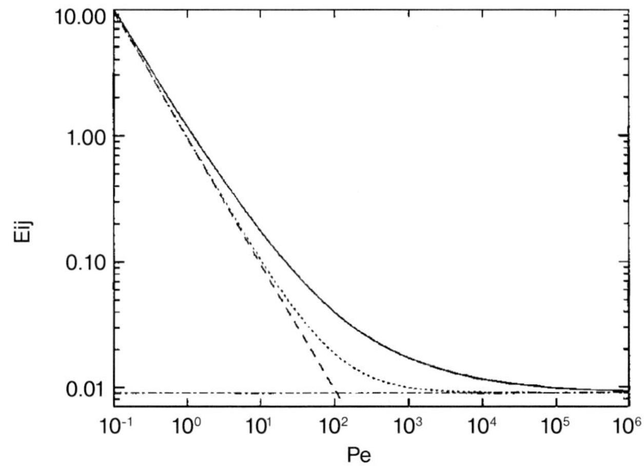

Fig. 11.5 Collision efficiency as a function of the relative Peclet number for a drop with $a_1/a_2 = 0.25$ and no inter-droplet forces. The solid line is the exact solution of Zinchenko and Davis [4], the dotted line represents the additivity approximation, the dashed line is the Brownian collision efficiency and the dashed-dotted line is the gravitational collision efficiency [6].

Expression (26) allows us to achieve an analytical solution of the PBE and derive some general conclusions regarding transition from Brownian to gravitational coagulation.

There is also an analytical expression for the collision frequency for the opposite case of large Peclet numbers when gravitational coagulation dominates [33]. The case of the weak gravity and strong Brownian aggregation has also been investigated [62]. The calculations in [33] for combined Brownian and gravitational coagulation were accomplished without taking account of hydrodynamic interactions.

11.5
Transition from Brownian to Gravitational Aggregation – Analytical Solution

There have been two analytical solutions of the PBE, which describe the role of the gravity in coagulation kinetics. Historically, the first was given by Dukhin [19, 20] and Melik and Fogler [32] in the mid-1980s. The second was suggested about 15 years later by Jung et al. [37]. These solutions are completely independent, which is why their agreement, as we will show below, is even more valuable.

11.5.1 Analytical Solution by Dukhin [19, 20]

This solution is valid for the "zone of uniform dispersed phase". The particle size distribution $\varphi(v,t)$ and its zero moment $N(t)$ were expanded in a series with respect to the small parameter Pe_0:

$$\varphi(v,t) = \varphi_1 + Pe_0\varphi_2 + \ldots \tag{27}$$

$$N(t) = N_1 + Pe_0 N_2 + \ldots \tag{28}$$

where

$$N_1 = \int_0^\infty \varphi_1(v,t)\mathrm{d}v \tag{29}$$

$$N_2 = \int_0^\infty \varphi_2(v,t)\mathrm{d}v \tag{30}$$

Here φ_1 and N_1 are terms corresponding to the Brownian aggregation only.

Substituting functions (27) and (28) into the population balance Eq. (11) yields a linear equation for $N_2(t)$. This equation contains functions $N_1(t)$ and $\varphi_1(v,t)$ that are known. The function $\varphi_1(v,t)$ can be found in a monograph [57] for different initial distributions. The solution of the linear equation for $N_2(t)$ requires the initial condition for the particle size distribution. The two different initial PSDs (exponential and monodisperse) were used by Dukhin [19].

The final result is given as a time function of the total number of particles during transition from Brownian to gravitational aggregation.

For exponential initial PSD:

$$N(t) = \frac{N_0}{1 + t/\tau_{\mathrm{sm}}}\left[1 - \frac{tPe^0(t/\tau_{\mathrm{sm}} + 2)}{4\tau_{\mathrm{sm}}(t/\tau_{\mathrm{sm}} + 1)}\right] \tag{31}$$

For monodisperse initial PSD:

$$N(t) = \frac{N_0}{1 + t/\tau_{\mathrm{sm}}}\left[1 - Pe^0\left(\frac{t}{4\tau_{\mathrm{sm}}} - \frac{\ln(1 + 2t/\tau_{\mathrm{sm}})}{8(1 + t/\tau_{\mathrm{sm}})^2}\right)\right] \tag{32}$$

The first interesting conclusion following from these equations is that transition from Brownian to gravitational aggregation is invariant with the shape of the initial particle size distribution. The terms proportional to the Peclet number differ only two-fold even at $t = \tau_{\mathrm{sm}}$. The difference in these term decays quickly with time, since

$$\lim(\tau \to \infty)\frac{\ln(1+2\tau)}{8(1+\tau)^2} = 0 \qquad (33)$$

Hence the heterogeneous system, in effect, "forgets" its initial state, and at the time of transition from Brownian aggregation to gravitational, initially different systems becomes identical.

This conclusion agrees with the idea of "self-preserving distribution" by Friedlander [16, 27] for Brownian aggregation. It turns out that this feature of the particle size distribution might have a much wider validity range.

11.5.2
Analytical Solution by Jung et al. [37]

There are several distinctive differences between this approach and the previous one. First, the authors consider only gravitational coagulation. They use collision frequency as defined by the Smoluchowski-Camp-Stein theory (Eq. 16), with collision efficiency that takes into account hydrodynamic interaction following Reed et al.'s work [51] (Eq. 24). We mentioned before that it is identical with Sutherland's approach [25] from our viewpoint.

They used the momentum method formulated by Lee [64]. He showed that the PBE would lead to three ordinary differential equations for the first three moments of particle size distribution. These three moments can be related to the parameters of the particle size distribution with a particular shape for a complete description of the kinetic process. They chose a log-normal distribution, defined as follows:

$$\varphi(v,t) = \frac{N(t)}{3\sqrt{2\pi}\ln\sigma(t)}\exp\left\{\frac{-\ln^2[v/v_g(t)]}{18\ln^2\sigma(t)}\right\}\frac{1}{v} \qquad (34)$$

where $v_g(t)$ is the geometric mean particle volume and $\sigma(t)$ is the geometric standard deviation based on the particle radius.

The set of differential equations for $v_g(t)$, $\sigma(t)$ and $N(t)$ does not offer any analytical solutions. This forced additional assumptions. In an attempt to derive an analytical solution, they disregard the volume of the smaller particle in comparison with the larger particle in the expression for collision efficiency:

$$\gamma_c^2|v^{\frac{2}{3}} - \tilde{v}^{\frac{2}{3}}| \cong b_0 v^{\frac{2}{3}}\tilde{v}^{\frac{2}{3}}$$

where b_0 is the correction coefficient, which is calculated using

$$\frac{d\varphi(v,t)}{dt} = \frac{1}{2}\int_0^v \beta^g(v', v-v')\varphi(v-v',t)\varphi(v')dv' - \int_0^v \beta^g(v,v')\varphi(v,t)\varphi(v',t)dv' \qquad (35)$$

$$b_0 = \frac{\iint \gamma_c^2 |v^{\frac{2}{3}} - \tilde{v}^{\frac{2}{3}}| \varphi(v,t)\varphi(\tilde{v},t)\mathrm{d}v\mathrm{d}\tilde{v}}{\iint v^{\frac{2}{3}}\tilde{v}^{\frac{2}{3}}\varphi(v,t)\varphi(\tilde{v},t)\mathrm{d}v\mathrm{d}\tilde{v}} \tag{36}$$

These assumptions allowed the author to derive analytical expressions for mean particle size, standard deviation and total number of particles versus time. We present here only the expression for the total number of particles, because it can be compared with the results of the previous analytical solution:

$$N(t) = N_0 \exp[9(Z - Z_0)] \left(\frac{1 - 2\exp 3Z_0}{1 - 2\exp 3Z} \right)^3 \tag{37}$$

where Z depends on the standard deviation: $Z = \ln^2\sigma$. The authors compared this analytical solution with the numerical solution and obtained very good agreement.

11.5.3
Comparison of Analytical Solutions and Following Conclusions

It is possible to compare predictions regarding the evolution of the total number of particles coming from both analytical solutions. The authors of the second solution presented their prediction in graphical format, reproduced in Fig. 11.6. They used the dimensionless time, which can be presented as follows:

$$\tilde{t} = t \frac{\pi Pe_0 \sqrt[3]{4\pi/3}}{3\tau_{sm}} \tag{38}$$

We can convert the expression from the first analytical solution (Eq. 34), to this dimensionless time. For times that exceed substantially the Smoluchowski times, this new expression would appear as

$$N(t) = \frac{N_0}{1 + t/\tau_{sm}} \left[1 - \left(\frac{4\pi}{3} \right)^{-\frac{4}{3}} \tilde{t} \right] \tag{39}$$

When the dimensionless time equals 1, gravity would reduce the total number of particles by 14.8% in addition to Brownian coagulation. This prediction of the first analytical solution agrees perfectly with the second analytical solution. We illustrate this statement with dashed lines in Fig. 11.6. It can be seen that the second analytical solution predicts this number reduction to be somewhere between 10 and 15%.

This agreement between two analytical solutions that are based on completely different approaches makes both of them much more reliable. We can use them for making certain predictions regarding the role of gravity in the coagulation process.

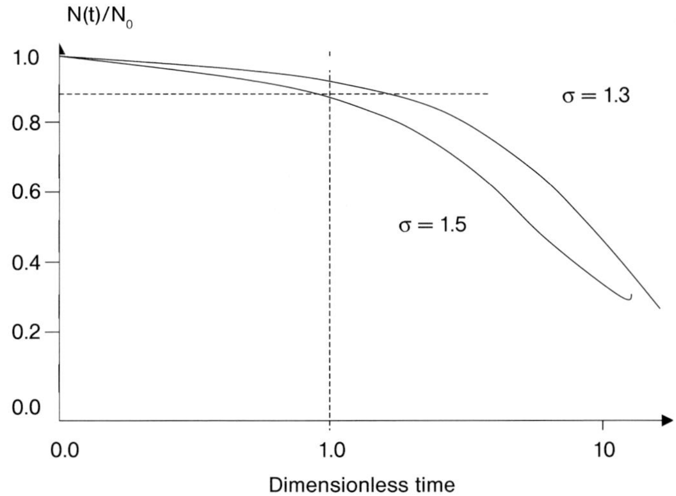

Fig. 11.6. Evolution of the total number of particles in time according to the analytical theory of Jung et al. [37].

For instance, we can introduce a critical time for this transient, which should be about the same for all heterogeneous systems. Terms that are proportional to the initial Peclet number become substantial when time approaches a certain critical value τ_{cr}. In order to define a definite time moment, we can follow the same principle as was used for introduction of the Smoluchowski time. The latter is defined as a time moment when the total number of particles reduces by half. Similarly, we can introduce τ_{cr} as a time moment when gravity reduces the total number of particles by half. This yields the following expression for this critical time:

$$\tau_{cr} = \frac{2\tau_{sm}}{Pe_0} = \frac{3\eta_m}{\Delta\rho g a d_0} \tag{40}$$

This parameter determines also valid time range of the Smoluchowski theory for Brownian aggregation.

One might argue that sedimentation instability would manifest itself prior to the changes described in this theory. In other words, it might be possible that the "zone of uniform dispersed phase" would disappear before this transition takes place. In order to resolve this issue, we could compare the critical time τ_{cr} with the time of sedimentation instability τ_{sd}:

$$\frac{\tau_{cr}}{\tau_{sd}} \approx \frac{d_0}{al} \tag{41}$$

The transition time is 100 times shorter than the sedimentation time for micron-sized particles at 1% volume fraction. For smaller particles this ratio is even greater.

Therefore, we can conclude that there is enough time for experimental observation of the transition from Brownian to gravitation aggregation in the "zone of uniform dispersed phase".

11.6
Transition from Brownian to Gravitational Aggregation – Numerical Solution

The most sophisticated numerical solution of the PBE was performed by Wang and Davis [5], extending their earlier work [7]. The purpose of this work was a description of the coalescence kinetics of emulsion droplets for a very wide range of droplet sizes. This would include both Brownian and gravitation coagulation mechanisms.

It is most important that they managed to avoid an "additivity approximation" for specifying collision frequency. They used a very detailed description of the collision frequency calculated by Zinchenko and Davis [4]. This work took into account complete hydrodynamic droplet interaction and did not apply an "additivity approximation" for combining Brownian and gravitational mechanisms of the relative particles motion.

The relationship between these mechanisms was characterized by the Peclet number, which they defined as follows:

$$Pe_{ij} = \frac{2\pi\Delta\rho g}{3kT} \frac{a_i(a_i + a_j)(a_i^2 - a_j^2)}{1 + \dfrac{a_i}{a_j}} \tag{42}$$

Their numerical procedure applied no restrictions on the Peclet number. They presented result in the range of Peclet numbers from 0.0001 to 1 000 000. They also used the Peclet number for the initial emulsion Pe_0 defined by Eq. (25).

It is important to appreciate the difference between the initial Peclet number Pe_0 and the Peclet number at a certain time moment $Pe_{ij}(t)$. For instance, some of the conclusions of Wang and Davis [5] are formulated for the initial Peclet number instead of the time-dependent Peclet number. In particular, gravity effects would eventually become important for any small initial Peclet number owing to gradual increase in the time-dependent Peclet number.

Regarding their calculation, it is difficult to convert the data into our symbols because their procedure of introducing dimensionless time was not described clearly enough. Fortunately, we have found definitions of all parameters in their previous paper [7].

It turns out that their characteristic time for Brownian coagulation (τ_b) is actually double the Smoluchowski time. If this assumption is valid, then Fig. 11.7 shows a plot of the average radius at $t = 100\,\tau_{sm}$ as a function of the initial Peclet number.

When $Pe_0 \to 0$ (Brownian effects alone), the calculated value is $a_{av} = 6.8\,a_o$, whereas the calculated values using the exact collision efficiencies are

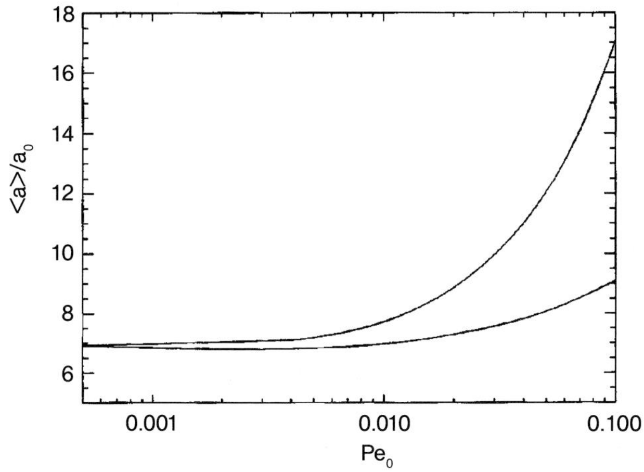

Fig. 11.7 The average drop radius at $t = 50\ \tau_b$ as a function of Pe_0 for a dispersion with an initial distribution having a standard deviation of 0.2. The upper curve is the result obtained using exact collision efficiencies and the lower curve is the result obtained using the additivity approximation [4].

$a_{av} = 17.1\,a_0$, $7.7\,a_0$ and $7.0\,a_0$ for $Pe_0 = 10^{-1}$, 10^{-2} and 10^{-3}, respectively. The calculated values using the additivity approximation are $a_{av} = 9.1\,a_0$, $7.0\,a_0$ and $6.8\,a_0$, respectively, for the same Peclet numbers.

Hence, neglecting gravitational effects or inaccurately accounting for them using the additivity approximation leads to a significant under-prediction of the droplet growth for $Pe > 0\ (10^{-2})$. This determines limits the analytical solution presented above that is based on the additivity approximation. Actually, coagulation evolves faster than predicted by analytical theory after the Peclet number exceeds roughly 0.01.

We can also compare the value of the critical time τ_{cr} established [19] with this numerical solution. In order to do this, we should define a similar critical time for the numerical solution τ_{cr}^{num}. We can assume that this time depends on the Smoluchowski time and the initial Peclet number in the same way as for the analytical solution, but with an unknown multiplier A:

$$\tau_{cr}^{num} = A\frac{\tau_{sm}}{Pe_0} \tag{43}$$

We can use some of Wang and Davis's conclusions for determining value of the A. For instance, they stated that gravitational effects become important at time $t = 50\tau_b = 100\,\tau_{sm}$ if the initial Peclet number $Pe_0 > 0.01$. This means that the coefficient A is approximately 1 and the value of the critical time for the transition from Brownian to gravitational coagulation according to the numerical solution equals

$$\tau_{\rm cr}^{\rm num} = \frac{\tau_{\rm sm}}{Pe_0} \tag{44}$$

It is only two times shorter than prediction of the analytical solution. We consider this to be good agreement, keeping in mind all assumptions involved in defining these parameters.

This research is the culmination of characterizing the influence of gravity on rapid coagulation.

There are several other studies that used various expressions for collision efficiency that were derived following the mathematical approach described above. It is possible to compare some of these results with the analytical solution [19].

For instance, Grant et al. [59] followed van Dongen and Ernst's path [31] and solved numerically the PBE for constant kernel and a sedimentation rate that increases linearly with the cluster size. They came to the conclusion that the total particle concentration decays as a power law of time during the transition period:

$$N(t) \sim t^a \quad a < 0 \tag{45}$$

This result does not agree with the analytical solutions.

This example shows the importance of the analytical solution of the reference point for verifying the underlying assumptions of the numerical solutions.

11.7
Experimental Data

There have been several studies of gravity-affected coagulation kinetics during last two decades. The first is the work of Folkersma et al. [35, 36], who used turbidity measurements and even performed experiments in space for observing the effect of a micro-gravity environment. They also used centrifugation for studying the effect of higher gravity.

Unfortunately, this work does not have direct relationship with this review because their particles were charged. The coagulation process that they observed does not follow route of the "fast coagulation". They observed a reduction in the coagulation rate with increasing gravity, instead of the increase expected for fast coagulation. Their explanation of the break-up of the aggregates due to convection was disproved later by Sun and Qiao [66]. We think that this effect can be explained by theoretical predictions [67]. In any case, the electric charge of the particles leaves this work beyond the scope of this review.

Verification of the fast kinetics coagulation theory requires proper selection of the heterogeneous system. For instance, aggregation of solid particles often leads to floc formation with fractal structures. These systems would be suitable for the verification of fractal theory.

In contrast, emulsion and aerosol drops can coalesce during aggregation and retain the spherical shape of the particles. This makes these systems much better candidates for verifying theories that assume a build-up of a compact aggregate. The list of these theories would include Smoluchowski's theory of Brownian aggregation and other theories describing the influence of gravity mentioned above.

One of the most detailed investigations of fractal coagulation kinetics was made by the Swiss Federal Institute of Technology (ETH) in Zurich, e.g. [38, 39]. In the first study they used small-angle light scattering and obscuration measurements for characterizing polystyrene latex coagulation. They arranged the conditions of flocs formation for studying fractal aspects of this phenomenon.

It turns out that after roughly 100 min the kinetic curve exhibits a deviation from the Smoluchowski law. The size of the flocs accelerates its growth with time. They explained this effect with floc sedimentation that would speed up their growth. In order to justify this assumption, they varied the density of water using heavy water (D_2O). This test confirmed that the observed deviation from the Smoluchowski law occurs due to gravity.

These experiments confirmed the 20-year-old prediction of the analytical theory formulated by Dukhin [19]. The decrease in the total number of particles compared with the Smoluchowski law is equivalent to increasing their size due to gravity-induced additional collisions.

There is a possibility of comparing this experiment with the theory. One can use Fuchs's procedure for calculating the critical size value for transition from Brownian coagulation to gravitational (see Section 11.4.3). Wu et al. [38] used this possibility without knowing that it had been suggested many years earlier by Fuchs.

Fuchs's procedure for calculating critical size simply assumes that $Pe = 1$ and uses this equation for calculating critical size. His calculation brought him to the conclusion that sedimentation becomes comparable to diffusion when the size is about 1 μm. The same calculation for latex by Wu et al. [38] yields a slightly larger size, 2.5 μm. The critical size increases owing to the much smaller density contrast. Fuchs assumed the density contrast to be about 1, whereas it is about 20 times smaller in the study by Wu et al. [38]. This value of the critical size that marks the transition to gravity-influenced coagulation is very close to the experimentally observed values [38].

This agreement between theory and experiment allows us to derive some conclusions regarding the hydrodynamic properties of these latex flocs. The expression for the Peclet number (Eq. 25) was derived assuming Stokes law for the hydrodynamic flow pattern generated by sedimenting flocs. This assumption is true only when there is no flow through the flocs, and their hydrodynamic permeability equals 0.

There have been other studies that used the same assumption for estimating gravity effects, e.g. [60]. Unfortunately, this assumption is not always valid. It is known that advection is an important feature of fractal flocs [22].

One of the most convincing pieces of evidence for advection was given by Li and Logan [54]. They investigated the process of small particle capture by settling porous fractal flocs. They prepared these flocs from red-dyed latex microspheres with an initial size of 2.85 mm. These flocs were introduced on the top of the vessel that contained smaller fluorescent latex microspheres with a size of 1.48 mm. Settling aggregates sweep smaller particles. They proved without doubt that advection is a very important feature of this interaction.

First, the measured settling velocities of the flocs were on average nearly three times faster than those from Stokes' law. Second, the measured collision efficiencies between settling fractal flocs and small particles were 10 times greater than predicted by the curvilinear collision model [53] for impermeable spheres. Although greater permeability of fractal aggregates increased the collision frequencies between the flocs and suspended particles, the observed β values were still two orders of magnitude lower than predicted by the rectilinear model. This is explained by larger macropores of the fractal flocs, which increase hydrodynamic permeability but decrease the probability of particle collisions with the interior walls.

It should be pointed out that the size of the primary particles in the above work is 10 times larger than that of the latex particles from the previously described experiment by Wu et al. [38]. This might explain why advection exists in the Li and Logan's experiments and not in the experiments of Wu et al.

The last experimental work illustrates very clearly all difficulties of the coagulation kinetic theory for fractal flocs. Internal hydrodynamic permeability makes hydrodynamic flow induced by sedimentation much more complicated for a quantitative description. In addition, the interior of the flocs might be accessible for particles. Very little is known regarding their probability of collection on the interior walls of the flocs. These complications makes rigid aggregates and especially emulsions much more attractive objects for verifying the kinetic theory of coagulation, including gravity effects.

We stated before that the main peculiarity and benefit of studying mini-emulsions is the way in which they build up an aggregate. It can be opposite to the fractal – a compact new particle or even a new drop. The latter occurs when droplet coalescence within the aggregate is rapid. This is the most desirable routes of aggregation for verification of the kinetic theory. It has been stated [68–71] that emulsions follow this path of aggregation when the coalescence time τ_c is shorter than the Smoluchowski time τ_{sm}:

$$\tau_{sm} \gg \tau_c \tag{46}$$

where τ_c is the lifetime of the thin film between emulsion droplets.

The droplets disappear within the agglomerate owing to coalescence that is more rapid than a new droplet attaching to the agglomerate when Eq. (46) is valid. Hence the fractals do not grow because the aggregates consist from two droplets all the time. The size of the droplets grows, but the number of individual particles within the aggregate does not [68–71].

In the opposite case, when the coalescence inside the agglomerate cannot prevent the growth of the number of particles, flocs may form, similarly to a dispersion or aerosols.

Condition (46) indicates the applicability boundary for the kinetics of emulsion destabilization, considered in this section. Unfortunately, the practical use of Eq. (46) is complicated, because there is no commercial device for measuring τ_c, and its calculation is known only for a few model systems.

The link between the chemical nature of an adsorption layer, its structure and the coalescence time is not yet quantified. A premise for such quantification is the theory of the lifetime of a foam bilayer [72–76]. The main notions of this theory are similar to those of the theory of Derjaguin and co-workers [77, 78]. However, the theory [72–76] is specified for amphiphile foam films, whereas it is elaborated in detail and proven by experiment with water-soluble amphiphiles, such as sodium dodecyl sulfate [79]. As the dependence of the rupture of the emulsion film on surfactant concentration is similar to that for a foam film, modification of the theory with respect to emulsions may be possible. In our opinion, the theory of the thermodynamic stability of thin emulsion films [72–79] is one of the main achievements in emulsion science.

There are two more potential problems that might complicate the use of emulsions for testing kinetic theories – mobility of their surface and deformability of the droplet. Fortunately, there are some ways of eliminating or avoiding this problem. It can be done either theoretically or by suitable treatment of the particular system.

For instance, the role of the droplet surface mobility on the short-range hydrodynamic interaction and the collision frequency has been investigated [3]. However, there is no need to use this more complicated theory. The droplet surface can be strongly retarded by adding surfactant. This would reduce a spherical droplet to a solid particle from a hydrodynamic viewpoint.

The peculiarity of emulsion droplets, their deformation, is more complicated and deserves special attention. Only about a decade ago was essential progress achieved in the understanding and quantification of the role of deformation [3, 80–82]. Yiantsios and Davis [3] devised a method (YD) for the investigation of the evolution of a thin emulsion film under the action of lubrication arising in gravitational coagulation. They derived an equation for the distance dependence at the early stage droplet deformation. They considered the film rupture to be more a natural evolution than instability. According to the YD theory, droplet deformation does not influence the grazing trajectory in mini-emulsions.

This justifies the introduction of a simple method to calculate the collision efficiency in mini-emulsions. However, with decreasing surface tension or increasing droplet dimensions (transition to macro-emulsions), the droplet deformation can influence the collision efficiency. As a general comment, the importance of droplet deformation in mini-emulsions was overestimated for many years. It was shown only relatively recently [3] that the transition to charged droplets and lower electrolyte concentrations prevents droplet deformation.

Taking into account the weak influence of droplet deformation on rapid coagulation, we can conclude that rapid Brownian and gravitational coagulation cannot be significantly affected by droplet deformation in mini-emulsions. Meanwhile, a droplet surface is homogeneous in contrast to a solid particle surface and droplets satisfy a spherical model much better than many solid particles.

Hence the mini-emulsion can be used as a model system for the evaluation of the state of the theories of coagulation. Their polydispersity is not a serious obstacle [23]. It is pure mathematical problem, which may be solved using the PBE.

The first step is verification of the collision efficiency in a mini-emulsion. Although extensive studies have been made on the impact of two drops at high velocity, there is, perhaps, only one experimental paper, devoted to the hydrodynamic interaction between two drops with not a large difference in droplet dimensions in low Reynolds number flow [83]. In spite of rather large droplet dimension (about 700 μm), the settling velocity of an isolated drop was sufficiently small and accordingly Re was <1, because of the small density difference of the drop phases and the matrix fluids and the high viscosity of the chosen matrix fluids. This large droplet dimension enables one to measure small droplet horizontal displacements when they approach each other during settling.

Experimental results demonstrated that hydrodynamic interactions significantly reduce the relative velocity of two nearby drops and cause two drops to flow around each other with curved trajectories. The theory predicts that the effect of hydrodynamic interactions increases as the drop separation decreases, the size ratio decreases and the viscosity ratio increases, which were also confirmed.

Close attention to collision efficiencies for droplets of almost equal dimensions has been paid in atmospheric science. It is believed [84] that this kind of collision plays a major role in the earlier stage of gravitational coagulation, when droplets grow up to 20 μm owing to condensation. Both the theory and a review of experimental data are given in [84].

Another source of relevant experimental data is micro-flotation kinetics [29, 85–87]. Monodisperse latex or glass spherical particles were used in the experiments. The radius ratio was very small. Thus, a_1 in Eqs. (19) to (22) corresponds to the latex particle dimension and a_2 to the bubble dimension. The dependence $E \sim a_2^{-2}$ was confirmed in [29, 85]. The dependence $E \sim a_1^{1.5}$ was established in [85–87]. The combination of Eq. (20) that describes long-range hydrodynamic interaction and Eq. (21) for the short-range hydrodynamic interaction leads to the dependence $E \sim a_1^{1.4}$.

The direct observation of the grazing trajectory of latex particles (0.9 μm) in the vicinity of the rising bubble with radius 15 μm was accomplished [86, 87]. The cell was attached to the microscope stage, which was capable of moving vertically at the same velocity as that of the rising bubbles in the cell. Hence the theory of short-range hydrodynamic interaction is confirmed by the investigations [29, 85–87].

The next step is verification of the kinetic theory that describes the evolution of the particle ensemble. To our knowledge, there has been only one experimental study devoted to this subject for emulsions [34]. There three different oils were used (toluene, dichlorodecane and isolating oil) for preparing oil-in-water emulsions. The mean droplet diameter was about 2 μm. Several volume fractions were tested in the range 0.4–1%. Turbidity measurements were used.

The time of the sedimentation instability for these emulsions exceeds the time of aggregation approximately 10-fold, assuming droplet size = 2 μm, α=0.01,

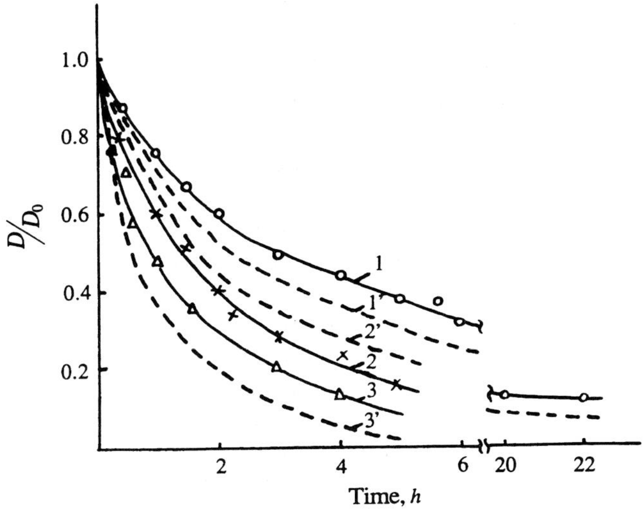

Fig. 11.8 Time dependence of the optical density of emulsions with different volume fraction of the transformation oil [34].

$l = 1$ cm and $\Delta\rho = 0.1$ g cm^{-3}. This means that the necessary condition for observing the transition from Brownian to gravitational coagulation is valid.

The authors [34] observed a deviation from the Smoluchowski law for perikinetic coagulation, as shown in Fig. 11.8. It can be seen that coagulation evolves faster than the Smoluchowski law for lower volume fractions of 0.4 and 0.7%. This is in agreement with the prediction of the theory.

The evolution of the most concentrated emulsion with 1% oil contradicts theory because it is slower than the Smoluchowski rate. However, the position of this experimental curve indicates that the basic underlying assumption regarding the relationship between optical properties and dispersed phase properties becomes invalid for this high concentration because it lies between curves with lower volume fractions.

It seems to us that this work can be considered at least a preliminary qualitative confirmation of the predictions for transition from Brownian to gravitational coagulation.

11.8
Conclusion

This review of many studies dedicated to fast coagulation kinetics allows us to conclude that substantial progress has been achieved in both theory and experiment. From the theoretical viewpoint, there are several different types of systems depending on the structure of aggregates. Historically, the first one corresponds to dense rigid aggregates or coalescing emulsion droplets. More recently, another type of system with porous aggregates that include substantial amount of dispersion media has attracted more attention, namely the area of fractal systems. In principle, the first type of system could be considered as an extreme fractal dispersion with fractal dimension 3. However, introduction of the dense aggregate as a new particle with the same density and other properties as primary particles leads to so many theoretical simplifications that this model certainly deserves an independent existence. Actually, practically all theoretical studies have been done for this model and not for the fractal aggregates model.

There is good agreement between several analytical and numerical solutions of coagulation kinetics. There are two critical parameters which mark the transition from Brownian to gravitational coagulation: critical particle size and time.

The value of the critical size is about 1 μm. This value is valid for a wide variety of systems, dense or fractal, and is almost independent of the particles density. The transition from Brownian to gravitational coagulation is very sharp on the particle size scale.

The estimated value of the critical time is given by the following equation:

$$\tau_{\text{cr}} \approx \frac{\tau_{\text{sm}}}{Pe_0}$$

All heterogeneous systems that undergo fast coagulation would accelerate their coagulation rate around this time owing to the increasing influence of gravity.

This conclusion was derived for the dense aggregates model and is not valid for fractal systems. It seems possible to generalize existing theories for fractal aggregates that do not exhibit advection – flow of liquid through the aggregate. The hydrodynamics of these fractal systems would be similar to those of dense aggregates, but with variable, size-dependent density and volume fraction.

There is also a strong motivation for the creation of a kinetic theory for such fractal systems from experiments. The most sophisticated experiment in this field [38] was performed for these fractals. Unfortunately, existing theories are not directly applicable to describe this experiment because they do not take into account variations of the aggregates' density and volume fraction with size. This experiment indicates that gravity might play an even earlier and more important role for fractals than for dense aggregates.

There have also been experimental studies on fractals with advection. Creating a kinetic theory for these fractals is a much more complicated task.

At the current stage, existing theory might be most useful for emulsion science. Modeling of the evolution of emulsions may result in quantification of the emulsion film stability, namely, establishment of the dependence of the coalescence time on the physico-chemical properties of the surfactant adsorption layer, its structure and the droplet dimensions. This quantification can form a basis for optimizing the selection of demulsifiers and synthesis for emulsion technology applications instead of the current empirical approach.

List of Symbols

A	Hamaker constant
a_1	Primary radius
a_i	Radius of the aggregate consisting of i primary particles
D	Diffusion coefficient
D_{ij}	Diffusion coefficient of the relative Brownian motion of aggregates containing i and j primary particles
d_1	Primary diameter
d_i	Diameter of the aggregate consisting of i primary particles
d_{max}	Maximum particle diameter
d_{f}	Fractal dimension
$d_{\mathrm{B\text{-}g}}$	Critical diameter of the transition from Brownian to gravitational coagulation
g	Earth's gravity acceleration
K_{f}	Constant in Smoluchowski time
k	Boltzmann constant
k_{f}	Fractal coefficient
I	Flux of particles
l	Depth of sedimentation
N_0	Number of primary particles per unit volume
$N(t)$	Number of particles per unit volume at time t
n_i	Number per unit volume of aggregates consisting of i primary particles
Pe	Peclet number
R_{ij}	Distance between centers of aggregates i and j
T	Absolute temperature
t	Time
x_{uni}	Boundary of the uniform dispersed phase zone
u_{s}	Sedimentation velocity
α	Volume fraction of dispersed phase
β_{ij}	Collision frequency of the aggregates containing i and j primary particles
β^{g}	Gravitational collision frequency
β^{d}	Diffusion (Brownian) collision frequency
δ	Frequency of break-up of aggregates
ε	Porosity of aggregate
η_{m}	Viscosity of dispersion medium
κ	Permeability of aggregate
v	Aggregate volume
v_{g}	Geometric mean volume
ρ_{m}	Density of dispersion medium
ρ_{p}	Density of particles
$\Delta\rho$	Density contrast
σ	Geometric standard deviation

τ_{sm}	Smoluchowski time
τ_{sd}	Sedimentation time
$\tau_{B\text{-}g}$	Critical time of the transition from Brownian to gravitational coagulation
$\phi(\nu)$	Volume fraction of the aggregates with volume ν

References

1 von Smoluchowski, M. *Phys. Z., 27,* 585 (**1916**).

2 Davis, R. H. *J. Fluid. Mech., 145,* 179 (**1984**).

3 Yiantsios, S. G., Davis, R. H. *J. Colloid Interface Sci., 144,* 412 (**1991**).

4 Zinchenko, A. Z., Davis, R. H. *J. Fluid. Mech., 280,* 119 (**1994**).

5 Wang, H., Davis, R. H. *J. Colloid Interface Sci., 178,* 47 (**1996**).

6 Zhang, X., Davis, R. H. *J. Fluid Mech., 230,* 479 (**1991**).

7 Wang, H., Davis, R. H. *J. Colloid Interface Sci., 108* (**1993**).

8 Derjaguin, B. V., Dukhin, S. S., Rulyov, N. N. Kinetic theory of flotation of small particles, in *Surface and Colloid Science,* Vol. 13, Matijevich, E. (ed.), Wiley-Interscience, New York, pp. 71–113 (**1983**).

9 Dukhin, S. S., Rulyov, N. N., Dimitrov, D. S. *Coagulation and Dynamics of Thin Films,* Naukova Dumka, Kiev (**1986**) (in Russian).

10 Derjaguin, B. V., Dukhin, S. S. *Trans. Inst. Min. Metall., 70,* 221–231 (**1960**).

11 Dukhin, S. S., Derjaguin, B. V. *Kolloidn. Zh., 20,* 326 (**1958**).

12 Rulyov, N. N. *Kolloidn. Zh., 40,* 898 (**1978**).

13 Rulyov, N. N. *Kolloidn. Zh., 40,* 1202 (**1978**).

14 Fuchs, N. A. *The Mechanics of Aerosols,* Pergamon Press, Oxford (**1964**).

15 Fuchs, N. A. *Uspekhi Mekhaniki Aerozoley,* Izd-vo Akad. Nauk SSSR, Moscow (**1961**) (in Russian).

16 Friedlander, Sh. K. *Smoke, Dust, and Haze,* Oxford University Press, Oxford (**2000**).

17 Camp, T. R., Stein, P. C. *J. Boston Soc. Civ. Eng., 30,* 219–237 (**1943**).

18 von Smoluchowski, M. *Z. Phys. Chem., 92,* 129 (**1917**).

19 Dukhin, A. S., *Kolloid. Z., 6,* 929–934 (**1988**).

20 Dukhin, A. S., *Kolloid. Z., 3,* 530–531 (**1986**).

21 Meakin, P. *Phys. Rev. Lett., 51,* 1119 (**1983**).

22 Meakin, P. *Adv. Colloid Interface Sci., 28,* 249–331 (**1988**).

23 Dukhin, A. S., Goetz, P. J. *Ultrasound for Characterizing Colloids,* Elsevier, Amsterdam (**2002**).

24 Melzak, Z. *Trans. Am. Math. Soc., 85,* 547 (**1957**).

25 Sutherland, K. L. *J. Phys. Chem., 58,* 394 (**1948**).

26 Derjaguin, B. V., Dukhin S. S. *Izv. Akad. Nauk SSSR, Otdel. Metall. Topl., 1,* 82 (**1959**).

27 Spielman, L. A., Fitzpatrick, J. A. *J. Colloid Interface Sci., 42,* 607 (**1973**).

28 Spielman, L. A., Fitzpatrick, J. A. *J. Colloid Interface Sci., 43,* 350 (**1973**).

29 Reay, D., Ratelif, G. A. *Can. J. Chem. Eng., 51,* 178 (**1973**).

30 Mandelbrot, B. B. *The Fractal Geometry of Nature,* Freeman, New York (**1982**).

31 Van Dongen, P. G. J., Ernst, M. H. *Phys. Rev. Lett., 54,* 1396 (**1985**).

32 Melik, D. H., Fogler, H. S. in *Encyclopedia of Emulsion Technology,* Vol. 3, Marcel Dekker, New York, pp. 3–78 (**1988**).

33 Wen, C.-S., Zhang, L., Lin, H. *J. Colloid Interface Sci., 142,* 258–265 (**1991**).

34 Mishchuk, N. A., Verbich, S. V., Dukhin, S. S., Oystein, H., Sjoblom, J. *J. Dispers. Sci. Technol., 18,* 517 (**1997**).

35 Folkersma, R., van Dieman, A. J. G., Stein, H. N. *J. Colloid Interface Sci., 206,* 482–493 (**1998**).

36 Folkersma, R., van Dieman, A. J. G., Stein, H. N. *J. Colloid Interface Sci., 206,* 494–505 (**1998**).

37 Jung, C.H., Park, S.H., Lee, K.W., Kuhlman, M.R. *Part. Sci. Technol., 18,* 89–102 (**2000**).

38 Wu, H., Lattuada, M., Sandkuhler, P., Sefcik, J., Morbidelli, M. *Langmuir, 19,* 10711–10718 (**2003**).

39 Sandkuhler, P., Lattuada, M., Wu, H., Sefcik, J., Morbidelli, M. *Adv. Colloid Interface Sci., 113,* 65–83 (**2005**).

40 Lyklema, J. *Fundamentals of Interface and Colloid Science, Vol. 1, Fundamentals,* Academic Press, London (**2000**).

41 Meakin, P., Viszek, T., Family, F. *Phys. Rev. B, 31,* 564 (**1985**).

42 Kilb, M. *Phys. Rev. Lett., 53,* 1653 (**1984**).

43 Viesek, T., Family, F. *Phys. Rev. Lett., 52,* 1669 (**1984**).

44 Scheludko, A. *Colloid Chemistry,* Elsevier, Amsterdam (**1967**).

45 Kumar, S., Narsimhan, G., Ramkrishna, D. *Ind. Eng. Chem. Res., 35,* 3155 (**1996**).

46 Spielman, L.A. *Annu. Rev. Fluid Mech., 9,* 297 (**1977**).

47 Goren, S.L., O'Neil, M.E. *Chem. Eng. Sci., 26,* 325 (**1971**).

48 Goren, S.L. *J. Fluid Mech., 41,* 613 (**1971**).

49 Dukhin, S.S., Sjoblom, J. Kinetics of Brownian and gravitational coagulation, in *Emulsion and Emulsion Stability,* Sjoblom, J. (ed.), Marcel Dekker, New York, pp. 41–180 (**1996**).

50 Jeffrey, D.J., Onishi, Y. *J. Fluid. Mech., 139,* 261 (**1984**).

51 Reed, L.D., Lee, K.W., Gieseke, J.A. *Nucl. Sci. Eng., 75,* 167–189 (**1980**).

52 Batchelor, G.K. *J. Fluid Mech., 119,* 379 (**1982**).

53 Han, M., Lawler, D.F. *J. Am. Water Works Assoc., 84* (10), 79–91 (**1992**).

54 Li, X.-Y., Logan, B.E. *Environ. Sci. Technol., 31,* 1229–1236 (**1997**).

55 Mazzolani, G., Stolzenbach, K.D., Elimelech, M. *J. Colloid Interface Sci., 197,* 334 (**1998**).

56 Voloshchuk, V.M., Sedunov, Yu.S. *Coagulation Processes in Disperse Systems,* Hydrometeoizdat, Leningrad (**1975**) (in Russian).

57 Voloshchuk, V.M. *Kinetic Theory of Coagulation,* Gidrometeoizdat, Leningrad (**1984**) (in Russian).

58 Petrov, V.G., Edissonov, I. *Biorheology, 33,* 353–364 (**1996**).

59 Grant, S.B., Kim, J.H., Poor, C. *J. Colloid Interface Sci., 238,* 238–250 (**2001**).

60 Grant, S.B., Poor, C., Relle, S. *Colloids Surf. A, 107,* 155–174 (**1996**).

61 Reddy, S.R., Melik, D.H., Fogler, H.S. *J. Colloid Interface Sci., 82,* 116 (**1981**).

62 Wang, Y.G., Wen, C.S. *J. Fluid. Mech., 214,* 599 (**1990**).

63 Simons, S., Williams, M.M.R., Cassel, J.S. *J. Aerosol.Sci., 17,* 789 (**1986**).

64 Lee, K.W. *J. Colloid Interface Sci., 92,* 315–325 (**1983**).

65 Sonntag, H., Strenge, K. *Coagulation Kinetics and Structure Formation,* VEB Deutscher Verlag der Wissenschaften, Berlin (**1987**).

66 Sun, Z.W., Qiao, R.L. *J. Colloid Interface Sci., 223,* 126–132 (**2000**).

67 Dukhin, A.S., *Kolloid. Z., 3,* 387–393 (**1988**).

68 Borwankar, R.P., Lobo, L.A., Wasan, D.T. *Colloids Surf., 69,* 135 (**1992**).

69 Dukhin, S.S., Sjoblom, J. *J. Dispers. Sci. Technol., 19,* 311 (**1998**).

70 Dukhin, S.S., Sjoblom, J., Wasan, D.T., Saether, O. *Colloids Surf., 180,* 223 (**2001**).

71 Dukhin, S.S., Saether, O., Sjoblom, J. in *Encyclopedic Handbook of Emulsion Technology,* Sjoblom, J. (ed.), Marcel Dekker, New York, Chapter 4 (**2001**).

72 Exerowa, D., Kashchiev, D. *Contemp. Phys., 27,* 429 (**1986**).

73 Exerova, D., Kashchiev, D., Platikanov, D. *Adv. Colloid Interface Sci., 40,* 201 (**1992**).

74 Exerova, D., Kruglyakov, P.M. *Foam and Foam Films,* Elsevier, Amsterdam (**1998**).

75 Kashchiev, D. *Nucleation: Basic Theory with Applications,* Butterworth-Heinemann, Oxford (**2000**).

76 Exerova, D. *Adv. Colloid Interface Sci., 96,* 75 (**2002**).

77 Derjaguin, B.V., Gutop, Yu.V. *Kolloidn. Zh., 24,* 431 (**1962**).

78 Prokhorov, A.V., Derjaguin, B.V. *J. Colloid Interface Sci., 125,* 11 (**1988**).

79 Kashchiev, D., Exerova, D. *J. Colloid Interface Sci., 203,* 146 (**1998**).

80 Klahn, J.K., Agterof, W.G.M., van Vurst, F,. Vader, R.D., Groot, F. Groeneweg, *Colloids Surf., 65,* 151 (**1992**).

81 Danov, K. D., Petsev, D. N., Denkov, N. D., Borwankar, R. *J. Chem. Phys., 99,* 7179 (**1993**).
82 Danov, K. D., Denkov, N. D., Petsev, D. N., Ivanov, I. B., Borwankar, R. *Langmuir, 9,* 1731 (**1993**).
83 Zhang, X.-G., Davis, R. H., Ruth, M. F. *J. Fluid. Mech., 249,* 227 (**1993**).
84 Lin, C. L., Lee, S. C. *J. Atmos. Sci., 32,* 1412 (**1975**).
85 Reay, D., Ratelif, G. A. *Can. J. Chem. Eng., 53,* 481 (**1975**).
86 Okada, K., Akagi, Y., Kogure, M., Yoshioka, N. *Can. J. Chem. Eng., 68,* 393 (**1990**).
87 Okada, K., Akagi, Y., Kogure, M., Yoshioka, N. *Can. J. Chem. Eng., 68,* 614 (**1990**).
88 Ivanov, I. B., Kralchevsky, P. A. *Colloids Surf., 128,* 155 (**1997**).
89 Danov, K. D., Kralchevsky, P. A., Ivanov, I. B. in *Encyclopedic Handbook of Emulsion Technology,* Sjoblom, J. (ed.), Marcel Dekker, New York, Chapter 26 (**2001**).
90 Dukhin, S. S. *Int. Tagung Grenzflach. Stoff. Berlin, 2,* 561 (**1975**).
91 Derjaguin, B. V., Dukhin, S. S., Rulyov, N. N. *Kolloidn. Zh., 39,* 1051 (**1977**).

Subject Index

Colloids and Interface Science Series, Vol. 2
Colloid Stability: The Role of Surface Forces, Part II. Edited by Tharwat F. Tadros

ISBN: 978-3-527-31503-1

q

r

s

Colloids and Interface Science Series

Colloid Stability
The Role of Surface Forces, Part I
Volume 1

2007
ISBN 13: 978-3-527-31462-1
ISBN 10: 3-527-31462-8

Colloid Stability
The Role of Surface Forces, Part II
Volume 2

2007
ISBN 13: 978-3-527-31503-1
ISBN 10: 3-527-31503-9

Colloid Stability and Applications in Pharmacy
Volume 3

2007
ISBN 13: 978-3-527-31463-8
ISBN 10: 3-527-31463-6

Colloids in Cosmetics and Personal Care
Volume 4

2007
ISBN 13: 978-3-527-31464-5
ISBN 10: 3-527-31464-4

Colloids in Agrochemicals
Volume 5

2007
ISBN 13: 978-3-527-31465-2
ISBN 10: 3-527-31465-2

Colloids in Paints
Volume 6

2008
ISBN 13: 978-3-527-31466-9
ISBN 10: 3-527-31466-0